| | hydroxyl | diol | carboxyl | carbonyl | amine |
|---|---|---|---|---|---|
| 2-methoxyethoxymethyl (MEM) | 74–77 | | 129–131 | | |
| methoxymethyl (MOM) | 68–71 | | 129–131 | | 218 |
| 1-methoxy-1-methylethyl | 84 | | | | |
| *p*-methoxyphenyl (PMP) | | | | | 224–225 |
| *p*-methoxyphenyldiphenylmethyl (MMTr) | 56 | | | | |
| 4-methoxytetrahydropyran-4-yl (MTHP) | 83–84 | | | | |
| methyl (Me) | 42–46 | | 121–125 | | |
| methylthiomethyl (MTM) | 72–74 | | 129–131 | | |
| *p*-nitrobenzyl (PNB) | | | 136–139 | | |
| *p*-nitrocinnamyloxycarbonyl (Noc) | | | | | 200 |
| *o*-nitrophenylsulfenyl (Nps) | | | | | 216–217 |
| ortho esters | | | 145–148 | | |
| 1,3-oxazolines | | | 148–149 | | |
| 9-phenyl-9-fluorenyl (PhFl) | | | | | 227–230 |
| phenylsulfonyl | | | | | 209–215 |
| 2-(phenylsulfonyl)ethoxycarbonyl (Psec) | | | | | 203–204 |
| 9-phenyl-9-xanthenyl (pixyl) | 57 | | | | |
| phthalimides | | | | | 186–189 |
| pivaloyl (Pv) | 22–28 | | | | |
| pivaloyloxymethyl (Pom) | | | 128 | | |
| 2,2,5,5-tetramethyl-1-aza-2,5-disilacyclopentane (STABASE) | | | | | 230–233 |
| tetrahydropyran-2-yl (THP) | 83–85 | | | | |
| 1,1,3,3-tetraisopropyldisiloxane-1,3-diyl (TIPDS) | | 112–114 | | | |
| thexyldimethylsilyl (TDS) | 41–42 | | | | |
| tosyl (Ts) | | | | | 212–215 |
| 2-tosylethoxycarbonyl (Tsoc) | | | | | 204 |
| 2-tosylethyl (TSE) | | | 134–136 | | |
| 2,2,2-trichloroethoxycarbonyl (Troc) | | | | | 207–209 |
| 2,2,2-trichloroethyl (TCE) | | | 131–132 | | |
| triethylsilyl (TES) | 31–33 | | 143–145 | | |
| trifluoroacetyl | | | | | 189–190 |
| triisopropylsilyl (TIPS) | 40–41 | | | | 232 |
| trimethylsilyl (TMS) | 29–31 | | 143–145 | | 230–233 |
| 2-(trimethylsilyl)ethanesulfonyl (SES) | | | | | 215–216 |
| 2-(trimethylsilyl)ethoxycarbonyl (Teoc) | | | | | 205–207 |
| 2-(trimethylsilyl)ethoxymethyl (SEM) | 80–82 | | 129–131 | | 218 |
| 2-(trimethylsilyl)ethyl (TMSE) | 80 | | 132–134 | | |
| 2,6,7-trioxabicyclo[2.2.2]octane (OBO) | | | 145–148 | | |
| 4,4′,4″-tris(benzoyloxy)trityl (TBTr) | 57 | | | | |
| trityl (Tr) | 54–59 | | | | 227–230 |

THIEME FOUNDATIONS OF ORGANIC CHEMISTRY SERIES

*Series Editors: D. Enders, R. Noyori, B. M. Trost*

# Protecting Groups

# Protecting Groups


**Philip J. Kocienski**

University of Southampton

Georg Thieme Verlag Stuttgart • New York 1994

Prof. Philip J. Kocienski

Department of Chemistry
University of Southampton
Highfield
Southampton SO9 5NH, UK

*Library of Congress Cataloging-in-Publication Data*

*Kocienski, Philip J.*:
Protecting groups / Philip J. Kocienski.
p. cm. -- (Thieme foundations of organic chemistry series )
Includes bibliographical references and Index.
ISBN 3-13-137001-7. -- ISBN 3-13-135601-4 (flexicover)
1. Organic compounds -- Synthesis. 2. Protective groups (Chemistry)
I. Title. II. Series.
QD262.K59 1994 94-17462
547'.2—dc20 CIP

*Die Deutsche Bibliothek – Cataloging-in-Publication Data*

*Kocienski, Philip J.*:
Protecting groups: Thieme foundations of organic chemistry series /ed. by D. Enders, R. Noyori, B. M. Trost –
Stuttgart; New York: Thieme, 1994
NE: Kocienski, Philip J.


Printed in Germany by Gutmann & Co GmbH, D-74388 Talheim

Georg Thieme Verlag, Stuttgart
ISBN 313 135601 4 Flexicover
ISBN 313 137001 7 Hardcover

Thieme Medical Publishers, Inc., New York
ISBN 0 86577 557 5 Flexicover
ISBN 0 86577 558 3 Hardcover

To Robert Cable, scholar and friend

# Foreword

Organic chemistry plays such a pivotal role in science ranging from biology to materials science that it becomes crucial to develop mechanisms to communicate to all the varied audiences. Textbooks have served as the major vehicle to introduce the subject to the novice. Monographs have largely evolved to focus on a specialized topic that is covered in depth *via* a multiauthor work coordinated by an editor. Such works typically become accounts of the state-of-the-art for a particular area at the time. On the other hand, a pedagogical treatment of a broad subject intended for the advanced students, either in graduate schools or industrial or governmental laboratories, is most valuable in transferring knowledge rather than presenting data.

Each book in this series is intended to cover an area where new developments are occurring that are having a broad impact on the science. As a result, each book should become essential reading for all students of organic chemistry and related disciplines. The series is intended to capture the creative energy of new discovery in order to inspire the interest and enthusiasm of the reader. The author should develop the topic conceptually and didactically, putting it into the context of organic chemistry as a whole in order to make it understandable for the broad readership intended.

The series will not provide an encyclopedic presentation of the field, which is the role of reference works, nor of the data, which is the role of data bases. It is intended to emphasize ideas, methodological thinking, and underlying principles. The coverage of the area should be authoritative and complete in terms of concepts. The concepts should emerge from the presentation of selective data that is critically evaluated; they should, however, transcend the current data available so that they remain equally valid even as new data is obtained and the field progresses. As appropriate for an advanced work, significant references to the primary literature and more complete referencing of the secondary literature should be provided. While the work is intended to be a scholarly treatment of the broad subject, its prime function is didactic.

D. Enders, Aachen
R. Noyori, Nagoya
B. M. Trost, Stanford

# Preface

Theodora Greene's landmark book *Protective Groups in Organic Synthesis* (1981) contained over 500 protective groups for 5 common functional groups. By the time the second edition by Greene and Wuts appeared in 1991, the number of protecting groups had expanded by a further 206. The increase reflects several factors including the need to circumvent inherent limitations in the existing repertoire of protecting groups, the more stringent demands of recent complex targets, and the need to adapt protecting groups to new technologies such as solid phase synthesis. However, the comprehensive approach of Greene and Wuts brought an inevitable increase in bulk because of the inclusion of protecting groups that satisfy the esoteric requirements of rather specialised niches of marginal interest to the synthetic community at large. We felt the need for a critical survey of the subject, that focuses on the most widely used protecting groups, for the most common functional groups, used by most organic chemists, most of the time. By these utilitarian criteria, we narrowed the field to about 50 protecting groups which have earned an honoured place in the synthetic repertoire.

Our survey is organised by functional group and special emphasis is placed on *deprotection* conditions applied to complex structures where selectivity is a prime issue. Discussion of each protecting group is divided into three sections beginning with deprotection conditions, followed by methods of formation, and ending with some cursory NMR data to aid the reader in identifying the background clutter contributed by a protecting group. Obviously, the chemical shifts and coupling data can vary significantly from those quoted. At the end of each chapter is a list of reviews which further amplifies topics covered in the individual chapters. An asterisk in the text indicates that a pertinent review can be found at the end of the chapter.

Over 500 schemes are provided to aid visual retrieval. No attempt is made to analyse protecting group issues in any one class of compounds comprehensively or systematically; rather, the illustrations span a wide domain of organic synthesis including alkaloids, terpenoids, polyketides, $\beta$-lactams, polyether antibiotics, macrolides, peptides, carbohydrates, glycolipids, glycopeptides, and nucleosides. We give priority to examples gleaned from the recent literature (covered up to the end of 1992) that are amply documented in full papers. Wherever possible, transformations are accompanied by key experimental details such as solvent, temperature, reaction time, stoichiometry, yield, and scale. Transformations in which the scale has been specified have been abstracted from papers providing detailed experimental procedures. A hazard warning has been included for some reagents (e.g. phosgene and haloalkyl ethers) which require special precautions. However, the reader cannot assume that the absence of a warning for any reagent, solvent, or product implies the need for any less vigilance or care.

I would like to thank the members of my research group who made valuable suggestions for improving the "user friendliness" of the book and for providing spectroscopic information. Dr. Stanislaw Marczak helped separate the wheat from the chaff during the early planning stages; Professor Ekkehard Winterfeldt suggested some examples and Dr. Ian Stevens provided valuable mechanistic information. Dr. Joe Richmond of Georg Thieme Verlag, Dr. Krzysztof Jarowicki, Dr. John Mellor and Dr. Georges Hareau read and corrected the entire manuscript and Sharon Casson kindly checked the references. Special thanks go to Professor Reinhard Hoffmann for providing some fascinating background information to Chapter 7 and Professor Horst Kunz for contributing many valuable insights, corrections, and encouragement.

By way of apology for errors of commission or omission, I would like to offer the following words from the Preface of Samuel Johnson's Dictionary (London, 1755):

> *A large work is difficult because it is large, even though all its parts might singly be performed with facility; where there are many things to be done, each must be allowed its share of time and labour, in the proportion only which it bears to the whole; nor can it be expected, that the stones which form the dome of a temple, should be squared and polished like the diamond of a ring.*

PHILIP J. KOCIENSKI,
*Southampton University*
*April, 1994*

# Contents

# Abbreviations

## Protecting Groups

Ac acetyl
Aloc allyloxycarbonyl
Bn benzyl
benzostabase 2,3-dihydro-1*H*-2,1,3-benzazadisilole
Boc *tert*-butoxycarbonyl
BOM benzyloxymethyl
Bz benzoyl
Cbz (or Z) benzyloxycarbonyl
DEIPS diethylisopropylsilyl
DMB 3,4-dimethoxybenzyl
DMTr di(*p*-methoxyphenyl)phenylmethyl
DTBS di-*tert*-butylsilylene
EE 2-ethoxyethyl
Fm 9-fluorenylmethyl
Fmoc 9-fluorenylmethoxycarbonyl
IPDMS isopropyldimethylsilyl
MBS *p*-methoxybenzenesulfonyl
MEM 2-methoxyethoxymethyl
MMTr *p*-methoxyphenyldiphenylmethyl
MOM methoxymethyl
MTHP 4-methoxytetrahydropyran-4-yl
MTM methylthiomethyl
Noc *p*-nitrocinnamyloxycarbonyl
Nps *o*-nitrophenylsulfenyl
OBO 2,6,7-trioxabicyclo[2.2.2]octyl
PhFl 9-phenyl-9-fluorenyl
pixyl 9-phenyl-9-xanthenyl
PMB *p*-methoxybenzyl
PMBM *p*-methoxybenzyloxymethyl
PMP *p*-methoxyphenyl
PNB *p*-nitrobenzyl
Pom pivaloyloxymethyl
Psec 2-(phenylsulfonyl)ethoxycarbonyl
Pv pivaloyl
SEM 2-(trimethylsilyl)ethoxymethyl
SES 2-(trimethylsilyl)ethanesulfonyl
STABASE 2,2,5,5-tetramethyl-1-aza-2,5-disilacyclopentane
TBS (or TBDMS) *tert*-butyldimethylsilyl
TBDPS *tert*-butyldiphenylsilyl
TBTr 4,4′,4″-tris(benzoyloxy)trityl
TCE 2,2,2-trichloroethyl
TDS thexyldimethylsilyl
Teoc 2-(trimethylsilyl)ethoxycarbonyl
TES triethylsilyl
thexyl 1,1,2-trimethylpropyl
THP tetrahydropyran-2-yl
TIPDS 1,1,3,3-tetraisopropyldisiloxane-1,3-diyl
TIPS triisopropylsilyl
TMOB 2,4,6-trimethoxybenzyl
TMS trimethylsilyl
TMSE 2-(trimethylsilyl)ethyl
TMSEC 2-(trimethylsilyl)ethoxycarbonyl
Tr trityl = triphenylmethyl
Troc 2,2,2-trichloroethoxycarbonyl
Ts tosyl = *p*-toluenesulfonyl
TSE 2-(*p*-toluenesulfonyl)ethyl
Tsoc 2-(*p*-toluenesulfonyl)ethoxycarbonyl

# Reagents and Solvents

| | |
|---|---|
| 9-BBN | 9-borabicyclo[3.3.1]nonane |
| CAN | ceric ammonium nitrate |
| CSA | camphorsulfonic acid |
| DABCO | 1,4-diazabicyclo[2.2.2]octane |
| DBN | 1,5-diazabicyclo[4.3.0]non-5-ene |
| DBU | 1,8-diazabicyclo[5.4.0]undec-7-ene |
| DCC | dicyclohexylcarbodiimide |
| DDQ | 2,3-dichloro-5,6-dicyano-1,4-benzoquinone |
| DEAD | diethyl azodicarboxylate |
| DHP | 3,4-dihydro-2*H*-pyrane |
| DIAD | diisopropyl azodicarboxylate |
| DIBALH | diisobutylaluminium hydride |
| DMAP | 4-dimethylaminopyridine |
| DME | 1,2-dimethoxyethane |
| DMF | *N,N*-dimethylformamide |
| DMPU | 1,3-dimethyl-3,4,5,6-tetrahydro-2(1*H*)-pyrimidinone |
| DMSO | dimethyl sulfoxide |
| EECE | electric eel cholinesterase |
| HMDS | 1,1,1,3,3,3-hexamethyldisilazane |
| HMPA | hexamethylphosphoramide |
| HOBT | 1-hydroxybenzotriazole |
| Im | imidazol-1-yl |
| LDA | lithium diisopropylamide |
| mcpba | *m*-chloroperoxybenzoic acid |
| MS | molecular sieves |
| NBS | *N*-bromosuccinimide |
| NCS | *N*-chlorosuccinimide |
| NIS | *N*-iodosuccinimide |
| PCC | pyridinium chlorochromate |
| PDC | pyridinium dichromate |
| PLE | pig liver esterase |
| PPL | pig pancreatic lipase |
| PPTS | pyridinium *p*-toluenesulfonate |
| Ppyr | 4-pyrrolidinopyridine |
| PTSA | *p*-toluenesulfonic acid |
| Pyr | pyridine |
| TBAF | tetrabutylammonium fluoride |
| TBDPSCl | *tert*-butyldiphenylsilyl chloride |
| TBSCl | *tert*-butyldimethylsilyl chloride |
| TBSOTf | *tert*-butyldimethylsilyl triflate |
| TESCl | triethylsilyl chloride |
| TESOTf | triethylsilyl triflate |
| Tf | trifluoromethanesulfonyl |
| TFA | trifluoroacetic acid |
| TFAA | trifluoroacetic anhydride |
| TfOH | triflic acid |
| THF | tetrahydrofuran |
| TIPSCl | triisopropylsilyl chloride |
| TIPSOTf | triisopropylsilyl triflate |
| TMEDA | *N,N,N′N′*-tetramethylethylenediamine |
| TMSBr | trimethylsilyl bromide |
| TMSCl | trimethylsilyl chloride |
| TMSI | trimethylsilyl iodide |
| TMSOTf | trimethylsilyl triflate |
| TPAP | tetra-*n*-propylammonium perruthenate |
| Ts | *p*-toluenesulfonyl |

# Chapter 1 Protecting Groups: An Overview

*Protection is not a principle, but an expedient*
*Benjamin Disraeli, 17 March, 1845*

## 1.1 Death, Taxes, and Protecting Groups

The problem of functional group incompatibility in the synthesis of complex organic structures has persisted since the pioneering research of Emil Fischer on the synthesis of carbohydrates. One of Fischer's enduring contributions to the development of organic chemistry was the notion that an otherwise reactive functional group could be temporarily rendered inert by appending a suitable protecting group which could then be later removed. Despite an intervening century of fabulous progress in synthetic methodology, the proliferation of protecting groups is a tacit acknowledgment that selectivity in functional group transformations remains a central and unsolved problem in organic synthesis. The problem is especially acute in the design and construction of polyfunctional molecules such as peptides, oligosaccharides, glycopeptides, glycolipids, nucleotides, and polyketides which often require a scaffold of protecting groups comparable in mass to the target itself.

There are 7 tactical considerations which define how effectively a protecting group will best fulfil its assigned strategic role of shielding a functional group from destruction (or reaction with another functional group):

1. It should be cheap or readily available.
2. The protecting group should be easily and efficiently introduced.
3. It should be easy to characterise and avoid such complications as the creation of new stereogenic centres.
4. It should be stable to the widest possible range of reaction and work-up conditions.
5. It should be stable to the widest possible range of techniques for separation and purification, such as chromatography.
6. It should be removed selectively and efficiently under highly specific conditions.
7. The by-products of the deprotection should be easily separated from the substrate.

If we accept that the business of organic synthesis is to promote reactivity and not prevent it, protecting groups — however elegant in conception — will necessarily excite disdain. Each protecting group lengthens a synthesis by at least two steps with the inevitable reduction in yield and increase in cost. Protecting groups add no value to a synthesis — they can only detract — yet the synthesis of a molecule of even modest complexity can seldom avoid their assistance. Contrary to popular belief, organic synthesis is not a mature science but an adolescent one and the continuing invention of new protecting groups is a symptom of our plight with regard to reaction selectivity. Like death and taxes, protecting groups have become a consecrated obstruction which we cannot elude; we will continue to depend on them for the forseeable future and we can admire the ingenuity that is invested in their design, but it is a wise practitioner who holds that "protection is not a principle, but an expedient".

## 1.2 Deprotection: The Concept of Orthogonal Sets

A typical synthetic plan to a complex natural product is a matrix of several interdependent and parallel strategies encompassing such issues as fragment synthesis, fragment linkage, stereochemistry, functional group interconversion, and protecting groups. Failure in any one can lead to expensive and wasteful modification — or defeat. Since a protection strategy is usually governed by the timing, sequence, and conditions of *deprotection*, we need to classify the repertoire of protecting groups into sets based on orthogonal deprotection modes. An orthogonal set is *ideally* a group of protecting groups whose removal is accomplished in any order with reagents and conditions which do not affect protecting groups in other orthogonal sets[1,2]. Unfortunately, the boundaries between orthogonal sets and the gradations of lability within orthogonal sets are not always well-defined leading to diminished selectivity. Nevertheless, as an organising principle, the concept of orthogonal sets is useful.

The protecting groups in this book are divided into 12 orthogonal sets which will be briefly described below. In the following chapters the common members of the orthogonal sets will be considered in greater detail.

## 1.2.1 Protecting Groups Cleaved by Basic Solvolysis

The acyl derivatives of thiols, hydroxyls (alcohols and carboxyls), and amino groups are amongst the oldest protecting groups still in standard use today. They are all easily prepared by standard methods from activated carboxylic acids but the relative ease of hydrolysis with base varies widely. Thiol esters are too susceptible to nucleophilic attack to offer sustained protection for the thiol group[3] but acetate, benzoate, and pivaloate esters (to name but a few) offer protection over a wide enough range of conditions to be synthetically useful. Acetates and benzoates especially are prized because they can be removed with $K_2CO_3$ or $NH_3$ in MeOH. Furthermore, the ease of cleavage can be tuned by taking advantage of steric and electronic effects. Thus pivaloate esters, with their greater steric hindrance, react with $NH_3$ in MeOH so slowly that acetates can be removed selectively[4] whereas trifluoroacetates are so reactive they hydrolyse at pH 7[5]. Some measure of the range of reactivity available by electronic tuning can be gleaned from the following relative rate data: acetate (1) chloroacetate (760) dichloroacetate (16, 000) trichloroacetate (100, 000)[6].

One problem associated with the use of esters as protecting groups in polyfunctional systems is the tendency for intramolecular transesterification leading to migration of the acyl function to a neighboring alcohol — a problem which is prominent with acetates but less significant with benzoates[7]. Nevertheless, migration of benzoates is observed when there is a driving force as illustrated by the benzoate migration [Scheme 1.1] *en route* to *N*-acetyl neuraminic acid[8,9]. In this case the migration was thermodynamically driven by the greater stability of the equatorial benzoate in the product.

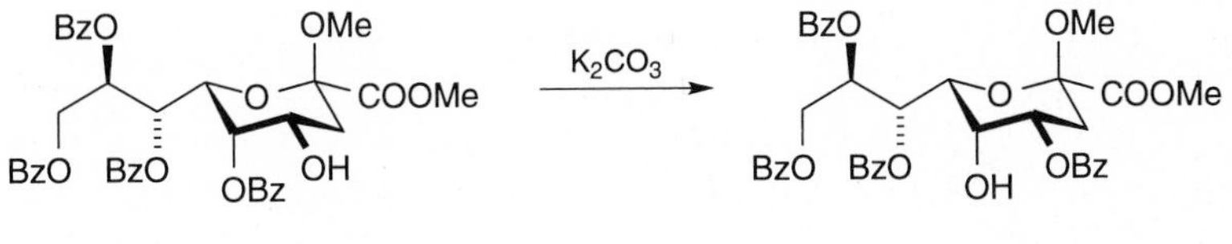

**Scheme 1.1**

In contrast to esters, amide hydrolysis usually requires rather forcing conditions; hence, amide protection of amines is not as common as gentler alternatives. Notable exceptions include trifluoroacetamides[10] and phthalimides. Trifluoroacetamides are so labile that they can be removed with $K_2CO_3$ in MeOH under conditions which preserve some methyl esters.

Phthalimides are a special case in the first orthogonal set because their cleavage by hydrazine in MeOH or EtOH is not strictly a solvolysis reaction. The reason for using hydrazine is apparent in the mechanism outlined in Scheme 1.2. The carbonyl groups of the imide function are much more susceptible to nucleophilic attack than an amide and the first step simply involves cleavage of the imide function by the usual addition–elimination mechanism. In the absence of an *internal nucleophile*, the resultant benzamide would be resistant ot further reaction but the presence of the neighboring hydrazide confers the advantage of high effective molarity leading to speedy cleavage with formation of the free amine.

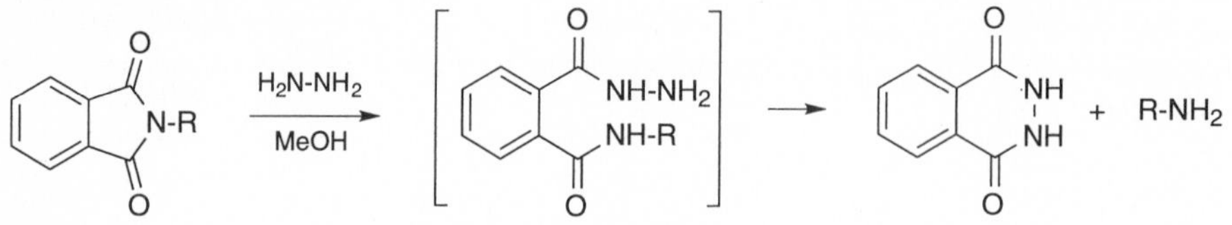

$H_2N\text{-}NH_2$, MeOH

**Scheme 1.2**

### 1.2.2 Protecting Groups Cleaved by Acid

The acid-labile protecting groups are the most difficult to classify into an orthogonal set because virtually all protecting groups can be cleaved by acid albeit under conditions which may be brutal. Nevertheless, in the synthesis of polyfunctional molecules, certain protecting groups have come to be valued by custom and practice for their lability under acidic conditions and these groups are sufficiently tolerant of protecting groups in other orthogonal sets to be useful.

The acid-labile orthogonal set can be roughly divided into two subsets. In the first subset heterolytic scission of a C–O bond in tertiary alkyl or benzylic ethers, esters, and urethanes is promoted by formation of a carbocation stabilised by induction or resonance [Scheme 1.3]. The reaction does not occur at ordinary temperatures unless provoked by prior coordination of a proton or Lewis acid to an oxygen atom. The range of conditions is remarkably broad; for example HBr in HOAc is required to deprotect benzylic esters and ethers whereas trityl ethers are so labile that dilute acetic acid is sufficient. *tert*-Butyl ethers, esters, and urethanes are typically cleaved with trifluoroacetic acid in dichloromethane. Formation of a stabilised carbocation intermediate is not a prerequisite for Lewis acid mediated cleavage of a C–O bond: provided a good nucleophile is present (bromide, iodide, thioether), even a robust methyl ether can be deprotected using $Me_3SiI$[11], $BBr_3$, or $BF_3$–thiolane as illustrated in Scheme 1.4[12] and phenolic isopropyl ethers can be deprotected in the presence of methyl ethers using $BCl_3$ [Scheme 1.5][13].

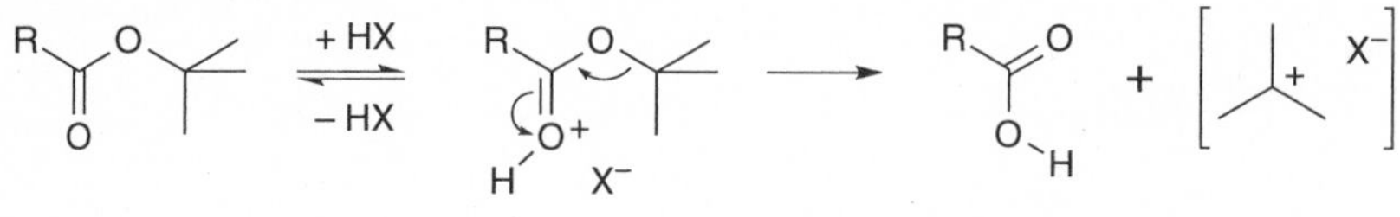

+ HX / – HX

**Scheme 1.3**

NaI, $Me_3SiCl$, MeCN, r.t., 68%

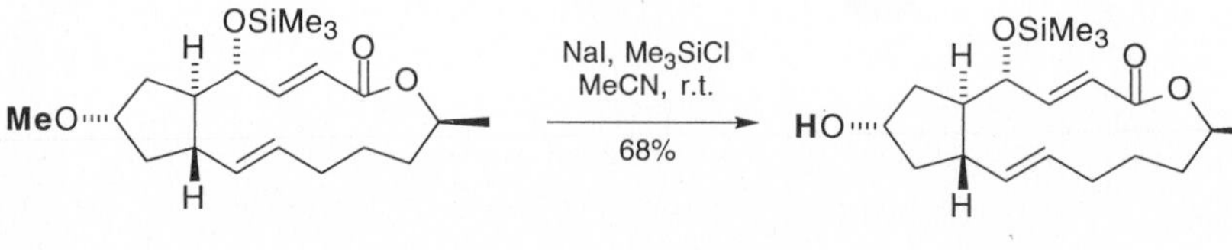

**Scheme 1.4**

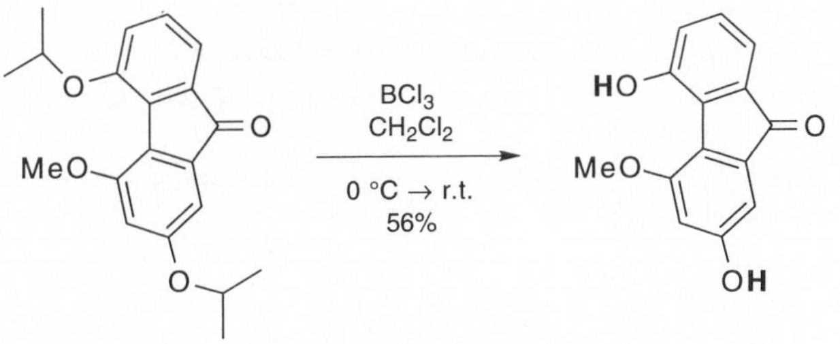

$BCl_3$
$CH_2Cl_2$
0 °C → r.t.
56%

Scheme 1.5

The second subset of acid labile protecting groups consists of *O,O*-acetals. Like the protecting groups in the first subset, heterolysis of an *O,O*-acetal is induced by protic acids and Lewis acids resulting in formation of a resonance stabilised intermediate — an oxonium ion [Scheme 1.6]. In the presence of nucleophiles (e.g., water), the highly electrophilic oxonium ion reacts further to produce a hemiacetal which can then collapse to give two alcohols (or a diol) and and a carbonyl derivative. The conditions required to deprotect *O,O*-acetals vary widely according to structure. Acyclic acetals, isopropylidene derivatives, and tetrahydropyranyl (THP) ethers are the most labile requiring dilute acetic acid or pyridinium *p*-toluenesulfonate (PPTS) in MeOH to accomplish deprotection. On the other hand, *O,O*-acetals lacking substituents at the acetal carbon such as methoxymethyl (MOM) and 2-methoxyethoxymethyl (MEM) ethers require dilute mineral acid at elevated temperature for deprotection. Benzyloxymethyl (BOM) and 2-(trimethylsilyl)ethoxymethyl (SEM) ethers are similar in reactivity to MOM and MEM ethers towards acid; however, alternative milder methods are available to cleave BOM (hydrogenolysis) and SEM (fluoride-induced fragmentation) groups (*vide infra*).

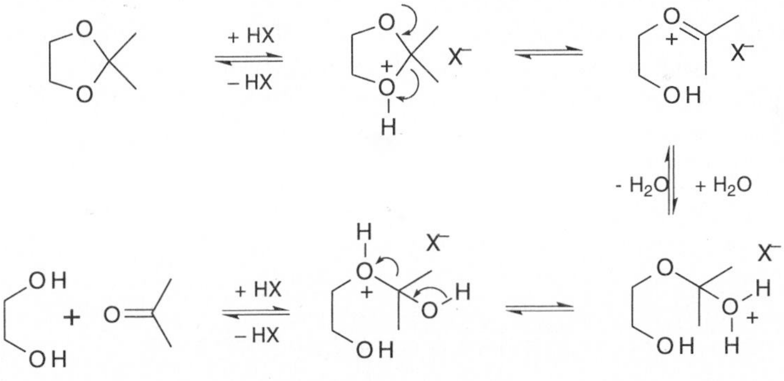

+ HX
– HX
$- H_2O$
$+ H_2O$

Scheme 1.6

### 1.2.3 Protecting Groups Cleaved by Heavy Metal Catalysis

Despite their similarity to the *O,O*-acetals, the *O,S*- and *S,S*-acetals are classified as a separate orthogonal set because they are virtually indestructable by protic acids but hydrolyse with the assistance of heavy metal catalysts such as Ag(I) and Hg(II); hence, *O,O*-Acetals are easily deprotected in the presence of *S,S*-acetals and *vice versa*. The reaction is usually buffered to consume the two equivalents of acid liberated in the hydrolysis according to the mechanism outlined in Scheme 1.7. The acceleration arises from soft–soft interactions with the sulfur and strictly does not constitute catalysis since the metal remains attached to the thiol product. Alternative methods of cleavage include *S*-alkylation in the presence of water or oxidative cleavage with *N*-bromosuccinimide (NBS) or iodine. Typical members of the set are methylthiomethyl (MTM ethers), dithiolanes, and dithianes.

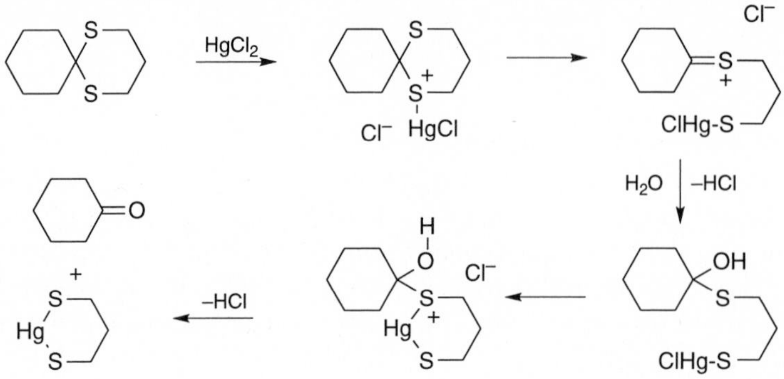

**Scheme 1.7**

### 1.2.4 Fluoride-Induced Cleavage of Si–O and Si–C Bonds

Silyl ethers are one of the most significant orthogonal sets to emerge in the last twenty years. All of the common trialkylsilyl ether protecting groups are labile to acid or base hydrolysis to widely varying degrees and their stability and ease of deprotection can be finely tuned by adjusting the substitution on silicon. However, it is the high thermodynamic affinity of silicon for fluorine which is especially advantageous in deprotection since the usual reagents — tetra-*n*-butylammonium fluoride (TBAF) in THF or HF in acetonitrile — are compatible with a wide range of functional groups and protecting groups in other orthogonal sets. Deprotection with fluoride proceeds via formation of a pentavalent fluorosiliconate intermediate [Scheme 1.8]. The most popular of the trialkylsilyl ethers used as protecting groups for alcohols are the triethylsilyl (TES), *tert*-butyldimethylsilyl (TBDMS or TBS), *tert*-butyldiphenylsilyl (TBDPS), and triisopropylsilyl (TIPS) ethers. Diols can be protected as the cyclic di-*tert*-butylsilylene (DTBS) and 1,1,3,3-tetraisopropyldisiloxanylidene (TIPDS) derivatives.

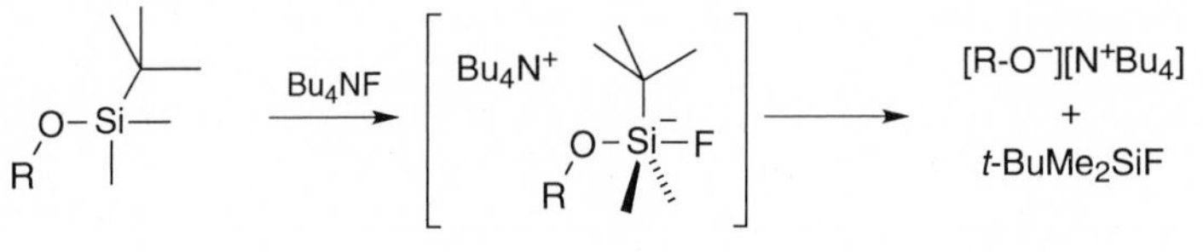

**Scheme 1.8**

Trialkylsilyl protection of carboxylic acids and amines is rare owing to hydrolytic lability. Nevertheless, synthetically useful silicon protecting groups have been developed for these functional groups in which the requisite stability is achieved by incorporating the silicon atom into a 2-(trimethylsilyl)ethyl (TMSE) substituent. The principal is illustrated [Scheme 1.9] by the reaction of 2-(trimethylsilyl)ethyl esters with TBAF: the pentavalent siliconate intermediate fragments with loss of ethylene and fluorotrimethylsilane[14,15] to liberate a carboxylic acid as its tetra-*n*-butylammonium salt.

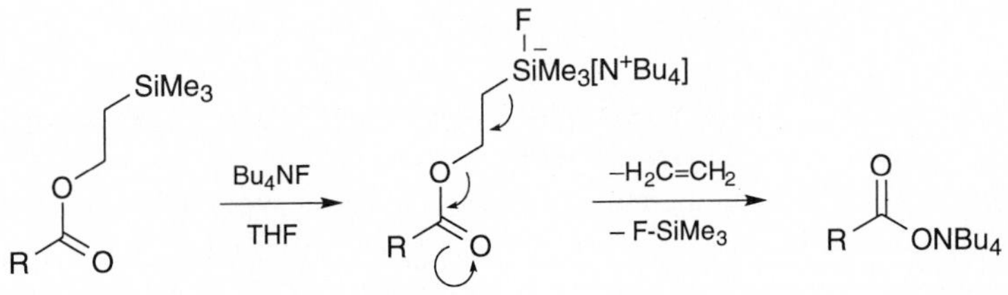

Scheme 1.9

Fluoride-induced fragmentation of 2-(trimethylsilyl)ethyl derivatives can also be used to release alcohols protected as 2-(trimethylsilyl)ethoxymethyl (SEM) ethers [Scheme 1.10][16] or 2-(trimethylsilyl)ethyl carbonates (RO-CO-$OCH_2CH_2SiMe_3$, TMSEC)[17]. By analogy, 2-(trimethylsilyl)ethyl carbamates (RNH-CO-$OCH_2CH_2SiMe_3$, Teoc) offer protection for amines[18] since the carbamate resulting from fragmentation readily loses carbon dioxide. All of these fluoride-induced fragmentation reactions may be considered as a special case of a β-elimination reaction in which the silicon is a surrogate for a proton and fluoride is acting as the base. There are two further orthogonal sets in which deprotection proceeds by an elimination process which differ in the method for initiating the elimination reaction.

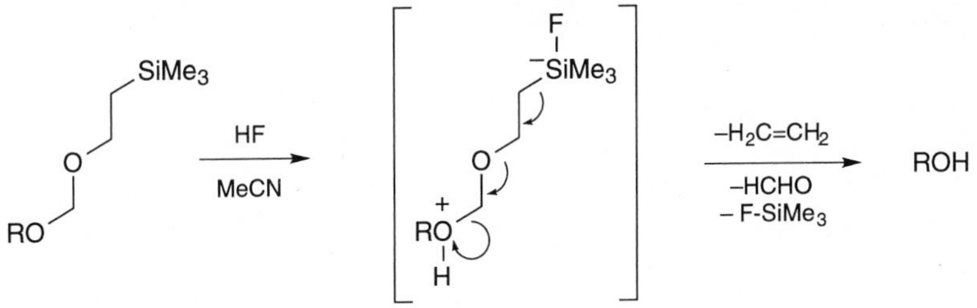

Scheme 1.10

## 1.2.5 Protecting Groups Cleaved by Reductive Elimination with Zinc

The second reaction which is akin to a *β*-elimination involves reductive elimination of 2,2,2-trichloroethyl (TCE) esters on treatment with Zn in acetic acid or Zn-Cu couple in DMF to give 1,1-dichloroethylene [Scheme 1.11][19]. The same principle can be applied to the protection of alcohols as 2,2,2-trichloroethoxymethyl ethers[20] or 2,2,2-trichloroethyl carbonates[21] and amines as 2,2,2-trichloroethoxycarbonyl (Troc) derivatives.

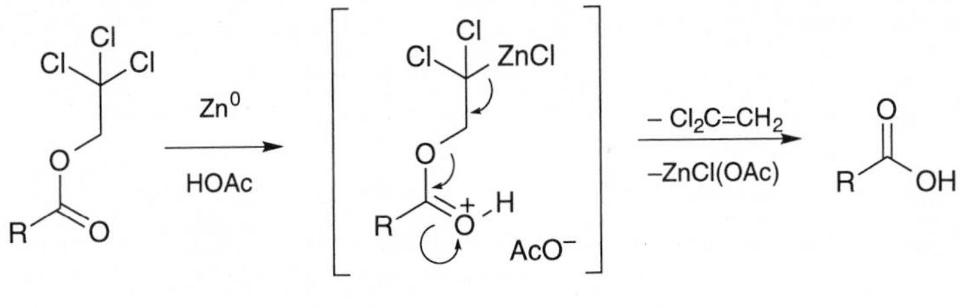

Scheme 1.11

### 1.2.6 Protecting Groups Cleaved by *β*-Elimination Reactions

The sixth orthogonal set consists of groups which deprotect by an $E1_{cb}$ mechanism. An important example involves the mild base-catalysed deprotection of 9-fluorenylmethoxycarbonyl (Fmoc) groups resulting in the liberation of an amino group [Scheme 1.12][22]. The rapid deprotonation of the fluorene group, which is greatly facilitated by the aromatic nature of the resultant dibenzocyclopentadienide anion, is accomplished with piperidine or morpholine in DMF. In a subsequent slower step, elimination generates dibenzofulvene (itself an unstable species that rapidly adds nucleophiles) and a carbamate residue which then decomposes with loss of carbon dioxide to release the free amine. The same principal has been applied to the protection of carboxyl groups as 9-fluorenylmethyl (Fm) esters[23]. The Fmoc group has found a secure niche in peptide and glycopeptide synthesis but, in general, protecting groups that depend on base-catalysed elimination are rare because base sensitivity is a liability not easily accommodated in most syntheses.

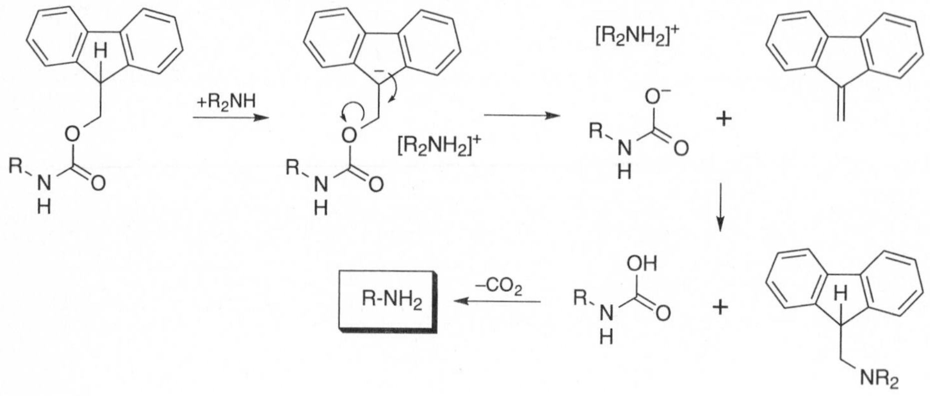

$+R_2NH$ $[R_2NH_2]^+$ $[R_2NH_2]^+$ $R\text{-}NH_2$ $-CO_2$ $NR_2$

Scheme 1.12

### 1.2.7 Protecting Groups Cleaved by Hydrogenolysis

An excellent method for cleaving benzylic ethers, esters, carbamates, and amines uses hydrogen in the presence of a transition metal catalyst such as Pd. Alternatively a process known as catalytic transfer hydrogenation can be employed which uses cyclohexadiene, cyclohexene, formic acid or ammonium formate as the hydrogen source[24]. The method is exceptionally mild and compatible with most functional groups devoid of unsaturation. Hydrogenolysis of benzyloxycarbonyl (Z or Cbz) groups of amines was a major advance in the synthesis of peptides and remains one of the cornerstones of peptide methodology[25]. Similarly, hydrogenolysis of benzyl ethers and benzylidene acetals is a powerful and common procedure in the elaboration of carbohydrate derivatives. Benzyl ethers can be cleaved in the presence of *p*-methoxybenzyl ethers using Raney nickel[26].

### 1.2.8 Protecting Groups Cleaved by Oxidation

Oxidative methods for removing protecting groups are rare. *p*-Methoxy- and 3,4-dimethoxybenzyl (PMB and DMB) ethers are conspicuous recent additions to the repertoire of standard protecting groups because they undergo easy electron transfer to 2,3-dichloro-5,6-dicyano-1,4-benzoquinone (DDQ) to generate an oxonium ion which can be captured by water [Scheme 1.13]. The corresponding acetals can also be deprotected under similar conditions[27]. Ceric ammonium nitrate (CAN) is an alternative oxidant which is superior to DDQ in some cases[28].

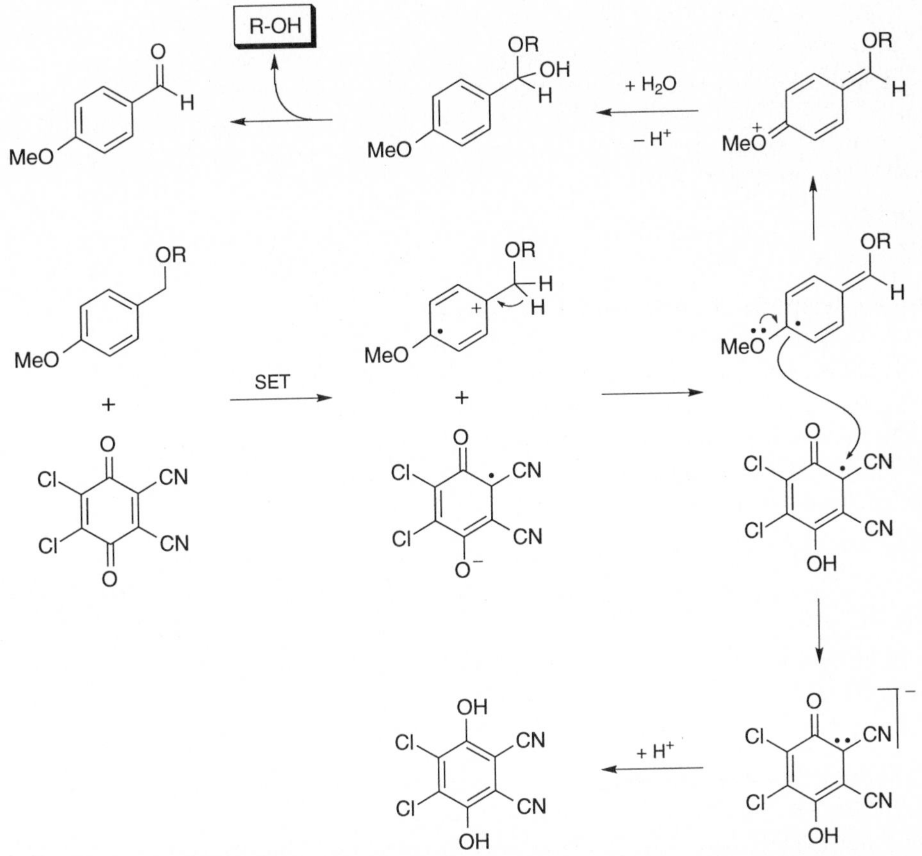

**Scheme 1.13**

### 1.2.9 Protecting Groups Cleaved Under Dissolving Metal Conditions

Sodium or lithium in liquid ammonia cleave benzylic ethers and esters in the presence of a proton source to provide a pentadienyl anion which expels an alkoxide or carboxylate leaving group [Scheme 1.14]. Aqueous acidic workup returns the carboxylic acid or alcohol and toluene. Many functional groups are unable to survive such powerful reducing conditions.

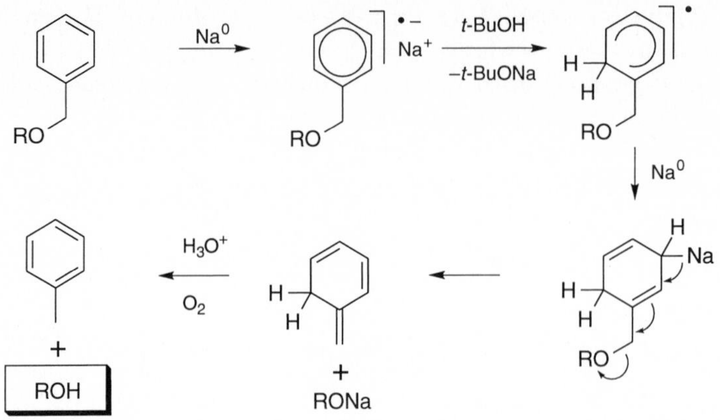

Scheme 1.14

## 1.2.10 Nucleophilic Cleavage of C–O Bonds

The resonance stabilisation of phenolate and carboxylate anions is just sufficient for them to serve as leaving groups in a classical bimolecular nucleophilic substitution [Scheme 1.15]. As a deprotection tactic, the reaction has comparatively narrow scope being limited to the scission of O-Me and O-Et bonds. Typical nucleophiles include chloride, iodide, cyanide, and thiophenolate in dipolar aprotic solvents at elevated temperature.

Scheme 1.15

## 1.2.11 Allylic Protecting Groups

The allyl protecting groups constitute an orthogonal set in their own right because the conditions required for their removal are mild and specific with a high tolerance of other functional groups and protecting groups. The allyl group has been used to protect alcohols (allyl ethers, allyl carbonates), carboxylic acids (allyl esters), and amines (allyl carbamates). The method of deprotection depends on the structure of the substrate. For allyl ethers which are stable to strong base, deprotection can be accomplished by a two step process involving isomerisation of the double bond of the allyl group using *t*-BuOK in DMSO. The resultant enol ether is a stable intermediate that can then be hydrolysed with mild acid[29,30]. Alternatively, rearrangement of the double bond can be accomplished using a transition metal complex such as Rh(I). A possible mechanism for the rhodium(I)-catalysed isomerisation is outlined in Scheme 1.16. The resonance stabilisation of carboxylate and carbamate anions activates allyl esters, carbonates and carbamates to nucleophilic attack under Pd(0)-catalysis in which cases $\pi$-allyl Pd complexes are intermediates (see section 2.4.6 for a mechanism). Pd(0)-catalysed transfer of allyl groups to a range of gentle nucleophiles has been perfected by Kunz and associates[31] for the deprotection of allyl esters and carbamates in acid-sensitive substrates.

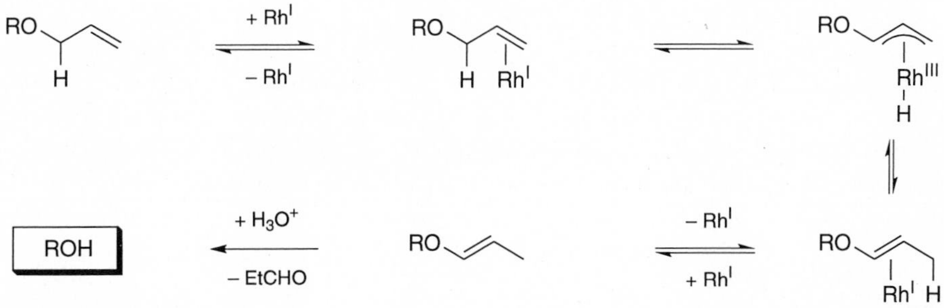

Scheme 1.16

### 1.2.12 Photodeprotection

A photoremoveable protecting group contains a chromophore which is sensitive to light but relatively stable to most chemical reagents[32-34]. For example, irradiation of *o*-nitrobenzyl ethers results in formation of *o*-nitrosobenzaldehyde and an alcohol according to the mechanism outlined in Scheme 1.17. The same principle has been applied to the photodeprotection of *o*-nitrobenzyl carbamates to form amines[35], *o*-nitrobenzyl carbonates to form alcohols[36], *o*-nitrobenzyl esters to form carboxylic acids[37], and *o*-nitrobenzylidene acetals to form 1,2- and 1,3-diols[38].

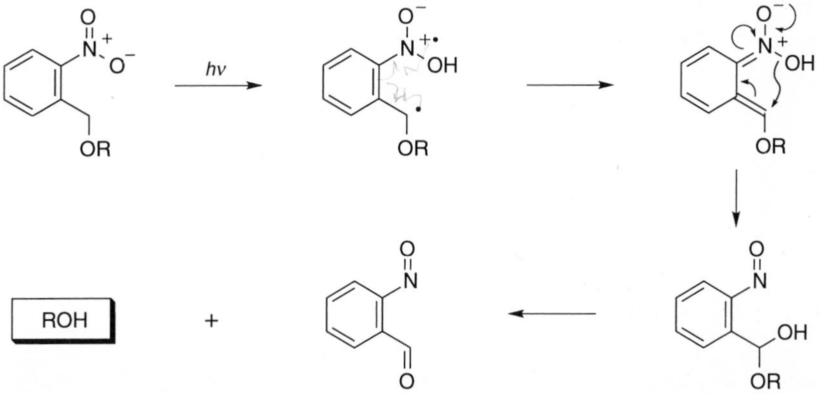

Scheme 1.17

## 1.3 Relay Deprotection

Relay deprotection is an artful dodge by which a protecting group that is hardy under a wide range of conditions is transformed by some chemical reaction into a new and labile group which is then cleaved under mild conditons. Perhaps the most significant example of relay deprotection to emerge in recent years is the allyl ether group and its variants which was introduced in section 1.2.11. However, relay deprotection strategies can be applied to protecting groups in other orthogonal sets as well. For example, the base sensitivity of groups in the $\beta$-elimination orthogonal set (section 1.2.6) can be circumvented by incorporating the C–H activating group in latent form as shown in Scheme 1.18[39].

Cleavage and reduction of a tetrahydropyranyl acetal in the presence of a sensitive dimethyl acetal was accomplished by conversion of the *β*-chloroethyl acetal **18.1** to a *β*-(phenylthio)ethyl acetal by nucleophilic displacement. On oxidation to the corresponding sulfone **18.2**, the molecule was primed for the ensuing base-catalysed elimination and reduction which was accomplished in a single operation using the basic metal hydride reducing agent sodium borohydride. Oxidation–elimination strategies of a similar nature have found application in carboxyl and amino protection.

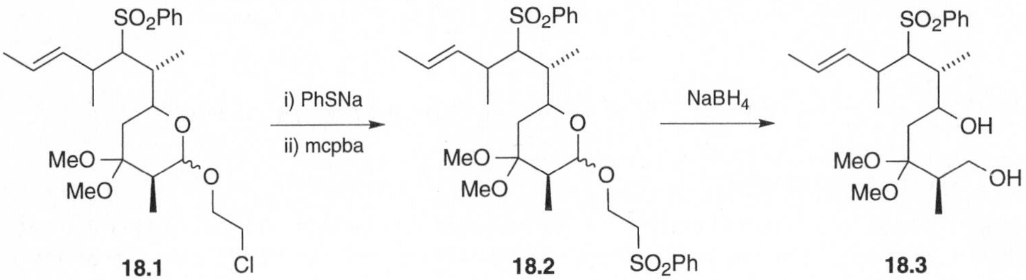

$SO_2Ph$ MeO MeO O O Cl **18.1** i) PhSNa ii) mcpba $SO_2Ph$ MeO MeO O O $SO_2Ph$ **18.2** $NaBH_4$ $SO_2Ph$ OH MeO MeO OH **18.3**

**Scheme 1.18**

## 1.4 Temporary Protection

A protecting group which must survive 10–20 steps or more had better be obdurate; but, what if we require protection for one or, at the most, two steps? Obviously, a frail protecting group might be used to advantage and it would be best if the protection, desired reaction(s), and deprotection were all done in one pot. An example of such *temporary* protection[40] involving a hypersensitive intermediate is shown in Scheme 1.19. During their synthesis of the Calicheamicin-Esperamicin core unit, Danishefsky and co-workers[41,42] were confronted with the problem of adding an alkynyl lithium to the less reactive ketone function of the oxo aldehyde **19.1.** By adding lithium *N*-methylanilide to the aldehyde function, the desired addition to the *α*-amino alkoxide **19.2** was accomplished and the protecting group removed on simple aqueous workup. Scheme 1.20 shows an example of temporary protection on a preparative scale and related reactions are known for the protection-functionalisation of pyridine[43,44] and indole derivatives[45].

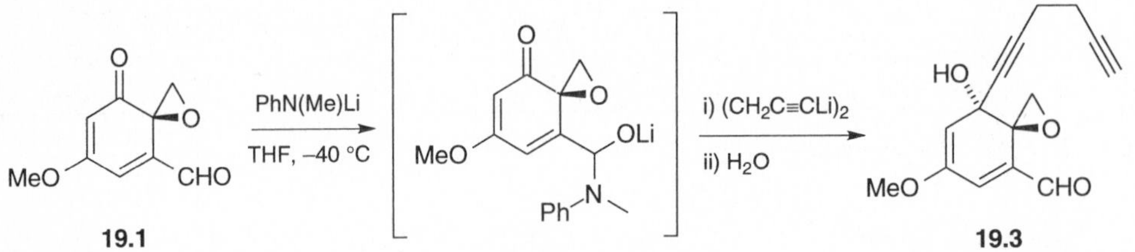

O O MeO CHO **19.1** PhN(Me)Li THF, –40 °C O O MeO OLi Ph N **19.2** i) $(CH_2C{\equiv}CLi)_2$ ii) $H_2O$ HO O MeO CHO **19.3** 50-60% overall

**Scheme 1.19**

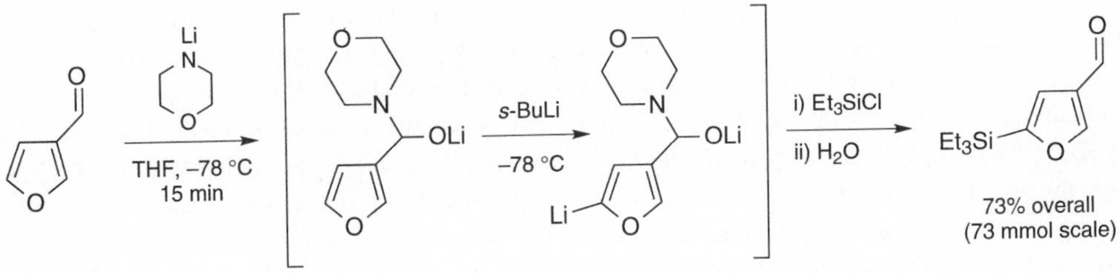

Li N O THF, −78 °C 15 min OLi s-BuLi −78 °C Li i) $Et_3SiCl$ ii) $H_2O$ $Et_3Si$ O 73% overall (73 mmol scale)

Scheme 1.20

# 1.5 Protecting Groups as Not-So-Innocent Bystanders

Ideally, a protecting group should be unobtrusive in every way. It should not alter the conformation of the host molecule; it should not influence the stereochemistry of any reactions at functional groups be they proximate or remote; nor should it participate in any reactions chemically. There are many examples where all of these strictures are violated. We will begin by showing how protecting groups participated unexpectedly in reactions intended for remote functional groups.

Attempts to effect a double nucleophilic displacement of the ditosylate **21.1** with NaCN in warm DMSO failed to return any of the desired dinitrile [Scheme 1.21]. Instead the rearranged product **21.3** was isolated which resulted from neighboring group participation of the dioxolane ring leading to formation of intermediate **21.2**[46].

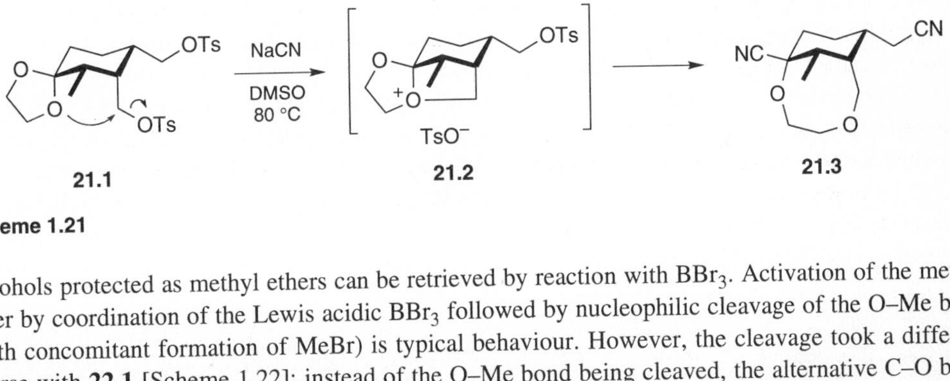

OTs NaCN DMSO 80 °C OTs $TsO^-$ NC CN

**21.1** **21.2** **21.3**

Scheme 1.21

Alcohols protected as methyl ethers can be retrieved by reaction with $BBr_3$. Activation of the methyl ether by coordination of the Lewis acidic $BBr_3$ followed by nucleophilic cleavage of the O–Me bond (with concomitant formation of MeBr) is typical behaviour. However, the cleavage took a different course with **22.1** [Scheme 1.22]: instead of the O–Me bond being cleaved, the alternative C–O bond cleaved owing to participation of the remote acetoxy group[47]. The formation of bromide **22.4** with *retention* of configuration is circumstantial evidence implicating dioxonium ion intermediate **22.3**.

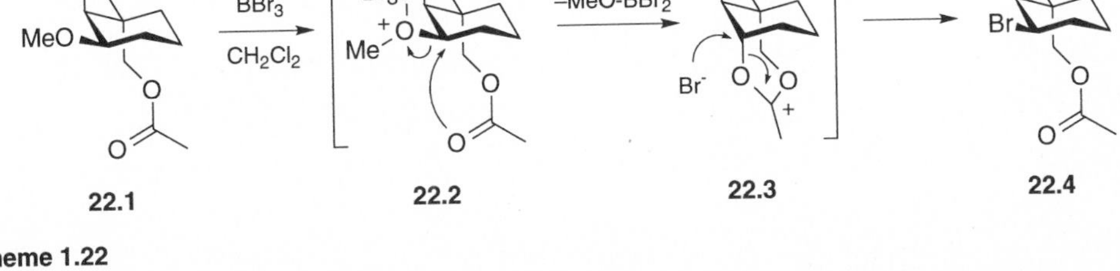

MeO $BBr_3$ $CH_2Cl_2$ $Br_3\bar{B}$ Me −MeO-$BBr_2$ $Br^-$ Br

**22.1** **22.2** **22.3** **22.4**

Scheme 1.22

Acyloxy participation has been used to advantage in controlling the stereochemistry of the Koenigs-Knorr glycosidation reaction [Scheme 1.23]. Treatment of the $\alpha$-glycosyl bromide **23.1** with $Ag^+$ in the presence of an alcohol results in the selective formation of the $\beta$-glycoside **23.3**. The reaction involves participation of the C-2 pivaloyloxy group and leads to formation of the *cis*-fused dioxonium ion **23.2** followed by alcohol attack at the anomeric centre (inversion) leading to the product. Pivaloyl ester groups, being more hindered, are less prone to competing capture of the dioxonium ion intermediate which can lead to ortho ester formation — a side reaction which affects the glycosidation of substrates protected with acetate esters[48].

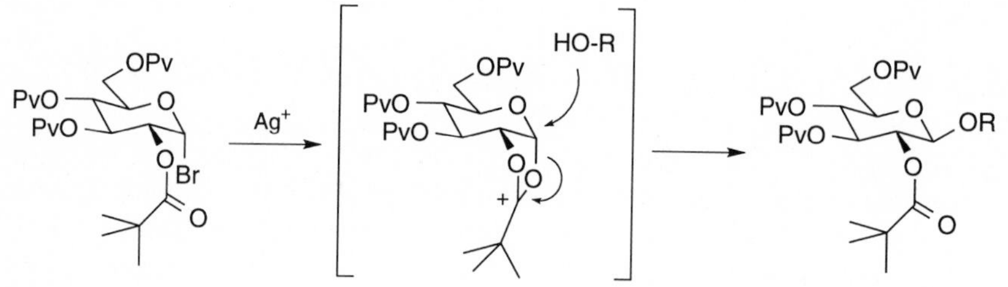

**Scheme 1.23**

Protecting groups can affect the ground state conformation of a molecule and therefore influence its chemistry. For example, the stereoselectivity of the nucleophilic addition of MeLi to the *cis*-fused decalone derivatives **24.1** and **24.2** depended on the steric bulk of the acetal protecting group [Scheme 1.24][49]. Thus, addition to the ketone function of dimethyl acetal **24.1** gave exclusively the oxo alcohol **24.4** after acetal hydrolysis. On the other hand, addition to the dioxolane-protected derivative **24.2** was less diastereoselective. The stereochemistry of the addition in the case of **24.2** was governed by the steric hindrance of the angular methyl group in the preponderant steroid conformation **24.5**. However, the reduced steric demand of the dioxolane ring results in a conformational equilibrium between the steroid and nonsteroid conformation **24.6** with nucleophilic addition to the latter being less stereoselective.

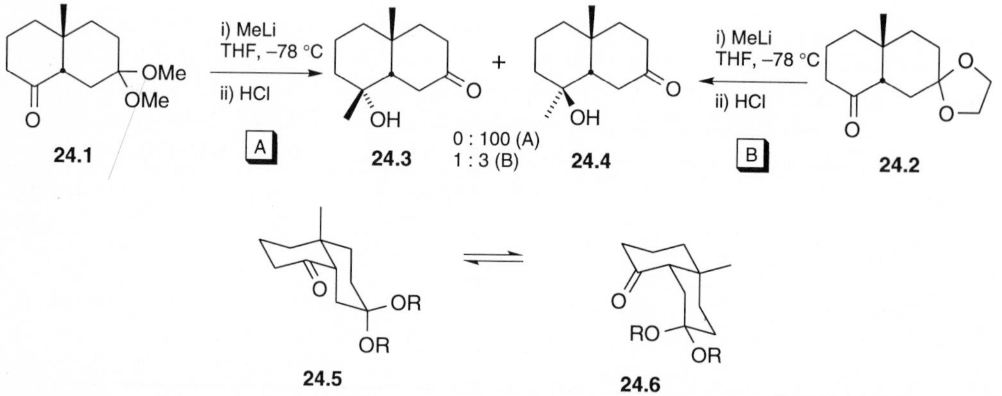

**Scheme 1.24**

Another example which shows how subtle changes in a protecting group can dramatically alter the ground state conformation of a molecule and hence its reactivity comes from the synthesis of dihydroerythronolide A by Stork and Rychnovsky[50]. Cyclisation of the hydroxy acid **25.1** [Scheme 1.25] to the 14-membered lactone ring of the target depended on the substituents on the 1,3-dioxane ring encompassing C9–C11. With **25.1a,b** cyclisation was thwarted completely but **25.1c** underwent macrolactonisation successfully. In the case of **25.1a,b** an axial Me group at the 2-position of the 1,3-dioxane ring incurs an unfavorable 1,3-diaxial interaction with the C9 substituent thereby destabilising the conformer best able to bring the reacting hydroxyl and carboxyl groups close to each other, whereas in **25.1c**, there is no such impediment.

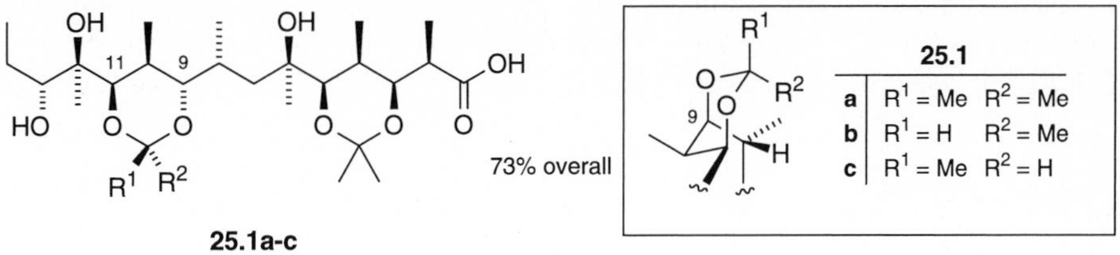

**25.1a-c**

73% overall

| 25.1 | | |
|---|---|---|
| a | $R^1$ = Me | $R^2$ = Me |
| b | $R^1$ = H | $R^2$ = Me |
| c | $R^1$ = Me | $R^2$ = H |

**Scheme 1.25**

Ether-type protecting groups are good Lewis bases which can coordinate to organometallic reagents and influence the stereochemistry of reactions at neighboring sites. Chelation-controlled 1,2-additions of organometallic reagents to $\alpha$- and $\beta$-benzyloxy aldehydes are well-known examples of the principle. By the same token, chelation control can be thwarted by switching to the less basic TBDMS or TBDPS ether protecting groups[51,52]. Recent theoretical studies indicate that the effect is principally electronic in origin[53]. Chelation control has also been implicated in conjugate addition reactions. For example, during a synthesis of the macrolide antibiotic 6-*epi*-erythromycin, Mulzer and co-workers[54] found that the stereochemistry at the anomeric centre (1′ position) of the THP protecting group had a profound effect on the stereochemistry of conjugate addition of $Me_2CuLi$ to the ynone **26.1** [Scheme 1.26]. Similarly, the high diastereoselectivity observed in the conjugate addition of MeLi to the $\alpha,\beta$-unsaturated sulfone **27.1** reflects the heteroatom-assisted delivery of MeLi by the MEM protecting group [Scheme 1.27][55].

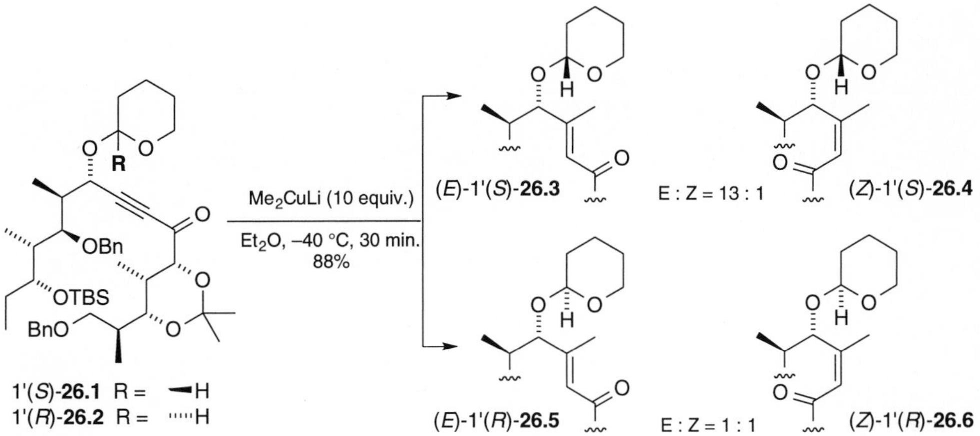

1′(*S*)-**26.1** R = H (wedge)
1′(*R*)-**26.2** R = H (hash)

$Me_2CuLi$ (10 equiv.), $Et_2O$, –40 °C, 30 min. 88%

(*E*)-1′(*S*)-**26.3** E : Z = 13 : 1 (*Z*)-1′(*S*)-**26.4**

(*E*)-1′(*R*)-**26.5** E : Z = 1 : 1 (*Z*)-1′(*R*)-**26.6**

**Scheme 1.26**

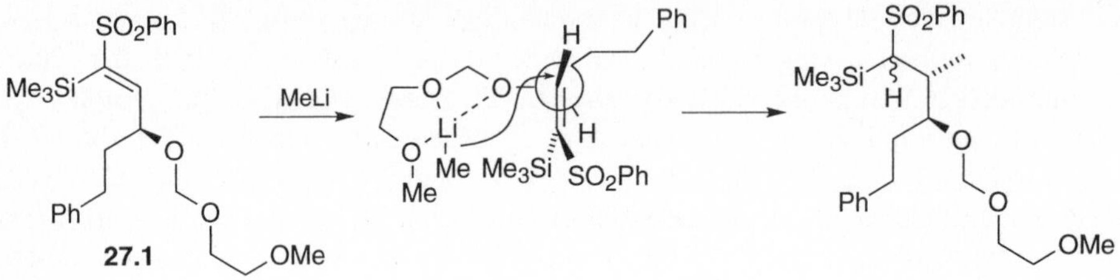

Scheme 1.27

Coordination effects can greatly facilitate displacement reactions too. Thus, the sulfur atom of a methylthiomethyl protecting group provided heteroatom-assisted displacement of the secondary tosylate **28.1** by organocuprates [Scheme 1.28][56].

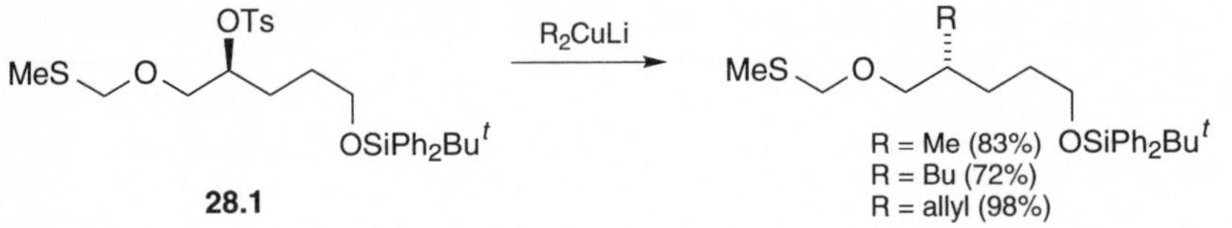

Scheme 1.28

Heteroatom-assisted metallation reactions are very common and especially useful for the selective *o*-functionalisation of aromatic compounds*. More recently, activation of non-aromatic substrates has attracted attention but only a few examples pertinent to protecting groups will be cited here. Kerrick and Beak[57] [Scheme 1.29] have shown that pyrrolidine protected with a *tert*-butoxycarbonyl (Boc) group **29.1** is enantioselectively deprotonated by *s*-BuLi activated by the homochiral alkaloid (–)-sparteine (**29.2**)[58]. Enantioselective deprotonations mediated by (–)-sparteine were first exploited by Hoppe and co-workers for the metallation of saturated[59] and allylic[60,61] alcohols using carbamates as activators*.

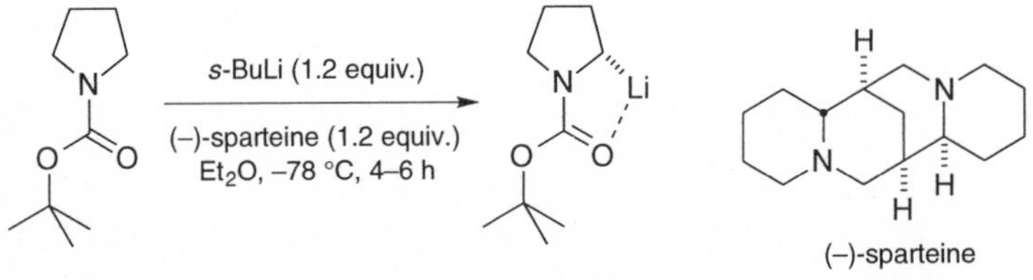

Scheme 1.29

The $\alpha$-lithiation of cyclic enol ethers is readily accomplished with strong bases such as *t*-BuLi[62] but the presence of heteroatoms can prevent or divert the desired course of reaction. For example, dihydropyran **30.1** protected with a *tert*-butyldimethylsilyl ether, underwent competing metallation of the Si-Me group [Scheme 1.30][63], but the problem is easily solved by switching to the more hindered triisopropylsilyl ethers[64].

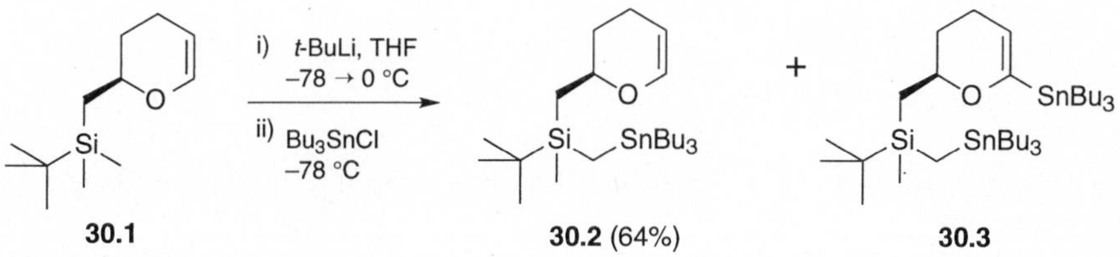

**Scheme 1.30**

We have seen that participation of a protecting group in a reaction intended for a functional group can be a boon or a nuisance depending on circumstances. We will end this section with an illustration of how the potential for neighboring group participation was built into a protecting group by *design*. Our example is the penultimate step of Fraser-Reid's synthesis[65] of the blood group substance B tetrasaccharide in which *n*-pentenyl glycosides* thrice served the dual and divergent roles of protection and activation of the anomeric centre in a crucial glycosidation reaction [Scheme 1.31]. Like any other typical glycoside, the *n*-pentenyl glycoside units in the monosaccharide components were installed at the outset, and being *O*-glycosides, they were stable under a wide range of conditions. However, the terminal alkene moiety strategically placed 5-atoms away from the glycosidic oxygen served as a trigger for the three glycosidation reactions required to assemble the target (only the last of which is shown here). Thus treatment of *n*-pentenyl glycoside **31.1** with *N*-iodosuccinimide and triethylsilyl triflate led to electrophilic attack on the alkene resulting in formation of the oxonium ion **31.2** which reacted with the the trisaccharide component **31.3** to give the desired tetrasaccharide in 91% yield along with 2-iodomethyltetrahydrofuran. The advantages of the strategy are obvious: once the appropriately protected monosaccharides were prepared, the chemical transformations were restricted to the coupling event followed by liberation of the pertinent hydroxyl group so that the next stage could proceed. Hence, protecting group adjustments in the precious coupled products were minimised.

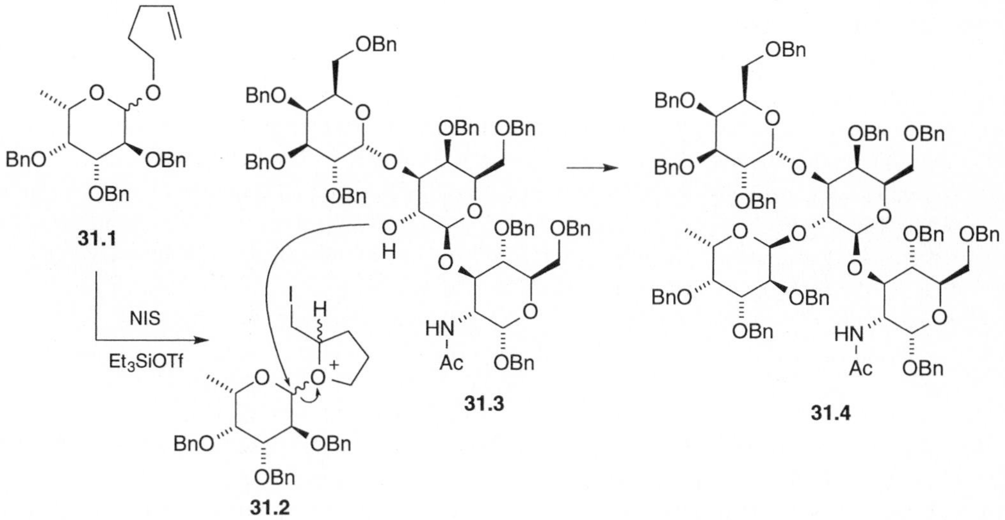

**Scheme 1.31**

# 1.6 Reviews

## 1.6.1 General Reviews on Protecting Groups


1 *Protective Groups in Organic Chemistry.* McOmie, J. F. W., Ed.; Plenum Press: New York, 1973.
2 *Protective Groups in Organic Synthesis*, 2nd ed. Greene, T. W.; Wuts, P. G. M.; Wiley-Interscience: New York, 1991.
3 Protecting Groups. Kunz, H.; Waldmann, H. In *Comprehensive Organic Synthesis* Trost, B. M.; Fleming, I., Eds.; Pergamon Press: Oxford, 1991; p 631.
4 Temporary Protection. Comins, D. L. *Synlett* **1992**, 615.
5 Photoremoveable Protecting Groups. Pillai, V. N. R. *Synthesis* **1980**, 1.
6 Photolytic Deprotection and Activation of Functional Groups. Pillai, V. N. R. *Org. Photochem.* **1987**, *9*, 225.
7 Applications of Photosensitive Groups in Carbohydrate Chemistry. Zehavi, V. *Adv. Carbohydr. Chem. Biochem.* **1988**, *46*, 179.
8 Synthetic Saccharide Photochemistry. Descotes, G. *Top. Curr. Chem.* **1990**, *154*, 39.


## 1.6.2 Reviews Concerning the Synthesis of Complex Targets Including Discussion of Protecting Group Problems and Strategies


1 The Chemical Synthesis of Oligonucleotides. Amarnath, V.; Broom, A. D. *Chem. Rev.* **1977**, *77*, 183.
2 The Chemical Synthesis of Oligo- and Poly-ribonucleotides. Reese, C. B. In *Nucleic Acids and Molecular Biology*; Eckstein F.; Lilley, D. M. J., Eds.; Springer-Verlag: Berlin, 1989; p 164.
3 Advances in the Synthesis of Oligonucleotides by the Phosphoramidite Approach. Beaucage, S. L.; Iyer, R. P. *Tetrahedron* **1992**, *48*, 2223.
4 The Functionalisation of Oligonucleotides via Phosphoramidite Derivatives. Beaucage, S.; Iyer, R. P. *Tetrahedron* **1993**, *49*, 1925.
5 The Synthesis of Modified Oligonucleotides by the Phosphoramidite Approach and Their Applications. Beaucage, S. L.; Iyer, R. P. *Tetrahedron* **1993**, *49*, 6123.
6 The Synthesis of Specific Ribonucleotides and Unrelated Phosphorylated Biomolecules by the Phosphoramidite Method. Beaucage, S.; Iyer, R. P. *Tetrahedron* **1993**, *49*, 10441.
7 Recent Developments in the Synthesis of *myo*-Inositol Phosphates. Billington, D. C., *Chem. Soc. Rev.* **1989**, *18*, 83.
8 Recent Advances in the Chemistry and Biochemistry of Inositol Phosphates of Biological Interest. Potter, B. L. V. *Nat. Prod. Rep.* **1990**, *7*, 1.
9 *Principles of Peptide Synthesis.* Bodanszky, M. Springer-Verlag: Berlin, 1984.
10 *The Practice of Peptide Synthesis.* Bodanszky, M.; Bodanszky, A. Springer-Verlag: Berlin, 1984.
11 *The Chemical Synthesis of Peptides.* Jones, J. Clarendon Press: Oxford, 1991.
12 Convergent Solid-Phase Peptide Synthesis. Lloyd-Williams, P.; Albericio, F.; Girault, E. *Tetrahedron* **1993**, *49*, 11065.
13 Advances in Selective Chemical Synthesis of Complex Oligosaccharides. Paulsen, H. *Angew. Chem. Int. Ed. Engl.* **1982**, *21*, 155.

14 Synthesis of Complex Oligosaccharide Chains of Glycoproteins. Paulsen, H. *Chem. Soc. Rev.* **1984**, *13*, 15.

15 Synthesis of Glycopeptides, Partial Structures of Biological Recognition Compounds. Kunz, H. *Angew. Chem. Int. Ed. Engl.* **1987**, *26*, 294.

16 New Methods for the Synthesis of Glycosides and Oligosaccharides — Are There Alternatives to the Koenigs-Knorr Method? Schmidt, R. R. *Angew. Chem. Int. Ed. Engl.* **1986**, *25*, 212.

17 *n*-Pentenyl Glycosides in Organic Chemistry: A Contemporary Example of Serendipity. Fraser-Reid, B.; Udodong, U. E.; Wu, Z.; Ottosson, H.; Merritt, J. R.; Rao, C. S.; Roberts, C.; Madsen, R. *Synlett* **1992**, 927.

18 Synthesis of Glycolipids. Gigg, J.; Gigg, R. *Top. Curr. Chem.* **1990**, *154*, 77.

19 Chemical Synthesis of Glycosphingolipids. Nicolaou, K. C. *Chemtracts* **1991**, *4*, 181.

20 *The Inositol Phosphates.* Billington, D. C. VCH: Weinheim, 1992.

21 Preparation of Selectively Alkylated Saccharides as Synthetic Intermediates. Stanek, J. *Top. Curr. Chem.* **1990**, *154*, 209.

22 Synthesis of Oligosaccharides Related to Bacterial O-Antigens. Bundle, D. R. *Top. Curr. Chem.* **1990**, *154*, 1.

23 Syntheses of Deoxy Oligosaccharides. Thiem, J.; Klaffke, W. *Top. Curr. Chem.* **1990**, *154*, 285.

24 The Synthesis of Carbohydrate Derivatives from Acyclic Precursors. Ager, D. J.; East, M. B. *Tetrahedron* **1993**, *49*, 5683.

25 *Antibiotics and Antiviral Compounds.* Krohn, K.; Kirst, H. A.; Maag, H., Eds; VCH: Weinheim, 1993.

26 *The Organic Chemistry of β-Lactams.* Georg, G. I., Ed.; VCH: Weinheim, 1992.

### 1.6.3 Heteroatom-Assisted Reactions

1 Heteroatom-Facilitated Lithiations. Gschwend, H. W.; Rodriguez, H. R. *Org. React.* **1979**, 26.

2 Stereo- and Regiocontrol by Complex Induced Proximity Effects: Reactions of Organolithium Compounds. Beak, P.; Meyers, A. I. *Acct. Chem. Res.* **1986**, *19*, 356.

3 Oxygen- and Nitrogen-Assisted Lithiation and Carbolithiation of Non-Aromatic Compounds; Properties of Non-Aromatic Organolithium Compounds Capable of Intramolecular Coordination to Oxygen and Nitrogen. Klumpp, G. W. *Rec. Trav. Chim. Pay-Bas* **1986**, *105*, 1.

4 Directed ortho Metallation. Tertiary Amide and *O*-Carbamate Directors in Synthetic Strategies for Polysubstituted Aromatics. Snieckus, V. *Chem. Rev.* **1990,** *90*, 879.

5 Metallated 2-Alkenyl Carbamates: Chiral Homoenolate Reagents for Asymmetric Synthesis. Hoppe, D.; Kraemer, T.; Schwark, J. R.; Zschage, O. *Pure Appl. Chem.* **1990**, *62*, 1999.

## References

1 Barany, G.; Merrifield, R. B. *J. Am. Chem. Soc.* **1977**, *99*, 7363.

2 Barany, G.; Albericio, F. *J. Am. Chem. Soc.* **1985**, *107*, 4936.

3 Wünsch, E. *Methoden Org. Chem. (Houben-Weyl)* **1974**, *15/1*, 735.

4 Griffin, B. E.; Jarman, M.; Reese, C. B. *Tetrahedron* **1968**, *24*, 639.

5 Cramer, F.; Bär, H. P.; Rhaese, H. J.; Sänger, W.; Scheit, K. H.; Schneider, G.; Tenigkeit, J. *Tetrahedron Lett.* **1963**, 1039.

6 Isaacs, N. S. *Physical Organic Chemistry*; Halstead Press/Wiley: New York, 1987; pp p. 470.

7 Haines, A. H. *Adv. Carbohydr. Chem. Biochem.* **1976**, *33*, 11.

8 Danishefsky, S. J.; De Ninno, M. P.; Chen, S.-h. *J. Am. Chem. Soc.* **1988**, *110*, 3929.

9 De Ninno, M. P. *Synthesis* **1991**, 583.

10 Weygand, F.; Geiger, R. *Chem. Ber.* **1956**, *89*, 647.
11 Jung, M. E.; Lyster, M. A. *Org. Synth. Coll. Vol. VI* **1988**, 353.
12 Le Drian, C.; Greene, A. E. *J. Am. Chem. Soc.* **1982**, *104*, 5473.
13 Wang, W.; Snieckus, V. *J. Org. Chem.* **1992**, *57*, 424.
14 Gerlach, H. *Helv. Chim. Acta* **1977**, *60*, 3039.
15 Sieber, P. *Helv. Chim. Acta* **1977**, *60*, 2711.
16 Lipshutz, B. H.; Pegram, J. J. *Tetrahedron Lett.* **1980**, *21*, 3343.
17 Gioelli, C.; Balgobin, N.; Josephson, S.; Chattopadhyaya, J. B. *Tetrahedron Lett.* **1981**, *22*, 969.
18 Carpino, L. A.; Tsao, J.-H.; Ringsdorf, H.; Fell, E.; Hettrich, G. *J. Chem. Soc., Chem. Commun.* **1978**, 358.
19 Woodward, R. B.; Heusler, K.; Gosteli, J.; Naegeli, P.; Oppolzer, W.; Ramage, R.; Ranganathan, S.; Vorbruggen, H. *J. Am. Chem. Soc.* **1966**, *88*, 852.
20 Jacobson, R. M.; Claser, J. W. *Synth. Commun.* **1979**, *9*, 57.
21 Windholz, T. B.; Johnson, D. B. R. *Tetrahedron Lett.* **1967**, 2555.
22 Carpino, L. A.; Sadat-Aalaee, D.; Beyermann, M. *J. Org. Chem.* **1990**, *55*, 1673.
23 Kessler, H.; Siegmeier, H. *Tetrahedron Lett.* **1983**, *24*, 281.
24 Entwhistle, I. D.; Johnstone, R. A. W.; Whitby, A. H. *Chem. Rev.* **1985**, *85*, 129.
25 Bergmann, M.; Zervas, L. *Chem. Ber.* **1932**, *65*,
26 Oikawa, Y.; Tanaka, T.; Horita, K.; Yonemitsu, O. *Tetrahedron Lett.* **1984**, *25*, 5397.
27 Oikawa, Y.; Yoshioka, T.; Yonemitsu, O. *Tetrahedron Lett.* **1982**, *23*, 889.
28 Johansson, R.; Samuelsson, B. *J. Chem. Soc., Perkin Trans. 1* **1984**, 2371.
29 Gigg, R.; Warren, C. D. *J. Chem. Soc. C* **1965**, 2205.
30 Gigg, R.; Warren, C. D. *J. Chem. Soc. C* **1968**, 1903.
31 Kunz, H. *Angew. Chem. Int. Ed. Engl.* **1987**, *26*, 294.
32 Pillai, V. N. R. *Synthesis* **1980**, 1.
33 Pillai, V. N. R. *Org. Photochem.* **1987**, *9*, 225.
34 Zehavi, V. *Adv. Carbohydr. Chem. Biochem.* **1988**, *46*, 179.
35 Amit, B.; Zehavi, U.; Patchornik, A. *J. Org. Chem.* **1974**, *39*, 192.
36 Cama, L. D.; Christansen, B. G. *J. Am. Chem. Soc.* **1978**, *100*, 8006.
37 Webber, J. A.; Van Heyningen, E. M.; Vasilieff, R. T. *J. Am. Chem. Soc.* **1969**, *91*, 5694.
38 Collins, P. M.; Munasinghe, V. R. N. *J. Chem. Soc., Perkin Trans. 1* **1983**, 921.
39 Isobe, M.; Kitamura, M.; Goto, T. *J. Am. Chem. Soc.* **1982**, *104*, 4997.
40 Comins, D. L. *Synlett* **1992**, 615.
41 Danishefsky, S. J.; Mantlo, N. B.; Yamashita, D. S.; Schulye, G. *J. Am. Chem. Soc.* **1988**, *110*, 6890.
42 Heseltine, J. N.; Cabal, M. P.; Mantlo, N. B.; Iwasawa, N.; Yamashita, D. S.; Coleman, R. S.; Danishefsky, S. J.; Schulte, G. K. *J. Am. Chem. Soc.* **1991**, *113*, 3850.
43 Kelly, T. R.; Kim, M. H. *J. Org. Chem.* **1992**, *57*, 1593.
44 Comins, D. L.; Baevsky, M. F.; Hong, H. *J. Am. Chem. Soc.* **1992**, *114*, 10971.
45 Ketcha, D. M.; Gribble, G. W. *J. Org. Chem.* **1985**, *50*, 5451.
46 Zalewski, Z. S.; Wendler, N. L. *Chem. Ind. (London)* **1975**, 280.
47 Danishefsky, S. J.; Schuda, P.; Kato, K. *J. Org. Chem.* **1976**, *41*, 1081.
48 Harreus, A.; Kunz, H. *Liebigs Ann. Chem.* **1986**, 717.
49 Kesselmans, R. P. W.; Wijnberg, J. P. B. A.; de Groot, A.; de Vries, N. K. *J. Org. Chem.* **1991**, *56*, 7232.
50 Stork, G.; Rychnovsky, S. *J. Am. Chem. Soc.* **1987**, *109*, 1565.
51 Keck, G. E.; Castellino, S. *Tetrahedron Lett.* **1987**, *28*, 281.
52 Kahn, S. D.; Keck, G. E.; Hehre, W. J. *Tetrahedron Lett.* **1987**, *28*, 279.
53 Shambayati, S.; Blake, J. F.; Wierschke, S. G.; Jorgensen, W. L.; Schreiber, S. L. *J. Am. Chem. Soc.* **1990**, *112*, 697.
54 Mulzer, J.; Mareski, P. A.; Buschmann, J.; Luger, P. *Synthesis* **1992**, 215.
55 Salomon, R. G.; Sachinvala, N. D.; Roy, S.; Basu, B.; Raychaudhuri, S. R.; Miller, D. B.; Sharma, R. B. *J. Am.Chem. Soc.* **1991**, *113*, 3085.
56 Hanessian, S.; Thavonekham, B.; DeHoff, B. *J. Org. Chem.* **1989**, *54*, 5831.
57 Kerrick, S. T.; Beak, P. *J. Am. Chem. Soc.* **1991**, *113*, 9708.
58 Gallagher, D. J.; Kerrick, S. T.; Beak, P. *J. Am. Chem. Soc.***1992**, *114*, 5872.
59 Hoppe, D.; Hintze, F.; Tebben, P. *Angew. Chem. Int. Ed. Engl.* **1990**, *29*, 1422.
60 Paulsen, H.; Hoppe, D. *Tetrahedron* **1992**, *48*, 5667.
61 Zschage, O.; Hoppe, D. *Tetrahedron* **1992**, *48*, 5657.
62 Boeckman, R. K.; Bruza, K. J. *Tetrahedron* **1981**, *37*, 3997.
63 Imanieh, H.; Quayle, P.; Voaden, M.; Conway, J.; Street, S. D. A. *Tetrahedron Lett.* **1992**, *33*, 543.
64 Friesen, R. W.; Sturino, C. F.; Daljeet, A. K.; Kolaczewska, A. *J. Org. Chem.* **1991**, *56*, 1944.
65 Udodong, U. E.; Srinivas Rao, C.; Fraser-Reid, B. *Tetrahedron* **1992**, *48*, 4713.

# Chapter 2 Hydroxyl Protecting Groups

**2.1 Introduction**

**2.2 Esters**

**2.3 Silyl Ethers**
2.3.1 Trimethylsilyl Ethers
2.3.2 Triethylsilyl Ethers
2.3.3 *tert*-Butyldimethylsilyl Ethers
2.3.4 *tert*-Butyldiphenylsilyl Ethers
2.3.5 Triisopropylsilyl Ethers
2.3.6 Less Common Trialkylsilyl Ethers

**2.4 Alkyl Ethers**
2.4.1 Methyl Ethers
2.4.2 Benzyl Ethers
2.4.3 *p*-Methoxybenzyl and 3,4-Dimethoxybenzyl Ethers
2.4.4 Trityl Ethers
2.4.5 *tert*-Butyl Ethers
2.4.6 Allyl Ethers and Allyloxycarbonyl Derivatives

**2.5 Alkoxyalkyl Ethers (Acetals)**
2.5.1 Methoxymethyl Ethers
2.5.2 Methylthiomethyl Ethers
2.5.3 (2-Methoxyethoxy)methyl Ethers
2.5.4 Benzyloxymethyl Ethers
2.5.5 *β*-(Trimethylsilyl)ethoxymethyl Ethers
2.5.6 Tetrahydropyranyl and Related Ethers

**2.6 Reviews**
2.6.1 General Reviews Concerning the Protection of Hydroxyls and Thiols
2.6.2 Reviews Concerning Enzyme-Mediated Esterification and Hydrolysis of Esters
2.6.3 Reviews Concerning Organosilicon and Organotin Chemistry Relevant to Hydroxyl Protection

**References**

## 2.1 Introduction

The hydroxyl group is nucleophilic, acidic ($pK_a$ 10–18), and easily oxidised by a wide range of reagents. Because it can participate in numerous transformations under mild conditions, a central problem in organic synthesis is to ensure that a specific hydroxyl function in a multifunctional molecule is protected from unwanted reactions altogether or until such time as its intrinsic reactivity is required. Over 150 hydroxyl protecting groups have been reported to date, but of these, only a comparatively small fraction are in common use; it is this small fraction which will be covered in the following survey.

## 2.2 Esters

**(i) Cleavage**

Esters provide a cheap and efficient means for protecting hydroxyl groups during oxidation, peptide coupling reactions or glycosidation. They are attacked by stronger nucleophiles such as Grignard reagents, metal hydride reducing agents, and the like but the limits of their stability are broad enough to ensure continued use in a wide range of applications. Our concern here is with esters that belong to orthogonal set 1 which are typically cleaved by base-catalysed solvolysis. Since esters straddle several orthogonal sets, they will be considered elsewhere in this book and most notably in Chapter 4 where their use in the protection of carboxylic acids is reviewed.

There are a very large number of different ester protecting groups available and we have selected only four of the more common representatives which exemplify the breadth of graduated hydrolytic lability that is available by simple steric or electronic tuning. Thus pivaloate esters, with their greater steric hindrance, react with $NH_3$ in MeOH so slowly that acetates can be removed selectively[1] whereas trifluoroacetates are so reactive they hydrolyse at pH 7[2]. Listed below is a relative reactivity order of esters in similar steric environments and in the ensuing discussion, we will focus on reactivity relative to acetate — probably the commonest of all the ester protecting groups:

$$t\text{-BuCO} < \text{PhCO} < \mathbf{MeCO} < \text{ClCH}_2\text{CO}$$

Acetates are generally cleaved under mildly basic conditions but they can be cleaved by acid-catalysed solvolysis (transesterification) as well; however, in the absence of water or alcohol, esters are fairly resistant to attack by acid. The point is illustrated by the transformations used in the closing stages of a recent synthesis of Peptide T, a partial sequence of the envelope glycoprotein of HIV-1 [Scheme 2.1][3]. The 8 *tert*-butyl protecting groups used to protect the *N*-terminal nitrogen (*tert*-butoxycarbonyl), the *C*-terminal carboxylic acid (*tert*-butyl ester), the phenolic OH of the tyrosine residue and 5 threonine residues (*tert*-butyl ethers) in substrate **1.1** were cleaved with trifluoroacetic acid. The $\alpha$-fucoside bond, which is ordinarily quite labile towards acid, remained intact due to the indirect protection afforded by the *O*-acetyl groups protecting the hydroxyl functions in the trisaccharide unit[4]. The *O*-acetyl groups were finally removed by gentle hydrolysis using $Et_2NMe$ in water (pH 9.0) to furnish the target glycopeptide **1.3**. Other basic catalysts which have been used in methanol or ethanol to remove acetates include $K_2CO_3$[5], $NH_3$, hydrazine[6], guanidine[7], and KCN[8].

One problem associated with the use of esters as protecting groups in polyfunctional systems, is the tendency for intramolecular transesterification leading to migration of the acyl function to a neighboring alcohol — a problem which is prominent with acetates but less significant with benzoates[9,10]. Nevertheless, migration of benzoates is observed when there is a driving force as illustrated by the benzoate migration [Scheme 2.2] *en route* to *N*-acetyl neuraminic acid[11,12]. In this case the migration was thermodynamically driven by the greater stability of the equatorial benzoate in the product; however, under acidic conditions (25% HF in MeCN, 50 °C, 24 h), migration of an axial benzoyl to an adjacent equatorial hydroxyl on a cyclohexane ring does not occur[13].

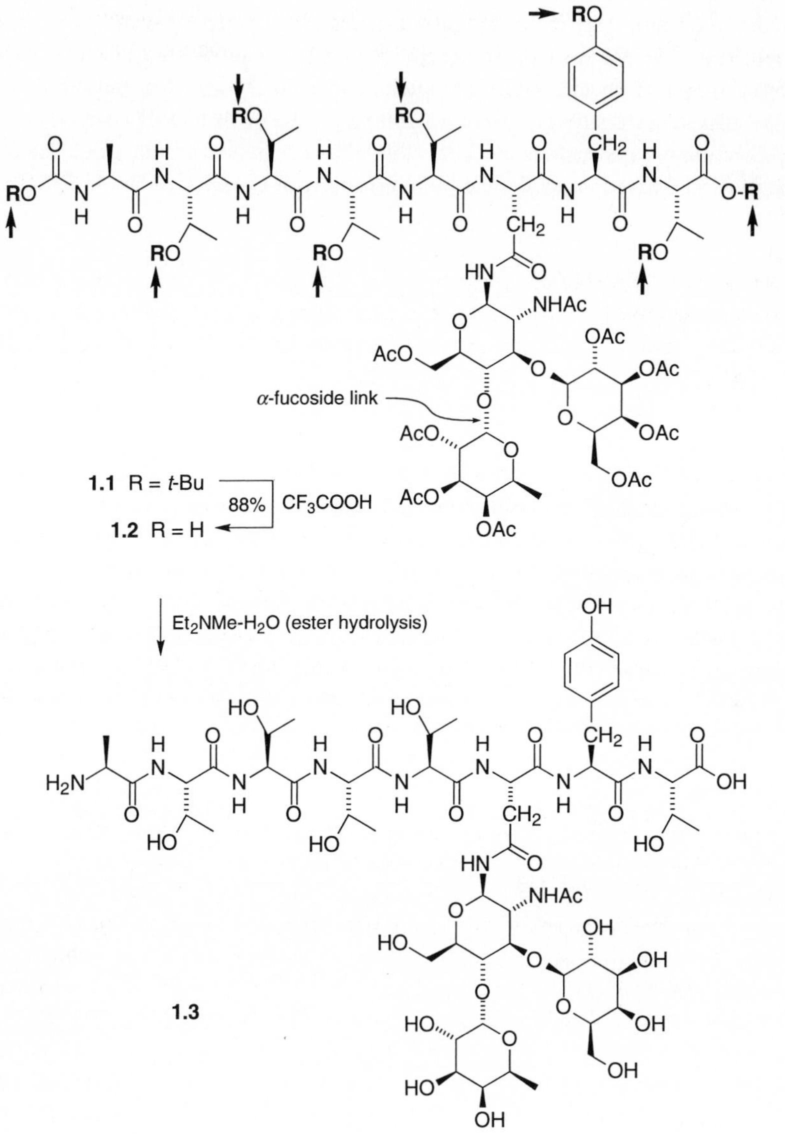

**Scheme 2.1**

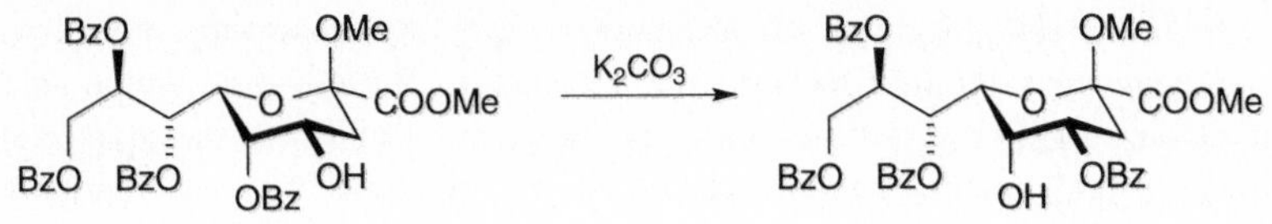

**Scheme 2.2**

Increased hydrolytic stability of the ester function is readily attained by shielding the carbonyl group from nucleophilic attack. We have already cited the example of pivaloates which are hydrolysed very slowly by $NH_3$ in MeOH under conditions which readily cleave acetates. Indeed, the cleavage of a pivaloate ester may require strongly basic reagents (e. g. KOH in MeOH) which are incompatible with other protecting groups such as the TBS (TBDMS) ethers (see section 2.3.3). In such instances the pivaloate may be cleaved in good yield by reduction with $LiAlH_4$[14], *i*-$Bu_2AlH$ [Scheme 2.3][15], or $KBHEt_3$[16].

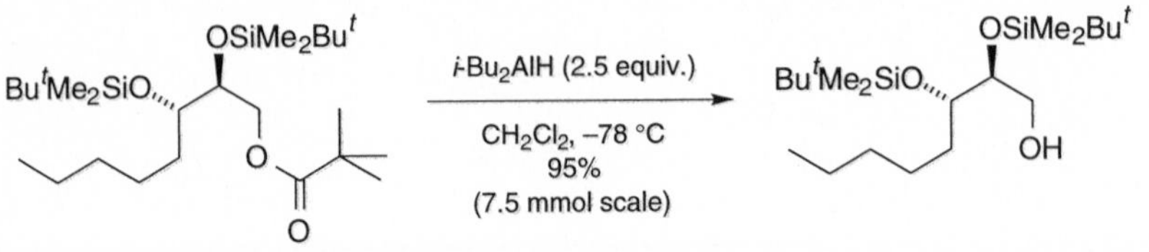

**Scheme 2.3**

The lability of esters can be easily enhanced by substituting the $\alpha$-position with electron-withdrawing groups. Thus, $\alpha$-methoxyacetate esters hydrolyse approximately 20 times faster than simple acetates and $\alpha$-phenoxyacetates hydrolyse 50 times faster[17]. More reactive still are $\alpha$-chloroacetates which cleave easily in the presence of acetates[18] as illustrated in Scheme 2.4[19]. Other mild reagents which have been deployed in the cleavage of $\alpha$-chloroacetates include thiourea[20], $NH_3$ in MeOH or toluene[21], aqueous pyridine (pH 6.7)[22], 2-mercaptoethylamine ($H_2NCH_2CH_2SH$), ethylene diamine ($H_2NCH_2CH_2NH_2$), and phenylene diamine[23]. Hydrazinedithiocarbonate ($H_2N$-NH-C(=S)SH, pre-

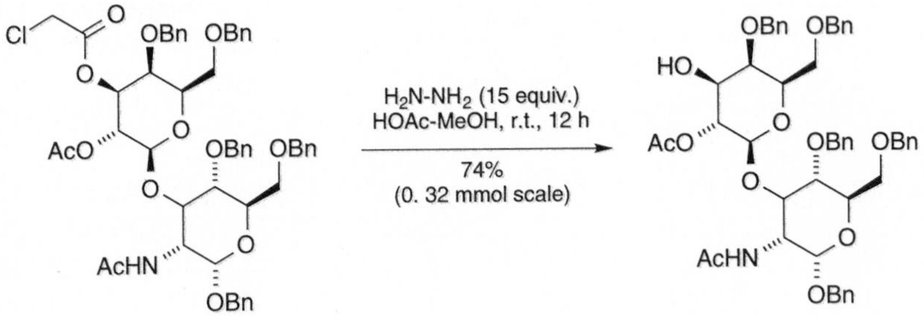

**Scheme 2.4**

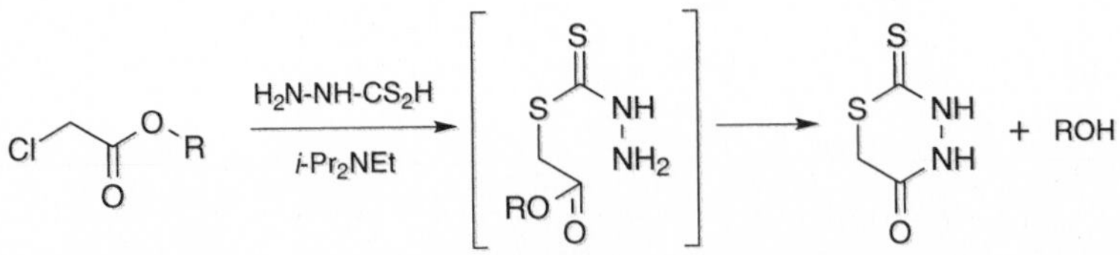

**Scheme 2.5**

pared by the reaction of hydrazine with carbon disulfide) in dioxane containing HOAc and $i$-$Pr_2NEt$ has been recommended as a much more reactive alternative to thiourea for the cleavage of chloroacetates in carbohydrate derivatives[24]. Like thiourea, the ease of cleavage by hydrazinedithiocarbonate depends on initial nucleophilic displacement according to the mechanism shown in Scheme 2.5.

Hydrazinedithiocarbonate was the only reagent of many tried which successfully accomplished the deprotection of three chloroacetate groups from the disaccharide fragment of Phyllanthostatin I [Scheme 2.6][25]. The reaction was complicated by acetate migration and low yields (*ca.* 40%) which were attributed to the limited stability of the hydrazinedithiocarbonate. The problems were eventually overcome by using triethylsilyl protecting groups in place of chloroacetate (see section 2.3.2).

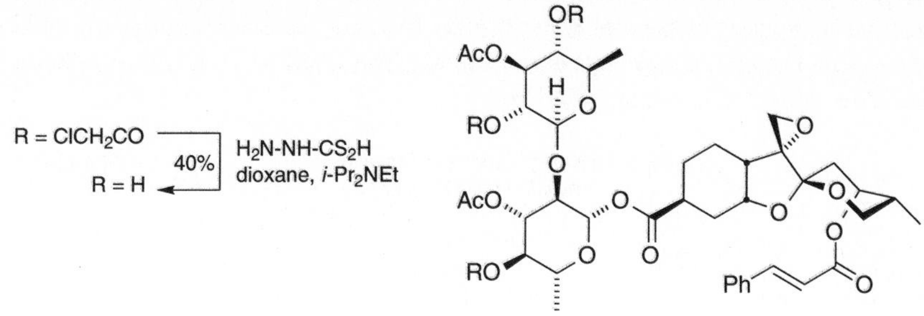

R = $ClCH_2CO$ → R = H: $H_2N$-NH-$CS_2H$, dioxane, $i$-$Pr_2NEt$, 40%

**Scheme 2.6**

Enzyme-catalysed reactions have an important role to play in organic chemistry and the intensity of activity in this area is reflected by the large number of reviews and monographs which have recently appeared (see section 2.6.2 at the end of this chapter). Esterases and dehydrogenases, in particular, have been explored because of their relevance to many of the standard functional group transformations required in organic synthesis. Many of the enzyme preparations are now available commercially and are effective on mole-scale transformations using organic solvents. Except for a pH meter, the equipment required is quite rudimentary. Much of the work to date on biotransformations has been stimulated by the ever-expanding need for relatively small enantiomerically pure building blocks. The mildness, high stereoselectivity, and substrate selectivity of enzymes has offered unique advantages which are also relevant in the context of protecting groups. The following examples illustrate how esterases (lipases) have been exploited for the endgroup differentiation of *meso*-diesters — transformations which would otherwise be difficult to achieve using "standard" synthetic methodology.

The homochiral prostaglandin precursor (1*S*,3*R*)-**7.1** was prepared [Scheme 2.7] by selective hydrolysis of the *meso*-diacetate of *cis*-4-cyclopentene-1,3-diol **7.2** using pig liver esterase (PLE)[26]

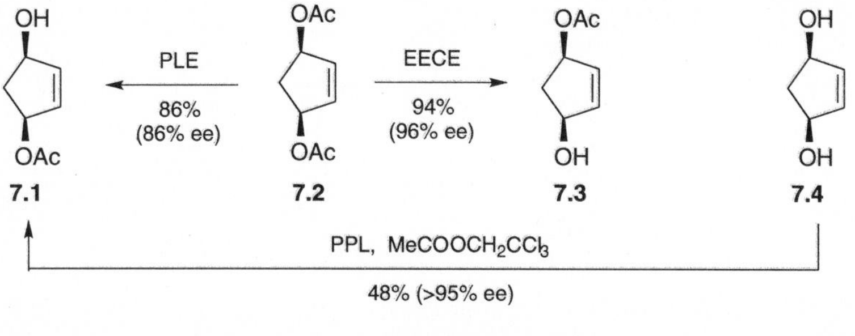

**7.2** → **7.1**: PLE, 86% (86% ee); **7.2** → **7.3**: EECE, 94% (96% ee); **7.4** → **7.1**: PPL, $MeCOOCH_2CCl_3$, 48% (>95% ee)

**Scheme 2.7**

whereas the antipode (1*R*,3*S*)-**7.3** was the product of electric eel cholinesterase (EECE) hydrolysis[27]. Since enzyme-catalysed reactions are reversible, transesterification can also be used to prepare esters. In the case at hand, *cis*-4-cyclopentene-1,3-diol (**7.4**) was transesterified using pig pancreatic lipase (PPL) and trichloroethyl acetate[28]. The trichloroethyl ester is used in order to influence the position of equilibrium since trichloroethanol is a better leaving group and weaker nucleophile than cyclopentene-diol[29].

Scheme 2.8 provides a further illustration of the power of enzymatic methods for the selective synthesis of either member of an enantiomeric pair on a preparative scale[30,31]. *Pseudomonas cepacia* lipase (PCL) hydrolysed the *meso*-diacetate **8.1** to give (2*S*,3*R*)-2,3-*O*-cyclohexylidene-erythritol monoacetate (**8.4**) in 91% chemical yield and ≥99% e.e. whereas transesterification using vinyl acetate accomplished the conversion of the diol **8.3** to the (2*R*,3*S*)-antipode **8.4** in diminished yield (78%) but comparable e.e. Vinyl or propenyl acetate ensures a favorable equilibrium owing to tautomerisation of the enol generated during the transesterification[32,33].

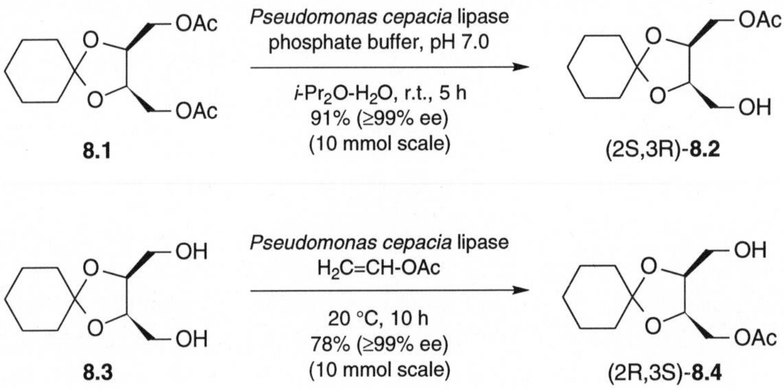

OAc, OAc; **8.1** — *Pseudomonas cepacia* lipase, phosphate buffer, pH 7.0; *i*-$Pr_2O$-$H_2O$, r.t., 5 h; 91% (≥99% ee); (10 mmol scale) → OAc, OH; (2S,3R)-**8.2**

OH, OH; **8.3** — *Pseudomonas cepacia* lipase, $H_2C$=CH-OAc; 20 °C, 10 h; 78% (≥99% ee); (10 mmol scale) → OH, OAc; (2R,3S)-**8.4**

**Scheme 2.8**

Biotransformations of *meso*-substrates to enantiomerically pure compounds are especially efficacious because the maximum theoretical conversion is 100%. However, racemic chiral alcohols can also be conveniently resolved via "enantioselective esterification" wherein an enzyme mediates acyl transfer to enantiomers of a given alcohol at different rates albeit with a maximum theoretical conversion of only 50%[34]. A practical application of such a kinetic resolution is shown in the PPL-mediated esterification of the racemic alcohol **9.1** [Scheme 2.9][35] and many other examples of equal efficiency are known[36-38].

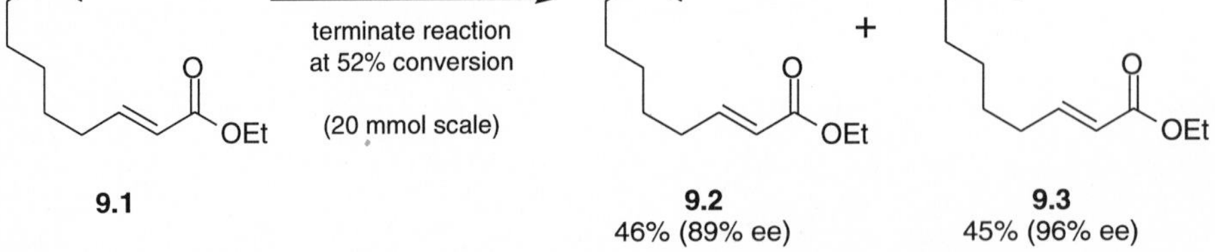

OH, O, OEt; **9.1** — PPL, *n*-$PrCOOCH_2CF_3$, $Et_2O$, 67 h; terminate reaction at 52% conversion; (20 mmol scale) → OCO-*n*-Pr, O, OEt; **9.2**, 46% (89% ee) + OH, O, OEt; **9.3**, 45% (96% ee)

**Scheme 2.9**

Two recent examples will suffice to demonstrate the use of lipases in the formation and hydrolysis of esters in circumstances which could be addressed only with great difficulty by standard synthetic operations. The first example [Scheme 2.10] arose from a search for analogues of Castanospermine as inhibitors for human immunodeficiency virus (HIV)[39] and it cogently illustrates the use of lipases in *organic solvents* pioneered by Klibanov's laboratory[40]. Castanospermine (**10.1**) underwent selective esterification of the C1 hydroxyl function using subtilisin in pyridine. A second selective acylation of the C7 hydroxyl in **10.2** was accomplished with the lipase isolated from *Chromobacterium viscosum* (CV) to give the 1,7-di-*n*-butanoyl derivative **10.3** from which the C1 acyl function was selectively removed using subtilisin in aqueous media.

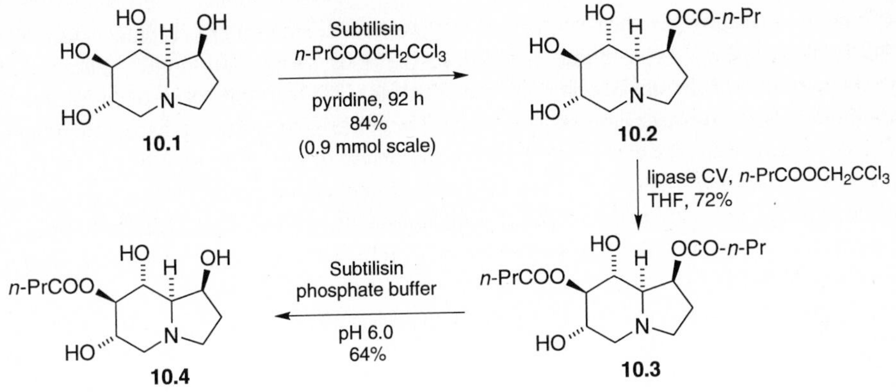

Subtilisin, n-$PrCOOCH_2CCl_3$, pyridine, 92 h, 84% (0.9 mmol scale): 10.1 → 10.2

lipase CV, n-$PrCOOCH_2CCl_3$, THF, 72%: 10.2 → 10.3

Subtilisin, phosphate buffer, pH 6.0, 64%: 10.3 → 10.4

**Scheme 2.10**

The second example [Scheme 2.11] shows how acyl migration, which hitherto has been considered a nuisance, can be put to good use. Lysofungin (**11.3**) is an antifungal phospholipid isolated in minute

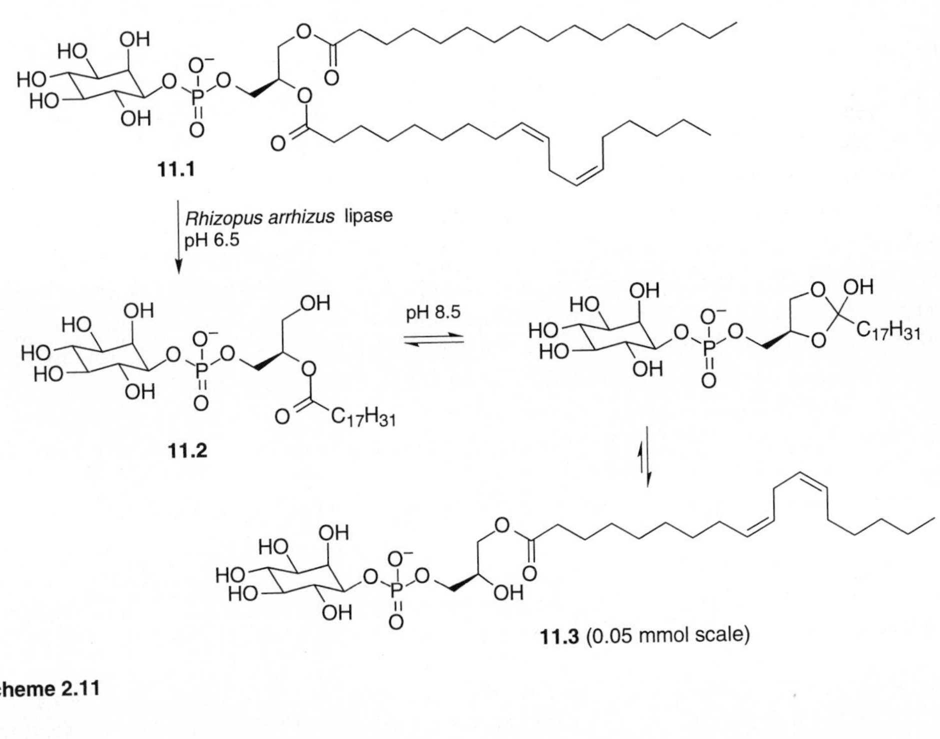

*Rhizopus arrhizus* lipase, pH 6.5: 11.1 → 11.2

pH 8.5: 11.2 ⇌ orthoester intermediate ($C_{17}H_{31}$) ⇌ **11.3** (0.05 mmol scale)

**Scheme 2.11**

yield from *Aspergillus fumigatus*. Faced with the prospect of a protracted total synthesis to secure adequate supplies of Lysofungin for biological evaluation, a Merck group[41] employed a highly selective hydrolysis of the stearoyl function from 1-stearoyl-2-linoleoyl-3-glycero-phosphatidyl-D-*myo*-inositol (**11.1**) which is readily available from soybeans. At pH 6.5, the conditions of the hydrolysis, the hydrolysis product **11.2** is stable; however, at pH 8.5 migration of the linoleoyl group to the terminus of the glycerol unit occurred to produce Lysofungin.

**(ii) Formation**

"Digressions, incontestably, are the sunshine; — they are the life, the soul of reading; —take them out of this book for instance, —you might as well take the book along with them." Laurence Sterne's apologia for rambling (*Tristram Shandy*, chapter 22, 1760–1767) might be invoked in favor of our small and limited detour through biotransformations. The subject is too important to ignore but too big to dwell upon. Fortunately, we can compensate for the length of our digression by the brevity of our treatment of the traditional methods for forming esters which are simple and few. In general reaction of an alcohol with the appropriate anhydride or acid chloride in pyridine at 0–20 °C is sufficient. In the case of tertiary alcohols acylation is very slow in which case a catalytic amount of 4-dimethyl-aminopyridine (DMAP) can be added to speed up the reaction. By the same token, selective acylation of a primary alcohol in the presence of one or more secondary alcohols is easily achieved with pivaloyl chloride [Scheme 2.12][15]. A more extensive discussion of the methods for forming and cleaving esters will appear in Chapter 4 when we consider protection of the carboxyl function.

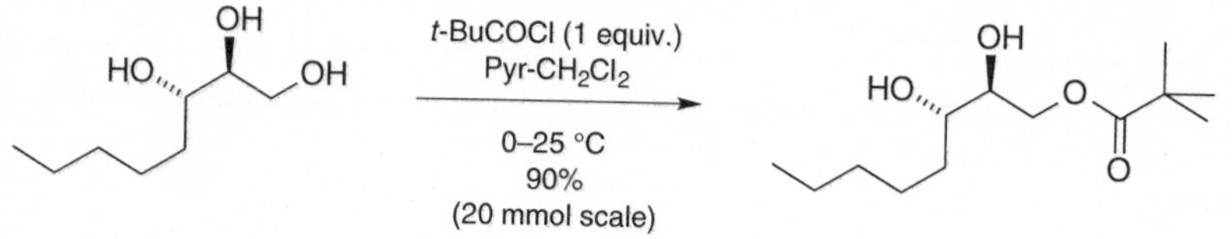

*t*-BuCOCl (1 equiv.)
Pyr-$CH_2Cl_2$
0–25 °C
90%
(20 mmol scale)

**Scheme 2.12**

## 2.3 Silyl Ethers

*O*-Silylation of alcohols was first introduced in the late 1950's to increase volatility and stability of polar compounds during gas chromatography and mass spectrometry. The synthetic potential of silyl ethers as protecting groups for the hydroxyl function was not widely appreciated until the early 1970's and it would be fair to say that now silyl protecting groups are probably used more than any other protecting group in organic synthesis. Silicon has also had a smaller but significant impact on the protection of other functional groups as well such as carboxylic acids, amines, and thiols. A number of useful reviews on the use of organosilicon reagents as protective groups in organic synthesis have appeared (see section 2.6.3 at the end of this chapter).

The reason for the wide popularity of silicon protecting groups is easy to understand: they are readily formed and cleaved under mild conditions and their relative stability can be finely tuned by simply varying the substituents on silicon. As a rule, the bulkier the substituents, the greater the stability towards acid and base hydrolysis, organolithium or Grignard reagents, oxidation, reduction, and column chromatography. However, stability is not solely a function of steric bulk since electronic effects play a role as well which can be exploited to differentiate stability under acidic or basic conditions. Thus, electron withdrawing groups on the Si (*e.g.* Ph) cooperate with the steric effect to enhance stability under acidic conditions (relative rate of hydrolysis of $Ph_3Si$ / $Me_3Si$ = 1 : 400) whereas under basic conditions the effects are opposed ($Ph_3Si:Me_3Si$ = 1). Kinetic data for the hydrolysis of silyl ethers has not been determined under standard conditions or even with the same substrates thereby precluding an absolute measure of relative stability. Nevertheless, a very rough order can be estimated from the data currently available in which the relative stability ($1/k_{rel}$) of $R_1R_2R_3SiO–R_4$ towards acid-catalysed solvolysis is $Me_3Si$ [1] < $Et_3Si$ [64] < *t*-$BuMe_2Si$ [20,000] < *i*-$Pr_3Si$ [700,000] < *t*-$BuPh_2Si$ [5,000,000] and towards base-catalysed solvolysis $Me_3Si$ [1] < $Et_3Si$ [10–100] < *t*-$BuMe_2Si$ = *t*-$BuPh_2Si$ [20,000] < *i*-$Pr_3Si$ [100,000].

Cunico and Bedell[42] measured the half lives of the more robust silyl ethers under a variety of conditions with the results shown in Table 1:

**Table 1** Half-lives in the cleavage of silyl ethers

| Protecting Group $SiR_3$ | *n*-Bu-O-$SiR_3$ | | *c*-$C_6H_{11}$-O-$SiR_3$ | | |
|---|---|---|---|---|---|
| | $H^{+\,a}$ | $OH^{-\,b}$ | $H^{+\,a}$ | $OH^{-\,b}$ | $F^{-\,c}$ |
| *t*-$BuMe_2Si$ | 1 min | 1 h | 4 min | 26 h | 76 min |
| *i*-$Pr_3Si$ | 18 min | 14 h | 100 min | 44 h | 137 min |
| *t*-$BuPh_2Si$ | 244 min | 4 h | 360 min | 14 h | - |

[a] 1% HCl / 95% EtOH, 22.5 °C; [b] 5% NaOH / 95% EtOH, 90 °C; [c] 2 equiv. TBAF / THF 22.5 °C

Silicon has a high affinity for fluorine and a practical consequence of the strength of the Si–F bond (142 kcal/mole *vs* 112 kcal/mole for Si–O) is the removal of silicon protecting groups under extremely mild and highly specific conditions using fluoride ion or HF — conditions which are compatible with most functional groups. In general, the order of cleavage of silyl ethers with basic fluoride reagents (*e.g.* tetra-*n*-butylammonium fluoride) parallels the order found for basic hydrolysis; similarly, slightly acidic fluorine-based reagents such as HF–MeCN parallel the order found for acid hydrolysis.

## 2.3.1 Trimethylsilyl Ethers

Trimethylsilyl ethers (abbreviated TMS) are widely used to derivatise polar and non-volatile compounds for gas chromatography or mass spectrometry but their use in synthesis is limited by their lability to hydrolysis and column chromatography.

**(i) Cleavage**

Owing to the sensitivity of TMS ethers, deprotection can usually be achieved under very mild conditions (e.g., HOAc or $K_2CO_3$ in MeOH). The rate of hydrolysis depends on both steric and electronic effects with hindered environments decreasing the rate and electron-withdrawing subsituents on the alcohol function increasing the rate.

**(ii) Formation**

Trimethylsilylation has been accomplished with a large number of reagents most of which are commercially available. The cheapest (TMSCl) and the most reactive (TMSOTf) rapidly silylate hydroxyl groups in the presence of a suitable base such as pyridine, $Et_3N$, $i$-$Pr_2NEt$, imidazole, or DBU but the resultant amine hydrochloride or triflate requires an aqueous workup to ensure complete removal. In some cases the insoluble salt may be removed by filtration without aqueous workup. A wide range of solvents can be used for the reaction such as $CH_2Cl_2$, acetonitrile, THF, or DMF. Care must be taken with TMSOTf since it will convert aldehydes and ketones to the corresponding enol silanes and it will open epoxides in a reaction which has preparative significance [Scheme 2.13][43]. Similar transformations can be accomplished with $t$-$BuMe_2SiOTf$ (TBSOTf) or $Et_3SiOTf$ (TESOTf)[44].

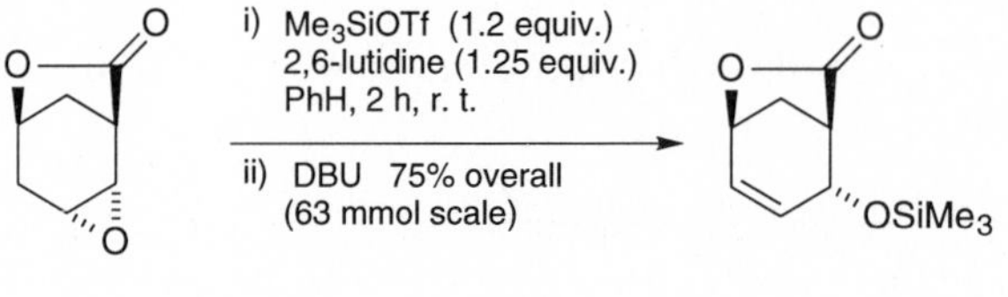

i) $Me_3SiOTf$ (1.2 equiv.)
2,6-lutidine (1.25 equiv.)
PhH, 2 h, r. t.
ii) DBU 75% overall
(63 mmol scale)
$OSiMe_3$

**Scheme 2.13**

In cases where aqueous workup or the presence of amine salts is undesirable, trimethylsilation can be achieved with a wide range of reagents whose by-products are volatile and therefore easily removed. Examples include $N,O$-bis(trimethylsilyl)acetamide (by-product $N$-trimethylsilylacetamide or acetamide) [Scheme 2.14][45]; hexamethyldisilazane (abbreviated HMDS, by-product ammonia) catalysed by TMSCl or, in hindered cases, TMSI [Scheme 2.15][46]; and 1-(trimethylsilyl)imidazole (abbreviated TMSIM, by-product imidazole) [Scheme 2.16][47].

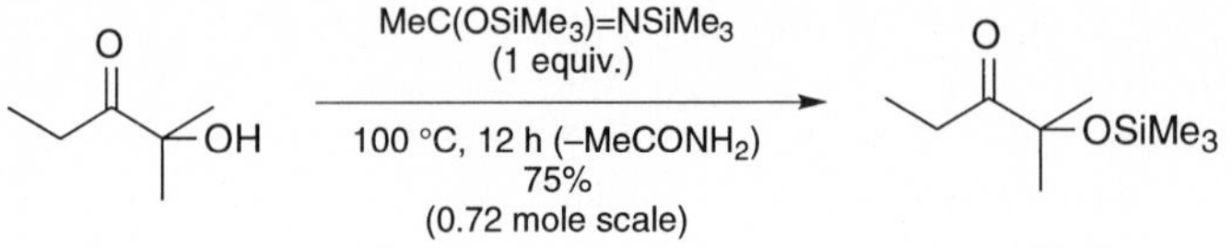

OH
$MeC(OSiMe_3)=NSiMe_3$
(1 equiv.)
100 °C, 12 h ($-MeCONH_2$)
75%
(0.72 mole scale)
$OSiMe_3$

**Scheme 2.14**

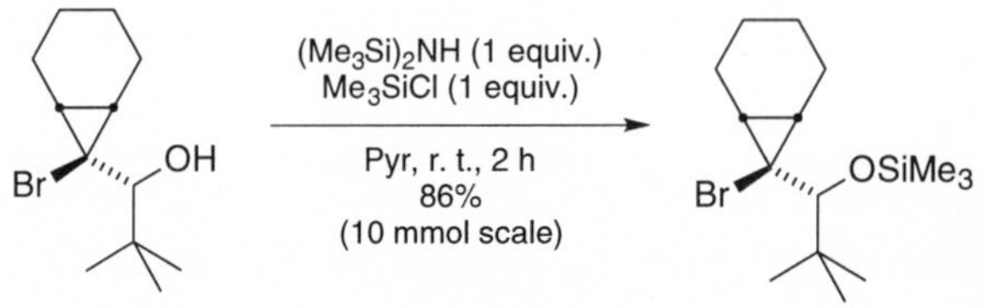

Br OH
$(Me_3Si)_2NH$ (1 equiv.)
$Me_3SiCl$ (1 equiv.)
Pyr, r. t., 2 h
86%
(10 mmol scale)
Br $OSiMe_3$

**Scheme 2.15**

COOMe
$COOBu^t$
$OSiMe_2Bu^t$
excess $Me_3Si$-Im
100 °C, 90 min
100%
$SiMe_3$
COOMe
$COOBu^t$
$OSiMe_2Bu^t$

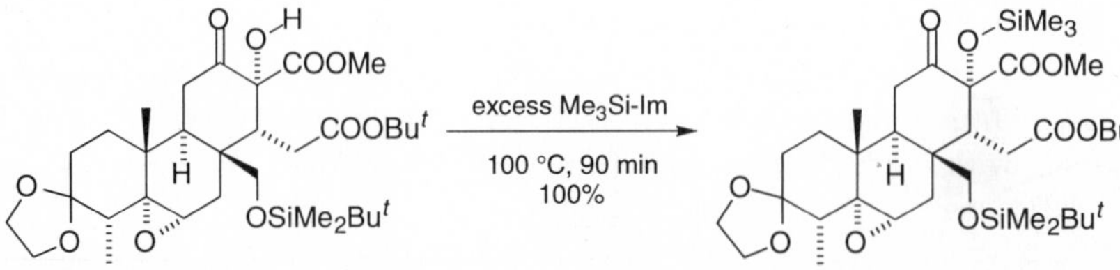

**Scheme 2.16**

**(iii) NMR Data**
$^1$H NMR: $\delta = 0.1$ (9H, s)
$^{13}$C NMR: $\delta = 3$ (3C, q)

## 2.3.2 Triethylsilyl Ethers

Triethylsilyl ethers (abbreviated TES) have only recently gained recognition as synthetically valuable protecting groups. The TES group is sufficiently stable to endure column chromatography and it is resistant to many oxidation, reduction, and organometallic reactions.

**(i) Cleavage**
The triethylsilyl group is 10–100 times more stable to hydrolysis or nucleophilic attack than the trimethylsilyl group but much more labile than the TBS group (see section 2.3.3). The relative lability of the TES group was a key strategic feature in the selective deprotection of an advanced intermediate in the Merck synthesis of the immunosuppressant FK-506 [Scheme 2.17][14] in which two TBS ethers and two triisopropylsilyl ethers survived intact using aqueous trifluoroacetic acid. Another structure of similar complexity is Cytovaricin whose synthesis by Evans' group[48] employed a diapason of silicon protection including 4 TES groups, 1 diethylisopropylsilyl, 1 TBS, and one di-*tert*-butylsilylene group which were cleaved in a single operation using HF in Pyr–THF [Scheme 2.45]. The TES group had survived a Grignard reagent addition, a Swern oxidation, a Horner–Wittig reaction, sulfone metallation using $Et_2NLi$, DDQ in aqueous dichloromethane (used to remove a *p*-methoxybenzyl ether), and Dess–Martin oxidation. The step in question will be given in full in the next chapter when we discuss silylene protection of diols.

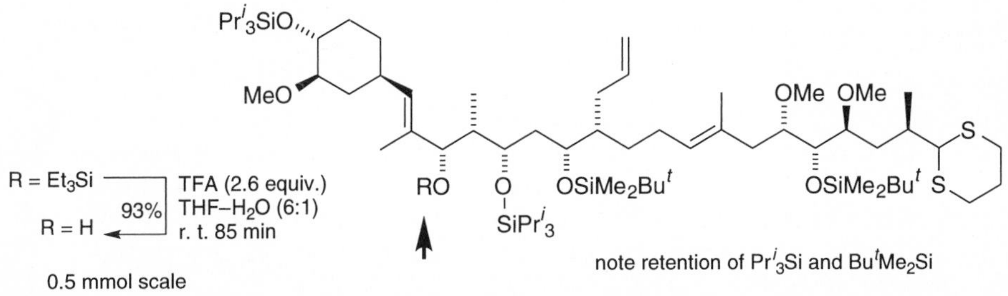

$Pr^i_3SiO$, MeO, OMe, OMe, S, S, RO, O, $SiPr^i_3$, $OSiMe_2Bu^t$, $OSiMe_2Bu^t$

R = $Et_3Si$ → R = H: TFA (2.6 equiv.), $THF–H_2O$ (6:1), r. t. 85 min, 93%

0.5 mmol scale

note retention of $Pr^i_3Si$ and $Bu^tMe_2Si$

**Scheme 2.17**

A mixture of $H_2O$, HOAc, and THF (3:5:11) at r.t. for 15 h has also been used to cleave a TES group in the presence of a TBS ether [Scheme 2.18][49]; similarly, selective cleavage of a TES group in the presence of a TBS ether has also been accomplished using 2% HF or HF–Pyr [Scheme 2.19][50].

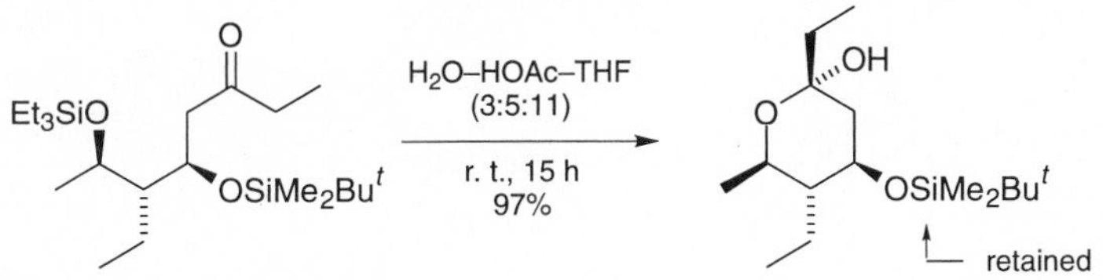

$Et_3SiO$, O, $OSiMe_2Bu^t$ → ($H_2O–HOAc–THF$ (3:5:11), r. t., 15 h, 97%) → O, OH, $OSiMe_2Bu^t$ (retained)

**Scheme 2.18**

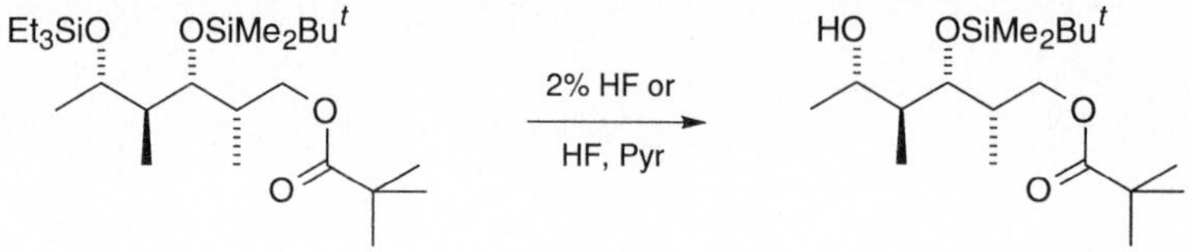
Et3SiO OSiMe2Bu^t
2% HF or
HF, Pyr
HO OSiMe2Bu^t

**Scheme 2.19**

The comparative lability of the TES ether group was a critical design feature of Smith and Rivero's synthesis of Phyllanthoside. In Scheme 2.6 we showed how the removal of chloroacetate from a disaccharide in the presence of acetate groups using hydrazinedithiocarbonate was compromised by low yields and acetate migration. A subsequent foray overcame the earlier problems by the use of TES groups in place of chloroacetates [Scheme 2.20][51]. The two TES groups in structure **20.1** were removed rapidly in quantitative yield without migration of acetate by simple treatment with aqueous acetic acid — conditions which left the two glycosidic links, a spiroacetal, and an epoxide intact.

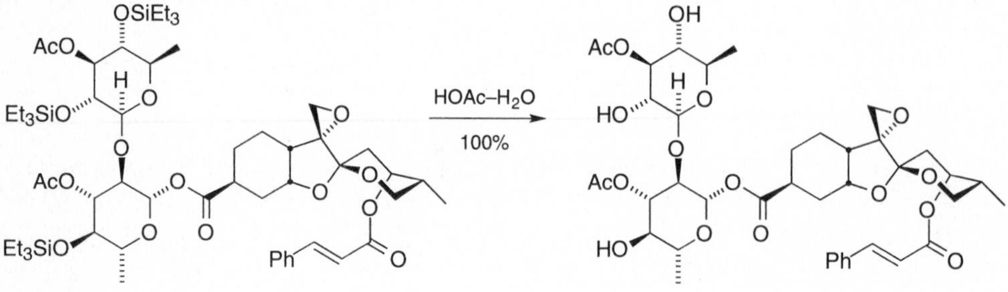
OSiEt3
AcO
Et3SiO
AcO
Et3SiO
Ph
HOAc–H2O
100%
OH
AcO
HO
AcO
HO
Ph

**20.1**

**Scheme 2.20**

**(ii) Formation**

TES ethers are prepared by the reaction of the alcohol with TESCl in the presence of a catalytic amount of imidazole [Scheme 2.21][52] or DMAP[53]. TESOTf in the presence of pyridine [Scheme 2.22][54] or 2,6-lutidine [Scheme 2.23][49] can be used to protect $\beta$-hydroxy aldehydes, ketones[55], and esters.

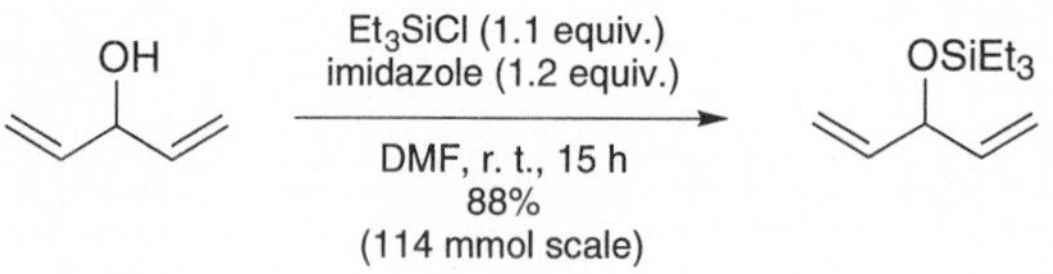
OH
Et3SiCl (1.1 equiv.)
imidazole (1.2 equiv.)
DMF, r. t., 15 h
88%
(114 mmol scale)
OSiEt3

**Scheme 2.21**

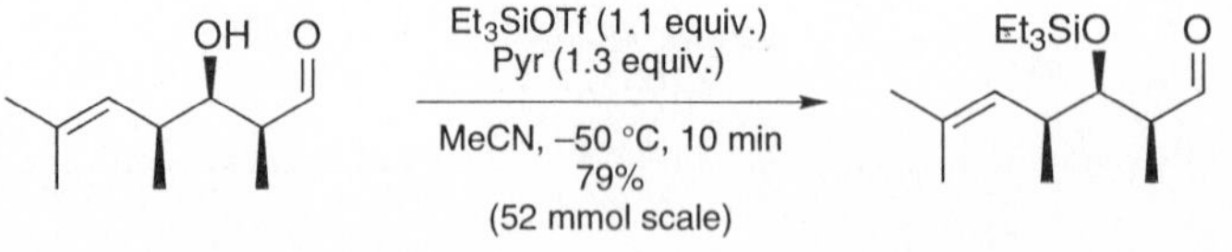
OH O
Et3SiOTf (1.1 equiv.)
Pyr (1.3 equiv.)
MeCN, –50 °C, 10 min
79%
(52 mmol scale)
Et3SiO O

**Scheme 2.22**

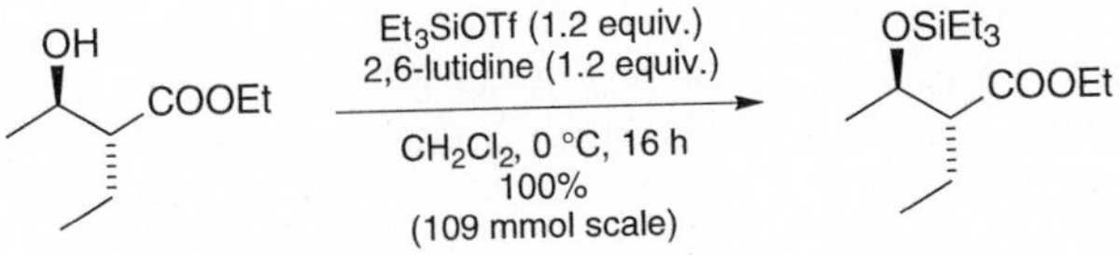

Scheme 2.23

**(iii) NMR Data**

$^{1}$H NMR: $\delta = 0.6$ (6H, q, $J = 7$–$8$ Hz), 0.9 (9H, t, $J = 7$–$8$ Hz)

$^{13}$C NMR: $\delta = 7$ (3C, q), 5 (3C, t)

### 2.3.3 *tert*-Butyldimethylsilyl Ethers

Since its introduction in 1972[56] the TBS group (abbreviated TBDMS or TBS) has become the most popular of the general purpose silicon protecting groups. TBS ethers are stable to chromatography and they are stable below 0 °C to strong non-protic bases such as *n*-alkyllithiums, Grignard reagents, enolates, and metallated sulfones; however, *t*-BuLi will deprotonate the Si–Me group[57,58] with preparatively useful efficiency in THF at about –10 °C. TBS ethers are comparatively stable to mild base but labile to mild acid and in the absence of Lewis acids, they are stable to metal hydrides such as $LiAlH_4$ but DIBALH cleaves them at room temperature[59]. Phenolic TBS ethers are less stable than those derived from simple alkanols but they have been used to protect phenolic hydroxyls during halogen-metal exchange of an aryl bromide using *n*-BuLi, tetrahydropyranyl ether formation using camphorsulfonic acid (CSA), acetate hydrolysis using $K_2CO_3$ in MeOH, and pyridinium dichromate oxidation of a primary alcohol to a carboxylic acid[60]. An important consideration in their use is the simplicity of their NMR spectra. Unfortunately *tert*-butyldimethylsilyl chloride (TBSCl, mp 86–89 °C) is expensive though it is readily prepared in high yield by the reaction of *t*-BuLi with dimethyldichlorosilane.

**(i) Cleavage**

Acetic acid removes TBS ethers at room temperature whilst leaving *tert*–butyldiphenylsilyl [Scheme 2.24][61] and triisopropylsilyl ethers[62] unscathed. Pyridinium tosylate (PPTS) in MeOH has also been

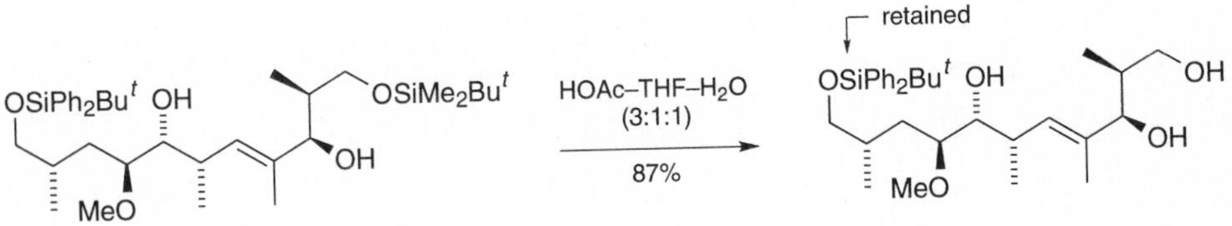

Scheme 2.24

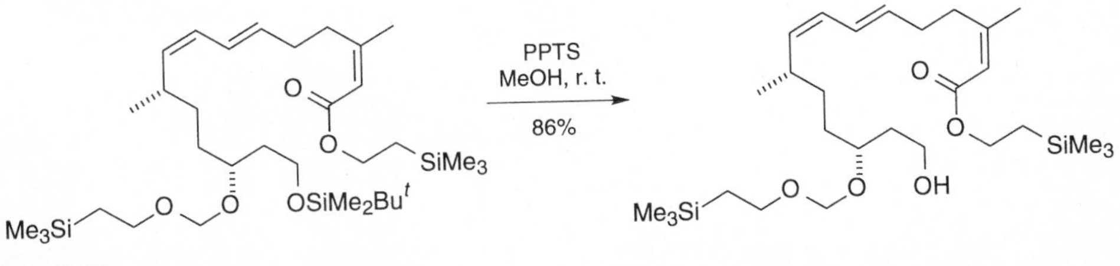

Scheme 2.25

used to cleave TBS ethers [Scheme 2.25][63], but *tert*–butyldiphenylsilyl ethers are impervious to attack under these conditions[64]. Scheme 2.26 shows that selective removal of the less hindered TBS ether of a 2′,5′-bis-*O*-(*tert*-butyldimethylsilyl)-3′-ketoadenosine derivative[65] is possible using aqueous $CF_3COOH$.

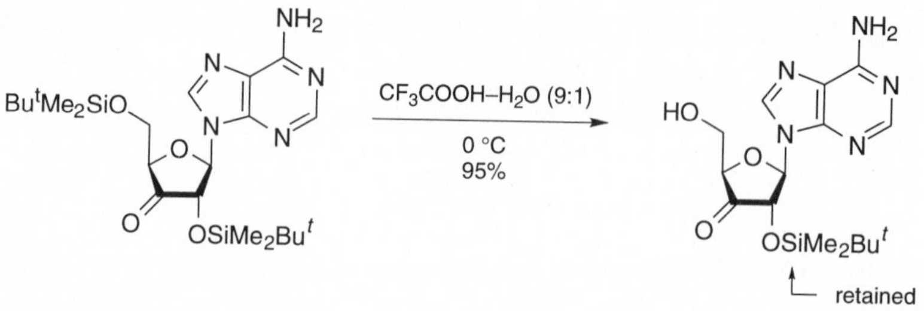

$CF_3COOH$–$H_2O$ (9:1), 0 °C, 95%

**Scheme 2.26**

The high affinity of Si for fluorine has recently proved fruitful in the deprotection of fragile advanced synthetic intermediates. During a synthesis of sensitive Prostaglandin D derivatives, Newton and co-workers were not able to deprotect a bis-TBS ether [Scheme 2.27][66] using aqueous acetic acid in the usual way but successful hydrolysis was accomplished using aqueous HF in acetonitrile — conditions which are now widely used in synthesis. HF ($pK_a$ 3.45) is only slightly more acidic than formic acid ($pK_a$ 3.75) and these conditions are mild enough to tolerate acetals, esters, and epoxides[11,67]. Selective removal of a primary allylic TBS group in the presence of a hindered secondary TBS group has been accomplished with HF in MeCN [Scheme 2.28][16].

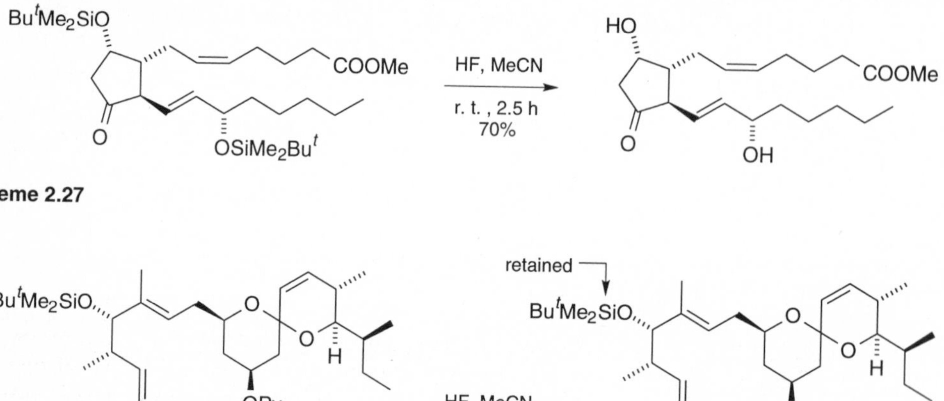

HF, MeCN, r. t., 2.5 h, 70%

**Scheme 2.27**

HF, MeCN, –20 °C, 85%

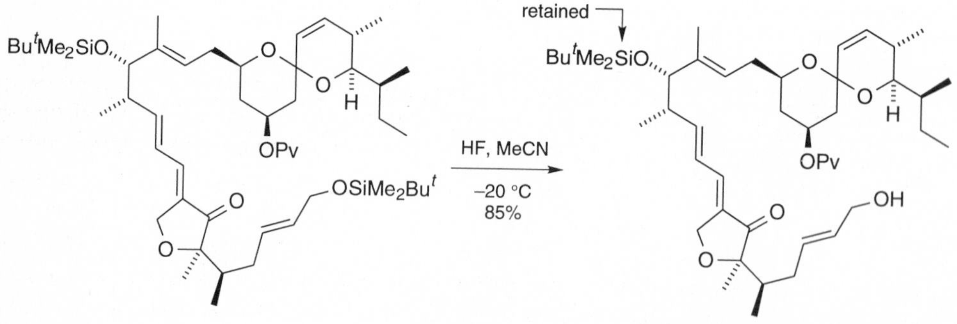

**Scheme 2.28**

A further advance in the selective deprotection of TBS ethers was stimulated by problems which arose during a synthesis of Tirandamycin [Scheme 2.29][68]. In the example, concomitant removal of the TBS ether and acid-catalysed closure of the bicyclic acetal was accomplished in high yield with excess HF in acetonitrile but the triisopropylsilyl (TIPS) group was also removed. However, treatment of the pyranone **29.1** with aqueous HF in the presence of a catalytic amount of fluorosilicic acid ($H_2SiF_6$) in acetonitrile resulted in selective removal of the TBS group and acetalisation to give **29.2**. Other acid-labile protecting groups such as THP, MEM, and acetonides survive these conditions. In a subsequent detailed study, the same laboratory investigated the relative rates of removal of the common silicon protecting groups[69,70]. The best conditions are as follows: the silyl ether (10 mmol) and $Et_3N$ (1.67 mmol) is dissolved in MeCN (100 mL) in a polyethylene bottle and cooled to 0 °C. Fluorosilicic acid (4.17 mmol), 1.54 mL of a 31% solution in water (2.50 fluoride equiv.) is then added and the mixture allowed to stir at room temperature. Under these conditions TBS ethers are removed in 20 min; TIPS ethers take 20 h and *tert*-butyldiphenylsilyl (TBDPS) ethers require 5 days.

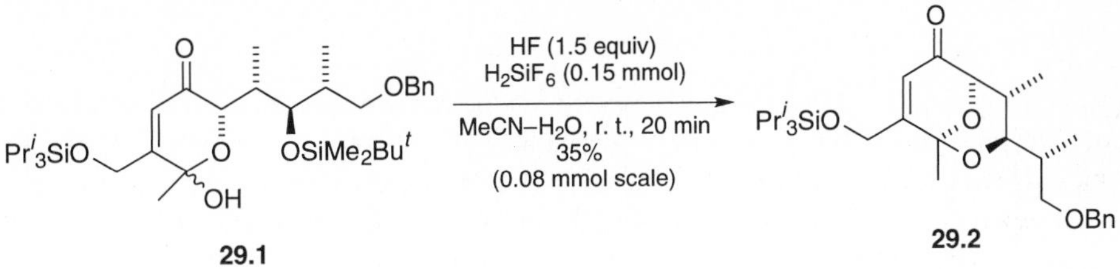

**Scheme 2.29**

HF•pyridine complex in MeOH has also been suggested for the removal of TBS ethers in acid- and base-sensitive substrates[15] and the method was used to remove three TBS ethers in one pot at the closing stages of a synthesis of Amphoteronolide [Scheme 2.30][71].

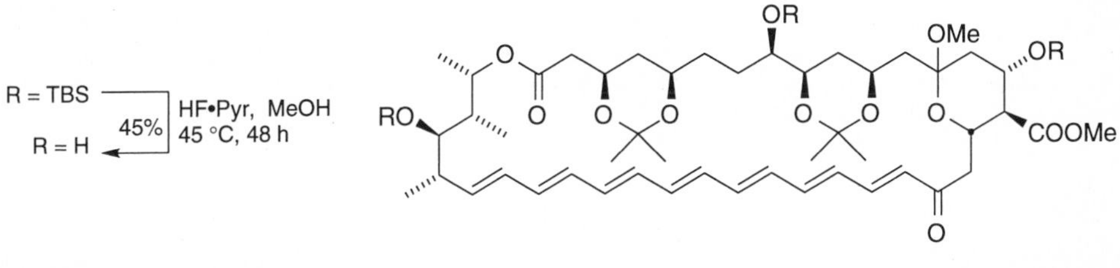

**Scheme 2.30**

Deprotection of TBS ethers is not limited to acidic reagents: fluoride under basic conditions is also effective and the most popular reagent is tetra-*n*-butylammonium fluoride (TBAF) which is available commercially as the trihydrate. Benzyltrimethylammonium fluoride has also been recommended in cases where TBAF has failed[72]. The solid trihydrate is highly hygroscopic and so it is usually dispensed as a solution in THF. The presence of water diminishes the effectiveness of TBAF owing to the tenacious hydration of fluoride ion; however, the water cannot be completely removed — nor should it — because anhydrous TBAF is unstable. Attempts to remove the water from the trihydrate by azeotropic distillation, for example, gives tri-*n*-butylamine. If necessary, a higher activity reagent can be obtained by running the deprotection reaction in THF in the presence of 4A molecular sieves. Schwesinger and co-workers have developed an anhydrous fluoride salt which has yet to make an impact on synthesis[73,74].

The principal detractions to TBAF are its high cost and its comparatively high basicity which can cause undesirable side reactions such as $\beta$-elimination[75] or enolate formation leading to further reactions as shown in Scheme 2.31[72]. During a recent synthesis of prostaglandin derivatives, Otera and co-workers[76] found that an undesirable $\beta$-elimination reaction caused by the basicity of TBAF could be suppressed by buffering the reaction with HOAc.

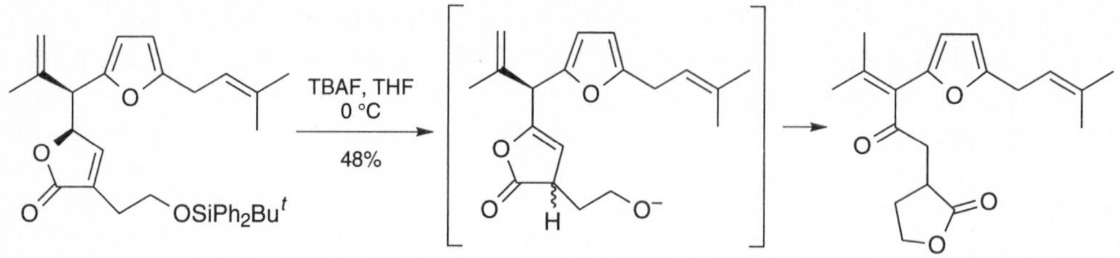

TBAF, THF
0 °C
48%

OSiPh2Bu$^t$

31.1 31.2

Scheme 2.31

Although TBAF generally leads to poorer selectivity in the removal of various silyl protecting groups compared with the acid-catalysed methods described above, some selective transformations have been described. For example, the phenolic TBS ether in **32.1** cleaved in preference to the primary alkyl TBS ether in the presence of one equivalent of TBAF to give phenol **32.3** whereas the alkyl TBS ether cleaved preferentially with HF–MeCN to give alcohol **32.2** [Scheme 2.32][77]. Alternatively, the selective desilylation of TBS ethers of phenols can be achieved using KF–$Al_2O_3$ in acetonitrile with ultrasonic irradiation under conditions which do not affect TBS ethers of benzylic alcohols or $\beta$-(trimethylsilyl)ethoxymethyl ethers of phenols[78]. Excellent discrimination between two secondary TBS groups in different steric environments has been accomplished using TBAF [Scheme 2.33][79].

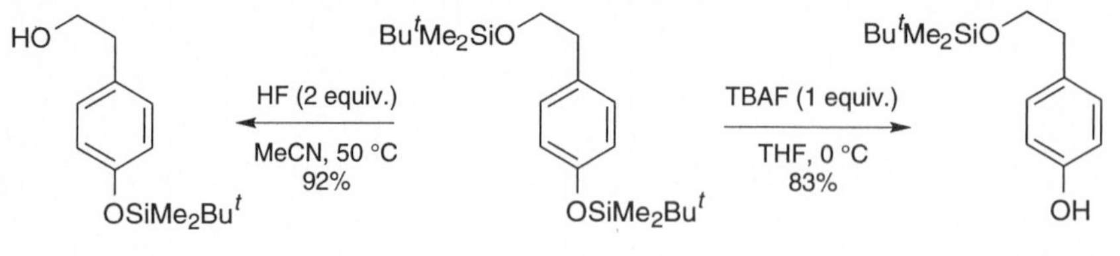

HO / OSiMe2Bu$^t$ — HF (2 equiv.), MeCN, 50 °C, 92% — Bu$^t$Me2SiO / OSiMe2Bu$^t$ — TBAF (1 equiv.), THF, 0 °C, 83% — Bu$^t$Me2SiO / OH

32.2 32.1 32.3

Scheme 2.32

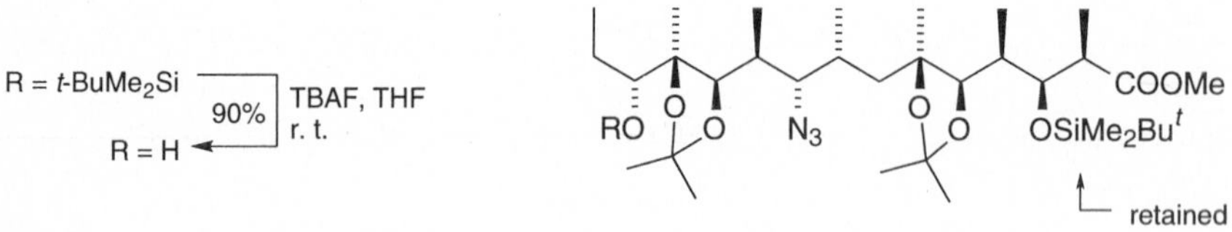

R = t-BuMe2Si → R = H, 90%, TBAF, THF, r. t.

RO, N3, COOMe, OSiMe2Bu$^t$ — retained

Scheme 2.33

Ammonium fluoride (10 equivalents) in refluxing MeOH for 5 h or at r.t. for 1–2 days has been proposed as an economic alternative to TBAF for the deprotection of TBS ethers as well as TBDPS ethers[80]. Isopropylidene derivatives and epoxides are stable to these conditions though acetate esters are cleaved. The method was recently applied on a minute scale to the selective deprotection of a TBS group protecting an allylic alcohol in a synthesis of Integerrimine [Scheme 2.34][81]. Note the survival of a TBS group protecting a tertiary alcohol in intermediate **34.2**.

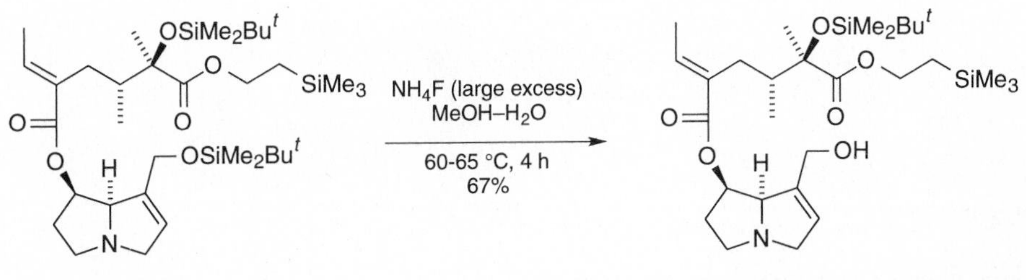
OSiMe2Bu^t
SiMe3
OSiMe2Bu^t
NH4F (large excess)
MeOH–H2O
60-65 °C, 4 h
67%
OH
34.1
34.2

**Scheme 2.34**

Cleavage of TBS ethers by catalytic hydrogenation is a promising and mild new technique which has yet to be adequately explored[82,83].

**(ii) Formation**

The steric bulk of the *tert*-butyl group significantly diminishes the rate of silylation with *t*-$BuMe_2SiCl$ so convenient rates are best achieved by the addition of basic activators such as imidazole or DMAP[84] and by using dipolar aprotic solvents such as dimethylformamide (DMF). Primary alcohols react much faster than secondary alcohols [Scheme 2.35][85] but tertiary alcohols are inert.

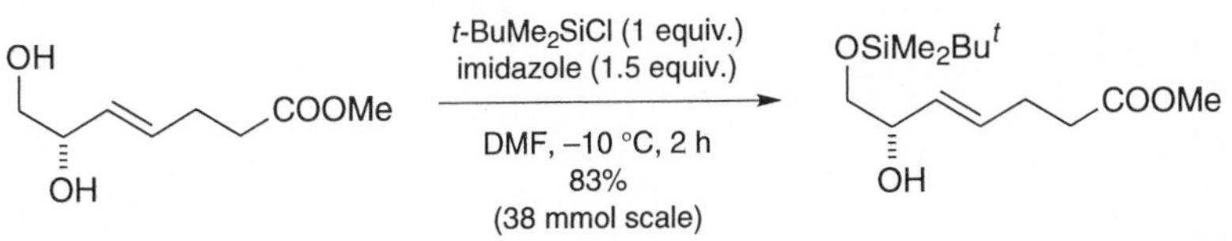
OH
COOMe
OH
t-BuMe2SiCl (1 equiv.)
imidazole (1.5 equiv.)
DMF, –10 °C, 2 h
83%
(38 mmol scale)
OSiMe2Bu^t
COOMe
OH

**Scheme 2.35**

Hindered secondary and tertiary alcohols can be silylated with *tert*-butyldimethylsilyl triflate (TBSOTf) using 2,6-lutidine as the base[86] as illustrated by the silylation of the secondary alcohol **36.1** in a recent synthesis of the diterpene Zoapatanol [Scheme 2.36][87] and the tertiary alcohol **37.1** [Scheme 2.37] originating in a synthesis of Erythronolide A[88]. The reactions can be run at molar concentrations using $CH_2Cl_2$ as a co-solvent and the recommended ratio of reactants is alcohol : TBSOTf : 2,6-lutidine = 1 : 1.5 : 2. TBSOTf is both expensive and very moisture sensitive and in our experience commercial samples do not store very well; consequently, it may be best to prepare it fresh from the reaction of *t*-$BuMe_2SiCl$ and triflic acid[86] or by the reaction of isopropenyltrimethylsilane with triflic acid[89]. For the silylation of *β*-hydroxy butyrolactones which are prone to dehydration during silylation with TBSOTf under the standard conditions, 2,6-di-*tert*-butylpyridine has been recommended as the base[90].

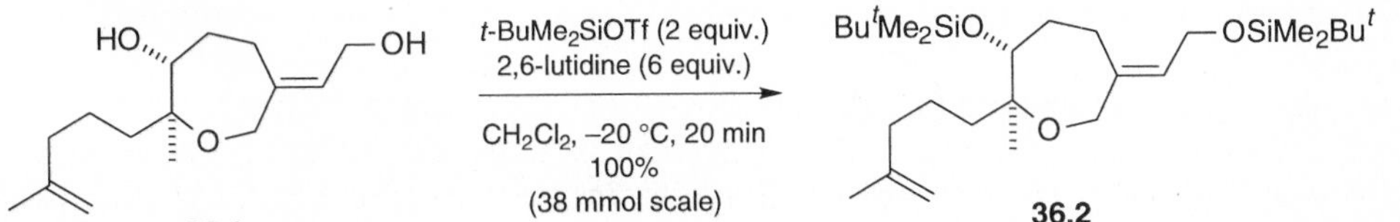
HO
OH
O
t-BuMe2SiOTf (2 equiv.)
2,6-lutidine (6 equiv.)
CH2Cl2, –20 °C, 20 min
100%
(38 mmol scale)
Bu^tMe2SiO
OSiMe2Bu^t
O

**36.1** **36.2**

**Scheme 2.36**

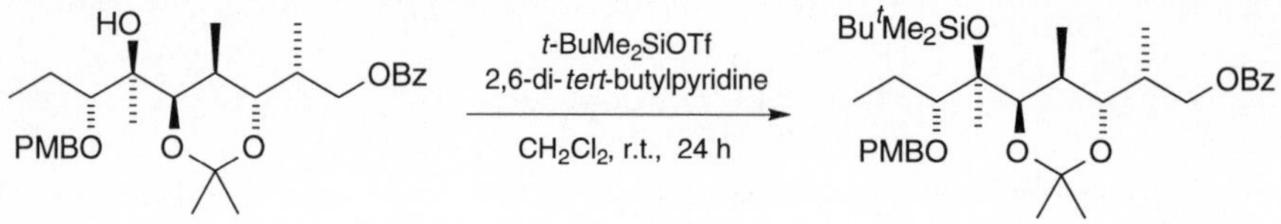

Scheme 2.37

*tert*-Butyldimethylsilyl triflate is a very powerful silylating agent and competing reactions can take place. For example, ketones are converted to the corresponding enol silyl ether under conditions similar to those used to protect hydroxyl functions[91] and the reaction has been recommended for the preparation of enol silyl ethers from sterically hindered ketones[92]. Fortunately, enol silyl ethers are hydrolysed more readily than alkyl silyl ethers and so the carbonyl function can be easily recovered whilst leaving the silyl ether intact. Another problem is the migration of silanes to proximate hydroxyl functions during silylation [Scheme 2.38][49] or under basic reaction conditions[93]. Yet another complication arises when epoxide rings are present: ring opening occurs in the presence of 2,6-lutidine to give allylic alcohols[44].

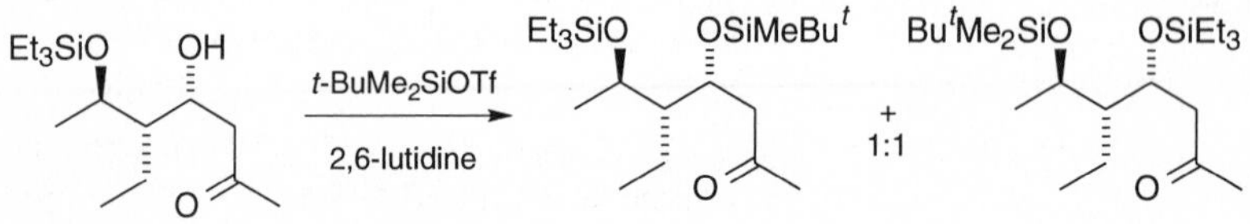

Scheme 2.38

A recent method which promises to provide a mild procedure for the preparation of TBS ethers of phenols and alcohols involves the reaction of *tert*-butyldimethylsilanol with the alcohol function under Mitsunobu conditions[94] [Scheme 2.39][95].

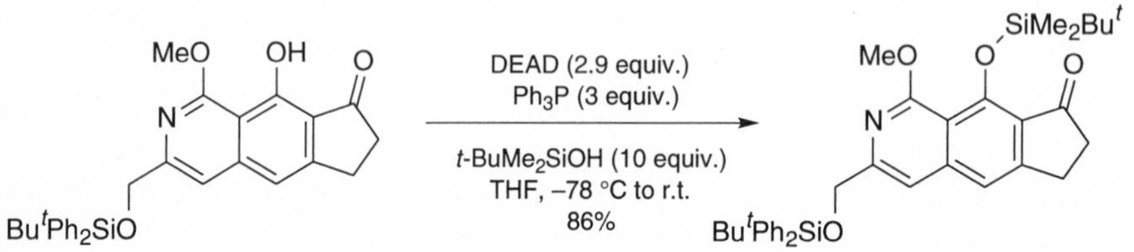

Scheme 2.39

**(iii) NMR Data**

$^{1}$H NMR: $\delta = 0.9$ (9H, s), 0.05–0.1 (6H, s)

$^{13}$C NMR: $\delta = 26$ (3C, q), 18 (1C, s), –4 [2C, q, $Si(CH_3)_2$]

## 2.3.4 *tert*-Butyldiphenylsilyl Ethers

The *tert*-butyldiphenylsilyl (abbreviated TBDPS) protecting group was first reported by Hanessian and Levallee in 1975[96]. Of the commonly used silyl protecting groups, the TBDPS group is the most stable to acid hydrolysis having a relative rate factor of 100–250 compared with the TBS group and 5–10 times compared with TIPS although it is cleaved at a comparable rate to the TBS group with TBAF.

The TBDPS group is also more stable towards carbanions than TBS. It survives DIBALH reductions, 80% HOAc (conditions that cleave trityl, THP, and TBS groups)[97,98]; aqueous trifluoroacetic acid in THF (conditions that cleave isopropylidene and benzylidene acetals)[99], $Me_2BBr$ in $CH_2Cl_2$ at –78 °C (conditions that cleave a MOM group)[100]; and catalytic amounts of NaOMe in MeOH at r.t. for 24 h (conditions which cleave acetate and benzoate esters). Methyl esters can be hydrolysed with NaOH (2 equiv.) in a mixture of *i*-PrOH and water (3:1) at 60 °C[101] and ketals can be cleaved[13] using mercaptoethanol in the presence of $BF_3{\bullet}OEt_2$. The TBDPS group is less prone to migrate to proximate hydroxyl groups under basic conditions than the TBS group but it may migrate under acidic conditions[13].

**(i) Cleavage**

TBDPS ethers are generally cleaved under the same conditions as those used for TBS ethers (i.e., TBAF–THF, HF–MeCN, or HF•pyridine–THF, *vide supra*) but longer reaction times are frequently necessary; consequently, selective removal of TBS groups in the presence of TBDPS groups is quite common.

**(ii) Formation**

TBDPS ethers are usually formed by the reaction of a primary or secondary alcohol with TBDPSCl in $CH_2Cl_2$ or DMF in the presence of imidazole [Scheme 2.40][102] or DMAP [Scheme 2.41][103]. Tertiary alcohols do not silylate. The higher steric bulk of the TBDPS group can result in much more selective protections of secondary hydroxyl groups than is possible with TBS groups. For example, monoprotection of the inositol derivative **42.1** [Scheme 2.42] occurred in 75% yield to give the C2 and C4 products **42.2** and **42.3** (20:1) whereas silylation with TBSCl under similar conditions only gave a 31% yield of the corresponding C2 and C4 TBS ethers in a ratio of 3.6:1[13].

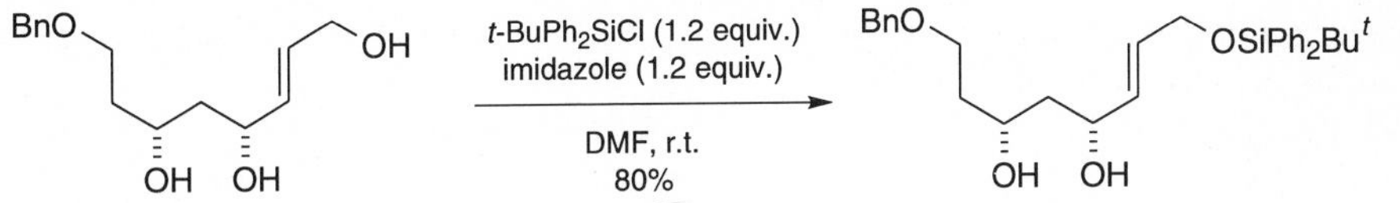

BnO OH OH OH

*t*-$BuPh_2SiCl$ (1.2 equiv.)
imidazole (1.2 equiv.)
DMF, r.t.
80%

BnO $OSiPh_2Bu^t$ OH OH

**Scheme 2.40**

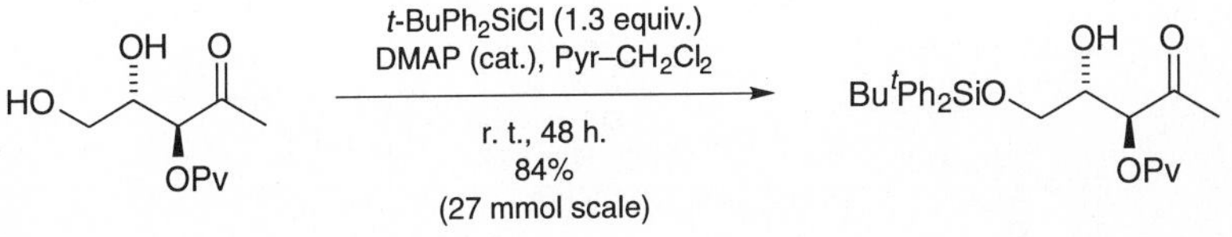

HO OH O OPv

*t*-$BuPh_2SiCl$ (1.3 equiv.)
DMAP (cat.), Pyr–$CH_2Cl_2$
r. t., 48 h.
84%
(27 mmol scale)

$Bu^tPh_2SiO$ OH O OPv

**Scheme 2.41**

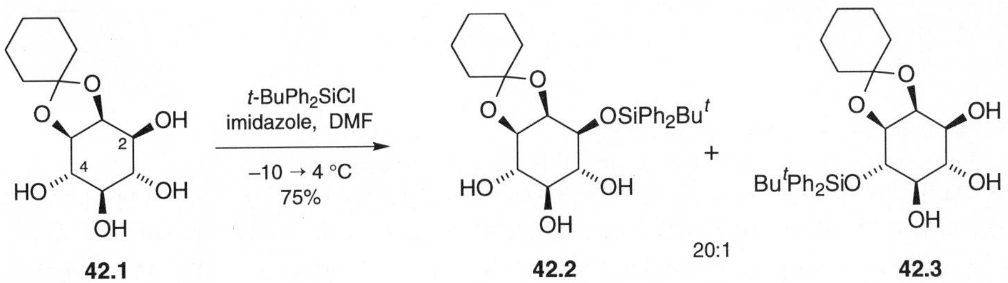

O O OH 2 4 HO OH OH

**42.1**

*t*-$BuPh_2SiCl$
imidazole, DMF
–10 → 4 °C
75%

O O $OSiPh_2Bu^t$ HO OH OH

**42.2**

\+

O O OH $Bu^tPh_2SiO$ OH OH

**42.3**

20:1

**Scheme 2.42**

**(iii) NMR Data**

$^1$H NMR: $\delta = 0.9–1.0$ (9H, s), 7.0–7.3 (10H, m)

$^{13}$C NMR: $\delta = 137$ (1C, s), 133 (1C, d), 129 (2C, d), 126 (2C), 27 (3C, q), 18 (1C, s)

## 2.3.5 Triisopropylsilyl Ethers

The large steric bulk of the triisopropylsilyl (abbreviated TIPS) group ensures high selectivity in the protection of primary hydroxyl groups over secondary and valuable stability under a wide range of reaction conditions[104]. TIPS groups are more stable than TBS or TBDPS groups towards basic hydrolysis and powerful nucleophiles though they are less stable than TBDPS groups towards acid hydrolysis. For example base hydrolysis of a methyl ester can be achieved in the presence of a TIPS ether whereas a TBS ether, under the same conditions, is destroyed[105]. Similarly, TIPS groups are inert towards powerful bases such as *t*-BuLi which is capable of metallating a Si–Me group[57] in TBS ethers.

**(i) Cleavage**

TIPS ethers are generally cleaved under the same conditions as those used for TBS ethers (i.e., TBAF–THF, HF–MeCN, or HF•pyridine–THF, *vide supra*) but longer reaction times are frequently necessary; consequently, TBS ethers can be removed selectively in many cases. The syntheses of FK-506 recently reported illustrate the remarkable stability of TIPS groups which survived many steps intact[14,106]. However, in the final step [Scheme 2.43] of Schreiber's synthesis[106], the removal of two TIPS and one TBS ether using HF in MeCN in standard laboratory glassware gave only 35% yield whereas the same reaction in polypropylene gave a 73% yield. These observations suggest that the well known reaction of HF with glass may generate species which can be deleterious to a fragile organic structure[107].

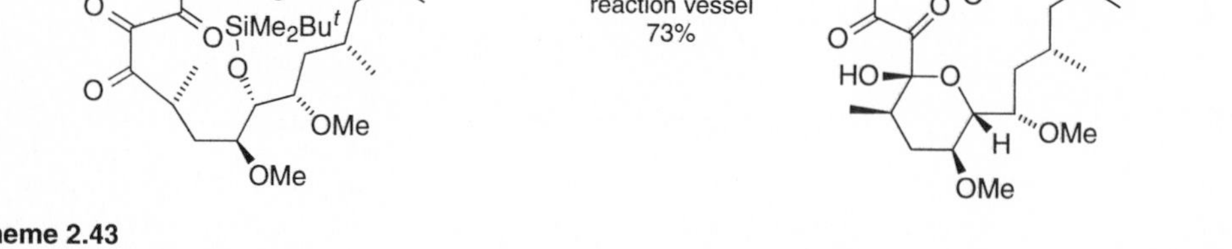

HF, MeCN

polypropylene reaction vessel 73%

**Scheme 2.43**

**(ii) Formation**

TIPS groups are formed under essentially the same conditions as TBS groups — i.e., primary or secondary alcohols are treated with TIPSCl in $CH_2Cl_2$ or DMF in the presence of imidazole or DMAP [Scheme 2.44][105]. For greater reactivity, TIPSOTf in the presence of 2,6-lutidine can be used[86]. TIPSOTf can be easily prepared on a large scale from triisopropylsilane and triflic acid in 97% yield.

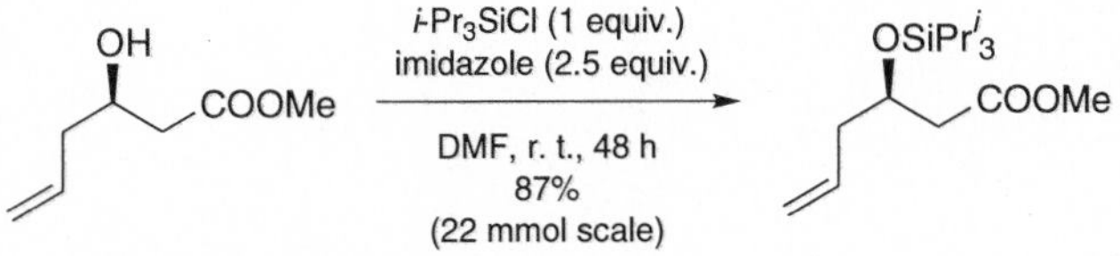

Scheme 2.44

### (iii) NMR Data
$^1$H NMR: $\delta = 0.9–1.1$ (18H, d), 0.9–1.1 (3H, septet)
$^{13}$C NMR: $\delta = 18$ (6C, q), 12 (3C, d)

## 2.3.6 Less Common Trialkylsilyl Ethers

The protecting groups cited above have been discussed in detail because they are commonly used but new groups are emerging under the pressure of functional and target complexity. For example, the synthesis of Cytovaricin[48] required a silicon protecting group which was more rugged than TES but which was labile in buffered HF•pyridine. These requirements were dictated by the extreme ease with which lactol dehydration occurred in the last step [Scheme 2.45] and the diethylisopropylsilyl (DEIPS) solved the problem admirably. A similar problem prevailed during the synthesis of Elaiophylin in which an acid-sensitive target had to be released in the last step [Scheme 2.46] under conditions which would not have been possible with the TBS group and again DEIPS and isopropyldimethylsilyl (IPDMS) groups were effective in providing stability during a protracted synthetic sequence but which were labile to mild acid hydrolysis[75]. A recent report[83] suggests that DEIPS ethers can be removed in the presence of TBS ethers using catalytic hydrogenolysis but the cleavage is slow enough to allow selective deprotection of a benzyl ether in the presence of a DEIPS group.

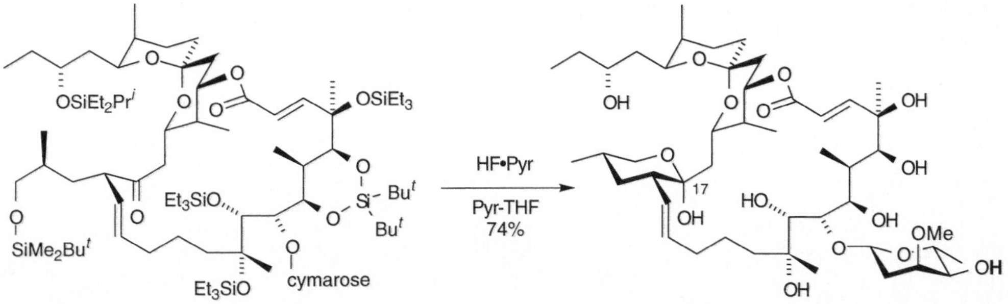

Scheme 2.45

Our final protecting group from the silyl ether orthogonal set is the thexyldimethylsilyl (TDS) group. Thexyldimethylsilyl chloride and the corresponding triflate have been recommended[108] as much cheaper substitutes for TBSCl and TBSOTf. Thexyldimethylsilyl ethers are formed at comparable rates and they are at least 2–3 times more stable than TBS ethers to acid and base hydrolysis but they introduce unwelcome clutter to NMR spectra. To date there have been comparatively few applications of TDS ethers. One indication of the promise they hold comes from the synthesis of the triquinane terpenoid Silphenene by the Franck–Neumann group[109]. A protecting group for a secondary hydroxyl was required which was resilient enough to survive some rather harsh conditions such as allylic oxidation using $SeO_2$ in dioxane at 101 °C, *O*-benzylation with *O*-benzyl trichloroacetimidate and

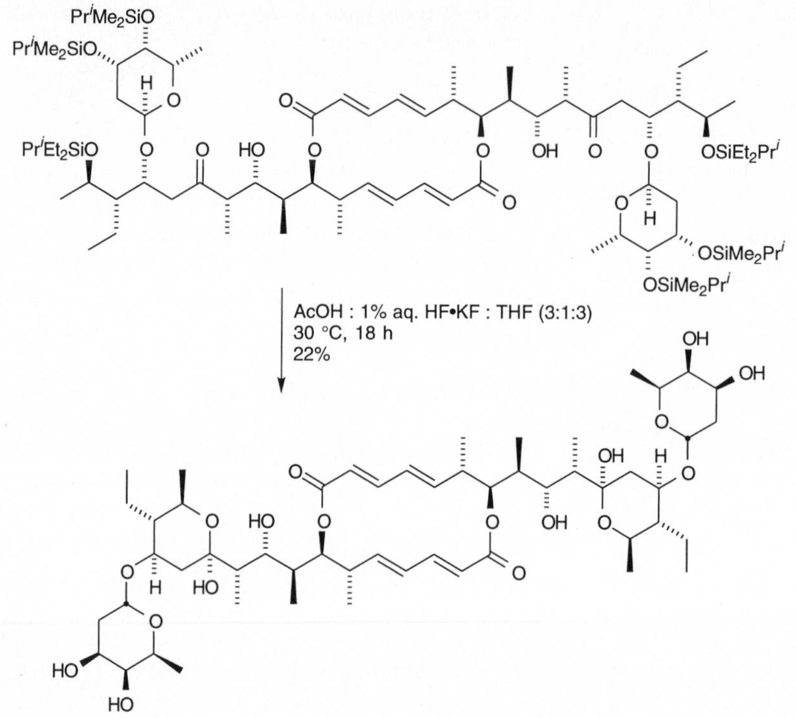

AcOH : 1% aq. HF•KF : THF (3:1:3)
30 °C, 18 h
22%

Scheme 2.46

$MeSO_3H$ at 40 °C, and — worst of all — the conditions required to effect the key silyl-assisted Nazarov cyclisation [Scheme 2.47]: $BF_3$•$OEt_2$ in refluxing PhEt (125 °C). A number of protecting groups were tried and failed but the thexyldimethylsilyl group was able to survive.

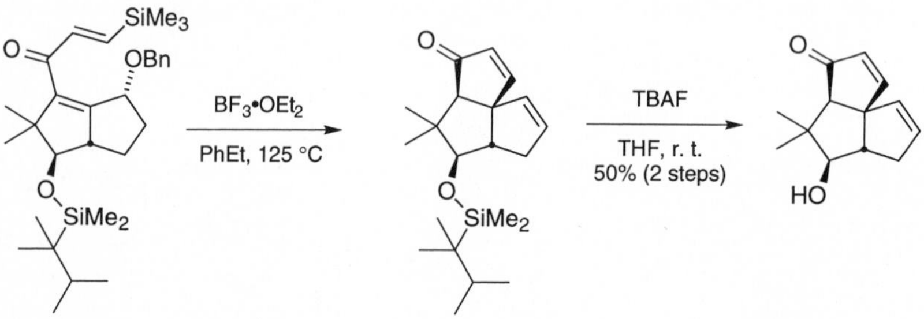

$BF_3$•$OEt_2$, PhEt, 125 °C; TBAF, THF, r. t., 50% (2 steps)

Scheme 2.47

## 2.4 Alkyl Ethers

### 2.4.1 Methyl Ethers

For sheer ruggedness and simplicity there are few protecting groups that can top the methyl ether. Of course there is a price to pay: the conditions required for deprotection are rather harsh and so comparatively few functional groups are compatible. Nevertheless, there is a *niche* for methyl ethers in the protection of hydroxyl groups which must survive strongly basic or acidic conditions. Phenol ethers cleave under conditions mild enough to be generally useful.

**(i) Cleavage**

Methyl ethers of aliphatic and alicyclic alcohols are usually cleaved with Lewis acids and in chapter 1 (section 1.2.2) we showed a cogent example of Lewis acid mediated methyl ether deprotection in Le Drian and Greene's synthesis of Brefeldin[110]. Typical reagents for cleaving methyl ethers include TMSI[111] in $CHCl_3$, $CH_2Cl_2$, or MeCN or $BBr_3$ in $CH_2Cl_2$. A practical advantage to TMSI is its easy preparation *in situ*[112-114] thereby ensuring a reagent free from protic acids which inevitably result when Lewis acids are exposed to adventitious water during storage and manipulation. The chemistry of TMSI has been reviewed[115-117]. $BF_3$•$OEt_2$ in the presence of ethanedithiol has also been employed[118,119].

$BBr_3$ is especially effective in the cleavage of phenol methyl ethers[120] and it has been recommended[121] as a superior reagent to TMSI. For stability and convenience of handling, the solid complex of $BBr_3$ and $Me_2S$ can be used[122]; the $Me_2S$ not only stabilises the complex but also provides a soft nucleophile which facilitates nucleophilic scission of the O–Me bond. The combination of a Lewis acid and a soft nucleophile is efficacious in many deprotection protocols which will be discussed later. A synthesis of the hexapeptide anti-tumour agent Deoxybouvaridin[123] provides a good example of the advantages of using ethanedithiol in conjunction with $AlCl_3$ for the cleavage of two phenol methyl ethers in a complex substrate [Scheme 2.48]. In the absence of ethanedithiol, only one O–Me bond was cleaved ($Me_b$) and that required 38 equivalents of $AlCl_3$ in $CH_2Cl_2$ at r.t. for 14 h. Another example of selectivity is shown in Scheme 2.49[124]. Methyl ethers ortho to acyl functions as in **49.1** generally cleave selectively with Lewis acids such as $BCl_3$ and $AlCl_3$[125].

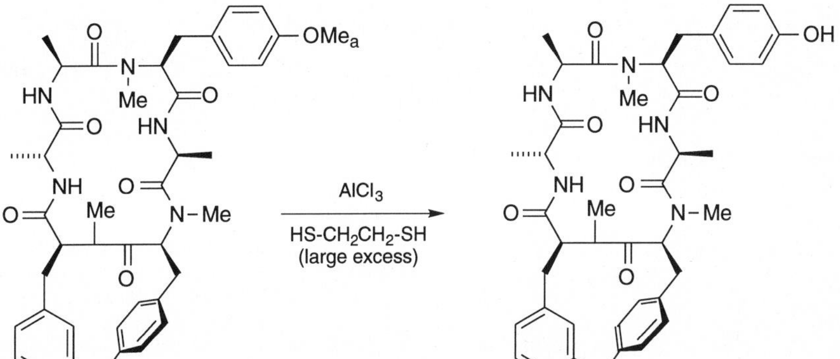

$AlCl_3$

HS-$CH_2CH_2$-SH
(large excess)

**Scheme 2.48**

i) $BCl_3$ , $CH_2Cl_2$, –20 °C

ii) aq. $Na_2CO_3$

94%

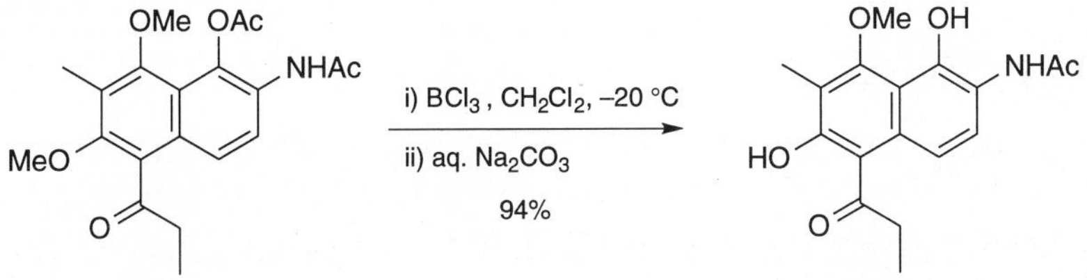

**49.1** **49.2**

**Scheme 2.49**

Phenol methyl ethers are susceptible to attack by powerful nucleophiles in dipolar aprotic solvents. Obviously, the resonance stabilisation of the phenolate anion affords sufficient weakening of the O–Me bond to make nucleophilic displacement feasible at elevated temperature. Scheme 2.50 illustrates the reaction in the last step of a synthesis of Milbemycin $\beta_3$[126] using EtSNa in refluxing DMF[127]. Phenol methyl ether deprotection can also be accomplished with LiCl under similar conditions[128].

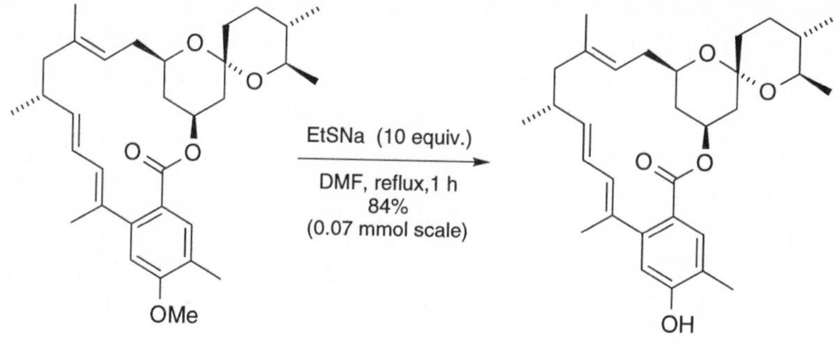

EtSNa (10 equiv.)
DMF, reflux,1 h
84%
(0.07 mmol scale)

OMe → OH

**Scheme 2.50**

**(ii) Formation**

Methyl ethers are usually prepared by some variant of the Williamson ether synthesis in which an alcohol reacts with either MeI, $(MeO)_2SO_2$, or $CF_3SO_2OMe$ (methyl triflate) in the presence of a suitable base. A word of caution: $(MeO)_2SO_2$ and $CF_3SO_2OMe$, like all powerful alkylating agents, are potentially carcinogenic and therefore should only be handled in a well-ventilated fume hood. For the *O*-methylation of phenols (p$K_a$ 10) a comparatively weak base such as $K_2CO_3$ in conjunction with $(MeO)_2SO_2$ is sufficient[127] whereas simple aliphatic alcohols require stronger bases such as NaH [Scheme 2.51][14] or $(TMS)_2NLi$ [Scheme 2.52][129]. The latter transformation is notable for the fact that *O*-methylation was accomplished without the complication of elimination.

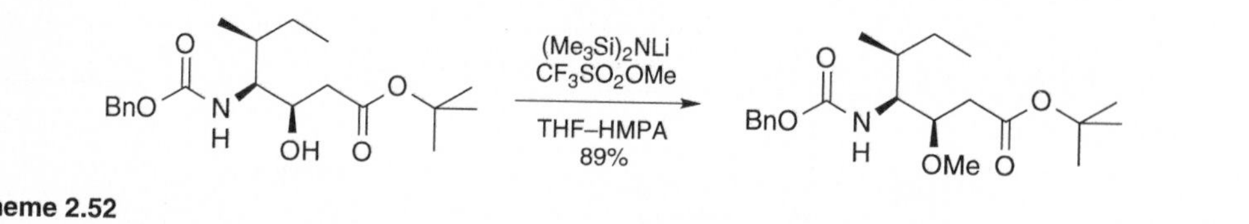

$Bu^tMe_2SiO$ OPMB OBn OH OH

NaH (5.2 mmol)
MeI (10.4 mmol)
THF, r.t., 36 h
98%
(1.73 mmol scale)

$Bu^tMe_2SiO$ OPMB OBn OMe OMe

**Scheme 2.51**

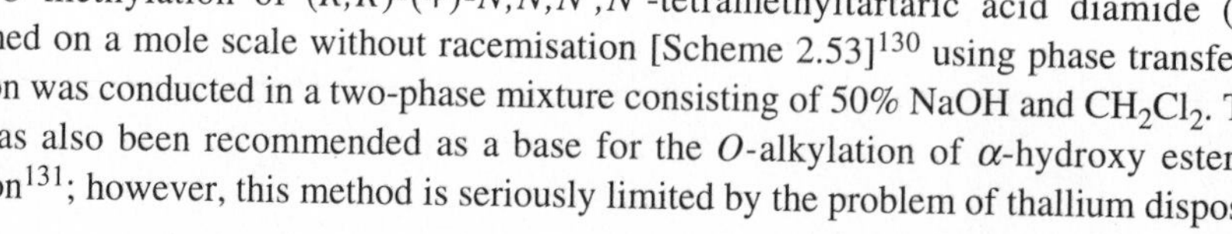

BnO N H OH O O

$(Me_3Si)_2NLi$
$CF_3SO_2OMe$
THF–HMPA
89%

BnO N H OMe O O

**Scheme 2.52**

A double *O*-methylation of (*R,R*)-(+)-*N,N,N′,N′*-tetramethyltartaric acid diamide (**53.1**) was accomplished on a mole scale without racemisation [Scheme 2.53][130] using phase transfer catalysis. The reaction was conducted in a two-phase mixture consisting of 50% NaOH and $CH_2Cl_2$. Thallium(I) ethoxide has also been recommended as a base for the *O*-alkylation of $\alpha$-hydroxy esters prone to racemisation[131]; however, this method is seriously limited by the problem of thallium disposal.

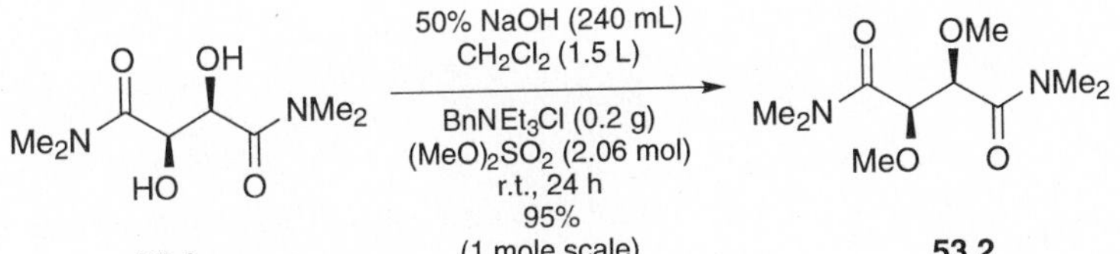

50% NaOH (240 mL)
$CH_2Cl_2$ (1.5 L)
$BnNEt_3Cl$ (0.2 g)
$(MeO)_2SO_2$ (2.06 mol)
r.t., 24 h
95%
(1 mole scale)

**53.1** **53.2**

**Scheme 2.53**

Base-sensitive substrates require some special precautions. With the very potent alkylating agent $CF_3SO_2OMe$, the highly hindered (and expensive) base 2,6-di-*tert*-butylpyridine can be used as shown in Scheme 2.54[14,132]. Alternatively, *O*-methylation using a combination of MeI and $Ag_2O$ as the base is effective for substrates harboring ester or lactone functions [Scheme 2.55][133] but appreciable conversions are best achieved only when MeI is used as the solvent and even then the reaction tends to be slow and the yields moderate. In some instances periodic sonication of the reaction appears to be beneficial[134].

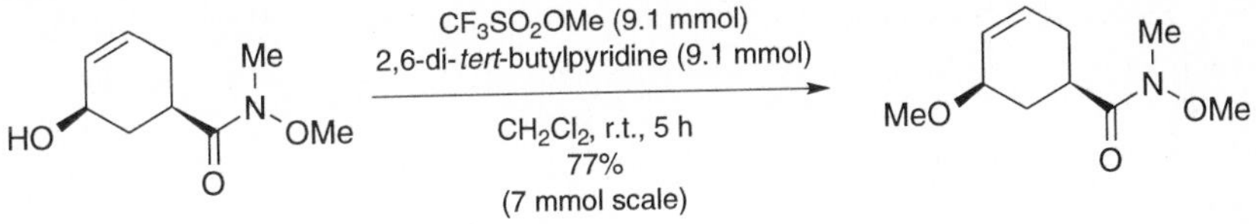

$CF_3SO_2OMe$ (9.1 mmol)
2,6-di-*tert*-butylpyridine (9.1 mmol)
$CH_2Cl_2$, r.t., 5 h
77%
(7 mmol scale)

**Scheme 2.54**

MeI (12 mL)
$CaSO_4$ (29.7 mmol)
$Ag_2O$ (12.9 mmol)
r.t., 18 h
70%
(8 mmol scale)

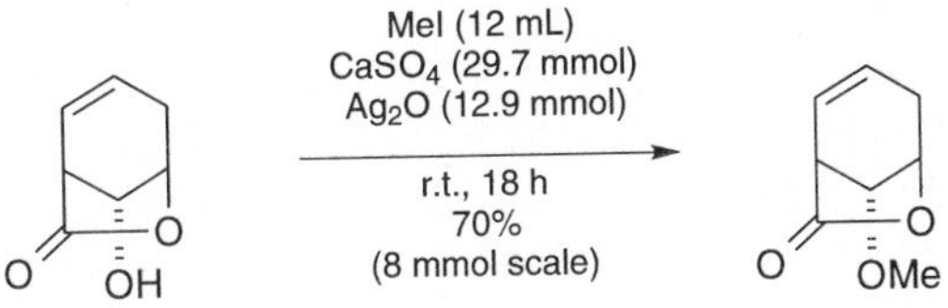

**Scheme 2.55**

Alcohols can be *O*-methylated with diazomethane in the presence of a protic acid such as $HBF_4$[135] or a Lewis acid. Meerwein and Hinz[136] showed that $ZnCl_2$ was an effective catalyst in 1930 and $BF_3$ later found favor[137]. Recently, a Japanese group[138] showed that an alcohol adsorbed onto neutral silica gel undergoes *O*-methylation. Despite the inconvenience and peril attending the preparation of $CH_2N_2$, the method can be used on a moderate scale [Scheme 2.56][139]. When $SnCl_2$ is used as the catalyst, selective mono-*O*-methylation of a 1,2-diol can be achieved [Scheme 2.57][140].

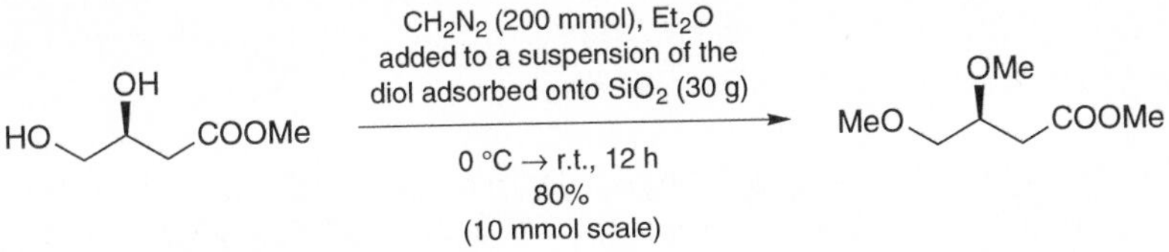

$CH_2N_2$ (200 mmol), $Et_2O$
added to a suspension of the
diol adsorbed onto $SiO_2$ (30 g)
0 °C → r.t., 12 h
80%
(10 mmol scale)

**Scheme 2.56**

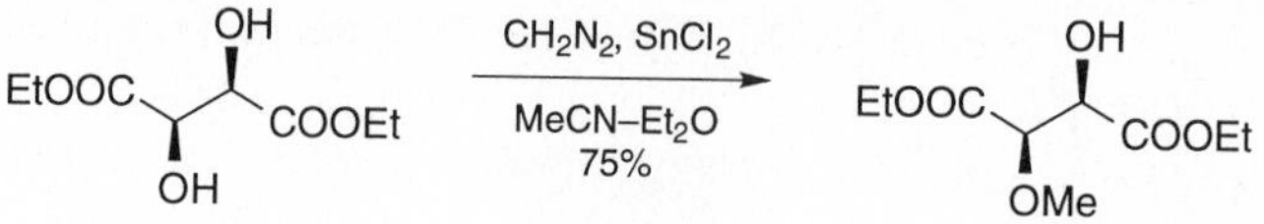

EtOOC, OH, OH, COOEt → EtOOC, OH, OMe, COOEt

$CH_2N_2$, $SnCl_2$

MeCN–$Et_2O$

75%

Scheme 2.57

**(iii) NMR Data**

$^{1}$H NMR: $\delta = 3.2$

$^{13}$C NMR: $\delta = 50–58$

## 2.4.2 Benzyl Ethers

Like methyl ethers, benzyl ethers (abbreviated Bn) are robust. They are stable to a wide range of aqueous acidic and basic conditions, and they are not readily attacked by most metal hydride reducing agents or mild oxidising agents (Swern, PCC, PDC, Dess–Martin periodinane, Jones, sodium periodate, lead tetraacetate, etc). At elevated temperature, and especially in the presence of Lewis acids, benzyl ethers may cleave with metal hydrides such as $LiAlH_4$. *n*-BuLi in THF may remove a benzylic proton from a benzyl ether (especially if an activator such as TMEDA or HMPA is present); *s*-BuLi and *t*-BuLi will almost certainly do so. As a group, the benzyl ethers straddle four orthogonal sets because they can be cleaved by Lewis acids, dissolving metal reduction, oxidation, and hydrogenolysis.

**(i) Cleavage**

Catalytic hydrogenolysis offers the mildest method for deprotecting benzyl ethers. The catalyst of choice is Pd–C in THF though others (e.g. Raney Ni[141], Rh–$Al_2O_3$[142]) can be used (*vide infra*). Platinum should be avoided because of competing reduction of the aromatic ring. The method is incompatible with the presence of most alkenes, alkynes, and alkyl halides — the prime exception being trisubstituted or hindered alkenes. A typical cause of vexation in the catalytic hydrogenolysis of benzyl ethers is catalyst poisoning which can occur when the substrate contains thioethers such as dithianes or methylthiomethyl ethers or even non-aromatic amines. Swern oxidations frequently introduce sulfur-containing contaminants which make a subsequent catalytic hydrogenolysis difficult if not impossible. Sulfur-containing impurities can sometimes be removed by treatment with Raney nickel.

Catalytic hydrogenolysis is particularly useful for deprotecting water-soluble alcohols since workup simply involves filtration to remove the catalyst (FIRE HAZARD : the catalyst may ignite filter paper when dry) and evaporation of the solvent. The yields are usually quantitative[143]. Scheme 2.58 illustrates the method[144]. Some measure of the dependability of catalytic hydrogenolysis can be gleaned from the final step of Schmidt's synthesis[145] of the octasaccharide fragment of the dimeric Le$^{X}$ antigen. In this synthesis no less than 17 benzyl groups were removed in one pot by catalytic hydrogenolysis to give the final product in 73% yield.

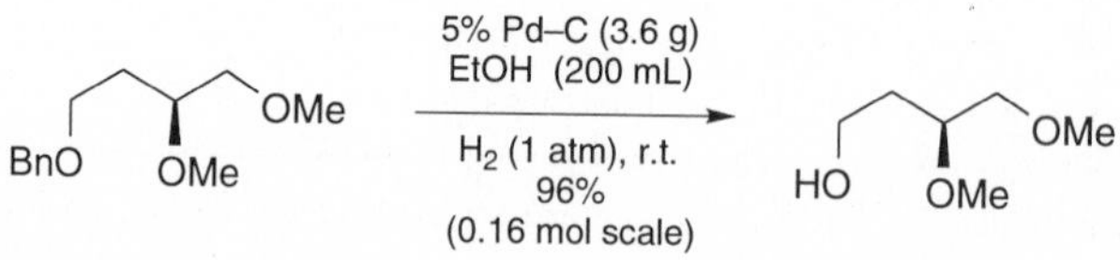

BnO, OMe, OMe → HO, OMe, OMe

5% Pd–C (3.6 g)

EtOH (200 mL)

$H_2$ (1 atm), r.t.

96%

(0.16 mol scale)

Scheme 2.58

An alternative procedure known as transfer hydrogenation[146,147] frequently has better functional group selectivity than the standard hydrogenation conditions and it is less susceptible to poisoning; indeed, in glacial acetic acid as solvent, transfer catalytic hydrogenation can be used to remove *O*-benzyl groups in the presence of methylthio ethers[148]. Palladium is the catalyst and refluxing MeOH or EtOH the usual solvent and cyclohexene[149], cyclohexadiene[150] or ammonium formate[151] can be used as the source of hydrogen rather than hydrogen gas. Benzylidene acetals, esters, and MOM, MEM, THP, TBS, and TBDPS groups survive. Scheme 2.59[152] illustrates the use of two sequential transfer hydrogenations whose success was in marked contrast to the failure of the usual methods. Note the survival of the tri-substituted alkene. These procedures are especially convenient for small scale work.

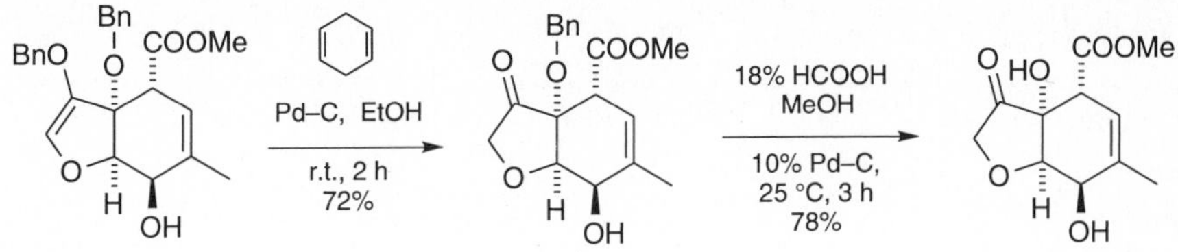

Pd–C, EtOH; r.t., 2 h; 72%. 18% HCOOH, MeOH; 10% Pd–C, 25 °C, 3 h; 78%

**Scheme 2.59**

*p*-Phenylbenzyl ethers have two advantages over ordinary benzyl ethers: first, they are more likely to be crystalline, and second, they undergo hydrogenolysis more rapidly. The latter feature was exploited by Mioskowski and Falck in their beautiful synthesis of Vineomycinone[153]. In this case, the *C*-glycoside ring in **60.1** [Scheme 2.60] had to remain intact (the ring oxygen itself is a benzylic oxygen) during the removal of three benzylic ether protecting groups. Attempts to use *p*-methoxybenzyl ethers failed because they could not be removed hydrogenolytically or by oxidative cleavage using DDQ (see section 2.4.3). The *p*-phenylbenzyl ethers, on the other hand, were readily removed by ordinary hydrogenolysis.

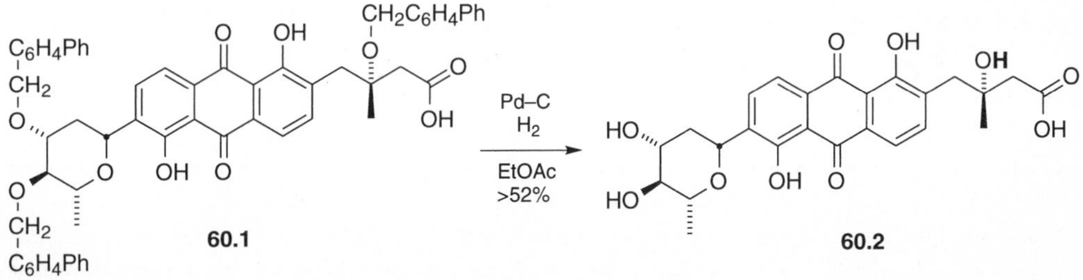

**60.1** → Pd–C, $H_2$; EtOAc; >52% → **60.2**

**Scheme 2.60**

Birch reduction involving sodium or lithium in liquid ammonia (bp –33 °C) in the presence of a proton source such as ethanol, *t*-BuOH, urea, or ammonium chloride will cleanly remove benzyl ethers. The conditions are brutal so most functional groups are incompatible. Isolated mono- and disubstituted alkenes are reduced very slowly with Na (but not with Li) whereas conjugated dienes are rapidly reduced to alkenes; reduction of alkynes is a highly stereoselective route to *trans*-alkenes but acetals and TBS ethers survive [Scheme 2.61][154]. Although unprotected carbonyls undergo reduction under these conditions, a carboxyl group is protected from reduction by its rapid conversion to the lithium carboxylate salt [Scheme 2.62][155]. Calcium in liquid ammonia offers a milder and more selective alternative to the standard Birch conditions which can be used to deprotect benzyl ethers in the presence of acetylenes and phenyl rings, furans, and cyclopropanes[156].

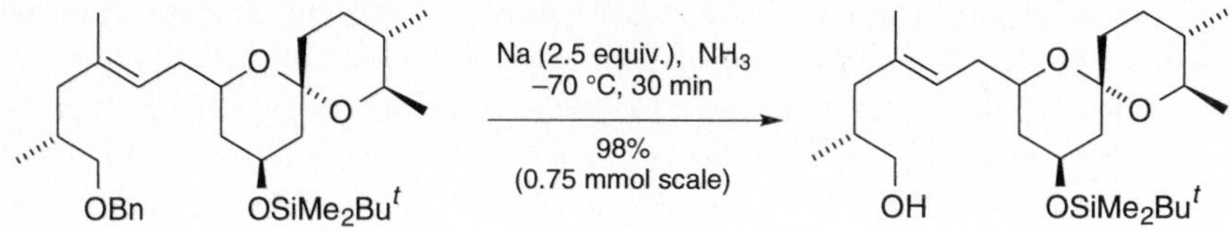

Na (2.5 equiv.), $NH_3$
–70 °C, 30 min
98%
(0.75 mmol scale)

OBn OSiMe$_2$Bu$^t$ → OH OSiMe$_2$Bu$^t$

**Scheme 2.61**

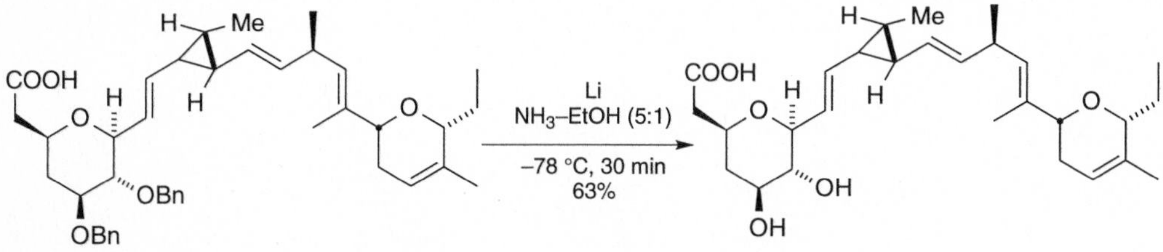

Li
$NH_3$–EtOH (5:1)
–78 °C, 30 min
63%

**Scheme 2.62**

Freeman and Hutchinson[157] recommend the use of the lithium di-*tert*-butylbiphenyl radical anion in THF at –78 °C to remove benzyl protecting groups as an alternative to the more usual Birch conditions; however, like the Birch reduction, few functional groups will survive the powerful reduction conditions. The method has recently been used by DeShong and co-workers[68,158] to remove a benzyl protecting group in the presence of an epoxide ring [Scheme 2.63].

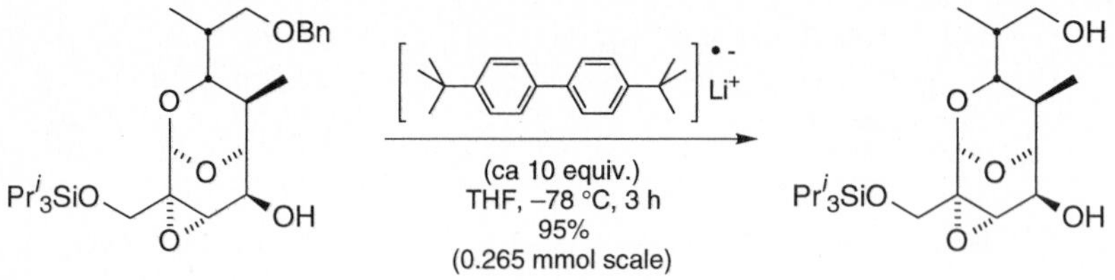

(ca 10 equiv.)
THF, –78 °C, 3 h
95%
(0.265 mmol scale)

**Scheme 2.63**

Various Lewis acids in $CH_2Cl_2$ or $ClCH_2CH_2Cl$, sometimes in combination with sulfur nucleophiles, have been used to cleave benzylic ethers under conditions which are mild enough to tolerate some functional groups. Typical reagents include TMSI[159], $SnCl_4$[160], $PhSSiMe_3$–$ZnI_2$[161] (tolerates ester functions), $BCl_3$[162] and $FeCl_3$[163]. The latter reagent cleaves benzyl ethers (15–30 min) and *p*-phenylbenzyl ethers (2–3 min) at room temperature in 70–85% yields and methyl ethers, benzoates and acetates are not affected. Scheme 2.64 shows the simultaneous cleavage of 2 phenol benzyl ethers and 2 *O*-benzyl carbamates from an intermediate in a synthesis of Biphenomycin[164] using TMSOTf and thioanisole in the presence of *o*-cresol.

Overman and Fukaya[165] used $BF_3$ in conjunction with EtSH[166] to remove an *O*-benzyl in the presence of an *N*-benzyl group without adversely affecting an ethyl ester function [Scheme 2.65].

2,3-Dichloro-5,6-dicyano-1,4-benzoquinone (DDQ) in the presence of water is a useful reagent combination for the cleavage of *p*-methoxybenzyl ethers (see section 2.4.3) but benzyl ethers can also be cleaved oxidatively using DDQ[167-169]. The method rescued a problematic deprotection during a synthesis of the antihelminthic agent Hikizimycin: Ikemoto and Schreiber[170] could not accomplish debenzylation of intermediate **66.1** [Scheme 2.66] under various reducing conditions owing to the

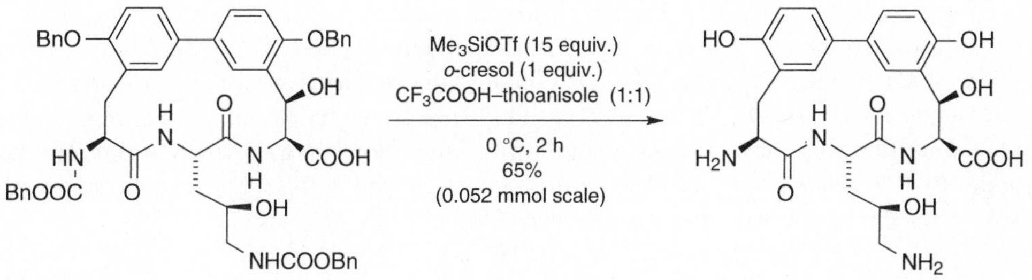
BnO
OBn
OH
HN
BnOOC
H
N
O
N
H
COOH
OH
NHCOOBn
Me3SiOTf (15 equiv.)
o-cresol (1 equiv.)
CF3COOH–thioanisole (1:1)
0 °C, 2 h
65%
(0.052 mmol scale)
HO
OH
H2N
COOH
NH2

**Scheme 2.64**

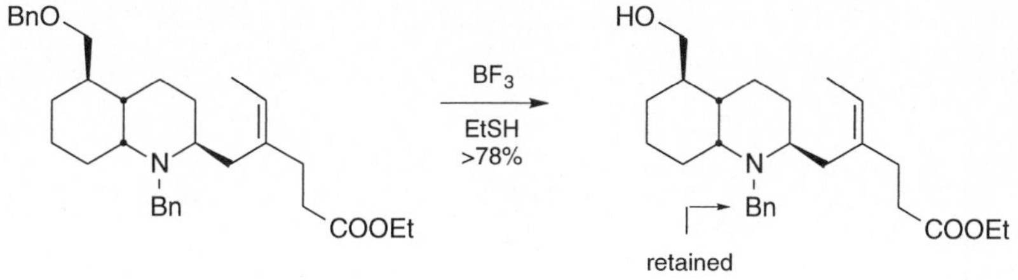
BnO
N
Bn
COOEt
BF3
EtSH
>78%
HO
retained

**Scheme 2.65**

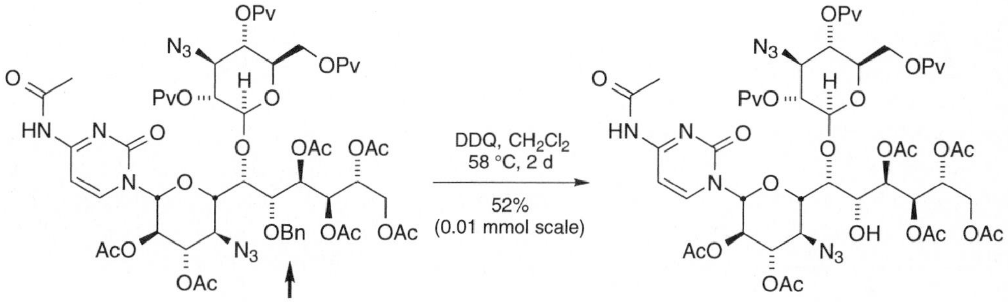
OPv
N3
PvO
H
O
HN
N
OAc
AcO
N3
OBn
DDQ, CH2Cl2
58 °C, 2 d
52%
(0.01 mmol scale)
OH

**66.1** **66.2**

**Scheme 2.66**

sensitivity of the cytosyl group. When DDQ and water were used, the debenzylation was slow and the DDQ decomposed giving acidic products causing deterioration of the substrate. However, *in the absence of water*[171], the reaction was clean giving a 52% yield of the debenzylated intermediate.

We have been rather cavalier in our treatment of the thiol group till now and our neglect will persist. However, whilst we contend with benzyl protecting groups it is worth pointing out that *S*-benzyl ethers remain one of the most frequently exploited means for thiol protection. *S*-Benzyl ethers are cleaved with Na in $NH_3$[172] or with HF in anisole[173] at r.t. Recently the *S*-2,4,6-trimethoxybenzyl (TMOB) group has been recommended as an efficacious acid-labile protecting group for cysteine thiol functions[174]; the trityl group is also effective (see next section).

**(ii) Formation**

It should come as no surprise that the methods for making benzyl ethers are analogous to those used to make methyl ethers (*vide supra*). Thus, *O*-alkylation of metal alkoxides with benzyl bromide or chloride (Williamson ether synthesis) is probably the most common method for preparing benzyl ethers. Since the metal alkoxides are usually generated with NaH or KH, this method is incompatible with base-sensitive functional groups. In the example [Scheme 2.67][144] a catalytic amount of KI or $Bu_4NI$ was added to accelerate the alkylation because iodide displaces bromide or chloride to give benzyl iodide *in situ* which is a much better alkylating agent. Iodide ion is then regenerated on alkylation with the benzyl iodide. Under these conditions even tertiary alcohols can be benzylated[175,176]. Alternatively, an alcohol can be protected with BnBr in the presence of silver oxide in DMF without affecting ester functions [Scheme 2.68][177,178].

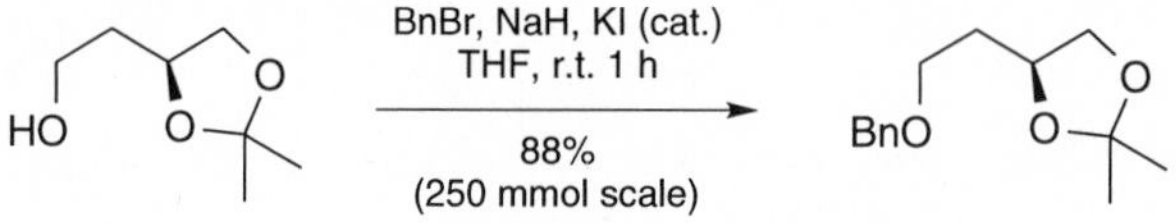
BnBr, NaH, KI (cat.)
THF, r.t. 1 h
88%
(250 mmol scale)
HO
O
O
BnO
O
O

**Scheme 2.67**

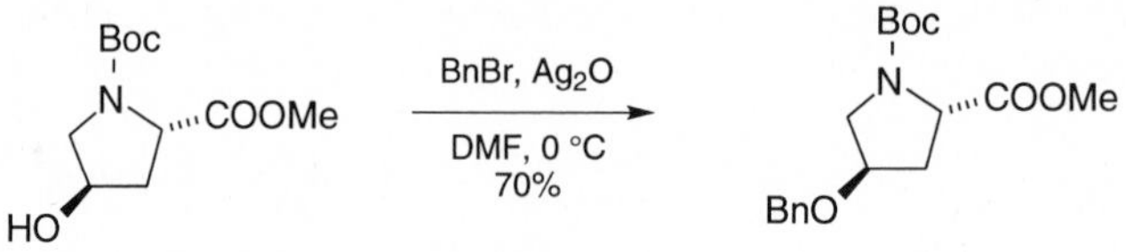
Boc
N
COOMe
HO
BnBr, Ag2O
DMF, 0 °C
70%
Boc
N
COOMe
BnO

**Scheme 2.68**

Benzyl ethers can also be prepared by nucleophilic cleavage of oxiranes by the alkali metal derivatives of benzyl alcohol. During a synthesis of inhibitors of inositol monophosphatase, a Merck group[179] obtained better yields and complete regioselectivity using a modification in which the oxirane and benzyl alcohol reacted in refluxing toluene in the presence of activated alumina [Scheme 2.69][180]. Another method which is useful for the selective protection of the equatorial hydroxyl of a cyclohexane-1,2-diol involves reaction with $(Bu_3Sn)_2O$ to give an intermediate stannoxane which is then simply heated with BnBr in an inert solvent such as toluene[181,182]. A closely related procedure will be covered in greater detail in section 2.4.6 when we discuss the formation of allyl ethers.

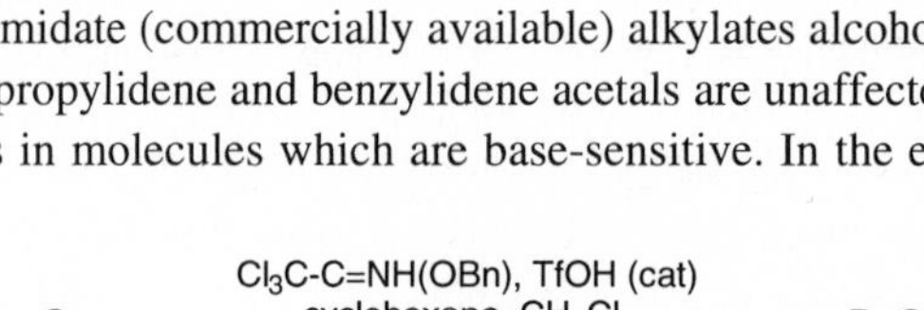
O
OPMB
BnOH, alumina
PhMe, Δ
OH
BnO
OPMB

**Scheme 2.69**

Benzyl 2,2,2-trichloroacetimidate (commercially available) alkylates alcohols in the presence of triflic acid[183]. Esters, imides, isopropylidene and benzylidene acetals are unaffected. This method allows the formation of benzyl ethers in molecules which are base-sensitive. In the example [Scheme 2.70][184],

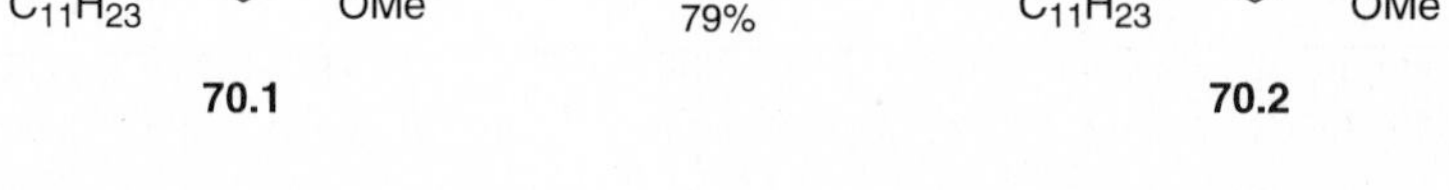
HO
O
C11H23
OMe
Cl3C-C=NH(OBn), TfOH (cat)
cyclohexane–CH2Cl2
r.t. 3 h
79%
BnO
O
C11H23
OMe

**70.1** **70.2**

**Scheme 2.70**

benzylation of the $\beta$-hydroxy ester **70.1** under the usual basic conditions would lead to retro-aldol reactions and/or elimination.

Scheme 2.71 exemplifies the application[185] of trichloroacetimidate activation to the synthesis of benzyl glycosides. Note the photochemical deprotection of an *o*-nitrobenzyl ether necessitated by the presence of acid- and base-sensitive groups (Fmoc) in the molecule.

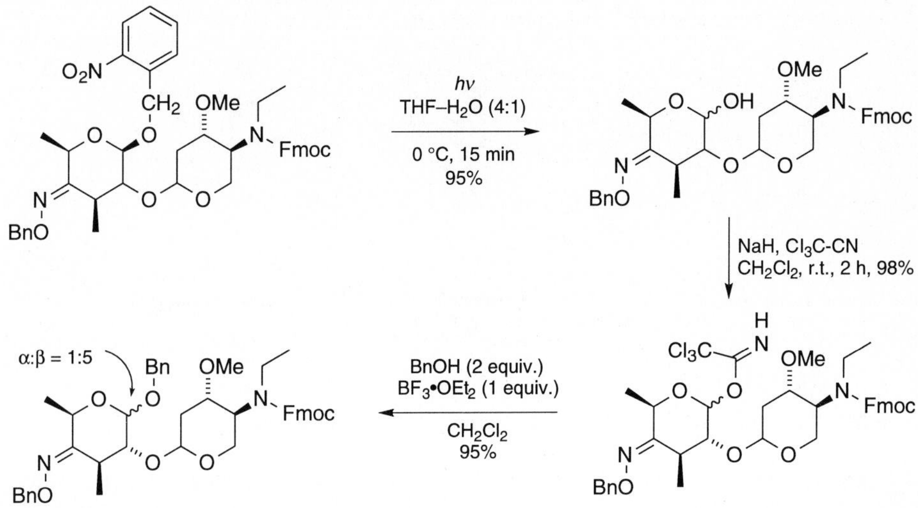

$h\nu$, THF–$H_2O$ (4:1), 0 °C, 15 min, 95%; NaH, $Cl_3C$-CN, $CH_2Cl_2$, r.t., 2 h, 98%; BnOH (2 equiv.), $BF_3$•$OEt_2$ (1 equiv.), $CH_2Cl_2$, 95%; α:β = 1:5

**Scheme 2.71**

Phenyldiazomethane[186] alkylates alcohols at –40 °C in the presence of a catalytic amount of $HBF_4$ in $CH_2Cl_2$ [Scheme 2.72][187]. The reaction conditions are sufficiently mild to alkylate tertiary alcohols; esters, acetals, and reactive halides are compatible. N-benzylation of primary and secondary amines occurs but much more slowly than alcohols.

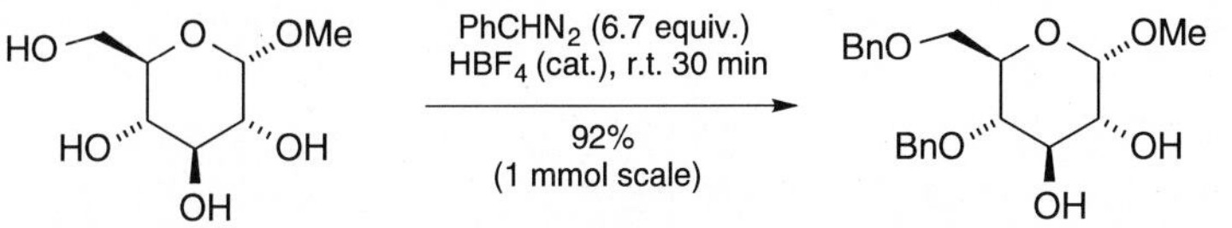

$PhCHN_2$ (6.7 equiv.), $HBF_4$ (cat.), r.t. 30 min, 92% (1 mmol scale)

**Scheme 2.72**

Unsymmetrically substituted benzylidene acetals can be cleaved regioselectively in some cases using Lewis acidic metal hydride reducing agents[188]. The cleavage results in the formation of a benzyl ether (more hindered position) and an alcohol. For example, cleavage of a benzylidene acetal with DIBALH in $CH_2Cl_2$ at r.t. for 96 h was used to elaborate the oxonane ring system of Obtusenyne [Scheme 2.73][189].

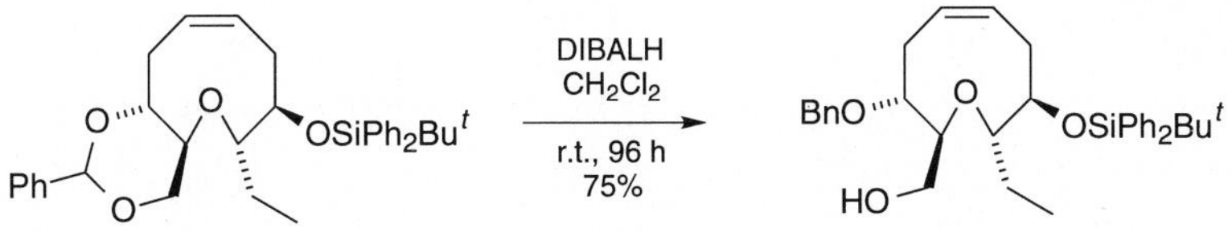

DIBALH, $CH_2Cl_2$, r.t., 96 h, 75%

**Scheme 2.73**

The enhanced acidity conferred upon *p*-acyl phenols is sufficient for participation in Mitsunobu displacements. Scheme 2.74 shows an example in which selective *O*-benzylation of the *p*-hydroxyl in the orcinol derivative **74.1** took place in preference to the *o*-hydroxyl[190].

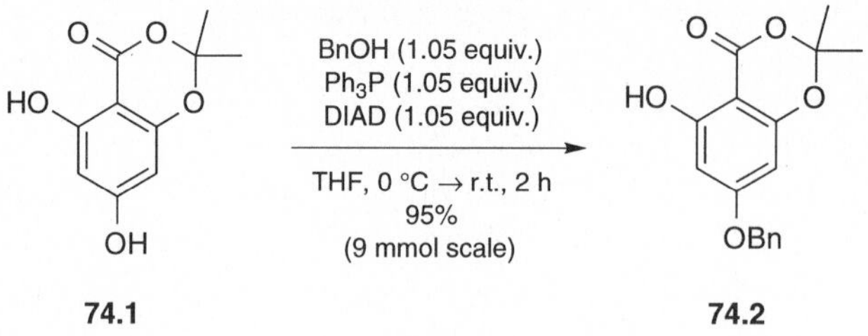

BnOH (1.05 equiv.)
$Ph_3P$ (1.05 equiv.)
DIAD (1.05 equiv.)
THF, 0 °C → r.t., 2 h
95%
(9 mmol scale)

**74.1** **74.2**

**Scheme 2.74**

**(iii) NMR Data**
$^1$H NMR: $\delta = 7.3$ (5H, m), 4.5 (2H, s or AB system if a stereogenic centre is present)
$^{13}$C NMR: $\delta = 137$ (1C, s), 129 (2C, d), 128 (2C, d), 127 (1C, s), 72 (1C, t)

### 2.4.3 *p*-Methoxybenzyl and 3,4-Dimethoxybenzyl Ethers

*p*-Methoxybenzyl ethers [abbreviated PMB or MPM (*p*-methoxyphenylmethyl)] and 3,4-dimethoxybenzyl ethers (abbreviated DMB or DMPM) are much less stable to acid than benzyl ethers. Aqueous mineral acid or CSA in methanol is sufficient to remove them. Their great virtue is that they can be removed under specific conditions which do not affect benzyl ethers or silyl ethers and therefore find use in the synthesis of functionally complex molecules where extensive selective protection–deprotection protocols are required.

**(i) Cleavage**
DMB and PMB ethers are especially susceptible to oxidative cleavage under essentially neutral conditions using 1.1–1.5 equiv of DDQ in $CH_2Cl_2$–water mixtures to give the desired alcohol and *p*–methoxybenzaldehyde or 3,4-dimethoxybenzaldehyde[191]. The reaction conditions are compatible with most other protecting groups (isopropylidene acetals, and MEM, MOM, BOM, THP, acetyl, TBS, benzyl, benzoyl, and tosyl groups) and functional groups (epoxides, alkenes, ketones). Scheme 2.75 shows that DMB ethers can be selectively removed in the presence of PMB ethers[141].

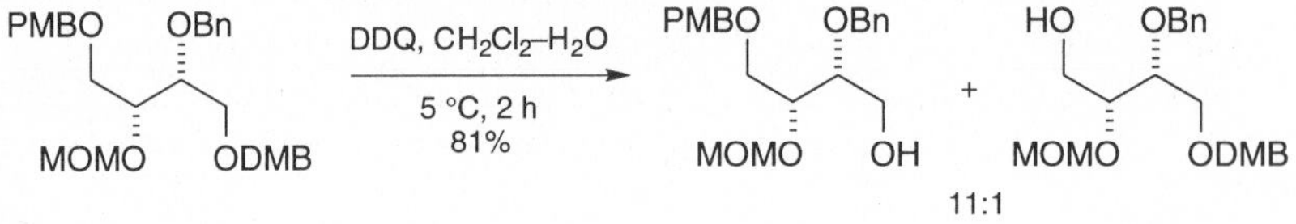

PMBO OBn / MOMO ODMB
DDQ, $CH_2Cl_2$–$H_2O$
5 °C, 2 h
81%
PMBO OBn / MOMO OH + HO OBn / MOMO ODMB
11:1

**Scheme 2.75**

Ceric ammonium nitrate (CAN) in aqueous acetonitrile has also been recommended[192] as a superior reagent for the oxidative cleavage of PMB ethers [Scheme 2.76][193,194]. Classon[195] used NBS or bromine as an alternative oxidant to remove PMB ethers in aminoglycosides when the usual DDQ failed.

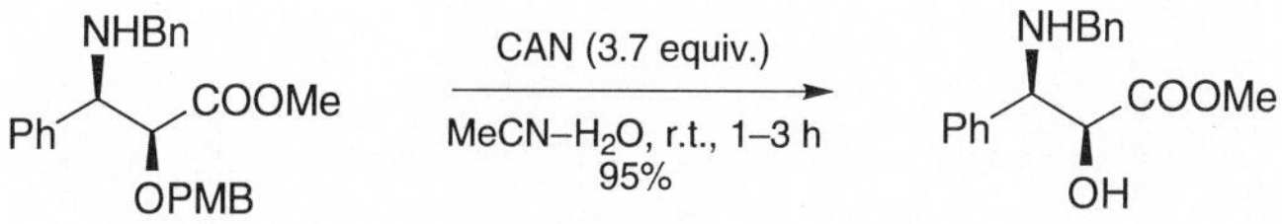
NHBn
COOMe
Ph
OPMB
CAN (3.7 equiv.)
MeCN–H2O, r.t., 1–3 h
95%
NHBn
COOMe
Ph
OH

**Scheme 2.76**

Catalytic hydrogenolysis is another mild option for the scission of PMB ethers, but under the usual conditions (Pd–C, $H_2$), selective deprotection of the various benzyl protecting groups is probably not possible. However, W-4 Raney nickel will selectively deprotect benzyl ethers in the presence of PMB and DMB ethers as illustrated in Scheme 2.77[141].

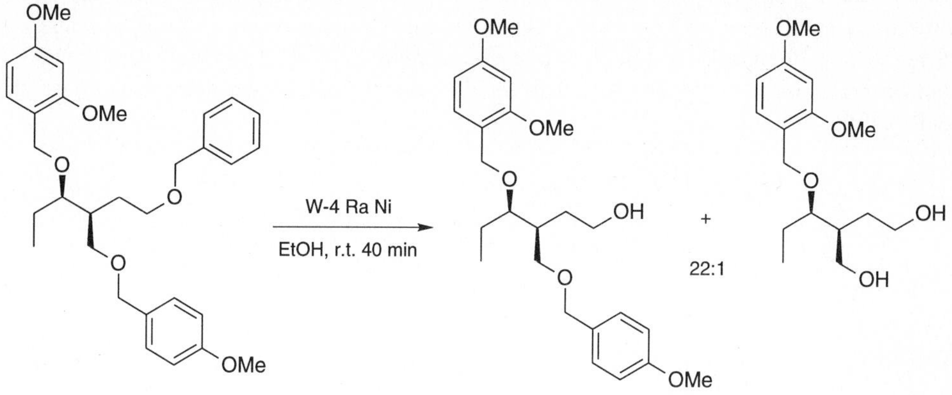
OMe
OMe
O
O
O
OMe
W-4 Ra Ni
EtOH, r.t. 40 min
OMe
OMe
O
OH
O
OMe
+
22:1
OMe
OMe
O
OH
OH

**Scheme 2.77**

Very few examples of Lewis acid mediated deprotection of PMB and DMB ethers have been reported despite the fact that the reaction should be much easier than the corresponding cleavage of benzyl ethers. Scheme 2.78 shows that the method holds some promise. Deprotection of the diolide **78.1** using $Me_2BBr$ occurred without 1,2-acyl migration[196] to give the hydroxymethyl-substituted diolide **78.2** in quantitative yield.

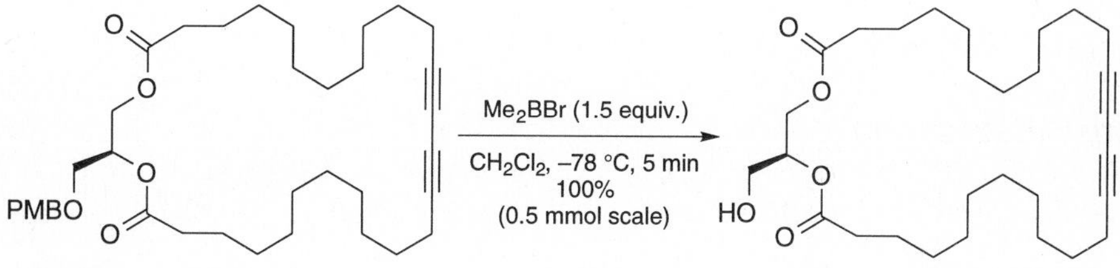
O
O
O
PMBO
O
Me2BBr (1.5 equiv.)
CH2Cl2, –78 °C, 5 min
100%
(0.5 mmol scale)
HO
O
O
O
O

**78.1** **78.2**

**Scheme 2.78**

**(ii) Formation**

The sodium alkoxides prepared from primary and secondary alcohols and NaH in DMF or DMSO react with PMBCl at r.t. to give the PMB ethers in good yield. Use of THF alone is not so satisfactory. PMBCl is moisture sensitive and deteriorates on standing. It should be freshly distilled *in vacuo* before use. DMB ethers are prepared by the same method. Scheme 2.79 illustrates the procedure[141].

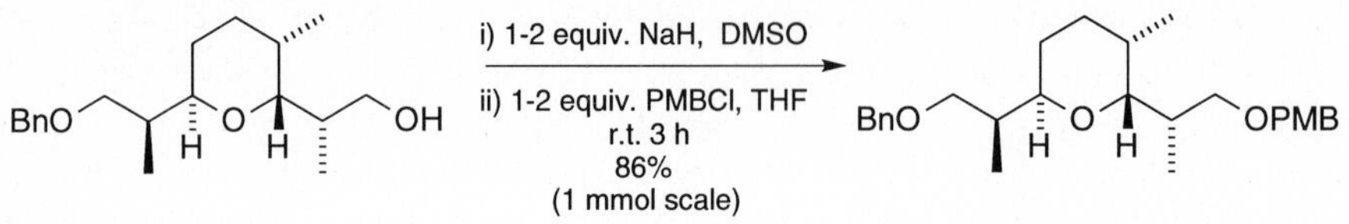

**Scheme 2.79**

PMB trichloroacetimidate (bp 135–137 °C / 0.7 mm Hg) is more reactive than its benzyl analogue and extremely sensitive to acids. It is best prepared immediately before use. Primary, secondary, or tertiary alcohols are simply treated with the trichloroacetimidate in ether at r.t. with 0.3 mol % triflic acid. The amount of acid is important: if too much is used, low yields and messy reactions are obtained. The conditions are compatible with epoxides, acetals, silyl groups, and esters. $\beta$-Hydroxy esters can be protected without elimination and racemisation of $\alpha$-stereogenic centres does not occur. Scheme 2.80 illustrates protection of a tertiary alcohol in the presence of an acid sensitive bis-spiroacetal ring system[197].

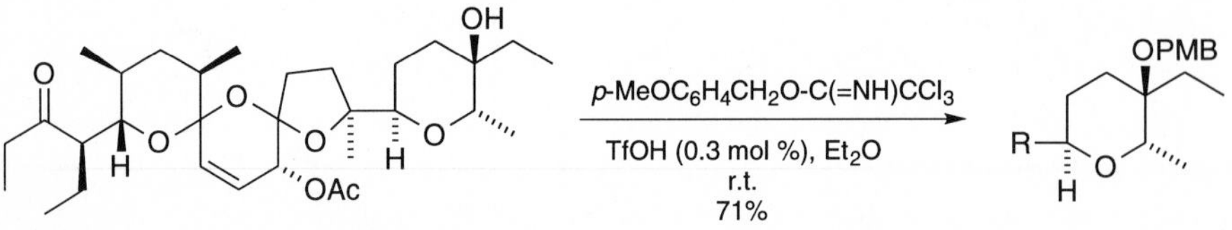

**Scheme 2.80**

Conversion of a 1,2- or 1,3-diol to the *p*-methoxybenzylidene acetal (*vide infra*) followed by regioselective reductive cleavage of the acetal can be used for the differential protection of a 1,2- or 1,3-diol [Scheme 2.81][48]. A related transformation has been applied to 3,4-dimethoxybenzylidene acetals[198].

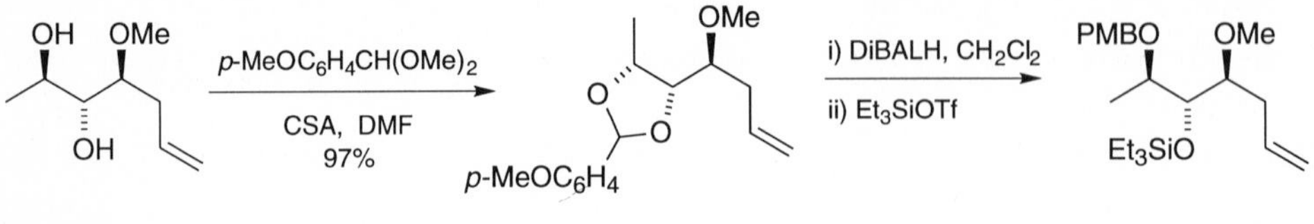

**Scheme 2.81**

**(iii) NMR Data (PMB ether)**

$^{1}$H NMR: $\delta = 7.3$ (2H, d, $J = 8$ Hz), 6.9 (2H, d, $J = 8$ Hz), 4.4 (2H, s), 3.8 (3H, s)

$^{13}$C NMR: $\delta = 159$ (1C, s), 131 (1C, s), 129 (2C, d), 114, 2C, d), 77 (1C, t), 55 (1C, q)

## 2.4.4 Trityl Ethers

In the late 1950's and early 1960's Khorana and co-workers prepared short, defined sequences of deoxyribopolynucleotides which were instrumental in deciphering the genetic code[199,200]. The chemical methodology used for the construction of these oligomers involved the condensation of a nucleoside having a free 3′-hydroxyl function and a 5′-hydroxyl bearing the acid-labile triphenylmethyl (trityl) group (abbreviated Tr) with a nucleoside 5′-phosphate in which the 3′-hydroxyl was protected with a base-labile group as illustrated in Scheme 2.82. The trityl group and its various derivatives were also to play a major role in the total syntheses of the genes encoding an alanine tRNA from yeast and a tyrosine suppressor tRNA from *E. coli*[201]. However, the trityl protecting group was hardly new: it had already been extensively exploited in classical carbohydrate chemistry because its hydrophobicity proved particularly valuable for the protection and manipulation of polar, water-soluble monosaccharides. Trityl ethers are stable to base and nucleophiles but easily removed under acidic conditions owing to the stability of the triphenylmethyl carbenium ion. To some extent the position of the trityl ether in the armoury of synthetic chemists has been usurped by the silicon protecting groups; nevertheless, trityl ethers continue to be used and are worthy of consideration.

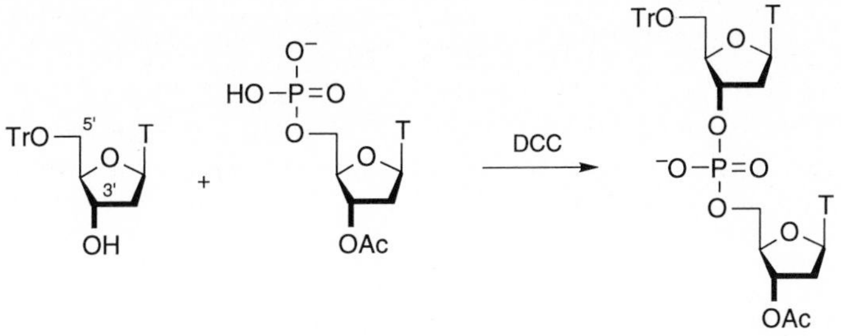

**Scheme 2.82**

**(i) Cleavage**

Trityl ethers are removed with protic acids such as formic acid in ether[202] — conditions mild enough to retain isopropylidene acetals and TBS ethers in some cases [Scheme 2.83]; alternatively, 80% acetic acid at reflux[203] or 1 M aqueous HCl in dioxane (4:96)[204] can be used. However, trityl ethers survive the formation of isopropylidene derivatives using PPTS and $CuSO_4$ in acetone at r.t.[79].

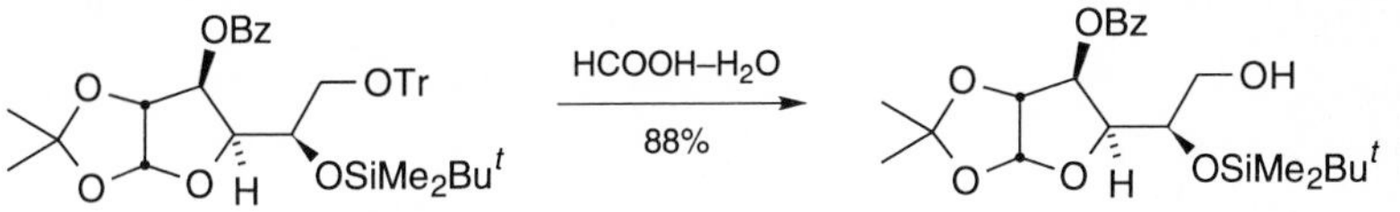

**Scheme 2.83**

The triphenylcarbenium cation may, in certain cases, attack sensitive substrates in which case it can be intercepted with a suitable nucleophile such as 2-methyl-2-butene[205] [Scheme 2.84]. Most Lewis acids such as $ZnBr_2$–MeOH, $EtAlCl_2$–$CH_2Cl_2$, $BF_3$•$Et_2O$–$HSCH_2CH_2SH$–$BF_3$ in methanol[206], or $FeCl_3$[207]

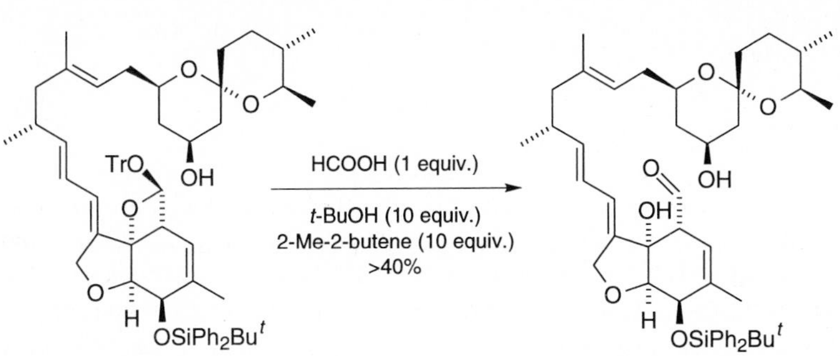

**Scheme 2.84**

will also cleave trityl ethers. Deprotection of a trityl ether of a glyceride derivative poses the problem of 1,2-acyl migration. A mild method which avoids migration simply involves passing a solution of the ether in hexane through a column of boric acid on silica. The deprotected alcohol is eluted from the column in good yield[208,209].

Trityl ethers are easily deprotected using Na in $NH_3$ [Scheme 2.85][210] but catalytic hydrogenolysis appears to be unreliable[211].

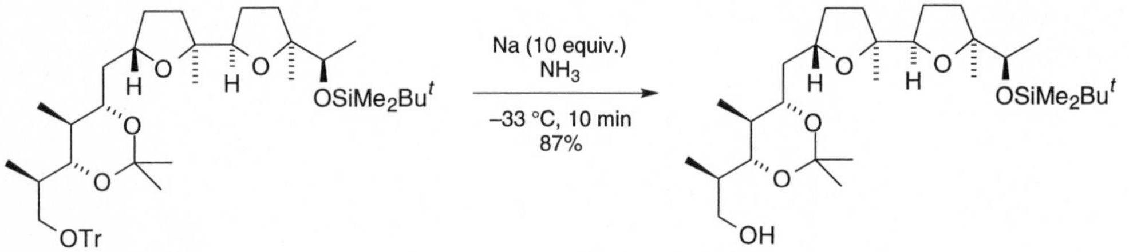

Na (10 equiv.)
$NH_3$
–33 °C, 10 min
87%

Scheme 2.85

*p*-Methoxy-substituted trityl ethers and related compounds have enhanced lability in acid which makes them useful[212] in cases where the quotidian trityl group is too stable [Scheme 2.86][213]. For each methoxy group introduced, acid-lability increases by about one order of magnitude[214]. Thus, monomethoxytrityl ethers (MMTr) of 5′-protected uridine derivatives hydrolyse in 2 h in 80% AcOH at r.t. whereas the parent trityl system requires 48 h. By contrast, the analogous dimethoxytrityl (DMTr) cleaves in 15 min. Obviously the enhanced lability of methoxy-substituted trityl ethers with protic acids reveals parallel behaviour with Lewis acids. For example $BF_3{\bullet}OEt_2$ in $CH_2Cl_2$–MeOH[215] and $ZnBr_2$ in $MeNO_2$[216] or $CH_2Cl_2$[217] have been successfully deployed in the cleavage of 5′-DMTr groups in nucleosides.

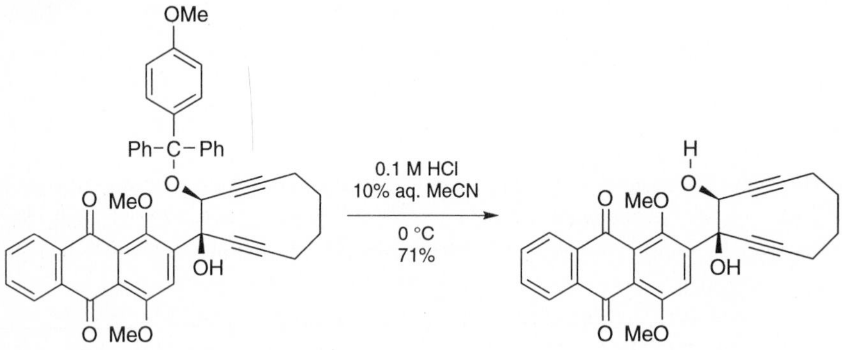

0.1 M HCl
10% aq. MeCN
0 °C
71%

Scheme 2.86

The high acid lability of 2,7-dibromo-9-phenylxanthen-9-yl ethers was a critical design feature in a synthesis of D-*myo*-inositol 1,4,5-triphosphate [Scheme 2.87][218]. Deprotection of the 3 phosphate ester moieties in **87.1** by treatment with tetramethylguanidine gave the tetramethylguanidinium salt **87.2** which, when dissolved in water, gave a pH of 6.0 — conditions which spontaneously removed the 6-*O*-(2,7-dibromo-9-phenylxanthen-9-yl) group. The triphosphate **87.3** was recovered from the guanidinium derivative by brief passage through an acid exchange resin and converted to the ammonium salt **87.4** with ammonia. The final deprotection of the cyclopentylidene acetal (see Chapter

3) was accomplished with aqueous acetic acid to give the target **87.5** in 33% overall yield. The related 9-(9-phenyl)xanthenyl ethers (pixyl ethers) have been used extensively for the protection of the 5′ hydroxyl function of nucleosides in cases where the trityl group proved too stable. They have the added advantage of higher crystallinity and lower UV detection limits to recommend them.

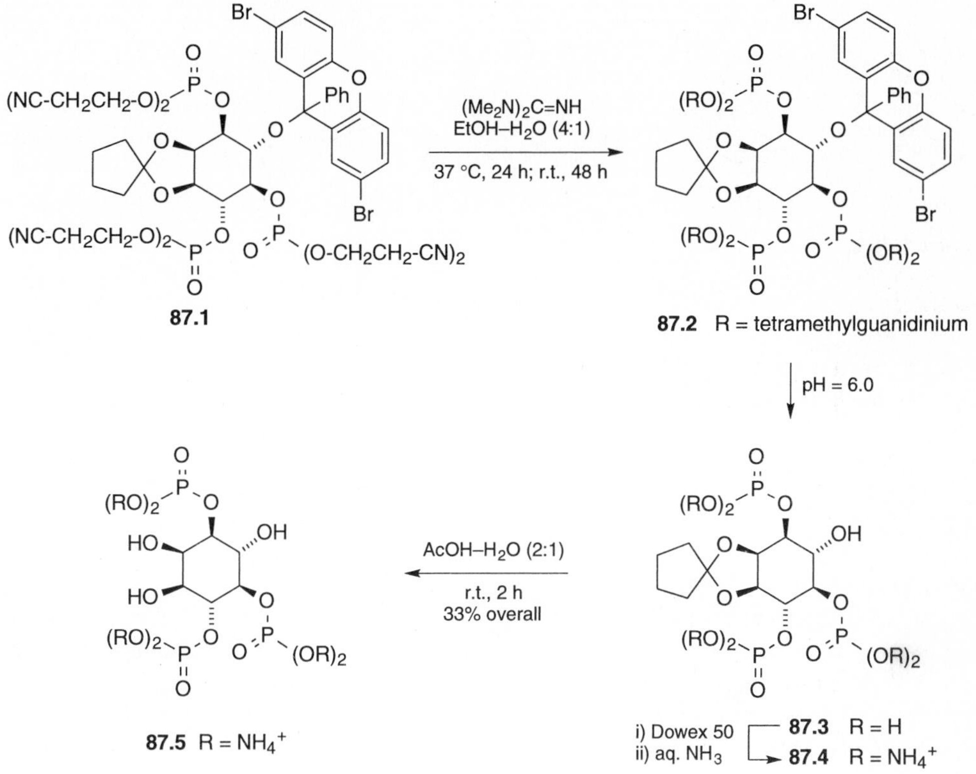

**Scheme 2.87**

Our discussion of the trityl family of protecting groups has so far centred on their membership in the acid-labile orthogonal set. To a lesser extent, the trityl family have also been modified for deployment in the base-labile orthogonal set. Thus, Sekine and Hata[219] showed that 4,4′,4″-tris-(benzoyloxy)trityl ethers (TBTr) of deoxythymidine were more resistant to acidic cleavage than the native trityl ethers owing to the electron-withdrawing nature of the three benzoyloxy groups. However, complete cleavage occurred within 10 min with 0.5 M NaOH according to the mechanism outlined in Scheme 2.88. The principal has been varied considerably and the relative lability finely tuned but, by-and-large, the base-labile trityl ethers have remained in the bailiwick of the oligonucleotide chemists. The interested reader should consult the recent review by Beaucage and Iyer[201] for a fuller discussion.

Before moving on to a discussion of the formation of trityl ethers, we should give brief mention of their use in the protection of thiols — a problem of commanding importance in peptide chemistry where the thiol group of cysteine must be tamed[220]. The *S*-trityl group can be cleaved using protic acids but a milder alternative exploits the high affinity of certain heavy metals such as Hg and Ag for sulfur[221]. The latter method[222] was used in an industrial synthesis of the penem antibacterial CP-70,429[223] [Scheme 2.89].

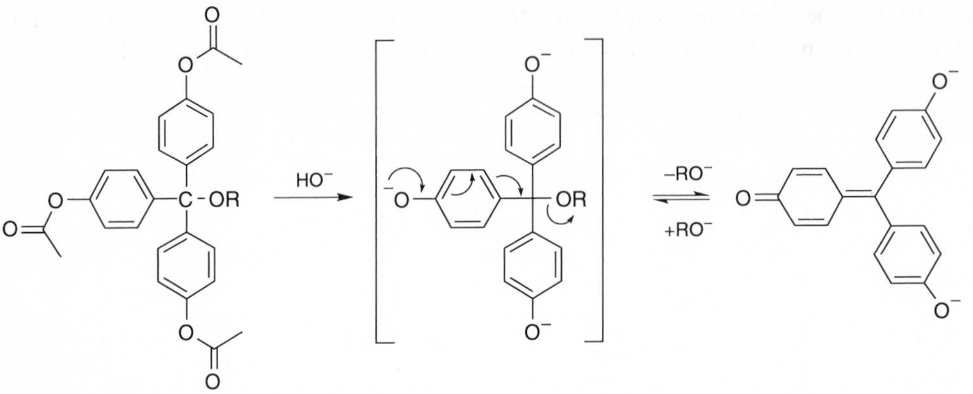

**Scheme 2.88**

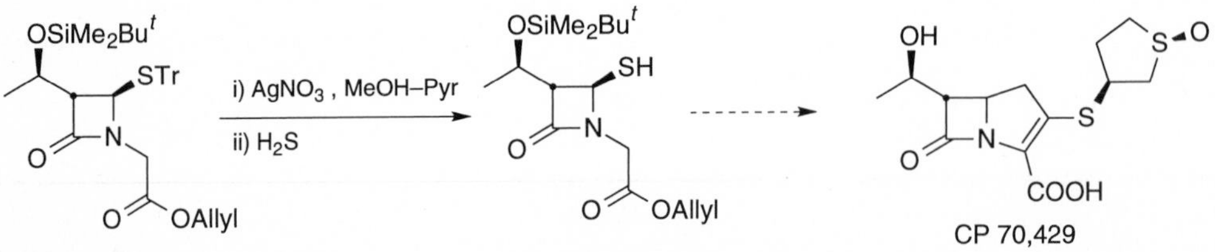

**Scheme 2.89**

**(ii) Formation**

The classical method for the introduction of the trityl group involves the reaction of a primary alcohol function (secondary alcohols react very slowly — if at all) with trityl chloride in pyridine [Scheme 2.90][204,211]. The reaction can be slow and a more convenient procedure using DMAP[84] or DBU[224] to accelerate the reaction has now been developed which is applicable to large scale as shown in the synthesis of a fragment of the macrolide antibiotic Erythronolide B [Scheme 2.91][225]. Alternatively, triphenylmethylpyridinium fluoroborate[226] has been used to achieve rapid and selective tritylation of primary alcohols[49].

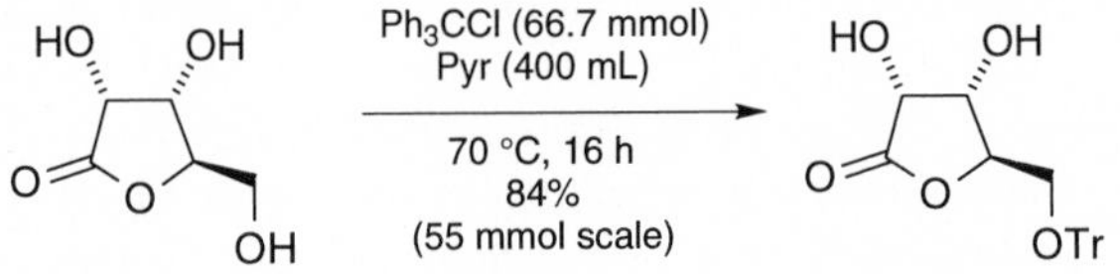

**Scheme 2.90**

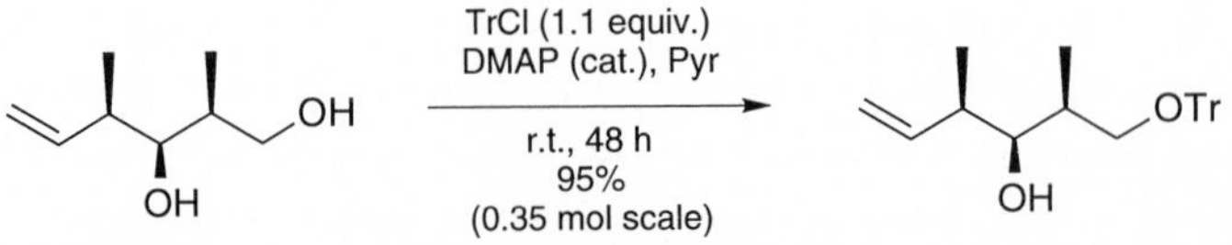

**Scheme 2.91**

During a synthesis of one of the Milbemycins, Hirama and co-workers[205] generated the extremely labile *β*-hydroxy aldehyde **92.1** which required protection before further crucial synthetic operations could be performed. The task was accomplished by protecting the $\beta$-hydroxy aldehyde as the trityl ether of the corresponding hemiacetal **92.2** using the potent tritylating agent TrOTf[227][Scheme 2.92].

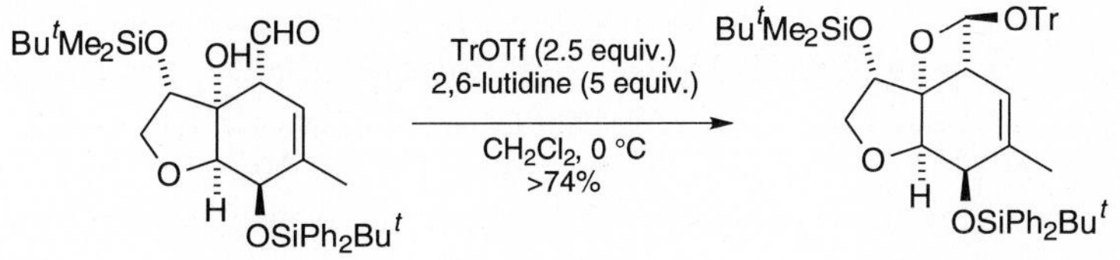

Bu$^t$Me$_2$SiO
OH
CHO
O
H
OSiPh$_2$Bu$^t$
TrOTf (2.5 equiv.)
2,6-lutidine (5 equiv.)
CH$_2$Cl$_2$, 0 °C
>74%
Bu$^t$Me$_2$SiO
O
OTr
O
H
OSiPh$_2$Bu$^t$


**92.1** **92.2**

**Scheme 2.92**

**(iii) NMR Data**
$^1$H NMR: $\delta = 7.3$ (15H, m)
$^{13}$C NMR: $\delta = 87$ (<u>C</u>Ph$_3$)

## 2.4.5 *tert*-Butyl Ethers

*tert*-Butyl ethers are stable to strongly basic conditions and will withstand attack by alkyllithium reagents which would attack most other protecting groups such as benzyl or allyl. They decompose in strong acid by an $E_1$ mechanism with loss of isobutene — a transformation which stems from the stability of the *tert*-butyl carbocation. A similar transformation takes place with *tert*-butyl esters and *N*-*tert*-butoxycarbonyl (Boc) derivatives (see section 6.2.4).

**(i) Cleavage**
The principal problem in the deployment of *tert*-butyl ethers is deprotection under conditions mild enough to accomodate acid-sensitive functional groups. Both protic and Lewis acids cleave *tert*-butyl

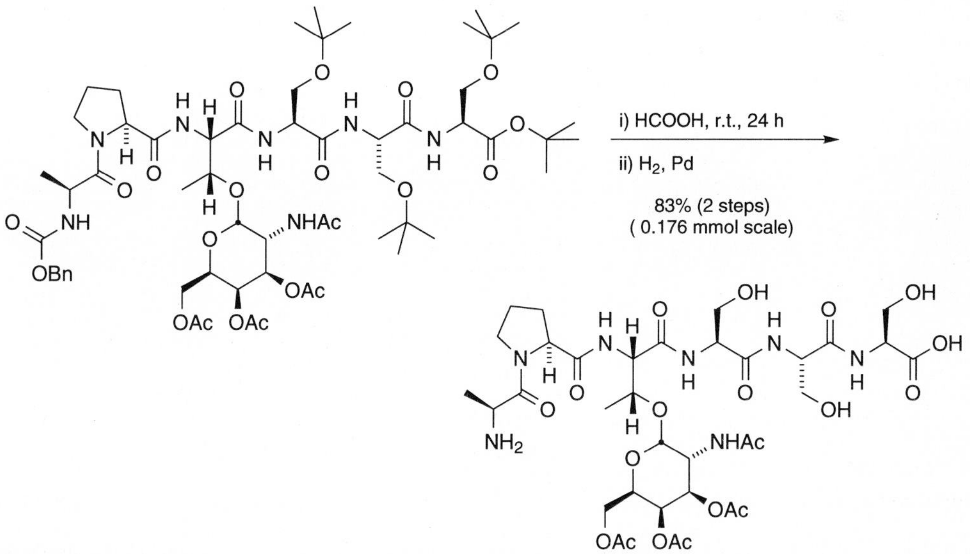

i) HCOOH, r.t., 24 h
ii) H$_2$, Pd
83% (2 steps)
( 0.176 mmol scale)
NH
OBn
NHAc
OAc
OAc OAc
NH$_2$
OH
OH
OH
NHAc
OAc
OAc OAc


**Scheme 2.93**

ethers rapidly. Typical reagents and conditions include anhydrous trifluoroacetic acid (1–16 h, 0–20 °C), HBr–HOAc (30 min, r.t.), 4 N HCl–dioxane, reflux, 3 h. However, the method is mild enough to be useful in peptide synthesis. For example, a synthesis of the *N*-terminal *O*-glycopeptide sequence of the immunomodulator Interleukin-2 [Scheme 2.93][228] made extensive use of *tert*-butyl ethers to protect the hydroxyl groups in serine and threonine residues. In the example shown, three *tert*-butyl ethers and one *tert*-butyl ester were deprotected simultaneously by simply stirring in neat formic acid for 24 h. Hydrogenolysis of the benzyloxycarbonyl derivative then unleashed the *N*-terminus to give the target. In the same paper a decapeptide derivative harboring 5 *tert*-butyl ethers and 3 *tert*-butyl ester functions was similarly deprotected in 89% yield by stirring in $CH_2Cl_2$–$CF_3COOH$ (2:1) at r.t. for 15 h.

A wide range of Lewis acids have been employed to cleave *tert*-butyl ethers but only a small selection of those used in various natural product syntheses will be cited here. Thus 10% anhydrous $FeCl_3$ in $Et_2O$ in the presence of 3–5 equiv. of acetic anhydride[229,230] was used to good effect to cleave a *tert*-butyl ether in good yield to give the corresponding acetate at a late stage in a synthesis of Didemnones A and B [Scheme 2.94][231]. $TiCl_4$ has also been used to give the alcohol [Scheme 2.95][232]. During a synthesis of 1$\alpha$,25-Dihydroxy-vitamin D [Scheme 2.96] a group at Hofmann–LaRoche[233] found that TMSI accomplished the deprotection of a *tert*-butyl ether in the presence of a secondary acetate ester. The reaction is usually performed[159] in $CCl_4$ or $CHCl_3$ at 25 °C for $\leq$ 10 min.

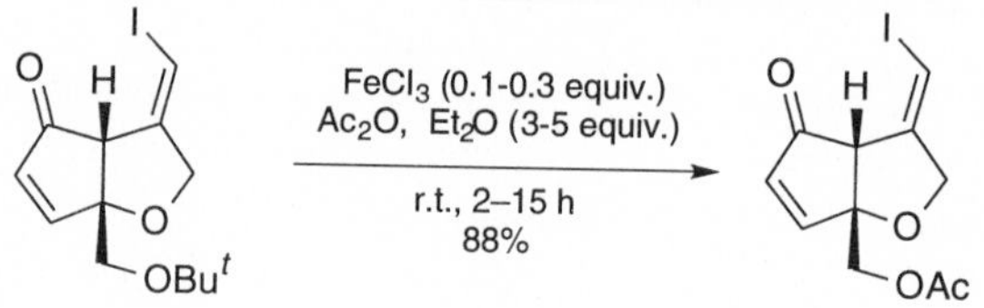

$FeCl_3$ (0.1-0.3 equiv.), $Ac_2O$, $Et_2O$ (3-5 equiv.), r.t., 2–15 h, 88%

**Scheme 2.94**

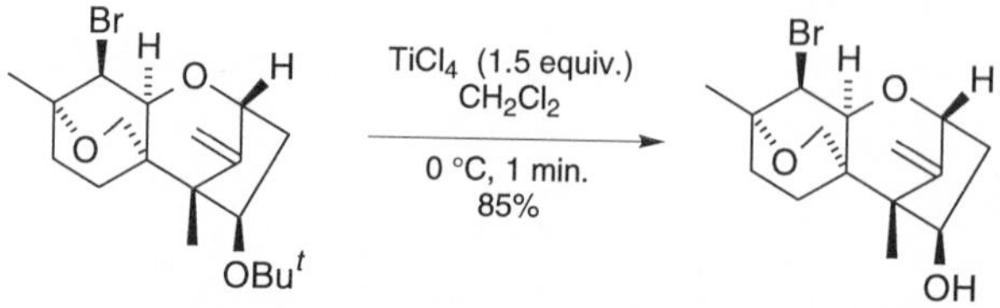

$TiCl_4$ (1.5 equiv.), $CH_2Cl_2$, 0 °C, 1 min., 85%

**Scheme 2.95**

$Me_3SiI$, $CCl_4$, 25 °C, 98%

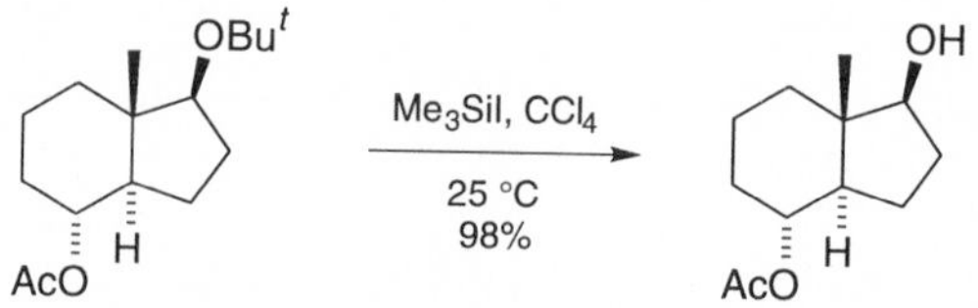

**Scheme 2.96**

**(ii) Formation**

The traditional method for preparing *tert*-butyl ethers involves reacting an excess of isobutene with a solution of the alcohol in $CH_2Cl_2$ in the presence of concentrated sulfuric acid and the method is effective for protecting the side chain hydroxyl functions of serine, threonine [Scheme 2.97], and tyrosine[234]. Milder alternative acid catalysts have been found; for example, a mixture of phosphoric acid and $BF_3$•$OEt_2$ was used by Hofmann–La Roche workers to prepare a *tert*-butyl ether and a *tert*-butyl ester in one pot from the homochiral $\beta$-hydroxy acid **98.1** as shown in Scheme 2.98[235]. A more convenient method involving use of Amberlyst H-15 resin in hexane as the acid catalyst deserves wider attention[229].

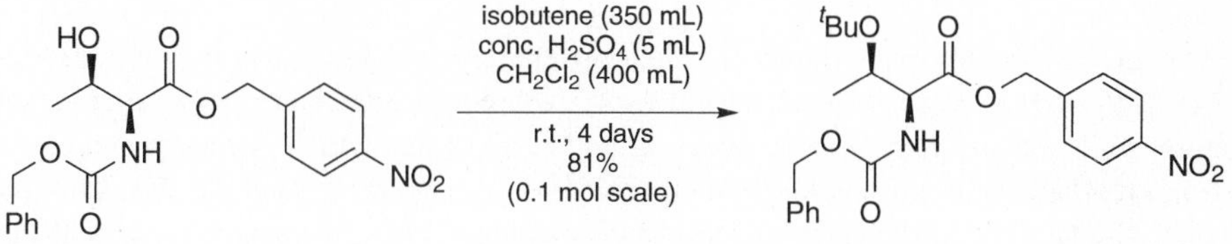

isobutene (350 mL)
conc. $H_2SO_4$ (5 mL)
$CH_2Cl_2$ (400 mL)
r.t., 4 days
81%
(0.1 mol scale)

**Scheme 2.97**

isobutene (70 mL)
$H_3PO_4$, $BF_3$•$OEt_2$
–72 °C, 3 h; 0 °C , 20 h
79%
(74 mmol scale)

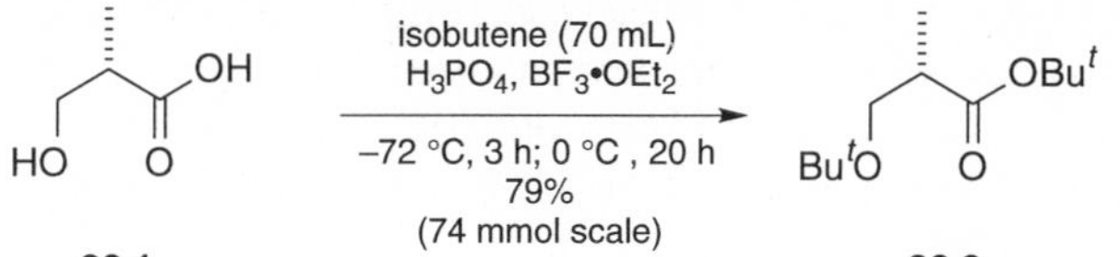

**98.1** **98.2**

**Scheme 2.98**

Perhaps the most versatile of the recent additions to the repertoire of methods for appending *tert*-butyl ethers involves alkylation of alcohols with *O-tert*-butyl trichloroacetimidate in the presence of a catalytic amount of $BF_3$ at r.t. for 16–21 h in cyclohexane–$CH_2Cl_2$ [Scheme 2.99][236]. *O-tert*-Butyl trichloroacetimidate is commercially available but it can be had more cheaply by synthesis from trichloroacetonitrile and *t*-BuOK (70% yield).

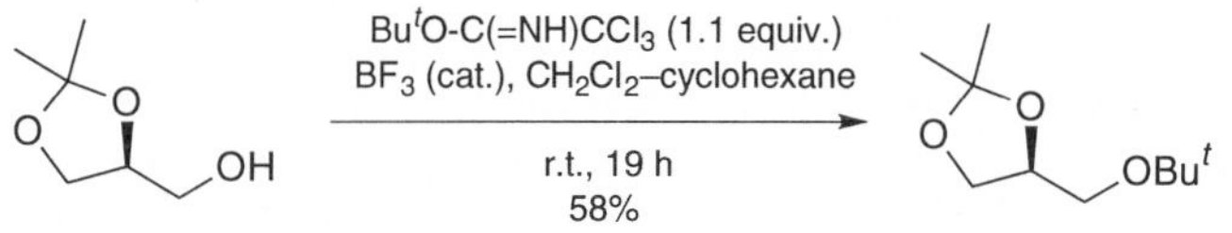

$Bu^tO$-C(=NH)$CCl_3$ (1.1 equiv.)
$BF_3$ (cat.), $CH_2Cl_2$–cyclohexane
r.t., 19 h
58%

**Scheme 2.99**

**(iii) NMR Data**
$^1$H NMR: $\delta = 1.1$ (9H, s)
$^{13}$C NMR: $\delta = 73$ (1C, s), 27 (3C, q)

## 2.4.6 Allyl Ethers and Allyloxycarbonyl Derivatives

Arguably the most significant new orthogonal set to emerge in recent years comprises the allyloxy protecting groups in their various guises. Allyl ethers were originally developed for the protection of hydroxyl groups in carbohydrates[237]. They are robust and therefore compatible with most reaction conditions (including glycosidation) required for the synthesis of oligosaccharides. Their limitations are obvious: strong electrophiles such as bromine are precluded as is catalytic hydrogenation. However, they are stable to both strong acid (M HCl, reflux, 10 h) and moderately strong base.

Allyloxycarbonyl (Aloc) derivatives have the advantage that they are often more easily prepared than the corresponding allyl ethers and they are more stable than ester protecting groups which find frequent use in carbohydrate chemistry. Since the development of milder conditions for deprotection, allyloxy protecting groups have become powerful tools for the construction of oligosaccharides and glycopeptides.

**(i) Cleavage**

Allyl ethers are prime exemplars of the principle of relay deprotection since they are quite resilient themselves and cannot be easily deprotected as such, but rearrangement of the double bond into conjugation with the oxygen atom provides a labile enol ether which is easily cleaved by acid hydrolysis or oxidation. The challenge then resides in the method used to effect double bond rearrangement. The Giggs' original procedure, developed nearly 30 years ago, used *t*-BuOK in DMSO at elevated temperature and these conditions, harsh as they are, are still occasionally used in robust substrates. For example, during a synthesis of the polyether antibiotic Monensin, the enol ether **100.2** was prepared by this method [Scheme 2.100] and then hydrolysed under Hg(II)-catalysis at neutral pH in order to preserve the labile SEM and isopropylidene groups[238].

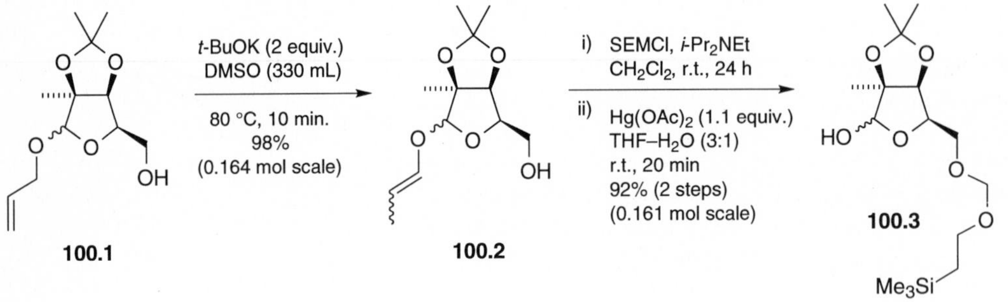

*t*-BuOK (2 equiv.)
DMSO (330 mL)
80 °C, 10 min.
98%
(0.164 mol scale)

i) SEMCl, *i*-$Pr_2NEt$
$CH_2Cl_2$, r.t., 24 h
ii) $Hg(OAc)_2$ (1.1 equiv.)
THF–$H_2O$ (3:1)
r.t., 20 min
92% (2 steps)
(0.161 mol scale)

**100.1** **100.2** **100.3**

**Scheme 2.100**

The true value of the allyloxy orthogonal set did not emerge until milder methods for effecting double bond rearrangement were found. Nowadays the rearrangement is usually achieved by transition metal catalysis and typical catalysts include Pd–C in MeOH, $[Ph_3P]_3RhCl$, DABCO–MeOH, or $Ir(COD)[PMePh_2]_2PF_6$[239] and it is noteworthy that allyl ethers isomerise at different rates depending on the conditions leading to the possibility for selectivity in deprotection: $[Ph_3P]_3RhCl$: allyl > methallyl > but-2-enyl; *t*-BuOK : but-2-enyl > allyl > methallyl. A possible mechanism for the Rh(I)-catalysed reaction was presented in Chapter 1 (Scheme 1.16).

A good example of the rearrangement–hydrolysis protocol appeared in a critical stage of Schmidt's synthesis [Scheme 2.101] of the octasaccharide fragment of the dimeric $Le^X$ antigen[145]. Attempts to isomerise and hydrolyse an allyl ether using Wilkinson's catalyst or *t*-BuOK–DMSO failed. However, Baudry's Ir complex[240] worked provided air and moisture were rigorously excluded. To complicate matters further, the resultant enol ether could not be hydrolysed using aqueous acid owing to the presence of the labile TBS ether. However, the enol ether was removed with HgO and $HgCl_2$ in aqueous acetone. The next step in the sequence is also included here and deserves special mention. Trichloroacetimidates can be used for more synthetically worthy goals than just putting on protecting groups. The glycosidation procedure devised by Schmidt[241,242] is a powerful and widely used variant of the classical Koenigs–Knorr glycosidation procedure which makes use of trichloroacetimidate activation[243].

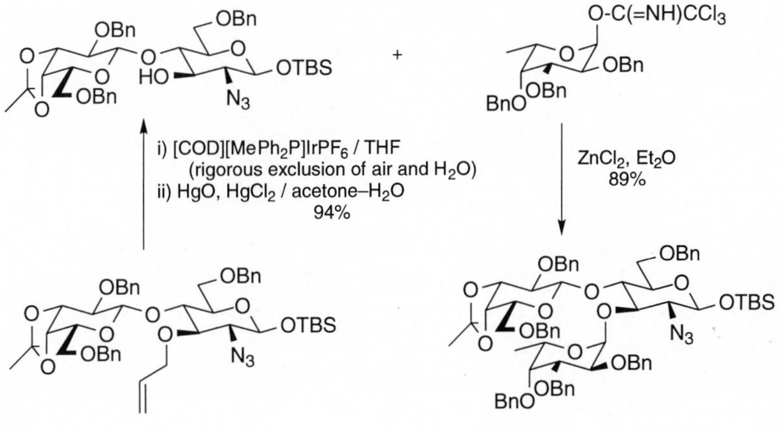

Scheme 2.101

The rearrangement of allyl ethers is not without its complications. For example, with less active catalysts which require elevated temperatures, adjacent azides can undergo [1,3]-dipolar cycloaddition on the alkene of the allyl ether function before isomerisation occurs [Scheme 2.102][244] but this can be avoided by using a more active Iridium catalyst which isomerises the alkene at r.t. Note that in this case, the enol ether was cleaved by oxidation with $OsO_4$. Iridium catalysts are also preferred over Rh owing to problems of competing reduction of the alkene which can occur with the latter[245-248].

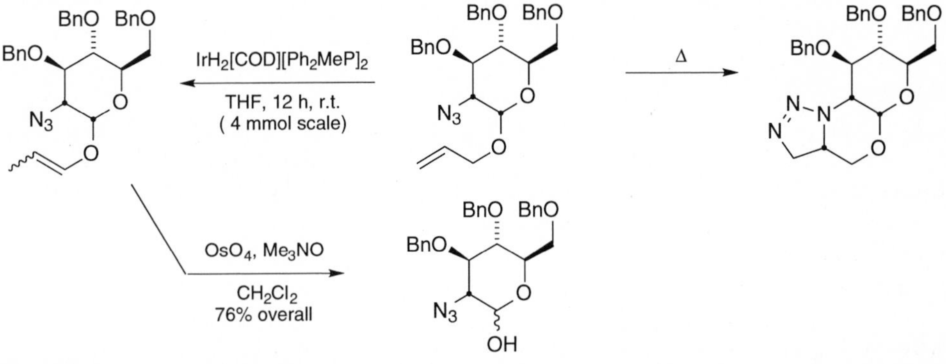

Scheme 2.102

During the final stages of Franck's route to Nogalomycin [Scheme 2.103][249], difficulties were encountered isomerising two allyl ether functions in **103.1** using $[Ph_3P]_3RhCl$ and other typical

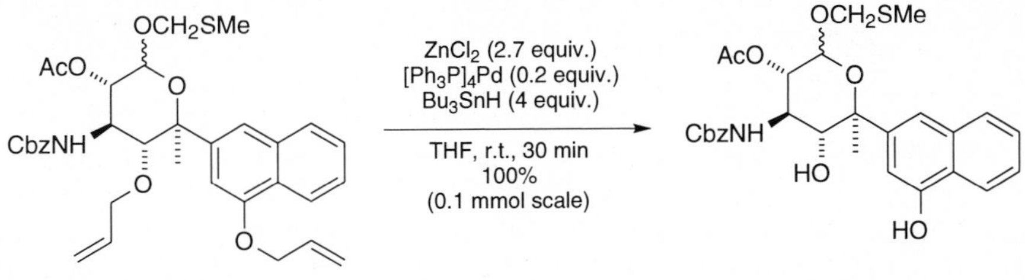

Scheme 2.103

catalysts and it was assumed that the catalysts were being poisoned by the presence of the methylthiomethyl ether. Eventually, a useful cleavage of the bisallyl ether to the diol was accomplished by Pd-catalysed $Bu_3SnH$ reduction[250] in 78% yield. A recent application of the method to the deprotection of phenol allyl ether **104.1** in an approach to the antibiotic Vancomycin demonstrates that aryl bromides and chlorides remain intact [Scheme 2.104][251].

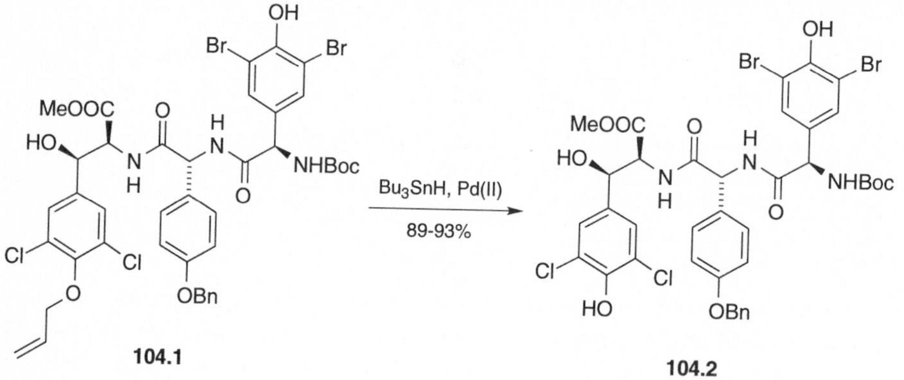

$Bu_3SnH$, Pd(II)
89-93%

**104.1** **104.2**

**Scheme 2.104**

The allyloxycarbonyl (Aloc) group is a precursor to allyl ethers (*vide infra*) but it can also be used as a protecting group in its own right though, compared with the corresponding allyl ether, the presence of the carbonate moiety imposes greater limits on the range of reaction conditions that can be tolerated. However, there are compensations in that the Aloc group is introduced more easily under the usual acylation conditions with allyl chloroformate. A striking example of the synthetic value of the Aloc group as a protecting group for alcohols as well as amines (see Chapter 6) comes from a recent synthesis of the glycopeptide Nephritogenoside [2.105][252]. A total of 11 Aloc groups were removed in a single operation in 55% yield by treating the substrate with $[Ph_3P]_4Pd$ in the presence of dimedone ($pK_a$ = 5.2) — conditions first developed by Kunz and Unverzagt[253]. The role of the dimedone is to intercept the intermediate π-allyl Pd complex before collapse to the allyl ether can occur. For a further discussion of the use of Pd-mediated *C*-alkylation as a deprotection tactic, see Chapter 4.

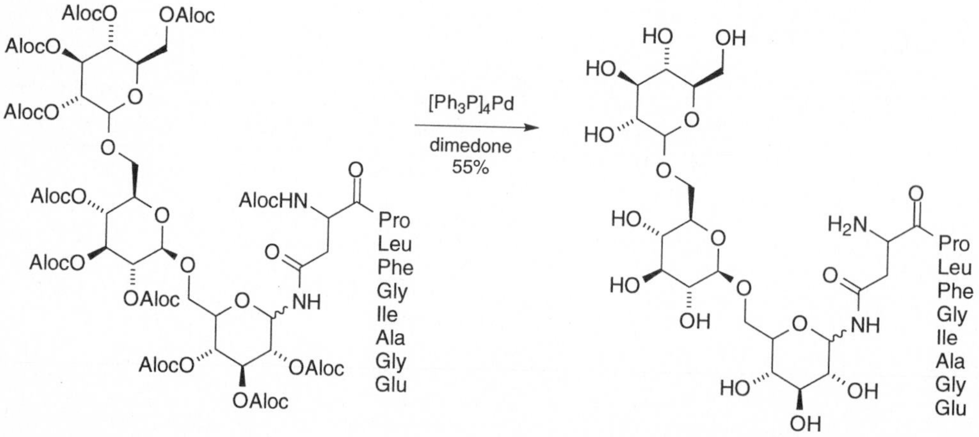

$[Ph_3P]_4Pd$
dimedone
55%

**Scheme 2.105**

Depending on the catalyst, allyl ethers and allyloxycarbonyl groups can be usefully distinguished. For example, an allyl ether can be isomerised preferentially in the presence of an Aloc group by using $[Ir(COD)(Ph_2MeP)_2]PF_6$ although the selectivity is not complete [Scheme 2.106][254]. Alternatively, the Aloc group can be selectively isomerised using Pd or Rh catalysts.

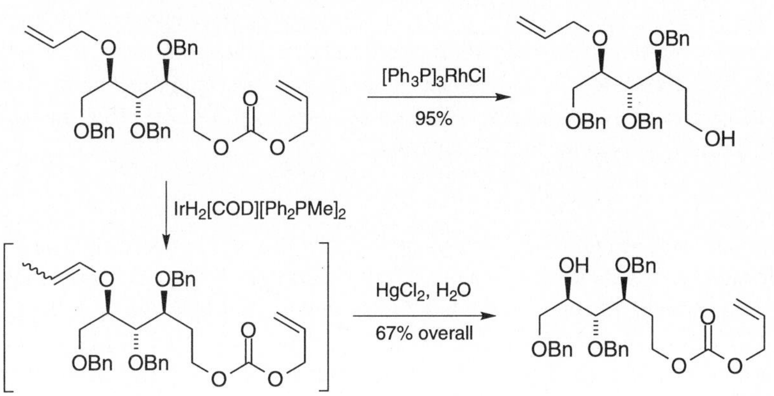

$[Ph_3P]_3RhCl$, 95%

$IrH_2[COD][Ph_2PMe]_2$

$HgCl_2$, $H_2O$, 67% overall

**Scheme 2.106**

A cogent illustration of the efficacy of allyl ether protection in sensitive substrates comes from the synthesis of Phyllanthoside [Scheme 2.107] recorded by Smith and Rivero[51]. Of immediate interest is the problem of how to release the anomeric hydroxyl from suitable precursors to give disaccharide

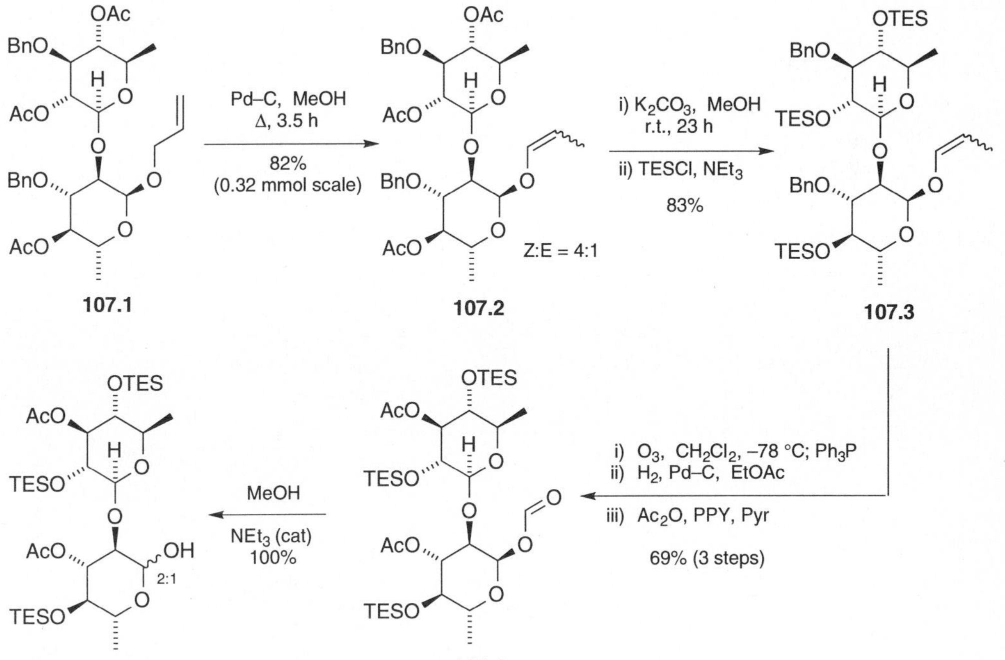

107.1 → 107.2: Pd–C, MeOH, Δ, 3.5 h, 82% (0.32 mmol scale); Z:E = 4:1

107.2 → 107.3: i) $K_2CO_3$, MeOH, r.t., 23 h; ii) TESCl, $NEt_3$; 83%

107.3 → 107.4: i) $O_3$, $CH_2Cl_2$, –78 °C; $Ph_3P$; ii) $H_2$, Pd–C, EtOAc; iii) $Ac_2O$, PPY, Pyr; 69% (3 steps)

107.4 → 107.5: MeOH, $NEt_3$ (cat), 100%; 2:1

**Scheme 2.107**

**107.5** in preparation for its union with Phyllanthocin, the aglycone of Phyllanthoside. The task required a suite of protecting groups which would ultimately enable selective deprotection of three hydroxyl functions without disturbing a glycosidic link or causing migration of two acetate functions (see section 2.3.2). The strategy adopted by Smith and Rivero required the conversion of **107.1** to **107.5** in 7 steps and began by Pd(0)-catalysed rearrangement of the allyl group in **107.1** to the enol ether **107.2**. After exchanging the acetate groups for TES groups, the enol ether function in **107.3** was ozonolysed to a very labile formate ester whose benzyl ether functions were converted to acetate esters to give **107.4**. The final step in the sequence unleashed the required anomeric hydroxyl by methanolysis in the presence of $NEt_3$. Another example of the value of allyl ethers can be found in van Boom's recent synthesis of *myo*-inositol phosphates[239].

**(ii) Formation**

Allyl bromide, like methyl iodide and benzyl bromide, is one of the triumvirate of common alkyl halides which alkylate metal alkoxides readily[255] but the strongly basic conditions limit the scope of the method. For example, attempts to alkylate the secondary and phenolic hydroxyl groups of the intermediate **108.1** in a synthesis of the antibiotic Nogalomycin using the traditional metal alkoxide alkylation failed owing to competing second-order Beckmann fragmentation of the isoxazoline ring [Scheme 2.108][249]. A successful double alkylation was eventually achieved to give the bisallyl ether **108.2** in 80% yield using potassium fluoride-impregnated alumina[256].

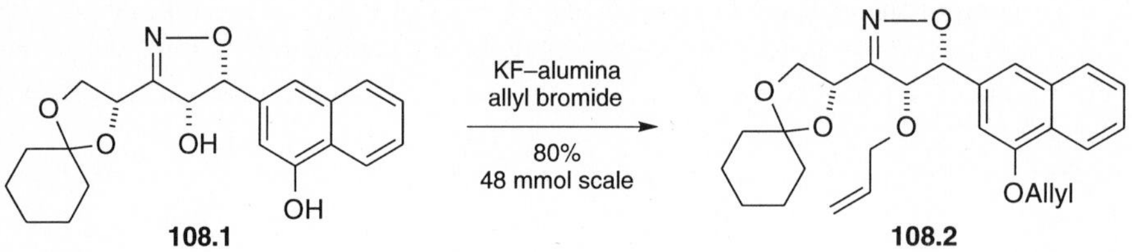

N—O
O
O
OH
OH
108.1
KF–alumina
allyl bromide
80%
48 mmol scale
N—O
O
O
O
OAllyl
108.2

**Scheme 2.108**

Acid-catalysed alkylation of an alcohol with the trichloroacetimidate derivative prepared from allyl alcohol and trichloroacetonitrile is readily accomplished as previously discussed for the preparation of benzyl and *tert*-butyl ethers[183]. However, these conditions are not compatible with many of the

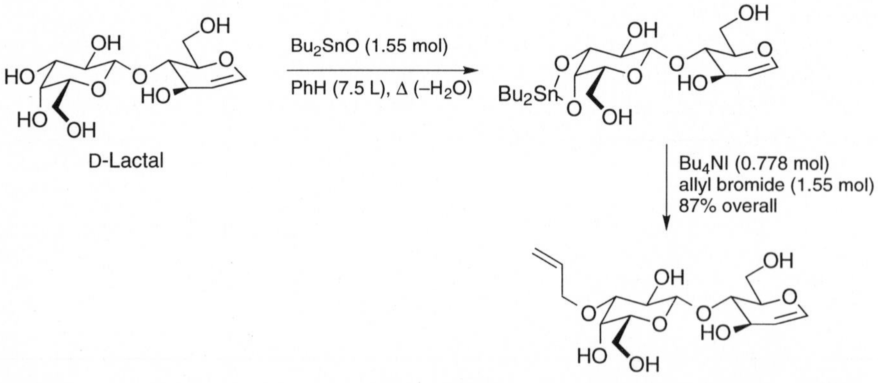

OH
OH
HO
O
O
HO
O
HO
OH
D-Lactal
$Bu_2SnO$ (1.55 mol)
PhH (7.5 L), Δ (–$H_2O$)
$Bu_2Sn$
$Bu_4NI$ (0.778 mol)
allyl bromide (1.55 mol)
87% overall

**Scheme 2.109**

protecting groups employed in oligosaccharide synthesis and so two new methods for *O*-allylation under essentially neutral conditions have been devised. The first method takes advantage of the mild conditions and regioselectivity of stannylene alkylations. Stannylenes are formed by reacting diols with $Bu_2SnO$ with removal of water. Cis-fused 5-membered stannylene rings form preferentially and the resultant stannylenes alkylate selectively at the equatorial position. The method is illustrated by manipulations on the disaccharide *O*-lactal which began with stannylene formation on an 0.8 mole scale [Scheme 2.109][257].

The regioselective manipulation of hydroxyl groups via stannylene derivatives is applicable to a range of other reactions such as oxidation, sulfonylation, and acylation and is noteworthy because it accomplishes the regioselective *activation* of a specific hydroxyl function[258]. The origin of the activation is not yet clear but structural evidence including X-ray data point to dimeric (or oligomeric) stannylene structures in which the tin atoms are in the centre of a trigonal bipyramid with the butyl groups occupying the two equatorial positions [Scheme 2.110]. The more electronegative of the two oxygen atoms occupies the apical position and is coordinated to only one tin atom whereas the less electronegative oxygen is ensconced in a $Sn_2O_2$ bridge and is therefore coordinated to two tin atoms[259]. Furthermore, it has been proposed[260] that it is the more electronegative of the two hydroxyl oxygen atoms that occupies the apical position. Thus the observed regioselectivity is a consequence of a cascade of effects beginning with the selection of a particular pair of hydroxyl functions for stannylene formation followed by orientation of the more electronegative oxygen in the apical position which is intrinsically more reactive.

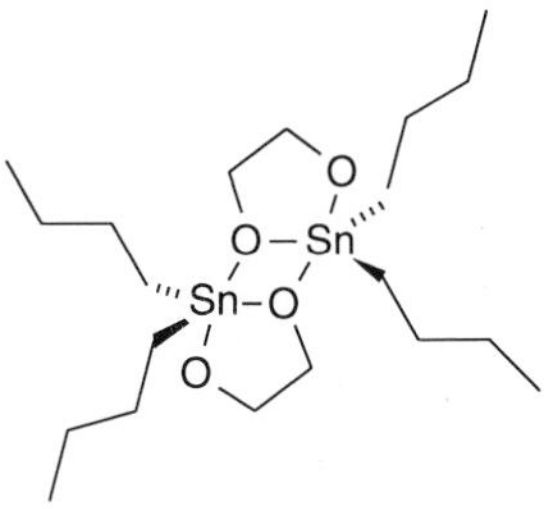

Dimeric structure of stannylene derivatives of 1,2-diols

**Scheme 2.110**

A second very mild procedure involves the Pd-catalysed extrusion of $CO_2$ from allyloxycarbonyl (Aloc) derivatives which, in turn, are simply made by acylation of an alcohol with allyl chloroformate in the usual way. Our example shows a double allylation in the presence of the sensitive 1,1,3,3-tetraisopropyldisiloxane-1,3-diyl group[261] [Scheme 2.111].

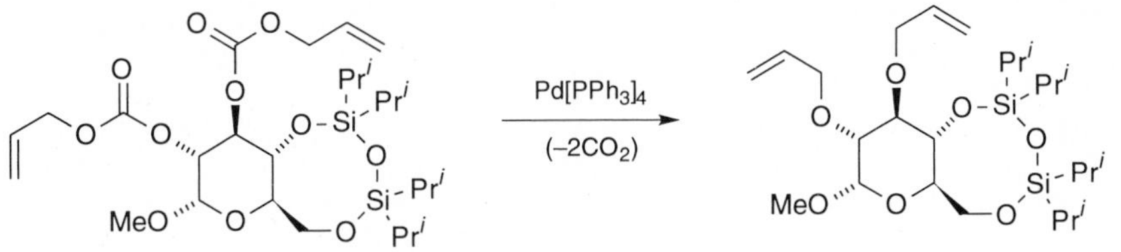

$Pd[PPh_3]_4$ $(-2CO_2)$

**Scheme 2.111**

**(iii) NMR Data**

$^1$H NMR: $\delta = 5.7$ (1H, ddt, $J_{cis} = 10$ Hz, $J_{trans} = 17$, $J_{vic} = 5$ Hz, **CH**=$CH_2$), 5.25 (1H, dd, $J_{trans} = 17$, $J_{gem} = 1$ Hz, CH=$CH_{cis}\mathbf{H}_{trans}$), 5.18 (1H, dd, $J_{cis} = 10$, $J_{gem} = 1$ Hz, CH=$\mathbf{CH}_{cis}H_{trans}$), 3.8 (2H, m)

$^{13}$C NMR: $\delta = 134$ (1C, d), 116 (1C, t), 73 (1C, t)

## 2.5 Alkoxyalkyl Ethers (Acetals)

Acetal-type protecting groups have long been principally associated with the protection of 1,2- and 1,3-diols and aldehydes and ketones. Their extension to the protection of alcohols is a comparatively recent development made popular by the work of Corey and his associates at Harvard[262]. Most of the protecting groups we will consider in this section are introduced by alkylation of a hydroxyl function with a chloroalkyl ether in the presence of a weak and hindered base such as $i$-$Pr_2NEt$ (Hünig's base). The reactions are much faster than similar alkylations with MeI, BnBr, or allyl bromide because the lone pairs on oxygen in ethers of the type RO-$CH_2$-X assist departure of the leaving group X giving an oxonium ion whose capture by a hydroxy group results in alkylation. Thus, the mechanism of the alkylation ($S_N1$) is fundamentally different from those discussed in the previous section. All alkylating agents are potentially carcinogenic but the haloalkylethers and thioethers require particular vigilance and should only be handled in a well-ventilated fume hood with protection of eyes and skin. These reagents should not be used without a proper and thorough safety assessment.

### 2.5.1 Methoxymethyl Ethers

Methoxymethyl ethers (abbreviated MOM) are the first and simplest of the acetal-type protecting groups whose prime virtues are ease of introduction. MOM ethers are the most robust of the alkoxymethyl ethers. With the exception of benzyl ethers, most other protecting groups are imperilled by the conditions required to remove MOM groups.

**(i) Cleavage**

Acidic hydrolysis of MOM ethers is not easy and rather harsh conditions are usually required such as a small amount of concentrated HCl in neat MeOH or in aqueous MeOH [Scheme 2.112][263]. Trifluoroacetic acid accomplished the deprotection of a hindered MOM group in a synthesis of Pleuromutilin[264].

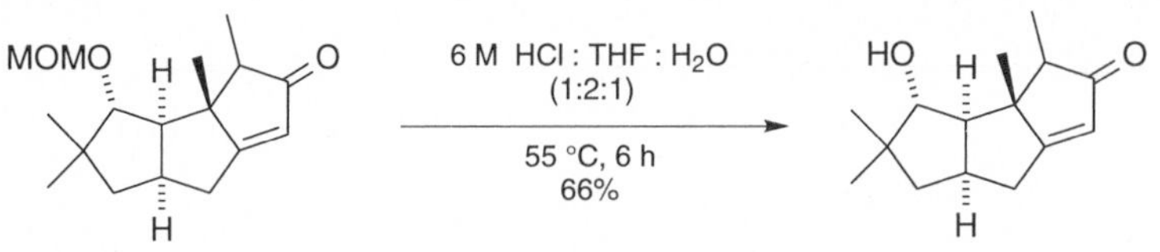

MOMO H O — 6 M HCl : THF : $H_2O$ (1:2:1), 55 °C, 6 h, 66% → HO H O

**Scheme 2.112**

Lewis acids offer opportunities to vary their reactivity by suitable choice of solvent, temperature, and additives which provide greater latitude in selecting conditions for cleaving MOM ethers in the presence of other protecting groups or functional groups. Thus MOM ethers are removed at –30 °C in $CH_2Cl_2$ in the presence of bromotrimethylsilane [Scheme 2.113][265,266]—conditions which preserve

TBDPS ethers but TBS ethers may cleave to some extent. Trityl, tetrahydropyranyl, and isopropylidene acetals cleave. The method has recently been employed in a synthesis of Nogalomycin[267].

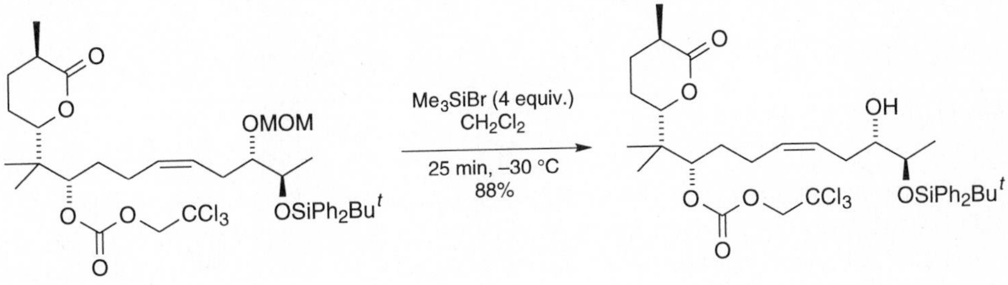

Me$_3$SiBr (4 equiv.), CH$_2$Cl$_2$, 25 min, –30 °C, 88%

**Scheme 2.113**

Boron trichloride will also remove MOM ethers but comparatively few functional groups will survive such harsh conditions[268]. Similarly, $BF_3 \bullet OEt_2$ in the presence of thiophenol removes MOM groups and this reagent combination was used by Corey and co-workers in a synthesis of Ginkolide A[269] and by Vedejs and Larsen in a synthesis of Fulvine and Crispatine [Scheme 2.114][270].

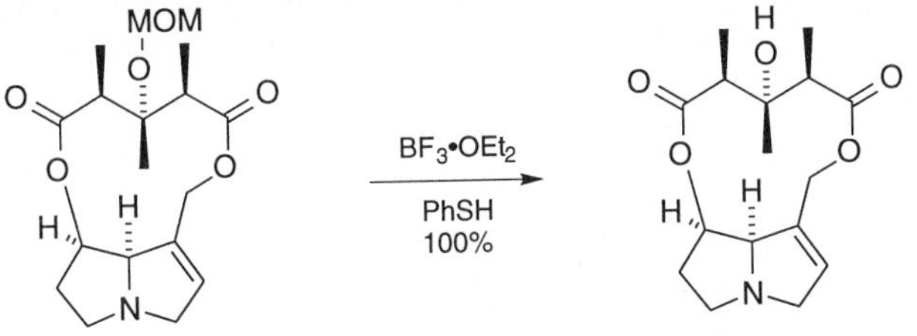

$BF_3 \bullet OEt_2$, PhSH, 100%

**Scheme 2.114**

Electron donating ligands attenuate the reactivity of Lewis acids thereby increasing their compatibility with a wider range of heteroatoms. The advantages of such reactivity moderation are apparent in the use of ($i$-PrS)$_2$B-Br cleavage of 2 MOM ether groups in a synthesis of Aplasmomycin [Scheme 2.115][271] and the use of Me$_2$BBr to remove a MOM group at the end of the synthesis of the cyclodepsipeptide Didemnin A [Scheme 2.116][62]. In the transformation shown in Scheme 2.115, the neighbouring hydroxyl function first reacts with the ($i$-PrS)$_2$B-Br to form an intermediate di-isopropylthioborate ester which coordinates to the MOM group and thereby facilitates its cleavage. Oxygen substitution also moderates the reactivity of boron-based Lewis acids in a useful way. For example, catechol boron bromide cleaves MOM and MEM ethers at comparable rates but *tert*-butoxycarbonyl, benzyloxycarbonyl, *tert*-butyl ethers, benzyl ethers, and TBS ethers are usually compatible[272].

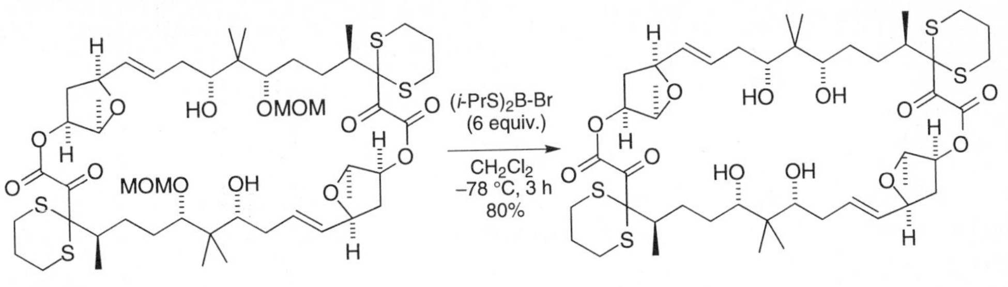

($i$-PrS)$_2$B-Br (6 equiv.), CH$_2$Cl$_2$, –78 °C, 3 h, 80%

**Scheme 2.115**

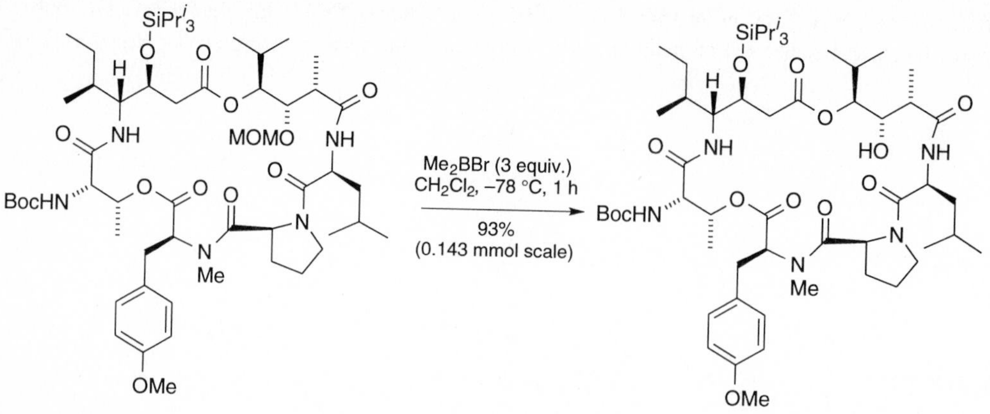

Scheme 2.116

In the foregoing discussion we have focussed on the cleavage of MOM ethers in their guise as protecting groups. Their Lewis-acid mediated cleavage can also be used in alkoxyalkylation reactions under conditions which are quite mild. For example, Linderman and co-workers[273] were able to preserve the reactive C–Sn bond in **117.2** during its $TiCl_4$-mediated alkylation with the ketene acetal derivative **117.1** [Scheme 2.117]. The resultant stannane **117.3** was then exploited in a novel synthesis of furanone derivatives.

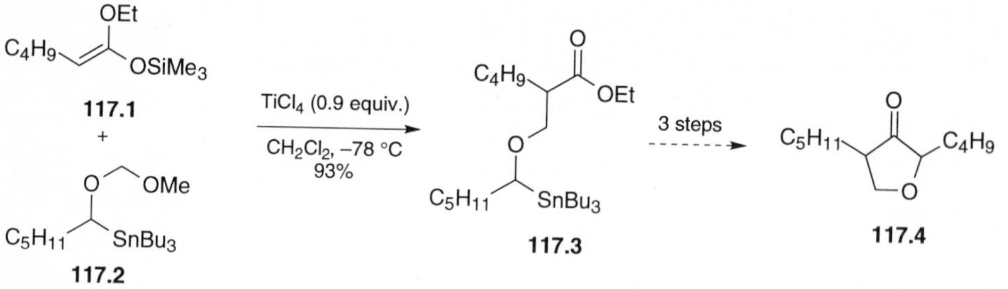

Scheme 2.117

**(ii) Formation**

The sodium or potassium alkoxides prepared from primary and secondary alcohols and NaH or KH in THF react with MOMCl to give the MOM ethers in good yield [Scheme 2.118][211,263,274]. An alternative milder and more common procedure involves alkylation of the alcohol with MOMCl or MOMBr[264] in the presence of $i$-$Pr_2NEt$ [Scheme 2.119][275,276]. This method can be adapted to the

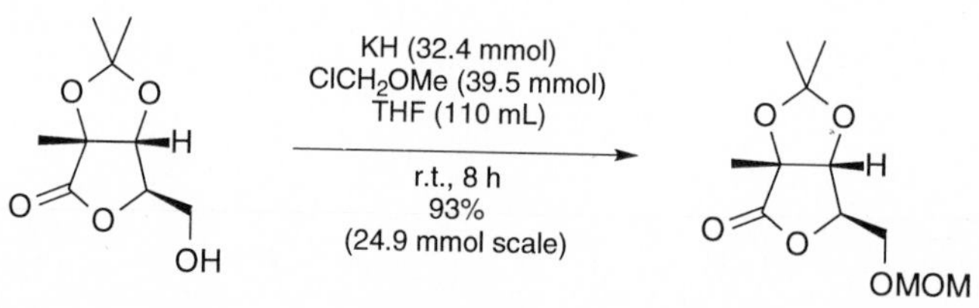

Scheme 2.118

protection of tertiary alcohols [Scheme 2.120][277] provided the MOMCl is further activated by conversion to the corresponding iodide *in situ* .

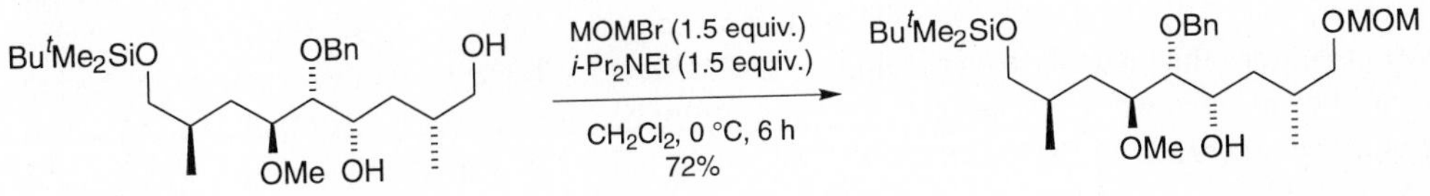
Bu$^t$Me$_2$SiO
OBn
OH
OMe OH
MOMBr (1.5 equiv.)
i-Pr$_2$NEt (1.5 equiv.)
CH$_2$Cl$_2$, 0 °C, 6 h
72%
Bu$^t$Me$_2$SiO
OBn
OMOM
OMe OH

**Scheme 2.119**

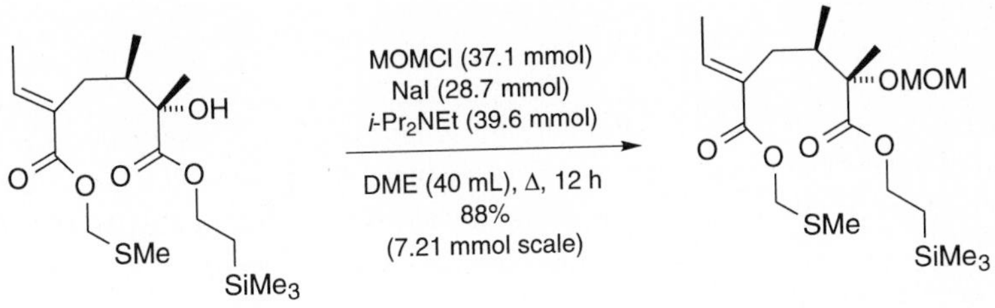
OH
O
O
O
O
SMe
SiMe$_3$
MOMCl (37.1 mmol)
NaI (28.7 mmol)
i-Pr$_2$NEt (39.6 mmol)
DME (40 mL), Δ, 12 h
88%
(7.21 mmol scale)
OMOM
SMe
SiMe$_3$

**Scheme 2.120**

Selective protection of the less hindered alcohol of a 1,3-diol can be achieved using the stannylene activation procedure used previously to prepare allyl ethers[278] [Scheme 2.121].

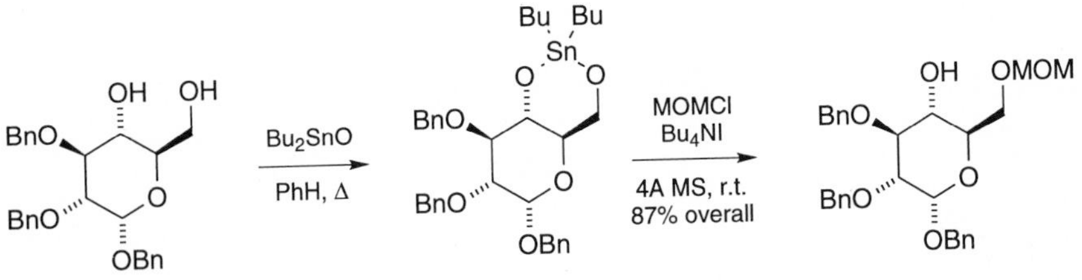
OH OH
BnO
BnO
O
OBn
Bu$_2$SnO
PhH, Δ
Bu Bu
Sn
O O
BnO
BnO
O
OBn
MOMCl
Bu$_4$NI
4A MS, r.t.
87% overall
OH OMOM
BnO
BnO
O
OBn

**Scheme 2.121**

Alcohols react with a large excess of methylal (formaldehyde dimethyl acetal) via an acetal exchange process at room temperature in the presence of acidic catalysts [Scheme 2.122][270,279-281].

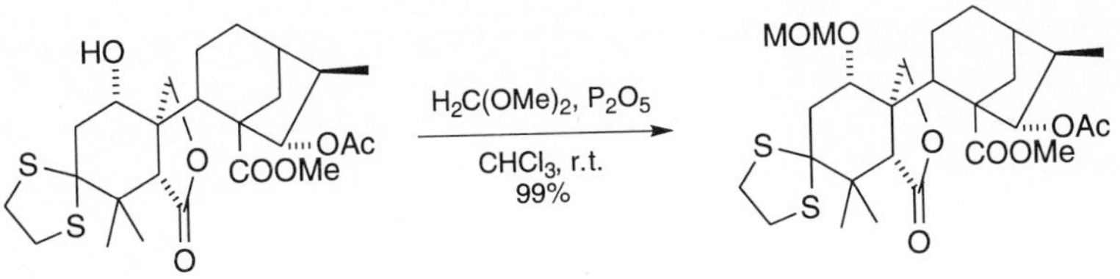
HO
OAc
O
COOMe
S
S
O
H$_2$C(OMe)$_2$, P$_2$O$_5$
CHCl$_3$, r.t.
99%
MOMO
OAc
O
COOMe
S
S
O

**Scheme 2.122**

**(iii) NMR Data**

$^1$H NMR: $\delta = 4.6$ (2H, s), 3.3 (3H, s)

$^{13}$C NMR: $\delta = 96$ (1C, t), 55 (1C, q)

## 2.5.2 Methylthiomethyl Ethers

Methylthiomethyl ethers (abbreviated MTM) are comparable in stability to alkoxymethyl ethers towards strongly basic conditions or mild acid. For example, MTM ethers survive aqueous acetic acid under conditions that hydrolyse dioxolanes or THP protectors. It can be removed under rather specific and mild conditions which do not affect most acetal-type protectors such as MEM, MOM, etc. (see below). The presence of the sulfur makes this group liable to oxidation by strong oxidants such as peracids, Cr(VI), NBS and it will poison Pd catalysts. The MTM group was first introduced by Corey and co-workers[282] and its virtues exploited by them in a synthesis of the antibiotics Erythronolide[283,284] and Brefeldin[285].

**(i) Cleavage**

MTM ethers, like most *O,S*-acetals, are stable to aqueous acid but they can be removed under essentially neutral conditions in the presence of heavy metal catalysts with a high affinity for sulfur such as Hg(II) or Ag(I). Typically, the substrate is treated with $HgCl_2$ in MeCN–$H_2O$ (4:1) at r.t. or slightly above[282]. With acid-sensitive substrates a buffer such as $CaCO_3$[282,286] [Scheme 2.123] or $CdCO_3$ can be added[287] to destroy HCl generated in the hydrolysis and it is noteworthy that dithianes survive these conditions intact. $AgNO_3$ in aqueous THF using 2,6-lutidine as the acid scavenger has also been recommended[288]. During a synthesis of the macrocyclic diolide Colletodiol, Keck and co-workers[289] found that the cleavage of an MTM ether from the substrate **124.1** using Hg(II) proved exceptionally difficult and capricious, but with Ag(I), the reaction was clean and efficient [Scheme 2.124]. The same conditions can be used to remove an MTM ether in the presence of a dithiane[288].

TBSO S MTMO S → $HgCl_2$, $CaCO_3$ / MeCN–$H_2O$ → TBSO S HO S

**Scheme 2.123**

O O O O COSEt OMTM **124.1** → $AgNO_3$ (5.52 mmol), 2,6-lutidine (2.76 mmol), THF–$H_2O$ (4:1, 5 mL) / Δ, 20 h, 79%, (0.37 mmol scale) → O O O O COOH OH **124.2**

**Scheme 2.124**

Thioethers are reasonably good nucleophiles and *S*-methylation with excess MeI in the presence of $NaHCO_3$ in aqueous acetone at r.t. or reflux affords dimethylsulfonium intermediates which hydrolyse under conditions which do not affect dioxolanes[290] or dioxanes [Scheme 2.125][283].

In a recent synthesis of the pyrollizidine alkaloid Integerrimine, an MTM ether protecting group was removed in the final step by hydride abstraction using triphenylcarbenium tetrafluoroborate [Scheme 2.126][291].

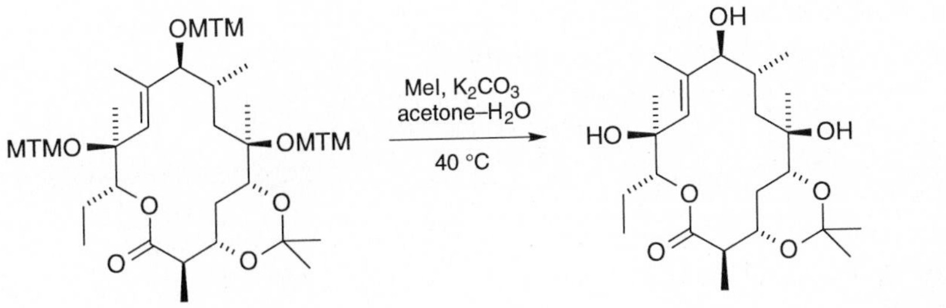
OMTM
MTMO
OMTM
MeI, $K_2CO_3$
acetone–$H_2O$
40 °C
OH
HO
OH

Scheme 2.125

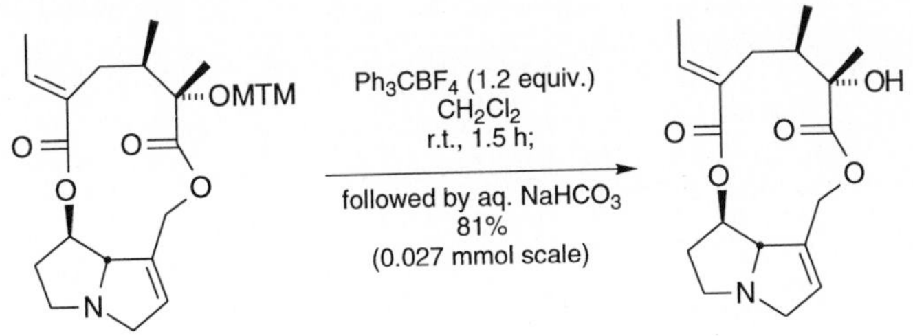
OMTM
$Ph_3CBF_4$ (1.2 equiv.)
$CH_2Cl_2$
r.t., 1.5 h;
followed by aq. $NaHCO_3$
81%
(0.027 mmol scale)
OH

Scheme 2.126

A relay deprotection strategy based on the selective hydrolysis of MTM ethers provides a way of accelerating the cleavage of 5′-acyl derivatives of nucleotides[292,293]. The principle is illustrated by the hydrolysis of 5′-protected thymidine derivative **127.1** [Scheme 2.127]. In concentrated ammonium hydroxide, **127.1** was converted into thymidine (**127.3**) with a half-life of around 6 h at 20 °C. However, hydrolysis of the MTM ether using $Hg(ClO_4)_2$ and 2,4,6-collidine gave the intermediate **127.2** within 3 h at 20 °C. Subsequent treatment of **127.2** with $K_2CO_3$ in aqueous THF released the thymidine within 30 s.

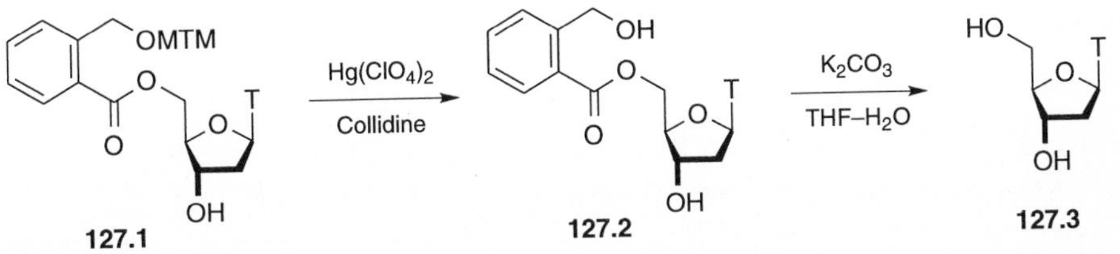
OMTM
$Hg(ClO_4)_2$
Collidine
OH
$K_2CO_3$
THF–$H_2O$
HO
T
OH
127.1
127.2
127.3

Scheme 2.127

**(ii) Formation**

The methods of preparing MTM ethers are analogous to those used to prepare MOM ethers. Thus, the sodium alkoxide prepared from primary and secondary alcohols and NaH in DME reacts with MTMCl in the presence of NaI to give the MTM ethers in good yield. Alternatively, the alcohol can be alkylated with MTMCl in the presence of triethylamine and silver nitrate in benzene or cyclohexane at 60–80 °C[294].

Alkylation of secondary and tertiary alcohols under 'modified Pummerer' conditions is an effective means for preparing MTM ethers. The reaction involves treatment of the alcohols with DMSO in the presence of acetic anhydride [Scheme 2.128][287,289,290]; alternatively, dimethyl sulfide in the presence of dibenzoyl peroxide can be used [Scheme 2.129][295].

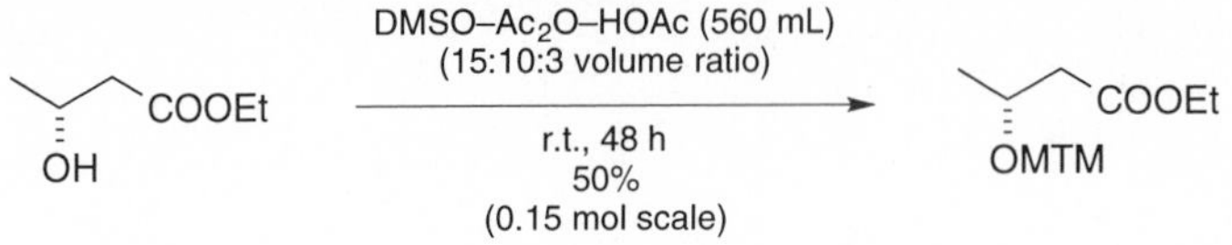

COOEt OH → DMSO–$Ac_2O$–HOAc (560 mL) (15:10:3 volume ratio); r.t., 48 h; 50% (0.15 mol scale) → COOEt OMTM

**Scheme 2.128**

OH O O COOMe → $Me_2S$ (8 equiv.) $(PhCOO)_2$ (4 equiv.); MeCN, 0 °C, 4 h; 86% → OMTM O O COOMe

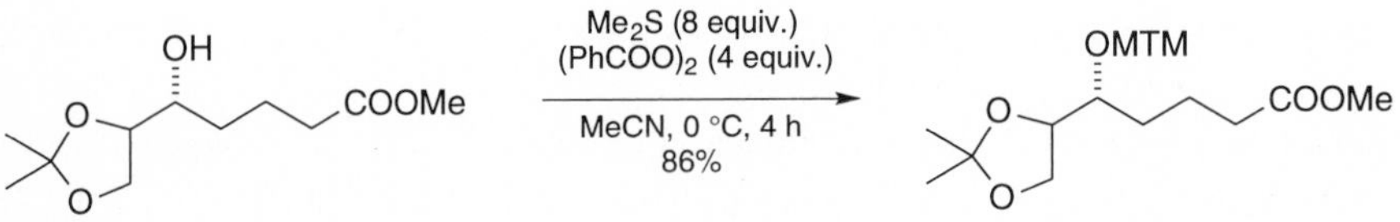

**Scheme 2.129**

**(iii) NMR Data**

$^1$H NMR: $\delta$ = 4.7–4.4 (2H, s), 2.1–2.4 (3H, s)

$^{13}$C NMR: $\delta$ = 72 (1C, t), 15 (1C, q)

## 2.5.3 (2-Methoxyethoxy)methyl Ethers

(2-Methoxyethoxy)methyl ethers (abbreviated MEM) were first described by Corey and co-workers[296]. They are roughly comparable in stability to MOM and SEM ethers towards protic acids though they decompose in the presence of Lewis acids more readily than MOM ethers. MEM ethers are stable to 0.05 equiv. of anhydrous *p*-toluenesulfonic acid (PTSA) in MeOH at 25 °C for 3–15 h or 3:1 AcOH–water at 35 °C for 4 h but not to HBr in HOAc or 2.0 M HCl. MEM ethers decompose slowly in $CF_3COOH$ but the reaction is slow enough to allow selective deprotection of *t*-Bu ethers or Boc groups.

**(i) Cleavage**

Williams and co-workers[297] used aq HBr in THF at r.t. for 72 h to remove three MEM groups simultaneously in the final stages of their synthesis of Breynolide [Scheme 2.130].

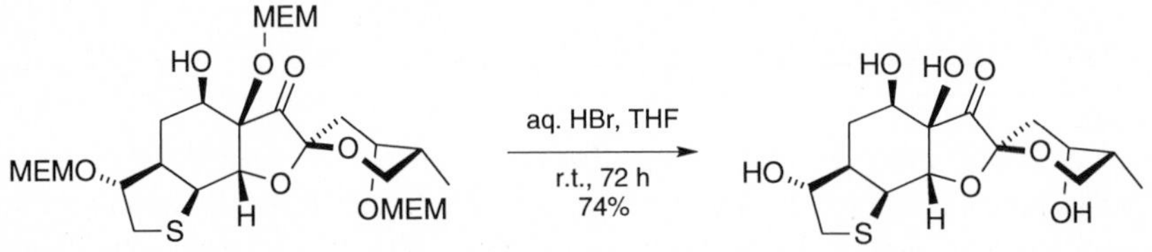

MEM O HO O MEMO S H O O OMEM → aq. HBr, THF; r.t., 72 h; 74% → HO HO O HO S H O O OH

**Scheme 2.130**

The additional oxygen suitably poised in a 1,2-diether arrangement serves as a bidentate ligand which enhances the lability of MEM ethers towards Lewis acids compared with MOM ethers. Corey claimed that anhydrous $ZnBr_2$ in $CH_2Cl_2$ at r.t. for 2–10 h cleaved MEM ethers but later reports[298] suggest that *wet* $ZnBr_2$ is better. $ZnCl_2$, $ZnI_2$, $SnCl_4$, $MgCl_2$, and $MgBr_2$ were less effective. Recently, a Merck group accomplished the removal of the MEM ether of a $\beta$-hydroxy ester without retroaldolisation, elimination, or destruction of a proximate TBDPS group [Scheme 2.131][101].

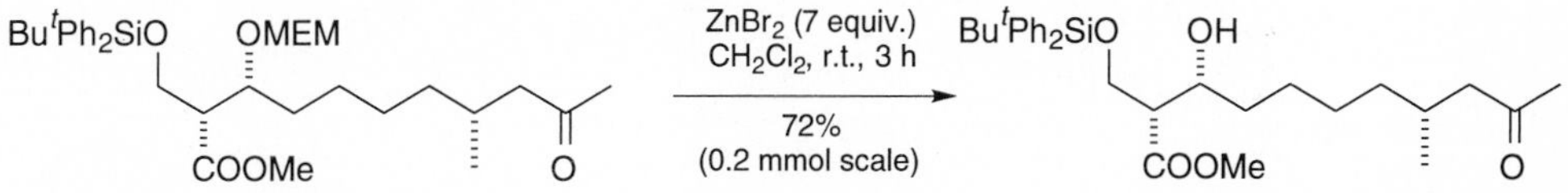

**Scheme 2.131**

The ability of the MEM group to coordinate to metals noted above can markedly influence the stereochemical course of organometallic reactions. In chapter 1 we showed how a MEM group assisted the diastereoselective conjugate addition of alkyllithiums to unsaturated sulfones [Scheme 1.26][299]. We now show another example taken from a synthesis of Taxusin[300]. A key step in the synthesis involved a nucleophilic acylation reaction in which $\alpha$-methoxyvinyllithium (an acetyl anion equivalent) added to the carbonyl group of the intermediate **132.1** [Scheme 2.132]. Given the chair-boat conformation of the cyclooctane ring in **132.1**, one would expect the organolithium to add from outside the ring system, i.e. away from the cleft caused by the fused six-membered ring. In the event, the reaction of **132.1** with 5 equivalents of $\alpha$-methoxyvinyllithium in THF at –13 °C for 12 h gave a 1:1 mixture of the diastereoisomers **132.2** and **132.3**. Thus it appeared that nucleophilic addition was indiscriminate despite the marked steric bias imposed by the conformation of the ring. Holton and co-workers postulated that the MEM ether was acting as a ligand and thus directing the lithium reagent to the more sterically hindered $\alpha$-face of the molecule. To prove the point, the solvent was changed from THF (a good ligand for lithium) to the poorly coordinating hexane which should amplify the ligand effect of the MEM ether. Indeed, the addition now occurred in 90% yield to give the $\alpha$-adduct **132.3** exclusively[301].

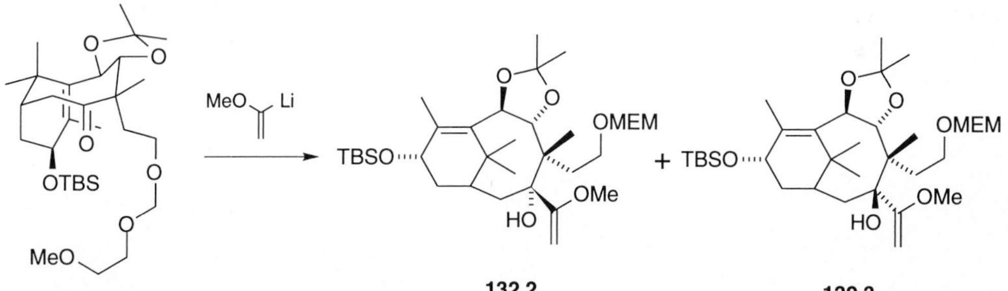

**Scheme 2.132**

A wide range of Lewis acids and conditions cleave MEM ethers and the choice of reagent will depend very much on functional and protecting group compatibility. We have already mentioned zinc halides which are especially mild but more vigorous reagents have also been employed including $TiCl_4$ in $CH_2Cl_2$ at 0 °C [Scheme 2.133][296,302,303] and TMSCl–NaI in MeCN at –10 °C[304]. During the synthesis of Taxusin alluded to above, Holton and co-workers were unable to cleave the MEM ether from the intermediate **134.1** using either $ZnBr_2$ or $TiCl_4$ [Scheme 2.134]. After many trials of various reagents and conditions, they found that a catalytic amount of anhydrous $FeCl_3$ in neat $Ac_2O$ at low temperature was effective[300,305]. A mixture of acetate and acetoxy methyl ether resulted which was then cleaved with $K_2CO_3$ in MeOH to give the desired alcohol **134.2** in 90% yield. Note the preservation of the dioxolane ring and TBS ether.

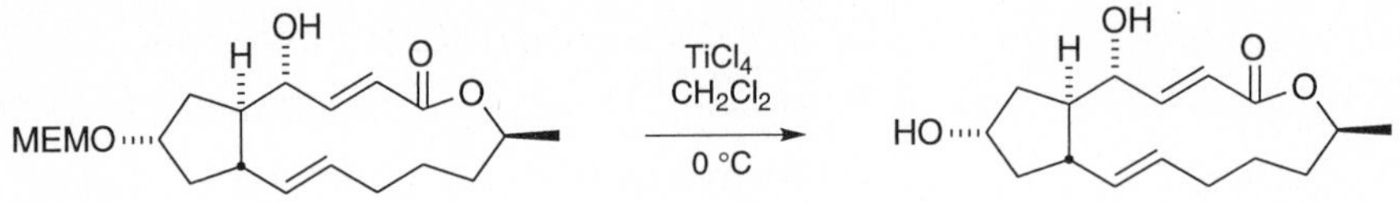

MEMO, OH, H, O, O → $TiCl_4$, $CH_2Cl_2$, 0 °C → HO, OH, H, O, O

**Scheme 2.133**

TBSO, O, O, OMEM, O → i) $FeCl_3$ (cat.), $Ac_2O$, –45 °C; ii) $K_2CO_3$, MeOH; 90% for 2 steps → TBSO, O, O, OH, O

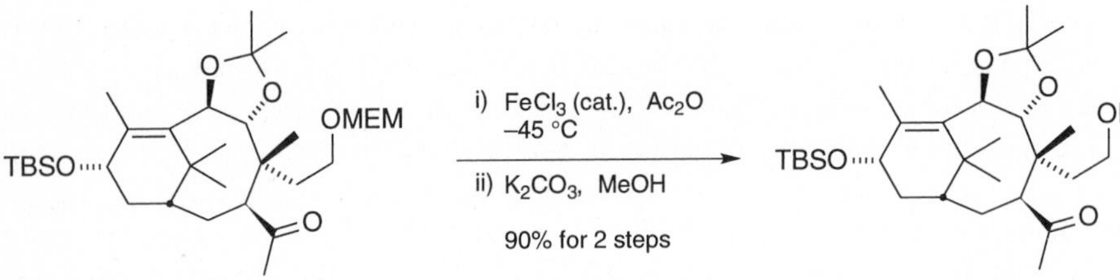

**134.1** **134.2**

**Scheme 2.134**

Boron halides, whose reactivity has been tuned by substitution, offer mild alternatives which have found favor. MEM, MOM, MTM and similar derivatives are cleaved at –78 °C by $Me_2BBr$ [Scheme 2.135][306] or $Ph_2BBr$[307]. However, a wide range of protecting groups are affected and there is little selectivity: dimethyl acetal > MOM ≈ MEM ≈ dimethyl ketal ≈ 1,3-dioxolane ≈ 1,3-dioxane ≈ acetonides >> THP, TBS, ROMe, ROBn > ArOMe >> alkenes, acetates, benzoates, ethyl ethers, TBDPS[308]. 2-Chloro-1,3,2-dithiaborolane[309] is another boron Lewis acid which cleaves MEM ethers in the presence of benzyl ethers, silyl ethers, THP ethers, acetals, ketals, acetates, and benzoates in $CH_2Cl_2$ at –78 °C for 1 h [Scheme 2.136][310]. Catechol boron bromide cleaves MEM and MOM ethers at comparable rates but *tert*-butoxycarbonyl, benzyloxycarbonyl, *tert*-butyl ethers, benzyl ethers, and TBS ethers are usually compatible[272].

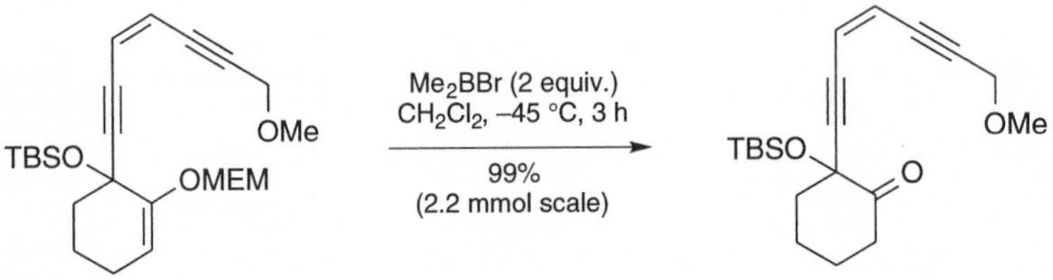

TBSO, OMe, OMEM → $Me_2BBr$ (2 equiv.), $CH_2Cl_2$, –45 °C, 3 h; 99% (2.2 mmol scale) → TBSO, OMe, O

**Scheme 2.135**

MEMO, O, $OSiPh_2Bu^t$ → Cl-B(S)S, $CH_2Cl_2$, –78 °C, 1 h, 87% → HO, O, $OSiPh_2Bu^t$

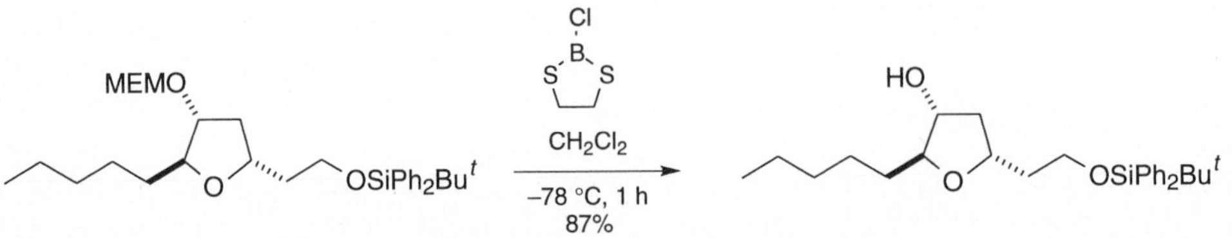

**Scheme 2.136**

Desperation can be an important stimulant for the development of new methodolgy and our next example exemplifies creative circumvention necessitated by the failure of more conventional methods for cleaving a MEM ether during a synthesis of Tirandamycic Acid. Ireland and co-workers[311] resorted to *n*-BuLi in heptane to generate a vinyl ether **137.2** (by elimination of LiOEt from **137.1**) which was subsequently hydrolysed with $Hg(OAc)_2$ [Scheme 2.137] and the method has been used by others[312,313].

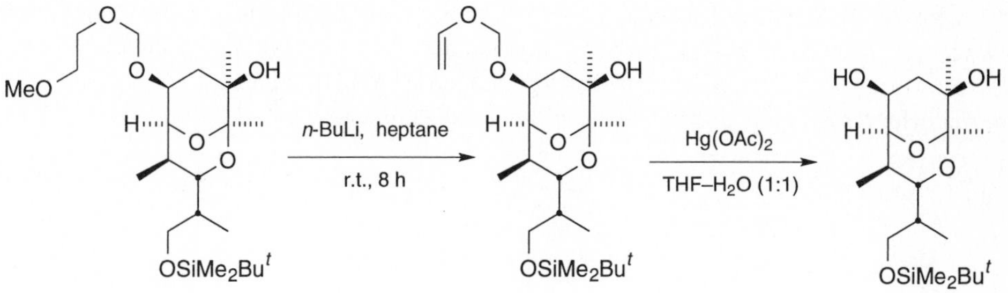

MeO O O OH H O O $OSiMe_2Bu^t$ **137.1**

*n*-BuLi, heptane; r.t., 8 h

O O OH H O O $OSiMe_2Bu^t$ **137.2**

$Hg(OAc)_2$; THF–$H_2O$ (1:1)

HO OH H O O $OSiMe_2Bu^t$ **137.3**

**Scheme 2.137**

### (ii) Formation

Alkylation of alcohols with (2-methoxyethoxy)methyl chloride (MEMCl) under an assortment of conditions pioneered by Corey still retains favor. These include reactions of the lithium or sodium alkoxides with MEMCl in THF or DME at 0 °C (10–60 min); reaction of the alcohol with MEMCl in $CH_2Cl_2$ in the presence of *i*-$Pr_2NEt$ at r.t. for 3 h; or reaction of the alcohol with MEMCl and $Et_3N$ in refluxing MeCN for 30 min. The latter conditions are neutral and therefore recommended for acid-sensitive substrates such as tertiary alcohols. For enhanced reactivity in sensitive substrates, [(2-methoxyethoxy)methyl]triethylammonium chloride (prepared from MEMCl and $Et_3N$) can be used [Scheme 2.138][314].

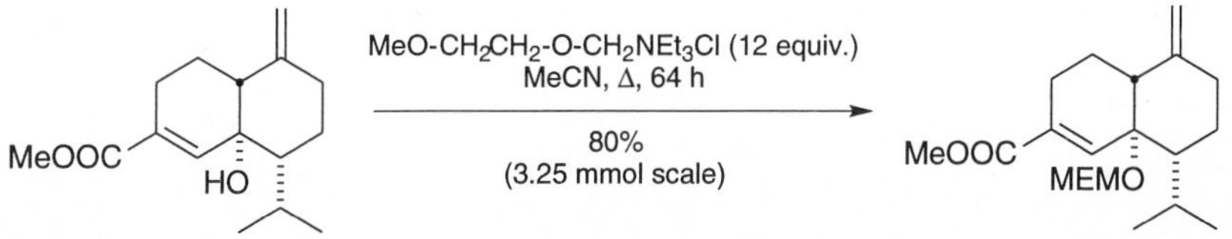

MeOOC HO

MeO-$CH_2CH_2$-O-$CH_2NEt_3Cl$ (12 equiv.); MeCN, Δ, 64 h; 80%; (3.25 mmol scale)

MeOOC MEMO

**Scheme 2.138**

### (iii) NMR Data

$^1$H NMR: $\delta = 4.7$ (2H, s), 3.6 (2H, m), 3.5 (2H, m), 3.3 (3H, s)

$^{13}$C NMR: $\delta = 94$ (1C, t, O-$CH_2$-O), 72 (1C, t), 67 (1C, t), 59 (1C, q)

## 2.5.4 Benzyloxymethyl Ethers

Benzyloxymethyl ethers (abbreviated BOM) are comparable in stability to MOM, MEM, and SEM ethers. However, like benzyl ethers they can be removed by hydrogenolysis or Birch reduction. The advantage BOM ethers have over benzyl ethers is their easier preparation and easier removal.

However, unlike benzyl ethers, they decompose in aqueous acid. The BOM group was introduced in 1975 by Stork and Isobe[315] but it has not been as widely applied as MOM or MEM groups though its virtues are gaining in appreciation.

**(i) Cleavage**

Preparative conditions for cleaving BOM protecting groups parallel the conditions used to cleave benzyl ethers. Thus, Birch reduction using sodium in liquid ammonia in the presence of EtOH at –60 °C removes BOM ethers [Scheme 2.139][316]. Under these conditions the BOM group can be removed in the presence of benzyl, furan, or vinyl iodides. Likewise, lithium in liquid ammonia has been used[317]. Provided the temperature is kept low and the reaction time short, a BOM group can even be removed with Na in liquid ammonia in the presence of an oxirane[318].

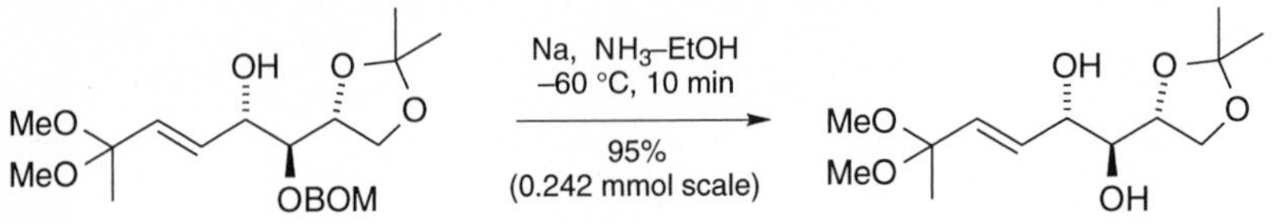

**Scheme 2.139**

BOM groups can easily be removed by catalytic hydrogenolysis. Typical catalysts include Pd–C[319] and $Pd(OH)_2$–C (Pearlman's catalyst) which tolerates the presence of tri-substituted alkenes[320]. For example a BOM group was successfully removed from the macrolide precursor **140.1** in the presence of three alkenes including di-substituted alkenes [Scheme 2.140][321]. With Ra-Ni and hydrogen, a BOM ether can be removed without detriment to a PMB ether [Scheme 2.141]. Hydrogenolysis of BOM ethers releases formaldehyde which can react with any amines present to result in reductive *N*-methylation. In such cases, glycinamide can be added to scavenge the formaldehyde[322].

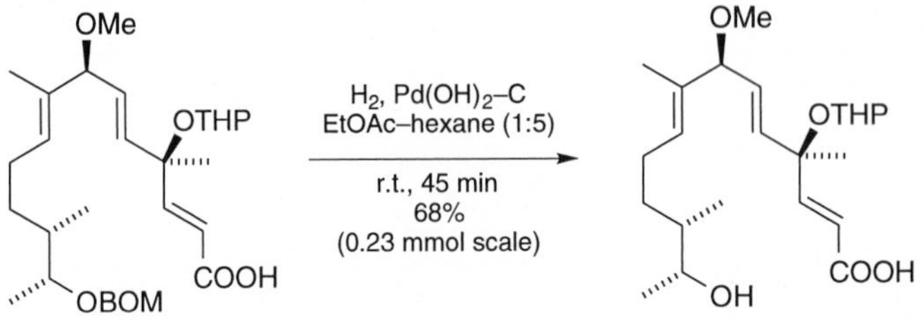

**Scheme 2.140**

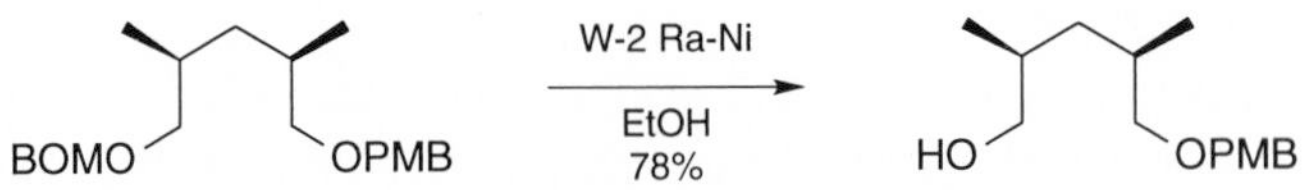

**Scheme 2.141**

Suzuki and co-workers[323] used a combination of $BF_3$•$OEt_2$ and PhSH to remove a BOM group in their synthesis of Protomycinolide [Scheme 2.142]. Conventional acid-catalysed hydrolysis of BOM ethers can also be used as shown by the cleavage of three phenolic BOM groups from the Olivomycin intermediate **143.1** [Scheme 2.143][324].

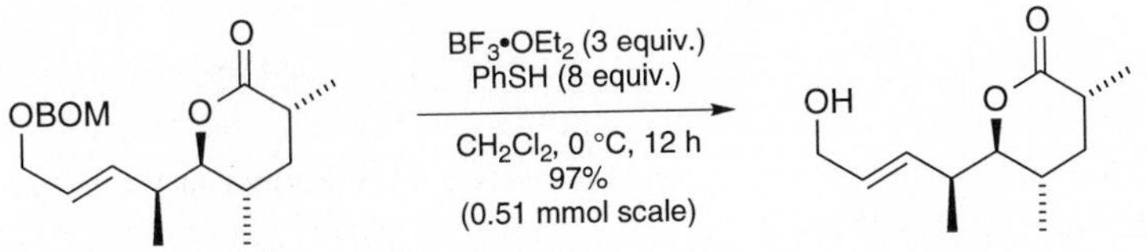

BF$_3$•OEt$_2$ (3 equiv.)
PhSH (8 equiv.)
CH$_2$Cl$_2$, 0 °C, 12 h
97%
(0.51 mmol scale)

**Scheme 2.142**

MeOH
Dowex 50W-X8
r.t., 5–6 d
90%

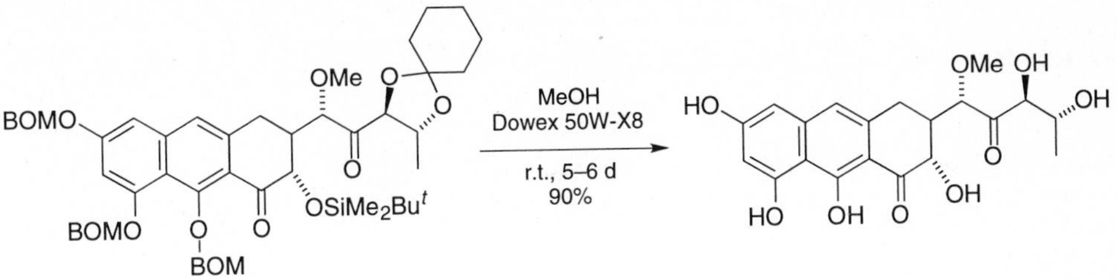

**143.1** **143.2**

**Scheme 2.143**

A *p*-methoxybenzyloxymethyl (PMBM) version of the BOM group has been described[325]. The alcohols are protected with *p*-methoxybenzyloxymethyl chloride (PMBMCl)[326] in the same way as BOM ethers. The PMBM ethers can be deprotected with DDQ and water (see PMB ethers).

**(ii) Formation**

Primary, secondary, and tertiary alcohols react with BOMCl in the presence of *i*-Pr$_2$NEt to give the BOM ethers[315] in good yields. If the reaction requires acceleration, some Bu$_4$NI (10 mol %) may be added to the reaction mixture [Scheme 2.144][176].

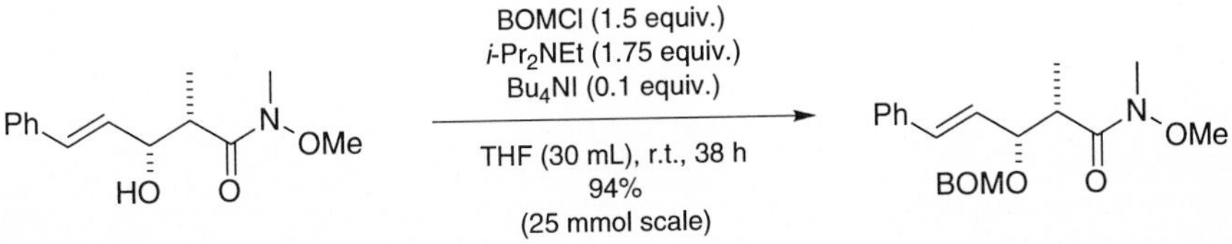

BOMCl (1.5 equiv.)
*i*-Pr$_2$NEt (1.75 equiv.)
Bu$_4$NI (0.1 equiv.)
THF (30 mL), r.t., 38 h
94%
(25 mmol scale)

**Scheme 2.144**

During an elegant synthesis of the polyether antibiotic X-206, Evans and co-workers[176] accomplished the protection of the secondary alcohol in **144.1** with BOMBr in the presence of "proton sponge" [1,8-bis(dimethylamino)naphthalene], without racemising the two stereogenic centres adjacent to the ketone function [Scheme 2.145].

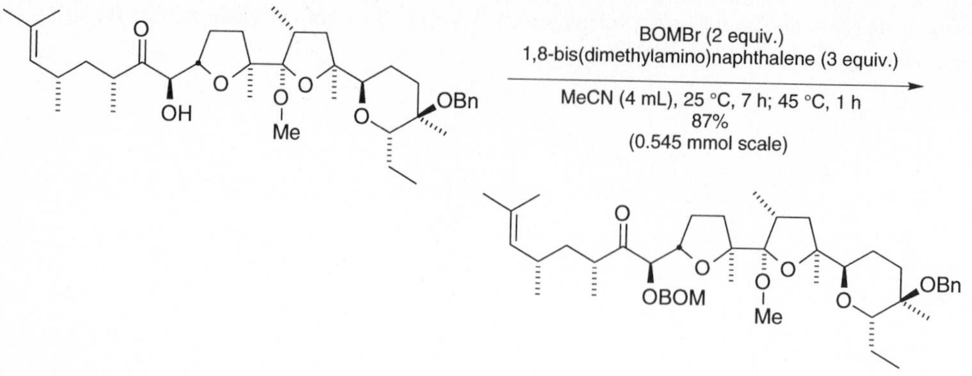

BOMBr (2 equiv.)
1,8-bis(dimethylamino)naphthalene (3 equiv.)
MeCN (4 mL), 25 °C, 7 h; 45 °C, 1 h
87%
(0.545 mmol scale)

Scheme 2.145

**(iii) NMR Data**

$^1$H NMR: $\delta$ = 7.4 (5H, m), 4.7 (2H, s), 4.6 (2H, s)

$^{13}$C NMR: $\delta$ = 138 (1C, s), 129 (2C, d), 128 (2C, d), 127 (1C, d), 95 (t, O-$CH_2$-O), 73–78 ($PhCH_2O$)

## 2.5.5 β-(Trimethylsilyl)ethoxymethyl Ethers

β-(Trimethylsilyl)ethoxymethyl ethers (abbreviated SEM) are more rugged than trialkylsilyl ethers but they can be deprotected using fluoride under forcing conditions. Silyl ethers can be deprotected in the presence of SEM ethers[327]. SEM ethers are stable to acidic conditions which cleave THP and TBS ethers.

**(i) Cleavage**

SEM ethers appear to be more labile than the corresponding MEM or MOM ethers to acid-catalysed hydrolysis. Concentrated HF in acetonitrile was used to deprotect a SEM ether in a synthesis of the marine toxin Latrunculin [Scheme 2.146][63] and it is noteworthy that a neighboring β-trimethylsilyl-ethyl ester was unscathed under the reaction conditions. Other acids may also be used. For example, Gadwood[328] successfully deprotected the SEM ether of a tertiary alcohol using 0.1 M HCl in MeOH without appreciable dehydration.

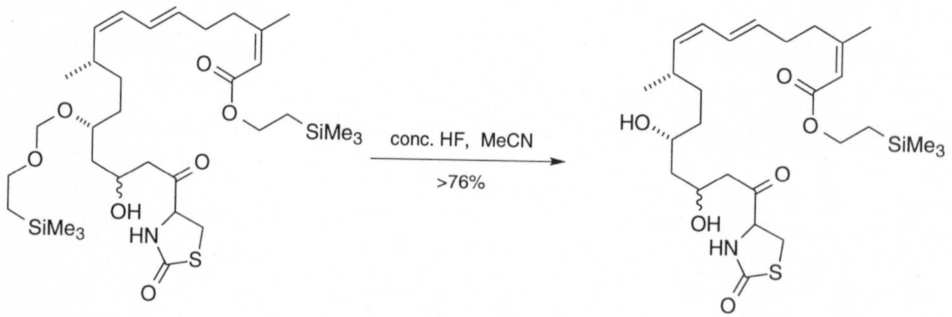

conc. HF, MeCN
>76%

Scheme 2.146

Lewis acid-mediated deprotection has seldom been exploited but the limited data available are encouraging. For example, $BF_3 \bullet OEt_2$ in $CH_2Cl_2$ cleaves a β-trimethylsilylethyl ether at 0–25 °C

[Scheme 2.147][329] and similar conditions should be applicable to SEM deprotection. Another example comes from the final step of a synthesis of the macrolide Methynolide [Scheme 2.148][330] in which $LiBF_4$ in a 1:1 mixture of MeCN and water at 72 °C for 5 h was used to remove a SEM protecting group[331,332]. Benzylidene acetals are also cleaved under these conditions.

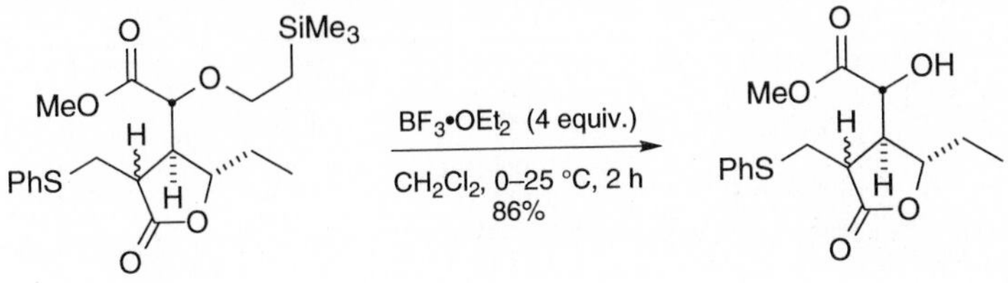

MeO SiMe3 PhS H H O O O O; BF3•OEt2 (4 equiv.), CH2Cl2, 0–25 °C, 2 h, 86%; MeO OH PhS H H O O O

**Scheme 2.147**

HO O O O OSEM; LiBF4, MeCN–H2O (1:1), 72 °C, 5 h, 88%; HO O O O OH

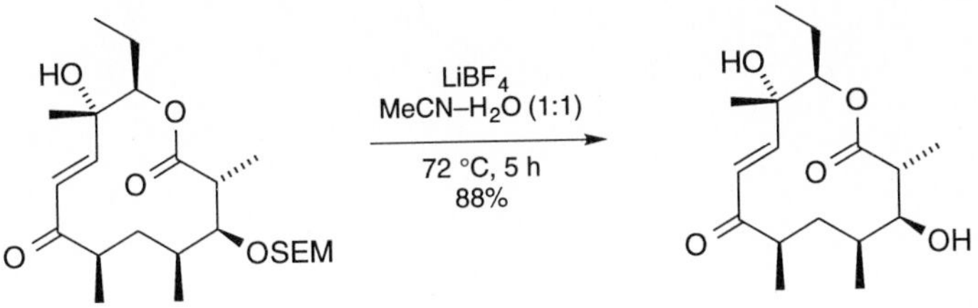

**Scheme 2.148**

Fluoride ions (typically TBAF) will induce a fragmentation reaction of SEM ethers resulting in loss of TMSF, ethylene, and formaldehyde to give the deprotected alcohol. However, compared with deprotection of ordinary silyl ethers, the reaction requires rather protracted reaction times, higher temperatures, or the presence of HMPA [Scheme 2.149][333]; indeed, TBDPS ethers can be removed selectively with TBAF in the presence of SEM ethers[297]. The toxicity of HMPA can be avoided by using dimethylpropylene urea (DMPU) as the solvent[334]. CsF in HMPA[238] or DMF[323] provides an alternative source of fluoride which effects deprotection of SEM ethers at elevated temperature.

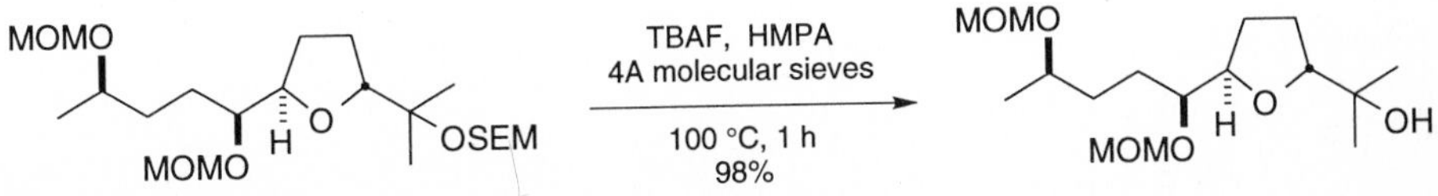

MOMO MOMO H O OSEM; TBAF, HMPA, 4A molecular sieves, 100 °C, 1 h, 98%; MOMO MOMO H O OH

**Scheme 2.149**

The additional resonance stabilisation intrinsic to a phenoxide anion is a key factor in accelerating the fragmentation of the SEM ethers of phenols as demonstrated in a synthesis of a Tetrahydrocannabinol precursor [Scheme 2.150][335].

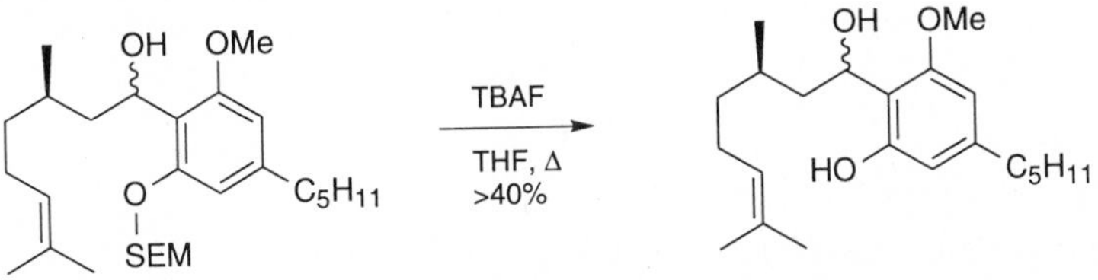

OH OMe O SEM C5H11; TBAF, THF, Δ, >40%; OH OMe HO C5H11

**Scheme 2.150**

During a synthesis of Milbemycin E, Thomas and co-workers needed to remove a SEM ether group without removing a TBS ether or a $\beta$-trimethylsilylethyl ester group. TBAF removes all three. These workers found that irradiation of the SEM group in the presence of 1 mole equivalent of iodine selectively removed the SEM in 92% yield [Scheme 2.151][336].

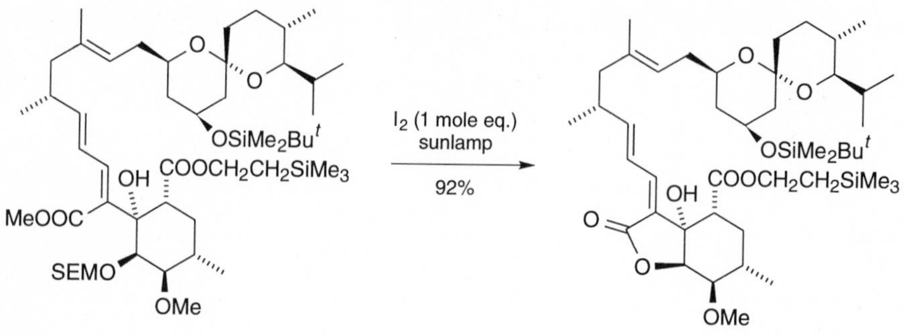

Scheme 2.151

### (ii) Formation

The sodium or potassium alkoxide prepared from primary and secondary alcohols and NaH or KH in THF reacts with $\beta$-(trimethylsilyl)ethoxymethyl chloride (SEMCl) to give the SEM ethers in good yield [Scheme 2.152][297]. Alternatively, the alcohol can be alkylated with SEMCl in the presence of *i*-$Pr_2NEt$ in $CH_2Cl_2$ at 40 °C [Scheme 2.153][327,337,338]. Conversion of the SEMCl to the corresponding iodide *in situ* using $Bu_4NI$ accelerates the reaction.

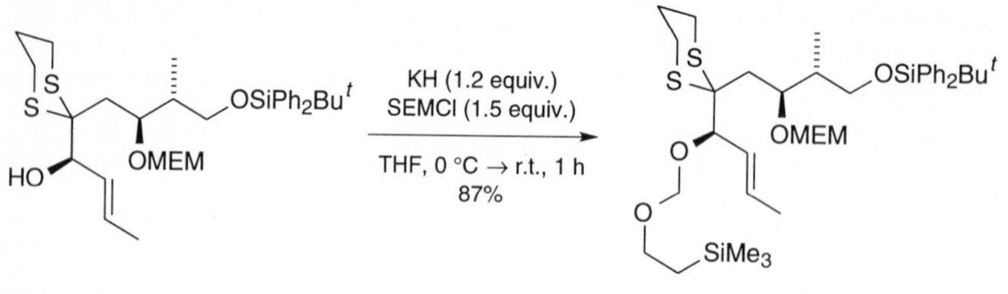

Scheme 2.152

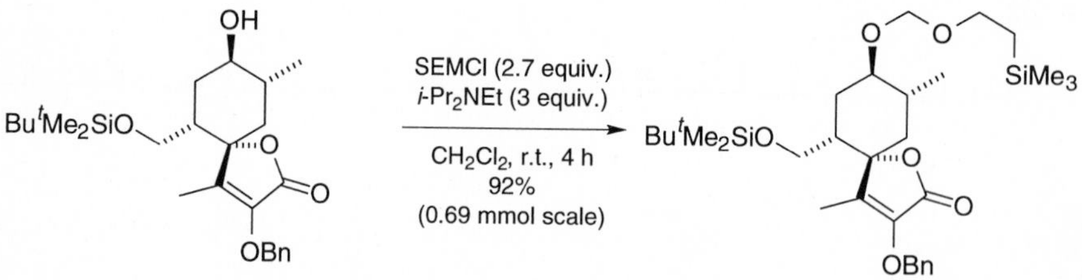

Scheme 2.153

### (iii) NMR Data

$^1$H NMR: $\delta = 4.7$ (2H, s), 3.6 (2H, m), 0.9 (2H, m), 0.03 (9H, s)

$^{13}$C NMR: $\delta = 95$ (1C, t, O-$CH_2$-O), 66 (1C, t, O-**$CH_2$**$CH_2$Si), 18 (1C, t, O-$CH_2$**$CH_2$**Si), –13 (3C, q).

## 2.5.6 Tetrahydropyranyl and Related Ethers

Tetrahydropyranyl ethers (abbreviated THP) were one of the first generally useful protecting groups for alcohols to be adopted[339] and they are still widely used today. The particular merits of the THP group are its ease of introduction, the low cost of dihydropyran, its stability (in the absence of acid) under a wide range of reaction conditions, and its ease of removal. Its two greatest detractions are the complexity of the NMR spectra and the fact that diastereoisomers are formed on reaction with chiral alcohols. THP ethers are stable to strongly basic conditions required for ester hydrolysis and they withstand attack by $LiAlH_4$ in the absence of Lewis acids and organometallic reagents (organocuprates, organolithiums, Grignard reagents) provided the temperature is maintained below 0 °C.

**(i) Cleavage**

THP ethers are labile to mild acid such as HOAc–THF–$H_2O$ (4:2:1) at 45 °C[340] — conditions which cleave TBS ethers but not MEM, MTM, or MOM ethers. If water is an objectionable component or low solubility a problem, THP removal can be accomplished by an acetal exchange process using MeOH or EtOH as solvent. Scheme 2.154 illustrates the deprotection of an acid-sensitive allylic alcohol using PTSA in MeOH[341]. PPTS in MeOH or EtOH (pH 3.0) can also be used but higher temperatures (45–55 °C) are required [Scheme 2.155][342,343]. Ion exchange resins in MeOH cleave THP ethers at r.t.[344,345].

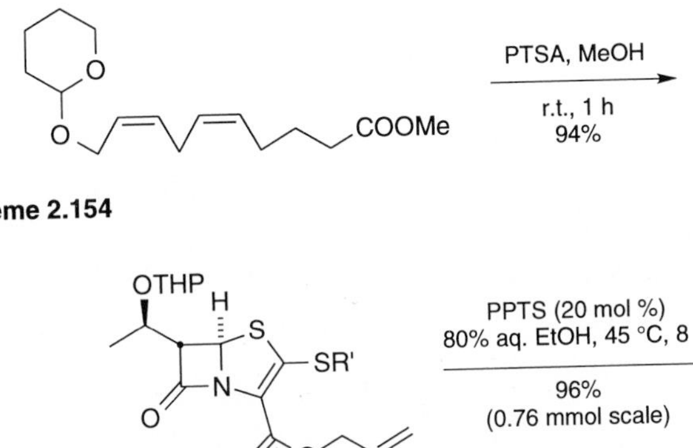

**Scheme 2.154**

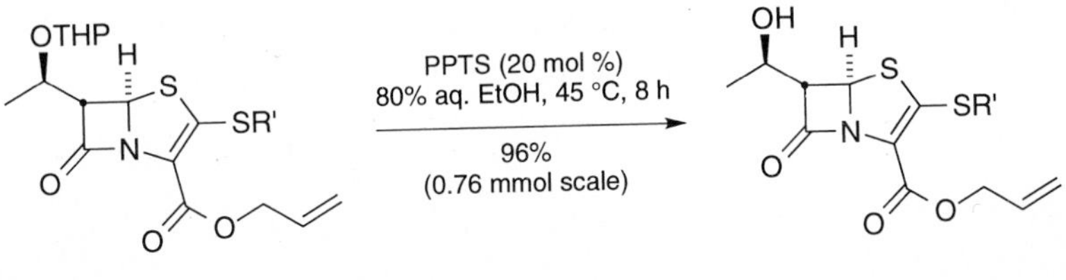

**Scheme 2.155**

Most Lewis acids will cleave THP ethers and conditions are mild enough to allow retention of protecting groups which would otherwise be labile in aqueous or alcoholic acid. For example, three equivalents of $MgBr_2$ in ether have been used to cleave THP ethers of primary and secondary alcohols[346] and these conditions are compatible with TBS and MEM ethers; however, MOM ethers cleave slowly. A further limitation is that tertiary and benzylic alcohols give the bromide. Ferric chloride bound to silica gel has been used to remove primary THP ethers[347].

Although the THP group is more labile towards hydrolysis than the other acetal type protecting groups we have discussed so far such as MOM, MEM, BOM, and SEM, there are many occasions when even greater lability is required. For example oligonucleotide synthesis imposes demands on ease of introduction and lability which are not always met by the traditional THP group. To that end Reese and co-workers[348] introduced 4-methoxytetrahydropyran-4-yl ethers (abbreviated MTHP) into the protecting group repertoire. 4-Methoxytetrahydropyranyl ethers are prepared the same way as THP ethers but they hydrolyse about 3 times faster than the corresponding THP ethers and they have the added bonus that their symmetrical structure is devoid of the problem of diastereoisomers. Unfortunately, 4-methoxydihydropyran is expensive. Reviews on the use of 4-methoxy–tetrahydro–pyranyl ethers and related labile protecting groups in oligonucleotide synthesis should be consulted for further examples of how protecting group design can be tailored to highly demanding synthetic problems[201,349,350].

The acid lability of 4-methoxy-tetrahydropyranyl ethers was a key design feature in a recent synthesis of 1,2-dipalmitoyl-*sn*-glycer-3-yl-D-*myo*-inositol-1-phosphate[351]. Hydrogenolysis of the phosphate phenol ether **156.1** [Scheme 2.156] released a phosphoric diester whose inherent acidity was sufficient to cleave the 4-methoxyltetrahydropyranyl ether function *in situ*. The desired product **156.2** was recovered as the sodium salt generated with the aid of an ion exchange resin.

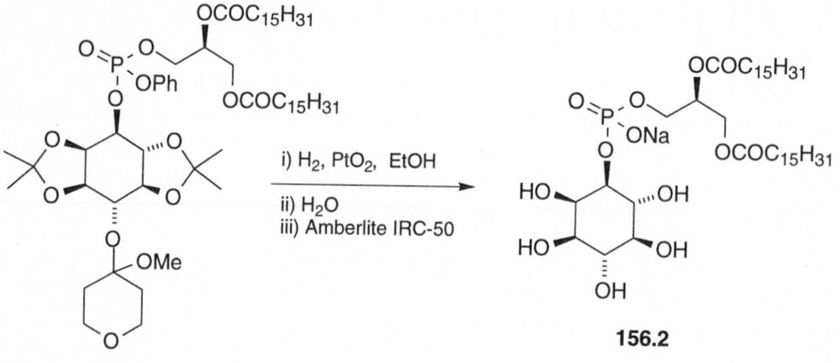

**Scheme 2.156**

1-Methoxy-1-methylethyl ethers are made by reacting alcohols with 2-methoxypropene which is cheap. However, the resultant acetals are quite labile (*ca.* 20 times faster hydrolysis than THP) and on occasion this lability can be put to good use. For example the Corey group has deprotected 1-methoxy-1-methylethyl ethers in the presence of allylic epoxides [Scheme 2.157][352,353].

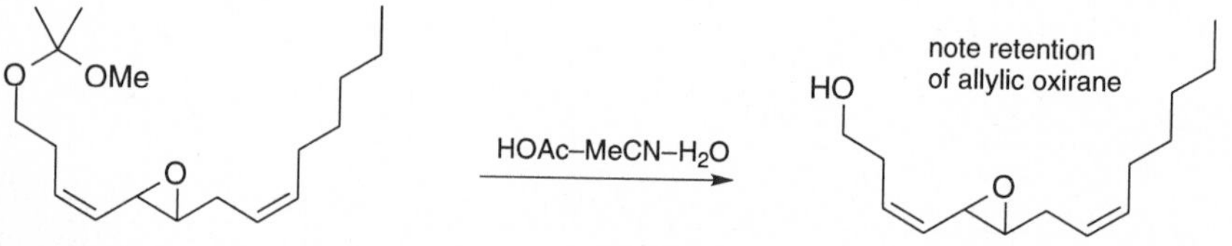

**Scheme 2.157**

1-Ethoxyethyl ethers (often abbreviated EE) are another variant of an acyclic acetal protecting group which is prepared from an alcohol and ethyl vinyl ether under acid catalysis. Being less highly substituted at the acetal centre, 1-ethoxyethyl ethers are more stable than the corresponding 1-methoxy-1-methylethyl ethers mentioned above but they can be removed selectively in the presence of 1,3-dioxolanes which are commonly used to protect 1,2-diols[354]. The ethers are strong enough to survive the conversion of an alkyl bromide to an organolithium[355]. 1-Ethoxyethyl ethers can be cleaved with $MgBr_2$ in $Et_2O$ — a property well exploited in Zwanenburg's synthesis of Pyrenophorol [Scheme 2.158][356].

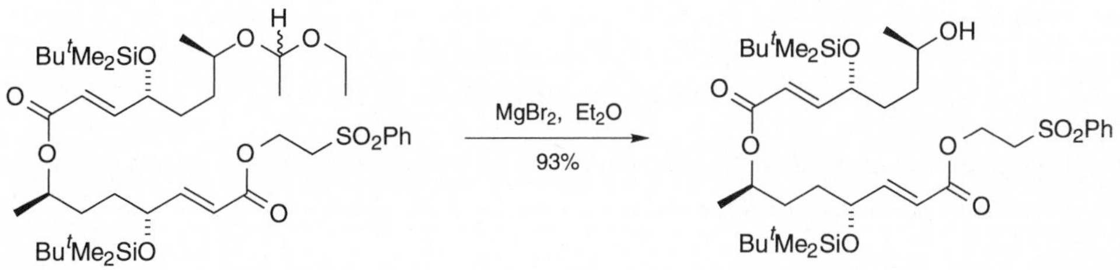

MgBr$_2$, Et$_2$O
93%

Scheme 2.158

### (ii) Formation

Acid-catalysed addition of primary, secondary, and tertiary alcohols to dihydropyran (DHP) in $CH_2Cl_2$ at r.t. is the only general method currently in use for preparing THP ethers and the variations cited below concern the choice of acid. The reaction proceeds by protonation of the enol ether carbon to generate a highly electrophilic oxonium ion which is then attacked by the alcohol. Yields are generally good. Favored acid catalysts include PTSA or CSA[340]. To protect tertiary allylic alcohols and sensitive functional groups such as epoxides, the milder acid PPTS has been employed [Scheme 2.159][321,343]. A variety of other acid catalysts have been used such as $POCl_3$, TMSI[357] and $(TMSO)_2SO_2$[358] but one cannot help but suspect that in all of these cases, the real catalyst is a proton derived from reaction of the putative catalysts with adventitious water.

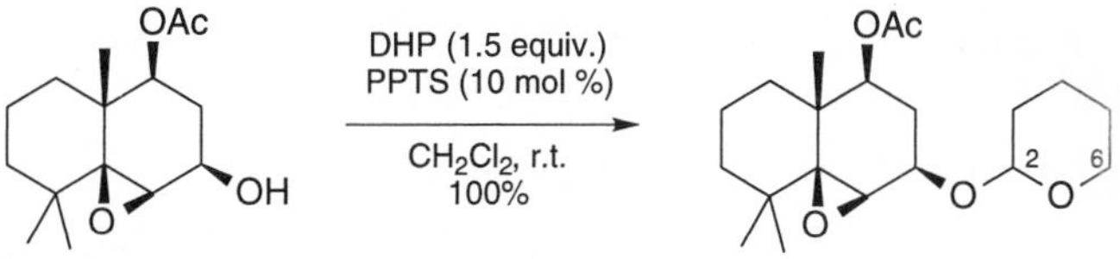

DHP (1.5 equiv.)
PPTS (10 mol %)
$CH_2Cl_2$, r.t.
100%

Scheme 2.159

In many instances the reaction of an alcohol with DHP (or ethyl vinyl ether or 2-methoxypropene) does not go to completion despite the addition of a large excess of the enol ether: as much as 20% of the starting material will be present at equilibrium. The equilibrium can be shifted toward product by adding excess finely powdered anhydrous $K_2CO_3$ and stirring the reaction mixture at r.t. As the acid concentration gradually diminishes, the reaction goes to completion.

In large scale processes the homogeneous reactions described above may not be convenient in which case a solid ion exchange resin such as Dowex 50 or Amberlyst H-15 can be used. The ion exchange resins consist of matrices of cross-linked styrene and divinylbenzene incorporating benzenesulfonic acid residues. The advantage these offer is that the beads are simply removed by decantation or filtration without the need for basification or aqueous workup[359]. A cross-linked polyvinylpyridine resin as its hydrochloride form (Reillex 425•HCl), which is effectively an insoluble form of PPTS, has also been recommended for THP ether formation[345].

### (iii) NMR Data

$^1H$ NMR: $\delta = 4.5$ (1H, m, O-CH-O), 3.8 (1H, m), 3.5 (1H, m), 1.9–1.4 (6H, m)

$^{13}C$ NMR: $\delta = 98$ (1C, d, C2), 31 (1C, t, C3), 19 (1C, t, C4), 25 (1C, t, C5), 62–75 (1C, t, C6)

# 2.6 Reviews

## 2.6.1 General Reviews Concerning the Protection of Hydroxyls and Thiols

For reviews concerning hydroxyl protecting groups in carbohydrate and nucleoside chemistry see section 1.6.2 at the end of Chapter 1.

1 Protection of Alcoholic Hydroxyl Groups and Glycol Systems. Reese, C. B. In *Protective Groups in Organic Chemistry*; McOmie, J. F. W., Ed.; Plenum: London, 1973; Chapter 3.
2 Schutzgruppen der alkoholischen Hydroxy-Funktion. Schaumann, E. In *Houben-Weyl*, 4th ed., Vol VI/1b; Kropf, H., Ed.; Thieme: Stuttgart, 1984; p. 737.
3 Protection for the Hydroxyl Group, Including 1,2- and 1,3-Diols. Greene, T. W.; Wuts, P. G. M. In *Protective Groups in Organic Synthesis*, 2nd ed.; Wiley: New York, 1991; Chapter 2.
4 Protection for Phenols and Catechols. Greene, T. W.; Wuts, P. G. M. In *Protective Groups in Organic Synthesis*, 2nd ed.; Wiley: New York, 1991; Chapter 3.
5 Protection of Thiols. Hiskey, R. G.; Rao, V. R.; Rhodes, W. G. In *Protective Groups in Organic Chemistry*; McOmie, J. F. W., Ed.; Plenum: London, 1973; Chapter 7.
6 Protection of the Thiol Group. Wolman, Y. In *The Chemistry of the Thiol Group;* Patai, S., Ed.; Wiley: New York, 1974; p. 669.
7 Schutzgruppen für Schwefel-Verbindungen. Gundermann, K.–D. in *Houben-Weyl*, 4th ed.; Vol. 11; Klamann, D., Ed.; Thieme: Stuttgart, 1985; p. 1624.
8 Protection for the Thiol Group. Greene, T. W.; Wuts, P. G. M. In *Protective Groups in Organic Synthesis*, 2nd ed.; Wiley: New York, 1991; Chapter 7.

## 2.6.2 Reviews Concerning Enzyme-Mediated Esterification and Hydrolysis of Esters

1 General Aspects and Optimisation of Enantioselective Biocatalysis in Organic Solvents – The Use of Lipases. Chen, C.–S.; Sih, C. J. *Angew. Chem. Int. Ed. Engl.* **1989**, *28*, 695.
2 Biotransformations in Organic Synthesis. Crout, D.H.G.; Christen, M. In *Modern Synthetic Methods;* Scheffold, R., Ed.; Springer: Berlin, 1989; Chapter 1.
3 *Biotransformations in Preparative Organic Chemistry.* Davies, H. G.; Green, R. H.; Kelly, D. R.; Roberts, S. M. Academic Press: New York, 1989.
4 Chiral Synthons by Ester Hydrolysis Catalysed by Pig Liver Esterase. Ohno, N.; Otsuka, M. *Org. Reactions* **1989**, *37*, 1.
5 Enzymic Catalysts in Organic Synthesis. Wong, C.–H. *Science* **1989**, *244*, 1146.
6 Resolution of Enantiomers via Biocatalysis. Sih, C. J.; Wu, S.–H. *Top. Stereochem.* **1989**, *19*, 63.
7 Recent Advances in the Use of Enzyme–Catalysed Reactions in Organic Synthesis. Turner, N. J. *Nat. Prod. Rep.* **1989**, *6*, 625.
8 Applications of Pig Liver Esterases in Asymmetric Synthesis. Zhu, L.–M.; Tedford, M. C. *Tetrahedron* **1990**, *46*, 6587.
9 Asymmetric Transformations Catalysed by Enzymes in Organic Solvents. Klibanov, A. M. *Acc. Chem. Res.* **1990**, *23*, 114.

10 Enzyme Catalysis in Synthetic Carbohydrate Chemistry. Drueckhammer, D. G.; Hennen, W. J.; Pederson, R. L.; Barbas, C. F., III; Gautheron, C. M.; Krach, T.; Wong, C.-H. *Synthesis* **1991**, 499.
11 Esterolytic and Lipolytic Enzymes in Organic Synthesis. Boland, W.; Frössl, C.; Lorenz, M. *Synthesis* **1991**, 1049.
12 Enzymic Methods in Preparative Carbohydrate Chemistry. David, S.; Augé, C.; Gautheron, C. *Adv. Carbohydr. Chem. Biochem.* **1991**, *49*, 176.
13 Enzymatic Protecting Group Techniques. Waldmann, H. *Kontakte* (Darmstadt), **1991** (2), 33.
14 The Biocatalytic Approach to the Preparation of Enantiomerically Pure Chiral Building Blocks. Santaniello, E.; Ferraboschi, P.; Grisenti, P.; Manzochi, A. *Chem. Rev.* **1992**, *92*, 1071.
15 *Selective Biocatalysis. A Synthetic Approach.* Poppe, L.; Novak, L. VCH: Weinheim, 1992.
16 *Biotransformations in Organic Chemistry.* Farber, K. Springer: Berlin, 1992.
17 Enzymatic Protecting Group Techniques in Bioorganic Synthesis. Reidel, A.; Waldmann, H. *J. Prakt. Chem.* **1993**, *335*, 109.

### 2.6.3 Reviews Concerning Organosilicon and Organotin Chemistry Relevant to Hydroxyl Protection

1 Synthetic Applications of Cyanotrimethylsilane, Iodotrimethylsilane, Azidotrimethylsilane, and Methylthiotrimethylsilane. Groutas, W. C.; Felker, D. *Synthesis* **1980**, 861.
2 Iodotrimethylsilane–A Versatile Synthetic Reagent. Olah, G. A.; Narang, S. C. *Tetrahedron* **1982**, *38*, 2225.
3 Use of Organosilicon Reagents as Protective Groups in Organic Synthesis. LaLonde, M.; Chan, T.–H. *Synthesis* **1985**, 817.
4 *Silylating Agents.* van Look, G. **1988**, Fluka.
5 Silyl Ethers as Protective Groups for Alcohols: Oxidative Deprotection and Stability under Alcohol Oxidation Conditions. Muzart, J. *Synthesis* **1993**, 11.
6 The Use of Tin Compounds in Carbohydrate and Nucleoside Chemistry. Blunden, S. J.; Cusack, P. A.; Smith, P. J. *J. Organomet. Chem.* **1987**, *325*, 141.

## References

1 Griffin, B. E.; Jarman, M.; Reese, C. B. *Tetrahedron* **1968**, *24*, 639.
2 Cramer, F.; Bär, H. P.; Rhaese, H. J.; Sänger, W.; Scheit, K. H.; Schneider, G.; Tenigkeit, J. *Tetrahedron Lett.* **1963**, 1039.
3 Kunz, H.; März, J. *Synlett* **1992**, 591.
4 Kunz, H.; Unverzagt, C. *Angew. Chem. Int. Ed. Engl.* **1988**, *27*, 1697.
5 Plattner, J. J.; Gless, R. D.; Rapoport, H. *J. Am. Chem. Soc.* **1972**, *94*, 8613.
6 Schultheiss-Reimann, P.; Kunz, H. *Angew. Chem. Int. Ed. Engl.* **1983**, *22*, 62.
7 Kunesch, N.; Miet, C.; Poisson, J. *Tetrahedron Lett.* **1987**, *28*, 3569.
8 Mori, K.; Tominaga, T.; Takigawa, T.; Matsui, M. *Synthesis* **1973**, 790.
9 Fromageot, H. P. M.; Reese, C. B.; Sulston, J. E. *Tetrahedron* **1968**, *24*, 3533.
10 Haines, A. H. *Adv. Carbohydr. Chem. Biochem.* **1976**, *33*, 11.
11 Danishefsky, S. J.; DeNinno, M. P.; Chen, S.-h. *J. Am. Chem. Soc.* **1988**, *110*, 3929.
12 De Ninno, M. P. *Synthesis* **1991**, 583.
13 Bruzik, K. S.; Tsai, M.-D. *J. Am. Chem. Soc.* **1992**, *114*, 6361.
14 Jones, T. K.; Reamer, R. A.; Desmond, R.; Mills, S. G. *J. Am. Chem. Soc.* **1990**, *112*, 2998.
15 Nicolaou, K. C.; Webber, S. E. *Synthesis* **1986**, 453.

16 Danishefsky, S. J.; Armistead, D. M.; Wincott, F. E.; Selnick, H. G.; Hungate, R. *J. Am. Chem. Soc.* **1989**, *111*, 2967.
17 Reese, C. B.; Stewart, J. C. M. *Tetrahedron Lett.* **1968**, 4273.
18 Reese, C. B.; Stewart, J. C. M.; van Boom, J. H.; de Leeuw, H. P. M.; Nagel, J.; de Rooy, J. F. M. *J. Chem. Soc., Perkin Trans. 1* **1975**, 934.
19 Udodong, U. E.; Srinivas Rao, C.; Fraser-Reid, B. *Tetrahedron* **1992**, *48*, 4713.
20 Claudemans, C. P. J.; Bertolini, M. J. In *Methods in Carbohydrate Chemistry*; R. J. Whistler and J. N. BeMiller, Ed.; Academic Press: New York, 1980; Vol. XIII; pp 272.
21 Flandor, J.; Garciá-Lopéz, M. T.; de Las Heras, F. G.; Méndez-Castrillón, P. P. *Synthesis* **1985**, 1121.
22 Johnson, F.; Starkovsky, N. A.; Paton, A. C.; Carlson, A. A. *J. Am. Chem. Soc.* **1964**, *86*, 118.
23 Cook, A. F.; Maichuk, D. T. *J. Org. Chem.* **1970**, *35*, 1940.
24 van Boeckel, C. A. A.; Beetz, T. *Tetrahedron Lett.* **1983**, *24*, 3775.
25 Smith, A. B.; Hale, K. J.; Vaccaro, H. A.; Rivero, R. A. *J. Am. Chem. Soc.* **1991**, *113*, 2112.
26 Laumen, K.; Reimerdes, E. H.; Schneider, M. P. *Tetrahedron Lett.* **1985**, *26*, 407.
27 Deardorff, D. R.; Matthews, A. J.; McMeekin, D. S.; Craney, C. L. *Tetrahedron Lett.* **1986**, *27*, 1255.
28 Theil, F.; Ballschuh, S.; Schick, H.; Haupt, M.; Hafner, B.; Schwarz, S. *Synthesis* **1988**, 540.
29 Therisod, M.; Klibanov, A. M. *J. Am. Chem. Soc.* **1986**, *108*, 5638.
30 Gais, H.-J.; Hemmerle, H.; Kossek, S. *Synthesis* **1991**, 169.
31 Pottie, M.; Van der Eycken, J.; Vandewalle, M. *Tetrahedron Asymmetry* **1991**, *2*, 329.
32 Degueil-Castaing, M.; De Jeso, B.; Drouillard, S.; Maillard, B. *Tetrahedron Lett.* **1987**, *28*, 953.
33 Wang, Y.-F.; Lalonde, J. J.; Momongan, M.; Bergbreiter, D.; Wong, C.-H. *J. Am. Chem. Soc.* **1988**, *110*, 7200.
34 Burgess, K.; Jennings, L. D. *J. Am. Chem. Soc.* **1991**, *113*, 6129.
35 Morgan, B.; Oehlschlager, A. C.; Stokes, T. M. *Tetrahedron* **1991**, *47*, 1611.
36 Laumen, K.; Breitgoff, D.; Schneider, M. P. *J. Chem. Soc., Chem. Commun.* **1988**, 1459.
37 Stokes, T. M.; Oehlschlager, A. C. *Tetrahedron Lett.* **1987**, *28*, 2091.
38 Belan, A.; Bolte, J.; Fauve, A.; Goucy, J. G.; Veschambre, H. *J. Org. Chem.* **1987**, *52*, 256.
39 Margolin, A. L.; Delinck, D. L.; Whalon, M. R. *J. Am. Chem. Soc.* **1990**, *112*, 2849.
40 Therisod, M.; Klibanov, A. M. *J. Am. Chem. Soc.* **1987**, *109*, 3977.
41 VanMiddlesworth, F.; Lopez, M.; Zweerink, M.; Edison, A. M.; Wilson, K. *J. Org. Chem.* **1992**, *57*, 4753.
42 Cunico, R. F.; Bedell, L. *J. Org. Chem.* **1980**, *45*, 4797.
43 Askin, D.; Angst, C.; Danishefsky, S. J. *J. Org. Chem.* **1987**, *52*, 622.
44 Fujiwara, S.; Smith, A. B. *Tetrahedron Lett.* **1992**, *33*, 1185.
45 Young, S. D.; Buse, C. T.; Heathcock, C. H. *Org. Synth. Coll. Vol. VII* **1990**, 381.
46 Hassig, R.; Siegel, H.; Seebach, D. *Chem. Ber.* **1982**, *115*, 1990.
47 Kerwin, S. M.; Paul, A. G.; Heathcock, C. H. *J. Org. Chem.* **1987**, *52*, 1686.
48 Evans, D. A.; Kaldor, S. W.; Jones, T. K.; Clardy, J.; Stout, T. J. *J. Am. Chem. Soc.* **1990**, *112*, 7001.
49 Seebach, D.; Chow, H.-F.; Jackson, R. F. W.; Sutter, M. A.; Thaisrivongs, S.; Zimmermann, J. *Liebigs Ann. Chem.* **1986**, 1281.
50 Boschelli, D.; Takemasa, T.; Nishitani, Y.; Masamune, S. *Tetrahedron Lett.* **1985**, *26*, 5239.
51 Smith, A. B.; Rivero, R. A.; Hale, K. J.; Vaccaro, H. A. *J. Am. Chem. Soc.* **1991**, *113*, 2092.
52 Oppolzer, W.; Snowden, R.; Simmons, D. P. *Helv. Chim. Acta* **1981**, *64*, 2002.
53 Roush, W. R.; Russo-Rodriguez, S. *J. Org. Chem.* **1987**, *52*, 598.
54 Heathcock, C. H.; Young, S. D.; Hagen, J. P.; Pilli, R.; Badertscher, U. *J. Org. Chem.* **1985**, *50*, 2095.
55 Masamune, S.; Lu, L. D.-L.; Jackson, W. P.; Kaiho, T.; Toyoda, T. *J. Am. Chem. Soc.* **1982**, *104*, 5523.
56 Corey, E. J.; Venkateswarlu, A. *J. Am. Chem. Soc.* **1972**, *94*, 6190.
57 Friesen, R. W.; Sturino, C. F.; Daljeet, A. K.; Kolaczewska, A. *J. Org. Chem.* **1991**, *56*, 1944.
58 Imanieh, H.; Quayle, P.; Voaden, M.; Conway, J.; Street, S. D. A. *Tetrahedron Lett.* **1992**, *33*, 543.
59 Corey, E. J.; Jones, G. B. *J. Org. Chem.* **1992**, *57*, 1028.
60 Kurokawa, N.; Ohfune, Y. *J. Am. Chem. Soc.* **1986**, *108*, 6041.
61 Marshall, J. A.; Sedrani, R. *J. Org. Chem.* **1991**, *56*, 5496.
62 Li, W.-R.; Ewing, W. R.; Harris, B. D.; Joullié, M. M. *J. Am. Chem. Soc.* **1990**, *112*, 7659.
63 White, J. D.; Kawasaki, M. *J. Am. Chem. Soc.* **1990**, *112*, 4991.
64 Prakash, C.; Saleh, S.; Blair, I. A. *Tetrahedron Lett.* **1989**, *30*, 19.
65 Robins, M. J.; Samano, V.; Johnson, M. D. *J. Org. Chem.* **1990**, *55*, 410.
66 Newton, R. F.; Reynolds, D. P.; Webb, C. F.; Roberts, S. M. *J. Chem. Soc., Perkin Trans. 1* **1981**, 2055.
67 Burke, S. D.; Cobb, J. E.; Takeuchi, K. *J. Org. Chem.* **1985**, *50*, 3421.
68 Shimshock, S. J.; Waltermire, R. E.; DeShong, P. *J. Am. Chem. Soc.* **1991**, *113*, 8791.
69 Pilcher, A. S.; Hill, D. K.; Shimshock, S. J.; Waltermire, R. E.; DeShong, P. *J. Org. Chem.* **1992**, *57*, 2492.
70 Pilcher, A. S.; DeShong, P. *J. Org. Chem.* **1993**, *58*, 5130.

71 Nicolaou, K. C.; Daines, R. A.; Chakraborty, T. K. *J. Am. Chem. Soc.* **1987**, *109*, 2208.
72 Paquette, L. A.; Doherty, A. M.; Rayner, C. M. *J. Am. Chem. Soc.* **1992**, *114*, 3910.
73 Schwesinger, R.; Link, R.; Thiele, G.; Rotter, H.; Honert, D.; Limbach, H. H.; Männle, F. *Angew. Chem. Int. Ed. Engl.* **1991**, *30*, 1372.
74 Seppelt, K. *Angew. Chem. Int. Ed. Engl.* **1992**, *31*, 292.
75 Toshima, T.; Tatsuta, K.; Kinoshita, M. *Bull. Chem. Soc. Jpn.* **1988**, *61*, 2369.
76 Otera, J.; Niibo, Y.; Nozaki, H. *Tetrahedron Lett.* **1992**, *33*, 3655.
77 Collington, E. W.; Finch, H.; Smith, I. J. *Tetrahedron Lett.* **1985**, *26*, 681.
78 Schmittling, E. A.; Sawyer, J. S. *Tetrahedron Lett.* **1991**, *32*, 7207.
79 Nakata, T.; Fukui, M.; Oishi, T. *Tetrahedron Lett.* **1988**, *29*, 2219.
80 Zhang, W.; Robins, M. J. *Tetrahedron Lett.* **1992**, *33*, 1177.
81 White, J. D.; Amedio, J. C.; Gut, S.; Ohira, S.; Jayasinghe, L. R. *J. Org. Chem.* **1992**, *57*, 2270.
82 Cormier, J. F. *Tetrahedron Lett.* **1991**, *32*, 187.
83 Toshima, K.; Yanagawa, K.; Mukaiyama, S.; Tatsuta, K. *Tetrahedron Lett.* **1990**, *31*, 6697.
84 Chaudary, S. C.; Hernandez, O. *Tetrahedron Lett.* **1979**, 99.
85 Suzuki, T.; Sato, E.; Unno, K.; Kametani, T. *J. Chem. Soc., Perkin Trans. 1* **1986**, 2263.
86 Corey, E. J.; Cho, H.; Rücker, C.; Hua, D. H. *Tetrahedron Lett.* **1981**, 3455.
87 Kocienski, P. J.; Love, C. J.; Whitby, R. J.; Costello, G.; Roberts, D. A. *Tetrahedron* **1989**, *45*, 3839.
88 Hikota, M.; Tone, H.; Horita, K.; Yonemitsu, O. *J. Org. Chem.* **1990**, *55*, 7.
89 Hudrlik, P. F.; Kulkarni, A. K. *Tetrahedron Lett.* **1985**, *26*, 1387.
90 Chamberlin, A. R.; Dezube, M.; Reich, S. H.; Sall, D. J. *J. Am. Chem. Soc.* **1989**, *111*, 6247.
91 Paquette, L. A.; Nitz, T. J.; Ross, R. J.; Springer, J. P. *J. Am. Chem. Soc.* **1984**, *106*, 1446.
92 Mander, L. N.; Sethi, S. P. *Tetrahedron Lett.* **1984**, *25*, 5953.
93 Howard, C.; Newton, R. F.; Reynolds, D. P.; Roberts, S. M. *J. Chem. Soc., Perkin Trans. 1* **1981**, 2049.
94 Mitsunobu, O. *Synthesis* **1981**, 1.
95 Clive, D. L. J.; Kellner, D. *Tetrahedron Lett.* **1991**, *32*, 7159.
96 Hanessian, S.; Levallee, P. *Canad. J. Chem.* **1975**, *53*, 2975.
97 Nicolaou, K. C.; Seitz, S. P.; Pavia, M. R.; Petasis, N. A. *J. Org. Chem.* **1979**, *44*, 4011.
98 Barrett, A. G. M.; Carr, R. A. E.; Attwood, S. V.; Richardson, G.; Walshe, N. D. A. *J. Org. Chem.* **1986**, *51*, 4840.
99 LeBlanc, Y.; Fitzsimmons, B. J.; Adams, J.; Perez, F.; Rokach, J. *J. Org. Chem.* **1986**, *51*, 789.
100 Hanessian, S.; Sumi, K. *Synthesis* **1991**, 1083.
101 Chiang, Y.-C. P.; Yang, S. S.; Heck, J. V.; Chabala, J. C.; Chang, M. N. *J. Org. Chem.* **1989**, *54*, 5708.
102 Nicolaou, K. C.; Pavia, M. R.; Seitz, S. P. *J. Am. Chem. Soc.* **1981**, *103*, 1224.
103 Ireland, R. E.; Obrecht, D. M. *Helv. Chim. Acta* **1986**, *69*, 1272.
104 Ogilvie, K. K.; Thompson, E. A.; Quilliam, M. A.; Westmore, J. B. *Tetrahedron Lett.* **1974**, 2865.
105 Bennett, F.; Knight, D. W.; Fenton, G. *J. Chem. Soc., Perkin Trans. 1* **1991**, 1543.
106 Nakatsuka, M.; Ragan, J. A.; Sammakia, T.; Smith, D. B.; Uehling, D. B.; Schreiber, S. L. *J. Am. Chem. Soc.* **1990**, *112*, 5583.
107 Ragan, J. A.; Standaert, R. F.; Schreiber, S. L. In *Strategies and Tactics in Organic Synthesis*; T. Lindberg, Ed.; Academic Press: San Diego, 1991; Vol. 3; pp 418.
108 Wetter, H.; Oertle, K. *Tetrahedron Lett.* **1985**, *26*, 5515.
109 Franck-Neumann, M.; Miesch, M.; Gross, L. *Tetrahedron Lett.* **1991**, *32*, 2135.
110 Le Drian, C.; Greene, A. E. *J. Am. Chem. Soc.* **1982**, *104*, 5473.
111 Jung, M. E.; Lyster, M. A. *Org. Synth. Coll. Vol. VI* **1988**, 353.
112 Ho, T.-L.; Olah, G. A. *Synthesis* **1977**, 417.
113 Olah, G. A.; Husain, A.; Balaram Gupta, B. G.; Narang, S. C. *Angew. Chem. Int. Ed. Engl.* **1981**, *20*, 690.
114 Jung, M. E.; Blumenkopf, T. A. *Tetrahedron Lett.* **1978**, 3657.
115 Olah, G. A.; Prakach, G. K. S.; Krishnamurthy, R. In *Advances in Silicon Chemistry*; G. L. Larson, Ed.; JAI Press: 1991; Vol. 1.
116 Olah, G. A.; Narang, S. C. *Tetrahedron* **1982**, *38*, 2225.
117 Groutas, W. C.; Felker, D. *Synthesis* **1980**, 861.
118 Node, M.; Hori, H.; Fujita, E. *J. Chem. Soc., Perkin Trans.1* **1976**, 2237.
119 Grieco, P. A.; Ferriño, S.; Vidari, G. *J. Am. Chem. Soc.*, **1980**, *102*, 7586.
120 McOmie, J. F. W.; West, D. E. *Org. Synth., Coll. Vol. V* **1973**, 412.
121 Vickery, E. H.; Pahler, L. F.; Eisenbraun, E. J. *J. Org. Chem.* **1979**, *44*, 4444.
122 Williard, P. G.; Fryhle, C. B. *Tetrahedron Lett.* **1980**, *21*, 3731.
123 Inaba, T.; Umezawa, I.; Yuasa, M.; Inoue, T.; Mihashi, S.; Itokawa, H.; Ogura, K. *J. Org. Chem.* **1987**, *52*, 2957.
124 Nagaoka, H.; Schmid, G.; Iio, H.; Kishi, Y. *Tetrahedron Lett.* **1981**, *22*, 899.
125 Parker, K. A.; Petraitis, J. J. *Tetrahedron Lett.* **1981**, *22*, 397.

126 Schow, S. R.; Bloom, J. D.; Thompson, A. S.; Winzenburg, K. N.; Smith, A. B. *J. Am. Chem. Soc.* **1986**, *108*, 2662.
127 Mirrington, R. N.; Feutrill, G. I. *Org. Synth., Coll. Vol. VI* **1988**, 859.
128 Bernard, A. M.; Ghiani, M. R.; Piras, P. P.; Rivoldini, A. *Synthesis* **1989**, 287.
129 Tomioka, K.; Kanai, M.; Koga, K. *Tetrahedron Lett.* **1991**, *32*, 2395.
130 Seebach, D.; Kalinowski, H.-O.; Langer, W.; Crass, G.; Wilka, E.-M. *Org. Synth., Coll. Vol. VII* **1990**, 41.
131 Kalinowski, H.-O.; Seebach, D.; Crass, G. *Angew. Chem. Int. Ed. Engl.* **1975**, *14*, 762.
132 Kocienski, P.; Stocks, M.; Donald, D.; Perry, M. *Synlett* **1990**, 38.
133 Pearlman, B. A. *J. Am. Chem. Soc.* **1979**, *101*, 6404.
134 Ley, S. V.; Anthony, N. J.; Armstrong, A.; Brasca, M. G.; Clarke, T.; Culshaw, D.; Greck, C.; Grice, P.; Jones, A. B.; Lygo, B.; Madin, A.; Sheppard, R. N.; Slawin, A. M. Z.; Williams, D. J. *Tetrahedron* **1989**, *45*, 7161.
135 Neeman, M.; Johnson, W. S. *Org. Synth., Coll. Vol. V* **1973**, 245.
136 Meerwein, H.; Hinz, G. *Liebigs Ann. Chem.* **1930**, *484*, 1.
137 Newman, M. S.; Beal, P. F. *J. Am. Chem. Soc.* **1950**, *72*, 5161.
138 Ohno, K.; Nishiyama, H.; Nagase, H. *Tetrahedron Lett.* **1979**, 4405.
139 Hoffmann, R. W.; Schlapbach, A. *Tetrahedron* **1992**, *48*, 1959.
140 Gateau-Olesker, A.; Cléophax, J.; Géro, S. D. *Tetrahedron Lett.* **1986**, *27*, 41.
141 Horita, K.; Yoshioka, T.; Tanaka, T.; Oikawa, Y.; Yonemitsu, O. *Tetrahedron* **1986**, *42*, 3021.
142 Oikawa, Y.; Tanaka, T.; Yonemitsu, O. *Tetrahedron Lett.* **1986**, *27*, 3647.
143 Freifelder, M. *Catalytic Hydrogenation in Organic Synthesis*; Wiley: New York, 1978, pp p. 109.
144 Willson, T. M.; Kocienski, P.; Jarowicki, K.; Isaac, K.; Hitchcock, P. M.; Faller, A.; Campbell, S. F. *Tetrahedron* **1990**, *46*, 1767.
145 Bommer, R.; Kinzy, W.; Schmidt, R. R. *Liebigs Ann. Chem.* **1991**, 425.
146 Brieger, G.; Nestrick, T. J. *Chem. Rev.* **1974**, *74*, 567.
147 Entwistle, I. D.; Johnstone, R. A. W.; Whitby, A. H. *Chem. Rev.* **1985**, *85*, 129.
148 Bodanszky, M.; Bodanszky, A. *The Practice of Peptide Synthesis*; Springer-Verlag: Berlin, 1984, pp p. 158.
149 Hanessian, S.; Liak, T. J.; Vanasse, B. *Synthesis* **1981**, 396.
150 Felix, A. M.; Heimer, E. P.; Lambros, T. J.; Tzougraki, C.; Meienhofer, J. *J. Org. Chem.* **1976**, *43*, 4194.
151 Bieg, T.; Szeja, W. *Synthesis* **1985**, 76.
152 Jung, M. T.; Street, L. J. *J. Am. Chem. Soc.* **1984**, *106*, 8327.
153 Bollitt, V.; Mioskowski, C.; Kollah, R. M.; Manna, S.; Rajapaksa, D.; Falck, J. R. *J. Am. Chem. Soc.* **1991**, *113*, 6320.
154 Kocienski, P.; Street, S. D. A.; Yeates, C.; Campbell, S. F. *J. Chem. Soc., Perkin Trans. 1* **1987**, 2171.
155 Kende, A. S.; Mendoza, J. S.; Fujii, Y. *Tetrahedron* **1993**, *49*, 8015.
156 Hwu, J. R.; Chua, V.; Schroeder, J. E.; Barrans, R. E.; Khoudary, K. P.; Wang, N.; Wetzel, J. M. *J. Org. Chem.* **1986**, *51*, 4731.
157 Freeman, P. K.; Hutchinson, L. L. *J. Org. Chem.* **1980**, *45*, 1924.
158 Ireland, R.; Smith, M. G. *J. Am. Chem. Soc.* **1988**, *110*, 854.
159 Jung, M. E.; Lyster, M. A. *J. Org. Chem.* **1977**, *42*, 3761.
160 Hori, H.; Nishida, Y.; Ohrui, H.; Meguro, H. *J. Org. Chem.* **1989**, *54*, 1346.
161 Nicolaou, K. C.; Pavia, M. R.; Seitz, S. P. *J. Am. Chem. Soc.* **1982**, *104*, 2027.
162 Williams, D. R.; Brown, D. L.; Benbow, J. W. *J. Am. Chem. Soc.* **1989**, *111*, 1923.
163 Park, M. H.; Takeda, R.; Nakanishi, K. *Tetrahedron Lett.* **1987**, *28*, 3823.
164 Schmidt, U.; Meyer, R.; Leitenberger, V.; Griesser, H.; Liebersknecht, A. *Synthesis* **1992**, 1025.
165 Overman, L. E.; Fukaya, C. *J. Am. Chem. Soc.* **1980**, *102*, 1454.
166 Fuji, K.; Ichikawa, K.; Node, M.; Fujita, E. *J. Org. Chem.* **1979**, *44*, 1661.
167 Tanaka, T.; Oikawa, Y.; Nakajima, N.; Hamada, T.; Yonemitsu, O. *Chem. Pharm. Bull.* **1987**, *35*, 2205.
168 Sviridov, A. F.; Ermolenko, M. S.; Yashunsky, D. V.; Borodkin, V. S.; Kochetkov, N. K. *Tetrahedron Lett.* **1987**, *28*, 3839.
169 Vedejs, E.; Buchanan, R. A.; Watanabe, Y. *J. Am. Chem. Soc.* **1989**, *111*, 8430.
170 Ikemoto, N.; Schreiber, S. L. *J. Am. Chem. Soc.* **1992**, *114*, 2524.
171 Naidu, M. V.; Venkama Rao, G. S.; Krishna Rao, G. S. *Synthesis* **1979**, 144.
172 Corrie, J. E. T.; Hlubucek, J. R.; Lowe, G. *J. Chem. Soc., Perkin Trans. 1* **1977**, 1421.
173 Sakakibara, S.; Shimonishi, Y.; Kishida, Y.; Okada, M.; Sugihara, H. *Bull. Chem Soc. Jpn.* **1967**, *40*, 2164.
174 Munson, M. C.; Garcia-Echeverria, C.; Albericio, F.; Barany, G. *J. Org. Chem.* **1992**, *57*, 3013.
175 Nicolaou, K. C.; Liu, J. J.; Hwang, C.-K.; Dai, W.-M.; Guy, R. K. *J. Chem. Soc., Chem. Commun.* **1992**, 1118.
176 Evans, D. A.; Bender, S. L.; Morris, J. *J. Am. Chem. Soc.* **1988**, *110*, 2506.
177 Mori, S.; Ohno, T.; Harada, H.; Aoyama, T.; Shioiri, T. *Tetrahedron* **1991**, *47*, 5051.
178 Van Hijfte, L.; Little, R. D. *J. Org. Chem.* **1985**, *50*, 3940.
179 Baker, R.; Leeson, P. D.; Liverton, N. J.; Kulagowski, J. J. *J. Chem. Soc., Chem. Commun.* **1990**, 462.

180 Posner, G. H.; Rogers, D. Z. *J. Am. Chem. Soc.* **1977**, *99*, 8208.
181 Chen, S.-H.; Horvath, R. F.; Joglar, J.; Fisher, M. J.; Danishefsky, S. J. *J. Org. Chem.* **1991**, *56*, 5834.
182 Cruzado, C.; Bernabe, M.; Martin-Lomas, M. *J. Org. Chem.* **1989**, *54*, 465.
183 Wessel, H.-P.; Iversen, T.; Bundle, D. R. *J. Chem. Soc., Perkin Trans. 1* **1985**, 2247.
184 Widmer, U. *Synthesis* **1987**, 568.
185 Nicolaou, K. C.; Schreiner, E. P.; Stahl, W. *Angew. Chem. Int. Ed. Engl.* **1991**, *30*, 585.
186 Creary, X. *Org. Synth.* **1985**, *64*, 207.
187 Liotta, L. J.; Ganem, B. *Tetrahedron Lett.* **1989**, *30*, 4759.
188 Lipták, A. *Carbohydr. Res.* **1978**, *53*, 69.
189 Curtis, N. R.; Holmes, A. B.; Looney, M. G. *Tetrahedron Lett.* **1992**, *33*, 671.
190 Dushin, R. G.; Danishefsky, S. J. *J. Am. Chem. Soc.* **1992**, *114*, 655.
191 Oikawa, Y.; Yoshioka, T.; Yonemitsu, O. *Tetrahedron Lett.* **1982**, *23*, 885.
192 Johansson, R.; Samuelsson, B. *J. Chem. Soc., Perkin Trans. 1* **1984**, 2371.
193 Georg, G. I.; Mashava, P. M.; Akgün, E.; Milstead, M. W. *Tetrahedron Lett.* **1991**, *32*, 3151.
194 Wang, Y.; Babirad, S. A.; Kishi, Y. *J. Org. Chem.* **1992**, *57*, 468.
195 Classon, B.; Garegg, P. J.; Samuelsson, B. *Acta Chem. Scand., Ser. B* **1984**, 419.
196 Hébert, N.; Beck, A.; Lennox, R. B.; Just, G. *J. Org. Chem.* **1992**, *57*, 1777.
197 Nakajima, N.; Horita, K.; Abe, R.; Yonemitsu, O. *Tetrahedron Lett.* **1988**, *29*, 4139.
198 Wanner, M. J.; Willard, N. P.; Koomen, G.-J.; Pandit, U. K. *Tetrahedron* **1987**, *43*, 2549.
199 Khorana, H. G. *Pure. Appl. Chem.* **1968**, *17*, 349.
200 Khorana, H. G. *Biochem. J.* **1968**, *109*, 709.
201 Beaucage, S. L.; Iyer, R. P. *Tetrahedron* **1992**, *48*, 2223.
202 Bessodes, M.; Komiotis, D.; Antonakis, K. *Tetrahedron Lett.* **1986**, *27*, 579.
203 Helferich, B. *Adv. Carbohydr. Chem. Biochem.* **1948**, *3*, 79.
204 García, M. L.; Pascual, J.; Borràs, L.; Andreu, J. A.; Fos, E.; Mauleón, D.; Carganico, G.; Arcamone, F. *Tetrahedron* **1991**, *47*, 10023.
205 Hirama, M.; Noda, T.; Yasuda, S.; Ito, S. *J. Am. Chem. Soc.* **1991**, *113*, 1830.
206 Prinz, H.; Six, L.; Ruess, K.-P.; Liefländer, M. *Liebigs Ann. Chem.* **1985**, 217.
207 Kim, K. S.; Song, Y. H.; Lee, B. H.; Hahn, C. S. *J. Org. Chem.* **1986**, *51*, 404.
208 Lok, C. M.; Ward, J. P.; van Dorp, D. A. *Chem. Phys. Lipids* **1976**, *16*, 115.
209 van Boeckel, C. A. A.; van Boom, J. H. *Tetrahedron* **1985**, *41*, 4545.
210 Hanessian, S.; Cooke, N. G.; Dehoff, B.; Sakito, Y. *J. Am. Chem. Soc.* **1990**, *112*, 5276.
211 Ireland, R. E.; Anderson, R. C.; Badoud, R.; Fitzsimmons, B. J.; McGarvey, G. J.; Thaisrivongs, S.; Wilcox, C. S. *J. Am. Chem. Soc.* **1983**, *105*, 1988.
212 Gaffney, P. R. J.; Changsheng, L.; Vaman Rao, M.; Reese, C. B.; Ward, J. G. *J. Chem. Soc., Perkin Trans. 1* **1991**, 1355.
213 Myers, A. G.; Dragovich, P. S. *J. Am. Chem. Soc.* **1992**, *114*, 5859.
214 Smith, M.; Rammler, D. H.; Goldberg, I. H.; Khorana, H. G. *J. Am. Chem. Soc.* **1968**, *84*, 430.
215 Engels, J. *Angew. Chem. Int. Ed. Engl.* **1979**, *18*, 148.
216 Mateucci, M. D.; Caruthers, M. H. *Tetrahedron Lett.* **1980**, *21*, 3243.
217 Kohli, V.; Blöcker, H.; Köster, H. *Tetrahedron Lett.* **1980**, *21*, 2683.
218 Reese, C. B.; Ward, J. G. *Tetrahedron Lett.* **1987**, *28*, 2309.
219 Sekine, M.; Hata, T. *J. Org. Chem.* **1983**, *48*, 3011.
220 Hiskey, R. G. In *The Peptides*; E. Gross and J. Meienhofer, Ed.; Academic Press: New York, 1981; Vol. 3; pp 137.
221 Hiskey, R. G. *J. Org. Chem.* **1966**, *31*, 1188.
222 Leanza, W. J.; DeNinno, F.; Muthard, D. A.; Wilkening, R. R.; Wildonger, K. J.; Ratcliffe, R. W.; Christensen, B. G. *Tetrahedron* **1983**, *39*, 2505.
223 Volkmann, R. A.; O'Neill, B. T. In *Strategies and Tactics in Organic Synthesis*; T. Lindberg, Ed.; Academic Press: Orlando, 1991; Vol. 3; pp 495.
224 Colin-Messager, S.; Girard, J.-P.; Rossi, J.-C. *Tetrahedron Lett.* **1992**, *33*, 2689.
225 Mulzer, J.; Kirstein, H. M.; Buschmann, J.; Lehmann, C.; Luger, P. *J. Am. Chem. Soc.* **1991**, *113*, 910.
226 Hanessian, S.; Staub, A. P. A. *Tetrahedron Lett.* **1973**, 3555.
227 Kobayashi, S.; Murakami, M.; Mukaiyama, T. *Chem. Lett.* **1985**, 1535.
228 Paulsen, H.; Adermann, K. *Liebigs Ann. Chem.* **1989**, 751.
229 Alexakis, A.; Gardette, M.; Colin, S. *Tetrahedron Lett.* **1988**, *29*, 2951.
230 Ganem, B.; Small, V. R. *J. Org. Chem.* **1974**, *39*, 3728.
231 Forsyth, C. J.; Clardy, J. *J. Am. Chem. Soc.* **1988**, *110*, 5911.
232 Schlessinger, R. H.; Nugent, R. A. *J. Am. Chem. Soc.* **1982**, *104*, 1116.
233 Baggiolini, E. G.; Iacobelli, J. A.; Hennessy, B. M.; Uskokovic, M. R. *J. Am. Chem. Soc.* **1982**, *104*, 2945.

234 Wünsch, E.; Jentsch, J. *Chem. Ber.* **1964**, *97*, 2490.
235 Cohen, N.; Eichel, W. F.; Lopresti, R. J.; Neukom, C.; Saucy, G. *J. Org. Chem.* **1976**, *41*, 3505.
236 Armstrong, A.; Brackenridge, I.; Jackson, R. F. W.; Kirk, J. M. *Tetrahedron Lett.* **1988**, *29*, 2483.
237 Gigg, J.; Gigg, R. *J. Chem. Soc. (C)* **1966**, 82.
238 Ireland, R. E.; Norbeck, D. W. *J. Am. Chem. Soc.* **1985**, *107*, 3279.
239 Dreef, C. E.; Tuinman, R. J.; Lefeber, A. W. M.; Elle, C. J. J.; van der Marel, G. A.; van Boom, J. H. *Tetrahedron* **1991**, *47*, 4709.
240 Baudry, D.; Ephritikhine, M.; Felkin, H. *J. Chem. Soc., Chem. Commun.* **1978**, 694.
241 Schmidt, R. R. *Angew. Chem. Int. Ed. Engl.* **1986**, *25*, 212.
242 Schmidt, R. R. *Pure Appl. Chem.* **1989**, 1257.
243 Vankar, Y. D.; Vankar, P. S.; Behrendt, M.; Schmidt, R. R. *Tetrahedron* **1991**, *47*, 9985.
244 Lamberth, C.; Bednarski, M. D. *Tetrahedron Lett.* **1991**, *32*, 7369.
245 Oltvoort, J. J.; van Boeckel, C. A. A.; de Koning, J. H.; van Boom, J. H. *Synthesis* **1981**, 305.
246 Warren, C. D.; Jeanloz, R. W. *Carbohydr. Res.* **1977**, *53*, 67.
247 Nishiguchi, T.; Tachi, K.; Fukuzumi, K. *J. Org. Chem.* **1975**, *40*, 237.
248 van Boeckel, C. A. A.; van Boom, J. H. *Tetrahedron Lett.* **1979**, 3561.
249 Yin, H.; Franck, R. W.; Chen, S.-L.; Quigley, G. J.; Todaro, L. *J. Org. Chem.* **1992**, *57*, 644.
250 Guibé, F.; Dangles, O.; Balavoine, G. *Tetrahedron Lett.* **1986**, *27*, 2365.
251 Evans, D. A.; Ellman, J. A.; DeVries, K. M. *J. Am. Chem. Soc.* **1989**, *111*, 8912.
252 Teshima, T.; Nakajima, K.; Takahashi, M.; Shiba, T. *Tetrahedron Lett.* **1992**, *33*, 363.
253 Kunz, H.; Unverzagt, C. *Angew. Chem. Int. Ed. Engl.* **1984**, *23*, 436.
254 Boullanger, P.; Chatelard, P.; Descotes, G.; Kloosterman, M.; van Boom, J. H. *J. Carbohydr. Chem.* **1986**, *5*, 541.
255 Corey, E. J.; Suggs, J. W. *J. Org. Chem.* **1973**, *38*, 3224.
256 Ando, T.; Yamawaki, J.; Kawate, T.; Sumi, S.; Hanafusa, T. *Bull. Chem. Soc. Jpn.* **1982**, *55*, 2504.
257 Schaubach, R.; Hemberger, J.; Kinzy, W. *Liebigs Ann. Chem.* **1991**, 607.
258 Blunden, S. J.; Cusack, P. A.; Smith, P. J. *J. Organometal. Chem.* **1987**, *325*, 141.
259 David, S.; Pascard, C.; Cesaria, M. *Nouveau J. Chim.* **1979**, *3*, 63.
260 David, S.; Hanessian, S. *Tetrahedron* **1985**, *41*, 643.
261 Oltvoort, J. J.; M., K.; van Boom, J. H. *Rec'l. Trav. Chim. Pays-Bas* **1983**, *102*, 501.
262 Corey, E. J.; Cheng, X.-M. *The Logic of Chemical Synthesis*; Wiley Interscience: New York, 1989, pp 436.
263 Wender, P. A.; Correira, C. R. D. *J. Am. Chem. Soc.* **1987**, *109*, 2523.
264 Gibbons, E. G. *J. Am. Chem. Soc.* **1982**, *104*, 1767.
265 Hanessian, S.; Deloure, D.; Dufresne, Y. *Tetrahedron Lett.* **1984**, *25*, 2515.
266 Ishihara, J.; Tomita, K.; Tadano, K.; Ogawa, S. *J. Org. Chem.* **1992**, *57*, 3789.
267 Kawasaki, M.; Matsuda, F.; Terashima, S. *Tetrahedron* **1988**, *44*, 5695.
268 Goff, D. A.; Harris, R. N.; Bottaro, J. C.; Bedford, C. D. *J. Org. Chem.* **1986**, *51*, 4711.
269 Corey, E. J.; Ghosh, A. K. *Tetrahedron Lett.* **1988**, *29*, 3205.
270 Vedejs, E.; Larsen, S. D. *J. Am. Chem. Soc.* **1984**, *106*, 3030.
271 Corey, E. J.; Hua, D. H.; Pan, B.-C.; Seitz, S. P. *J. Am. Chem. Soc.* **1982**, *104*, 6818.
272 Boeckman, R. K.; Potenza, J. C. *Tetrahedron Lett.* **1985**, *26*, 1411.
273 Linderman, R. J.; Graves, D. M.; Kwochka, W. R.; Ghannam, A. F.; Anklekar, T. V. *J. Am. Chem. Soc.* **1990**, *112*, 7438.
274 Kluge, A. F.; Untch, K. G.; Fried, J. H. *J. Am. Chem. Soc.* **1972**, *94*, 7827.
275 Stork, G.; Takahashi, T. *J. Am. Chem. Soc.* **1977**, *99*, 1275.
276 Askin, D.; Volante, R. P.; Reamer, R. A.; Ryan, K. M.; Shinkai, I. *Tetrahedron Lett.* **1988**, *29*, 277.
277 Narasaka, K.; Sakakura, T.; Uchimaru, T.; Guédin-Vuong, D. *J. Am. Chem. Soc.* **1984**, *106*, 2954.
278 David, S.; Thieffry, A.; Veyrières, A. *J. Chem. Soc., Perkin Trans. 1* **1981**, 1796.
279 Fuji, K.; Nakano, S.; Fujita, E. *Synthesis* **1975**, 276.
280 Gras, J.-L.; Kong Win Chang, Y.-Y.; Chang, L.; Guerin, A. *Synthesis* **1985**, 74.
281 Groziak, M. P.; Koohang, A. *J. Org. Chem.* **1992**, *57*, 940.
282 Corey, E. J.; Bock, M. *Tetrahedron Lett.* **1975**, 3269.
283 Corey, E. J.; Hopkins, P. B.; Kim, S.; Yoo, S.-e.; Nambiar, K. P.; Falck, J. R. *J. Am. Chem. Soc.* **1979**, *101*, 7131.
284 Schomburg, D.; Hopkins, P. B.; Lipscomb, W. N.; Corey, E. J. *J. Org. Chem.* **1980**, *45*, 1544.
285 Corey, E. J.; Wollenberg, R. H.; Williams, D. R. *Tetrahedron Lett.* **1977**, 2243.
286 Wachter, M. P.; Adams, R. E. *Synth. Commun.* **1980**, *10*, 111.
287 Yamada, K.; Kato, K.; Nagase, H.; Hirata, Y. *Tetrahedron Lett.* **1976**, 65.
288 Corey, E. J.; Hua, D. H.; Pan, B.-C.; Seitz, S. P. *J. Am. Chem. Soc.* **1982**, *104*, 6818.
289 Keck, G. E.; Boden, E. P.; Wiley, M. R. *J. Org. Chem.* **1989**, *54*, 896.
290 Pojer, P. M.; Angyal, S. J. *Aust. J. Chem.* **1978**, *31*, 1031.

291 Niwa, H.; Miyachi, Y.; Okamoto, O.; Uosaki, Y.; Kuroda, A.; Ishiwata, H.; Yamada, K. *Tetrahedron* **1992**, *48*, 393.
292 Brown, J. M.; Christodoulou, C.; Jones, S. S.; Modak, A. S.; Reese, C. B.; Sibanda, S.; Ubasawa, A. *J. Chem. Soc., Perkin Trans. 1* **1989**, 1735.
293 Brown, J. M.; Christodoulou, C.; Modak, A. S.; Reese, C. B.; Serafinowska, H. T. *J. Chem. Soc., Perkin Trans. 1* **1989**, 1751.
294 Suzuki, K.; Inanaga, J.; Yamaguchi, M. *Chem. Lett.* **1979**, 1277.
295 Medina, J. C.; Saloman, M.; Kyler, K. S. *Tetrahedron Lett.* **1988**, *29*, 3773.
296 Corey, E. J.; Gras, J.-L.; Ulrich, P. *Tetrahedron Lett.* **1976**, 809.
297 Williams, D. R.; Jass, P. A.; Tse, H.-L. A.; Gaston, R. D. *J. Am. Chem. Soc.* **1990**, *112*, 4552.
298 Guindon, Y.; Morton, H. E.; Yoakim, C. *Tetrahedron Lett.* **1983**, *24*, 3969.
299 Saloman, R. G.; Sachinvala, N. D.; Roy, S.; Basu, B.; Raychauduri, S. R.; Miller, D. B.; Sharma, R. B. *J. Am. Chem. Soc.* **1991**, *113*, 3085.
300 Holton, R. A.; Juo, R. R.; Kim, H. B.; Williams, A. D.; Harusawa, S.; Lowenthal, R. E.; Yogai, S. *J. Am. Chem. Soc.* **1988**, *110*, 6558.
301 Holton, R. A. In *Strategies and Tactics in Organic Synthesis*; T. Lindberg, Ed.; Academic Press: San Diego, 1991; Vol. 3; pp 165.
302 Ranganathan, D.; Ranganathan, S.; Mehrotra, M. M. *Tetrahedron* **1980**, *36*, 1869.
303 Yamaguchi, K. *Bull. Chem. Soc. Jpn.* **1987**, *60*, 2169.
304 Rigby, J. H.; Wilson, J. *Tetrahedron Lett.* **1984**, *25*, 1429.
305 Gross, R. S.; Watt, D. S. *Synth. Commun.* **1987**, 1749.
306 Magnus, P.; Carter, P.; Elliot, J.; Lewis, R.; Harling, J.; Pitterna, T.; Bauta, W. E.; Fortt, S. *J. Am. Chem. Soc.* **1992**, *114*, 2544.
307 Shibasaki, M.; Ishida, Y.; Okabe, N. *Tetrahedron Lett.* **1985**, *26*, 2217.
308 Guindon, Y.; Yoakim, C.; Morton, H. E. *J. Org. Chem.* **1984**, *49*, 3912.
309 Finch, A.; Pearce, J. *Tetrahedron* **1964**, *20*, 175.
310 Williams, D. R.; Sakdarat, S. *Tetrahedron Lett.* **1983**, *24*, 3965.
311 Ireland, R. E.; Wuts, P. G. M.; Ernst, B. *J. Am. Chem. Soc.* **1981**, *103*, 3205.
312 Martinez, G. R.; Grieco, P. A.; Williams, E.; Kanai, K.; Srinivasan, C. V. *J. Am. Chem. Soc.* **1982**, *104*, 1436.
313 Anderson, R. J.; Adams, K. G.; Chinn, H. R.; Henrick, C. A. *J. Org. Chem.* **1980**, *45*, 2229.
314 Williams, J. R.; Callahan, J. F. *J. Org. Chem.* **1980**, *45*, 4479.
315 Stork, G.; Isobe, M. *J. Am. Chem. Soc.* **1975**, *97*, 4745.
316 Pikul, S.; Raczko, J.; Ankner, K.; Jurczak, J. *J. Am. Chem. Soc.* **1987**, *109*, 3981.
317 Nagaoka, H.; Rutsch, W.; Schmid, G.; Ito, H.; Johnson, M. R.; Kishi, Y. *J. Am. Chem. Soc.* **1980**, *102*, 7962.
318 Still, W. C.; Mobilio, D. *J. Org. Chem.* **1983**, *48*, 4785.
319 Hirama, M.; Uei, M. *J. Am. Chem. Soc.* **1982**, *104*, 4251.
320 Still, W. C.; Murata, S.; Revial, G.; Yoshihara, K. *J. Am. Chem. Soc.* **1983**, *105*, 625.
321 Tanner, D.; Somfai, P. *Tetrahedron* **1987**, *43*, 4395.
322 Elie, C. J. J.; Muntendam, H. J.; Elst, v. d.; van der Marel, G. A.; Hoogerhout, P.; van Boom, J. H. *Rec'l Trav. Chim. Pays-Bas* **1989**, *108*, 219.
323 Suzuki, K.; Tomooka, K.; Katayama, E.; Matsumoto, T.; Tsuchihashi, G.-i. *J. Am. Chem. Soc.* **1986**, *108*, 5221.
324 Roush, W. R.; Michaelis, M. R.; Tai, D. F.; Chong, W. K. M. *J. Am. Chem. Soc.* **1987**, *109*, 7575.
325 Kozikowski, A. P.; Wu, J.-P. *Tetrahedron Lett.* **1987**, *28*, 5125.
326 Benneche, T.; Strande, P.; Undheim, K. *Synthesis* **1983**, 762.
327 Lipshutz, B. M.; Pegram, J. J. *Tetrahedron Lett.* **1980**, *21*, 3343.
328 Gadwood, R. C.; Lett, R. M.; Wissinger, J. E. *J. Am. Chem. Soc.* **1984**, *106*, 3869.
329 Burke, S.; Pacofsky, G. J. *Tetrahedron Lett.* **1986**, *27*, 445.
330 Ditrich, K. *Liebigs Ann. Chem.* **1990**, 789.
331 Lipshutz, B. H.; Pegram, J. J.; Morey, M. C. *Tetrahedron Lett.* **1981**, *22*, 4603.
332 Lipshutz, B. H.; Harvey, D. F. *Synth. Commun.* **1982**, *12*, 267.
333 Kan, T.; Hashimoto, M.; Yanagiya, M.; Shirahama, H. *Tetrahedron Lett.* **1988**, *29*, 5417.
334 Lipshutz, B. H.; Miller, T. A. *Tetrahedron Lett.* **1989**, *30*, 7149.
335 Marino, J. P.; Dax, S. L. *J. Org. Chem.* **1984**, *49*, 3671.
336 Karim, S.; Parmee, E. R.; Thomas, E. J. *Tetrahedron Lett.* **1991**, *32*, 2269.
337 Ireland, R. E.; Varney, M. D. *J. Org. Chem.* **1986**, *51*, 635.
338 Lipshutz, B. H.; Moretti, R.; Crow, R. *Tetrahedron Lett.* **1989**, *30*, 15.
339 Parham, W. E.; Anderson, E. L. *J. Am. Chem. Soc.* **1948**, *70*, 4187.
340 Bernardy, K. F.; Floyd, M. B.; Poletto, J.; Weiss, M. J. *J. Org. Chem.* **1979**, *44*, 1438.
341 Corey, E. J.; Niwa, H.; Knolle, J. *J. Am. Chem. Soc.* **1978**, *100*, 1942.
342 Gala, D.; Steinmann, M.; Jaret, R. S. *J. Org. Chem.* **1986**, *51*, 4488.

343 Miyashita, M.; Yoshikoshi, A.; Grieco, P. A. *J. Org. Chem.* **1977**, *42*, 3772.
344 Beier, R.; Mundy, B. P. *Synth. Commun.* **1979**, *9*, 271.
345 Johnston, R. D.; Marston, C. R.; Krieger, P. E.; Goe, G. L. *Synthesis* **1988**, 393.
346 Kim, S.; Ho Park, J. *Tetrahedron Lett.* **1987**, *28*, 439.
347 Fadel, A.; Salaun, J. *Tetrahedron* **1985**, *41*, 1267.
348 Reese, C. B.; Saffhill, R.; Sulston, J. E. *J. Am. Chem. Soc.* **1967**, *89*, 3366.
349 Reese, C. B. *Tetrahedron* **1978**, *34*, 3143.
350 Amarnath, V.; Broom, A. D. *Chem. Rev.* **1977**, *77*, 183.
351 Ward, J. G.; Young, R. C. *Tetrahedron Lett.* **1988**, *29*, 6013.
352 Corey, E. J.; Marfat, A.; Munroe, J. E.; Kim, K. S.; Hopkins, P. B.; Brion, F. *Tetrahedron Lett.* **1981**, *22*, 1077.
353 Corey, E. J.; Pyne, S. G.; Su, W.-g. *Tetrahedron Lett.* **1983**, *24*, 4883.
354 Nakatani, K.; Arai, K.; Hirayama, N.; Matsuda, F.; Terashima, S. *Tetrahedron* **1992**, *48*, 633.
355 Lutyen, M.; Keese, R. *Angew. Chem. Int. Ed. Engl.* **1984**, *23*, 390.
356 Dommerholt, F. G.; Thijs, L.; Zwanenburg, B. *Tetrahedron Lett.* **1991**, *32*, 1499.
357 Olah, G. A.; Husain, A.; Singh, B. P. *Synthesis* **1985**, 703.
358 Morizawa, Y.; Mori, I.; Hiyama, T.; Nozaki, H. *Synthesis* **1981**, 899.
359 Bongini, A.; Cardillo, G.; Orena, M.; Sandri, S. *Synthesis* **1979**, 618.

# Chapter 3 Diol Protecting Groups

## 3.1 Introduction

Over a century has lapsed since Emil Fischer's pioneering research in carbohydrate chemistry. As the first polyfunctional natural products whose chemistry was systematically and extensively investigated, the carbohydrates occupy a commanding place in the historical development of such familiar themes as the anomeric effect, the stereochemical consequences of the tetrahedral carbon, diastereoisomerism, and selective functional group transformations. Fischer deserves pride of place in our discussion of protecting groups since he was probably the first to consciously exploit protecting group strategies as a synthetic tool. Indeed the first protecting group to emerge in aid of selective synthetic transformations was the isopropylidene acetal[1] Fischer's signal contribution has stood the test of time since acetals remain the principal means of preserving 1,2- and 1,3-diols. A number of reviews dealing with the synthesis and reactions of acetals from a wider perspective than functional group protection are given in section 3.5.

## 3.2 Acetals

Acetals reveal virtually unlimited stability to basic conditions but they are quite fragile towards acid — a property which derives from participation of a lone pair on the adjacent oxygen atom in the cleavage of a protonated intermediate as depicted in Scheme 1.6. In this chapter our discussion of acetals will only embrace matters pertinent to the protection of diols and polyols. The subject will reappear in Chapter 4 when we consider the protection of carbonyls.

### 3.2.1 Benzylidene Acetals

Benzylidene acetals are frequently used in carbohydrate chemistry. They are stable to most strong bases, mild oxidants, and metal hydrides (in the absence of Lewis acids) but they are readily attacked by *N*-bromosuccinimide and ozone — a fact which can be usefully exploited (*vide infra*). Benzylidene acetals are not stable to strong bases such as alkyllithium reagents[2] and they are hydrogenolysed in the presence of Pd or Pt. Lewis acids will also decompose them. A further useful feature is that they can be reductively cleaved to give an alcohol and a benzyl ether.

**(i) Cleavage**

Catalytic hydrogenation offers the mildest and one of the most efficient means for cleaving benzylidene acetals to toluene and the corresponding diol and the limitations to the method are largely those already noted for the cleavage of benzyl ethers [Scheme 3.1][3]. The analogy with benzyl ethers also extends to the use of Birch reduction involving Na or Li in liquid ammonia in the presence of a proton source such as *t*-BuOH. Despite the obvious narrow range of functional groups compatible with such powerfully reducing conditions, Birch reduction is still widely used[4].

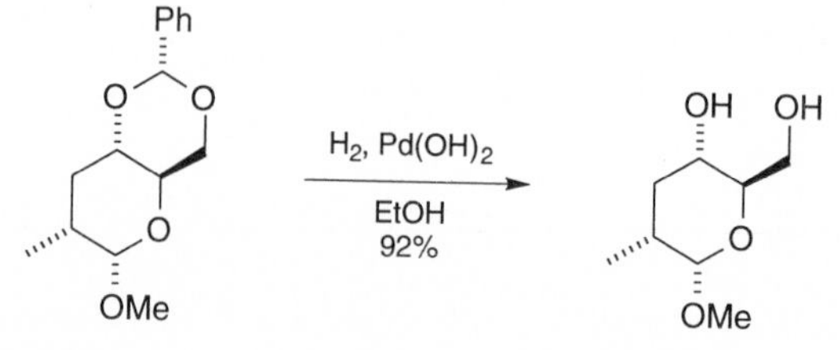

Ph, O, O, O, OMe → ($H_2$, $Pd(OH)_2$; EtOH; 92%) → OH, OH, O, OMe

**Scheme 3.1**

Acid-catalysed hydrolysis offers a cheap and obvious method [Scheme 3.2][5] for the cleavage of benzylidene acetals but for sensitive substrates transacetalisation using ethanethiol in the presence of a mild Lewis acid such as $Zn(OTf)_2$ is especially advantageous [Scheme 3.3][6].

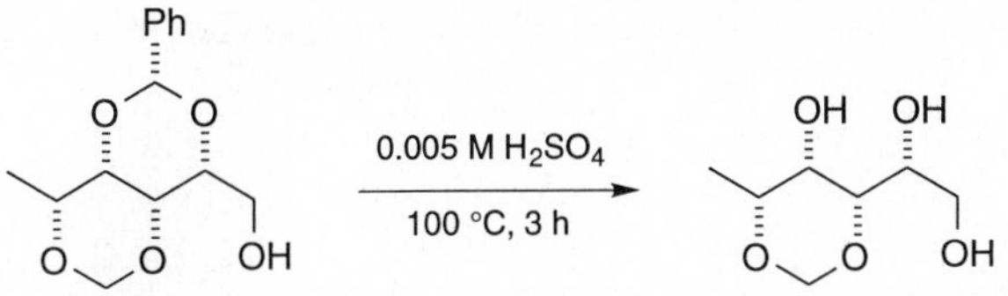

0.005 M $H_2SO_4$, 100 °C, 3 h

**Scheme 3.2**

EtSH (5 equiv.), $NaHCO_3$ (10 equiv.), $Zn(OTf)_2$ (3 equiv.), $CH_2Cl_2$, r.t., 5 h, 90%

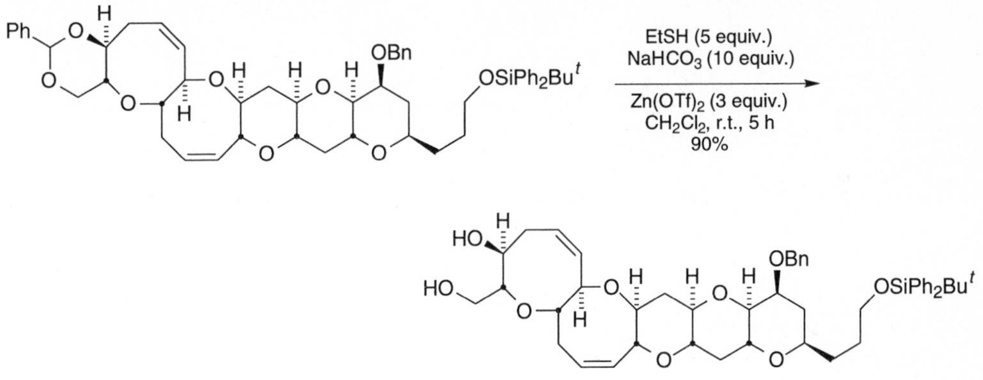

**Scheme 3.3**

Cleavage of the benzylidene protecting group can be effected with a range of other reagents which do not return the diol but derivatives thereof. For example, partial deprotection of unsymmetrical arylidene acetals is a useful procedure for selectively liberating only the less hindered hydroxyl for further manipulation and the method has found recent favor for the synthesis of polyketide chains. Chemical differentiation is accomplished by regioselective reductive cleavage of a benzylidene acetal with a Lewis acidic metal hydride such as DIBALH at room temperature or $LiAlH_4$–$AlCl_3$[7] to give the benzyl ether of the more hindered hydroxyl and a free hydroxyl group at the less hindered position. The reaction conditions are compatible with the presence of *tert*-butyldiphenylsilyl ethers [Scheme 3.4][8] and PMB ethers [Scheme 3.5][9]. With $NaBH_3CN$ and acid, benzylidene acetals cleave to give the benzyl ether of the less hindered alcohol[10]. Another combination of reagents which can be used to accomplish the selective cleavage of benzylidene acetals function is $BH_3$•$NMe_3$ and $AlCl_3$ as shown in Scheme 3.6[11,12].

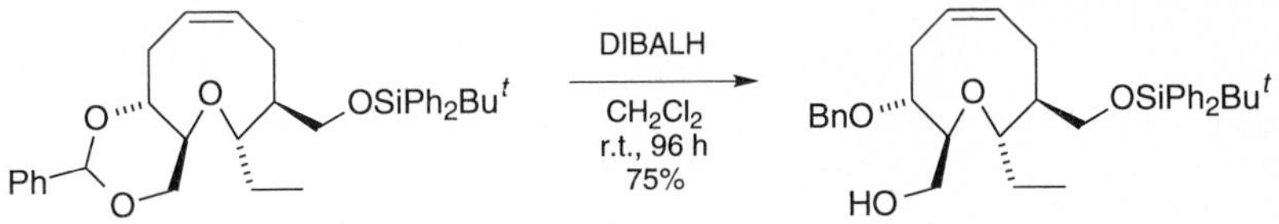

DIBALH, $CH_2Cl_2$, r.t., 96 h, 75%

**Scheme 3.4**

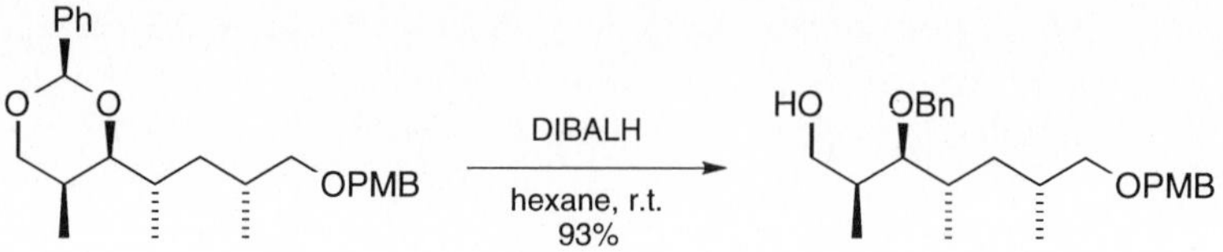

**Scheme 3.5**

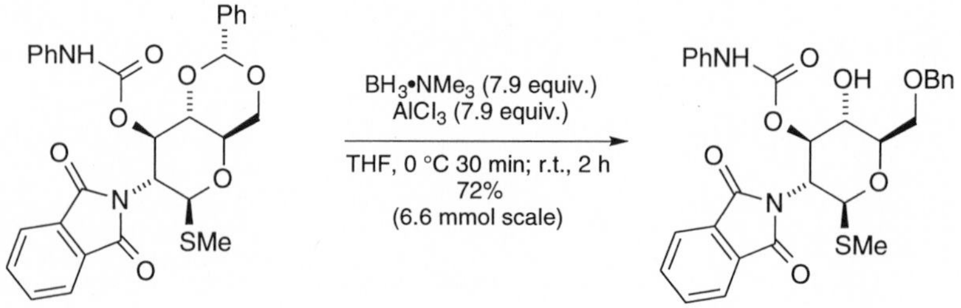

**Scheme 3.6**

The introduction of alkoxy groups onto the aromatic ring gives enhanced lability to the acetal function. Thus, *p*-methoxybenzylidene acetals, which hydrolyse approximately 10 times faster than the corresponding benzylidene acetals, may prove useful in acid-sensitive substrates[13]. *p*-Methoxy-benzylidene acetals also undergo easier regioselective reductive ring cleavage on treatment with DIBALH to give a mono-protected 1,3-diol derivatives in good yield [Scheme 3.7][14-16]. The transformation has been used effectively in the synthesis of FK-506 intermediates[17].

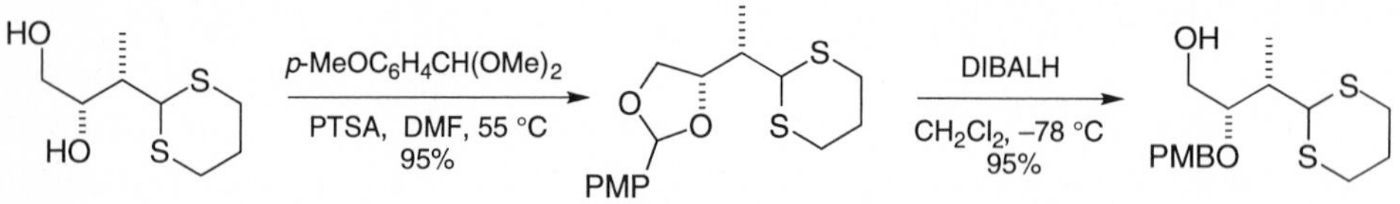

**Scheme 3.7**

The regiochemistry of reductive cleavage of *p*-methoxybenzylidene acetals depends on the reaction conditions and by suitable choice of solvent and electrophile, the distribution of regioisomers can be controlled. Thus, cleavage resulting in the PMB ether of the more hindered hydroxyl is accomplished with $LiAlH_4$–$AlCl_3$[18], $BH_3$•$NMe_3$–$AlCl_3$[19], $BH_3$ in THF at elevated temperature[20], or $NaBH_3CN$–TMSCl in MeCN [Scheme 3.8][21]. On the other hand, by using $NaBH_3CN$ in a mixture of $CF_3COOH$ and DMF, cleavage occurs in the opposite sense to provide the PMB ether of the less hindered hydroxyl [Scheme 3.8][21].

An added measure of versatility was added to the chemistry of arylidene acetals when it was discovered that the reductive cleavage process described above could be reversed under oxidative conditions. Thus, treatment of PMB ethers bearing proximate free hydroxyl groups with DDQ *in the absence of water* leads to *p*-methoxyphenyl dioxanes or dioxolanes [Scheme 3.9][22,23]. In the presence of water, methoxybenzylidene acetals undergo oxidative cleavage with ceric ammonium nitrate (CAN)[21] or DDQ[24,25] as previously witnessed in the analogous cleavage of PMB ethers (section 2.4.3).

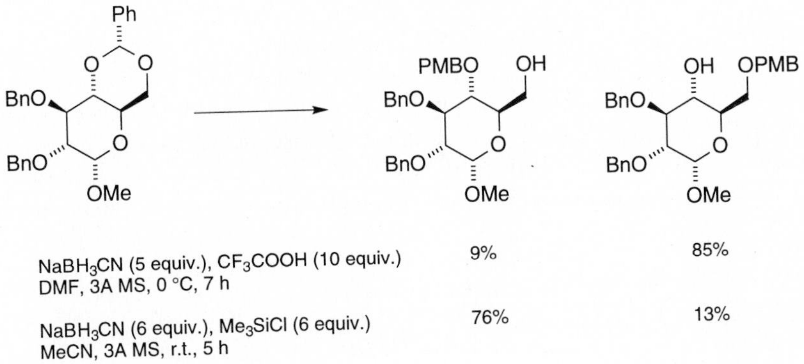
Ph
BnO
BnO
OMe
PMBO
OH
OPMB
NaBH3CN (5 equiv.), CF3COOH (10 equiv.)
DMF, 3A MS, 0 °C, 7 h
9%
85%
NaBH3CN (6 equiv.), Me3SiCl (6 equiv.)
MeCN, 3A MS, r.t., 5 h
76%
13%

**Scheme 3.8**

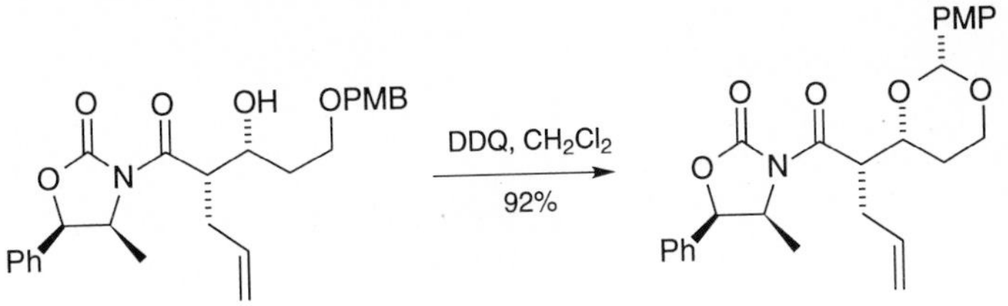
OH
OPMB
Ph
DDQ, CH2Cl2
92%
PMP

**Scheme 3.9**

Benzylidene acetals of 1,2- and 1,3-diols undergo ozonolytic cleavage rapidly at –78 °C at rates which compete with cleavage of alkenes [Scheme 3.10][26]. The reaction is not restricted to benzylidene acetals since dioxolane derivatives of aldehydes also cleave rapidly. The reaction was creatively exploited by Stork and Rychnovsky in the closing stages of their synthesis of dihydroerythronolide A[27] when it was used to unmask a 1,2-diol protected as a 2-methyl-1,3-dioxolane in the presence of a 2-methyl-1,3-dioxane. It is noteworthy that a 1,3-dioxane function cleaved very much slower [Scheme 3.11]. Similarly, benzylidene acetals can be oxidatively cleaved with *tert*-butyl hydroperoxide in the presence of catalytic amounts of $CuCl_2$[28] or $Pd(OAc)_2$[29] to give a benzoate ester and a free hydroxyl group but the regioselectivity of the reaction is poor.

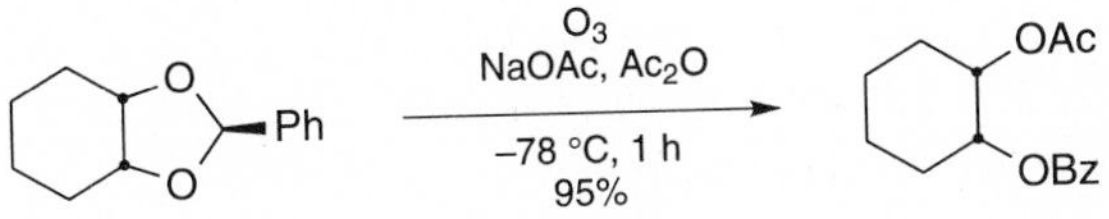
Ph
O3
NaOAc, Ac2O
–78 °C, 1 h
95%
OAc
OBz

**Scheme 3.10**

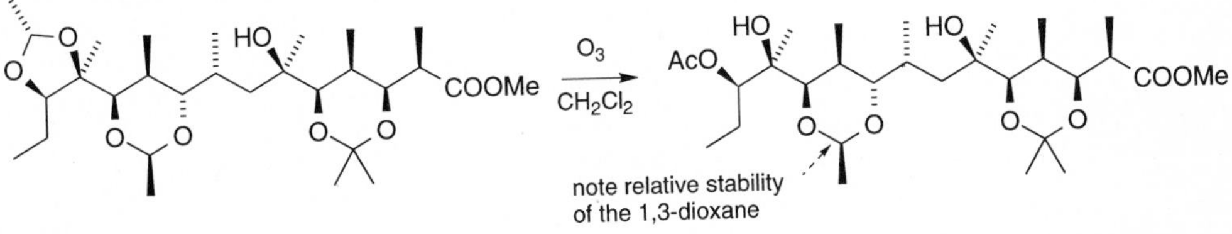
HO
COOMe
O3
CH2Cl2
AcO
note relative stability
of the 1,3-dioxane

**Scheme 3.11**

The fact that acyclic acetals could be cleaved by *N*-bromosuccinimide was discovered over 40 years ago[30]. The reaction was subsequently extended to cyclic derivatives[31] but its synthetic potential was not fully realised until Hanessian applied the method to the cleavage of carbohydrate benzylidene acetals to give bromo benzoate esters[32]. A noteworthy feature of the transformation is that it can be performed on a 100 g (or larger) scale[33]. The reaction has been exploited in the synthesis of a number of natural products such as Pseudomonic acid[34] [Scheme 3.12], Boromycin[35], Thienamycin[36], and Rifamycin[37]. A similar transformation has been accomplished using $BrCCl_3$ under photochemical conditions[38].

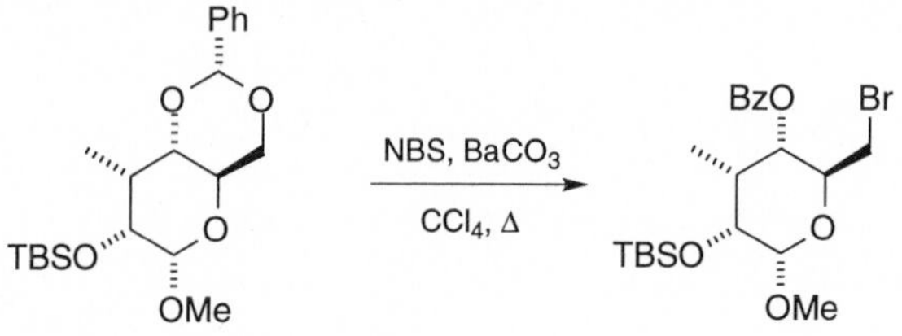

**Scheme 3.12**

Recently Williams and co-workers have shown that an intermediate generated on NBS oxidation of a 2-phenyl-1,3-dioxolane can be trapped by a proximate hydroxyl function to give a new synthesis of tetrahydrofurans [Scheme 3.13][39].

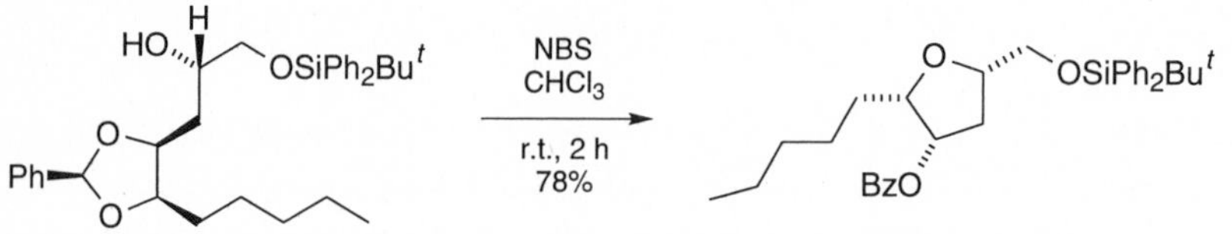

**Scheme 3.13**

Metallation of benzylidene acetals of 1,2-diols gives a benzylic carbanion which extrudes benzoate anion to give the alkene. The method has been applied to vicinal deoxygenation reactions and is exemplified by Whitham's synthesis of *trans*-cyclooctene [Scheme 3.14][40]. In some cases the cleavage can take a different course resulting in the formation of an enolate as exemplified in Scheme 3.15[41].

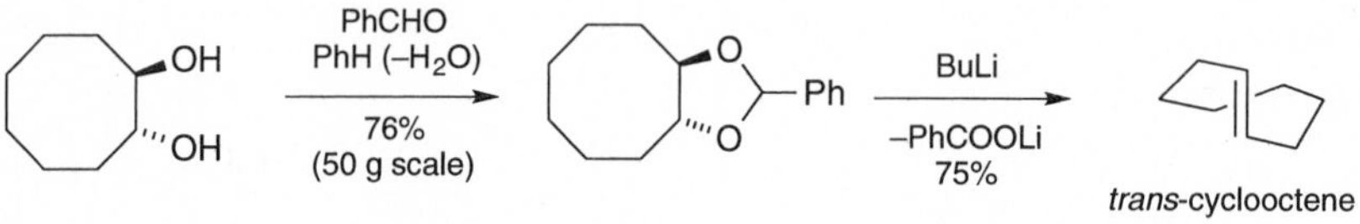

**Scheme 3.14**

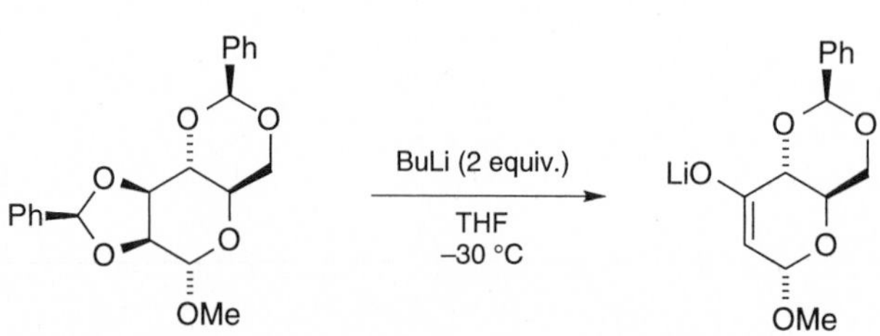

**Scheme 3.15**

## (ii) Formation

There are only two methods generally used to prepare benzylidene acetals: reaction of a diol with benzaldehyde in the presence of a protic acid or Lewis acid (usually $ZnCl_2$)[Scheme 3.16][42-44] — a reaction which is accelerated by ultrasonication[45] — or reaction of the diol with benzaldehyde dimethyl acetal (acetal exchange)[46] in the presence of an acid catalyst as shown in Schemes 3.17[47] and 3.18[48]. Both 1,3-dioxolanes or 1,3-dioxanes can be formed under these conditions.

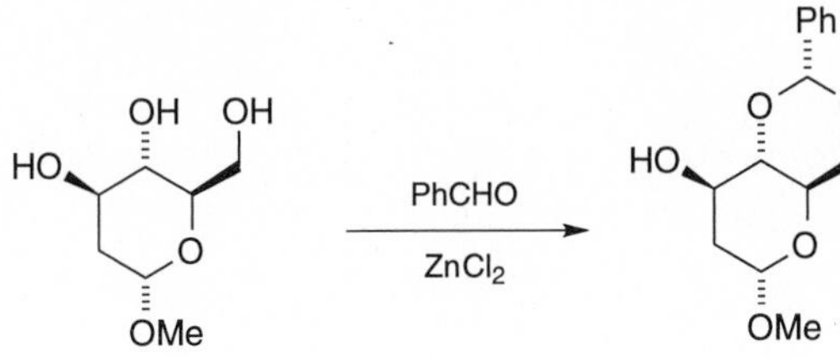
OH OH
HO
O
OMe
PhCHO
ZnCl2
Ph
O O
HO
O
OMe

Scheme 3.16

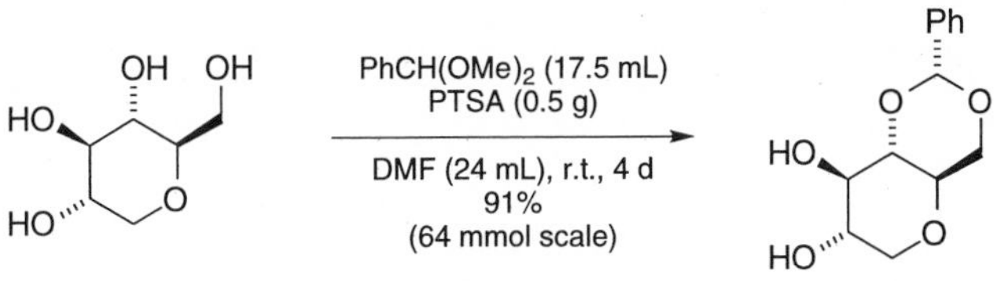
OH OH
HO
HO
O
PhCH(OMe)2 (17.5 mL)
PTSA (0.5 g)
DMF (24 mL), r.t., 4 d
91%
(64 mmol scale)
Ph
O O
HO
HO
O

Scheme 3.17

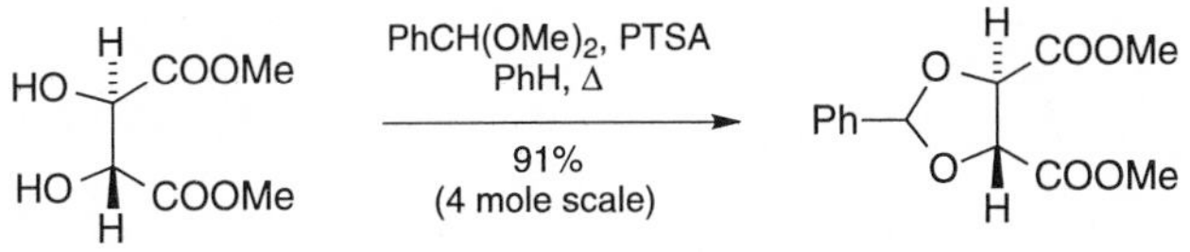
H
HO COOMe
HO COOMe
H
PhCH(OMe)2, PTSA
PhH, Δ
91%
(4 mole scale)
H
O COOMe
Ph
O COOMe
H

Scheme 3.18

*1,3-Dioxolane vs. 1,3-Dioxane Formation*

Treatment of a acyclic 1,2,3- or 1,2,4-trihydroxy alkanes with benzaldehyde or acetone and acid can result in the formation of a 1,3-dioxolane (5-membered ring) or a 1,3-dioxane (6-membered ring)*. In the absence of substituents, we would expect the dioxane ring to form preferentially at equilibrium since thermochemical studies indicate that it is more stable. However, the composition of the product will strongly depend on temperature, substituents and conditions. As a general rule *aldehydes (e.g. benzaldehyde, acetaldehyde) tend to give 6-membered rings and ketones (acetone, cyclohexanone) tend to give 5-membered rings.* Ketones give 5-membered dioxolane rings because the axial substituent at the acetal centre destabilises the corresponding 1,3-dioxane. In benzylidene acetals of 1,3-diols, the phenyl group occupies the equatorial position preferentially whereas the conformational preferences of the dioxolane derivatives is much less pronounced. Where added steric bias is required, either in fixing the conformation of a 1,3-dioxane ring or, in controlling 1,3-dioxane formation at the expense of the 1,3-dioxolane in 1,2,4-triols, mesitylmethylene acetals can be used[49].

Baggett and co-workers[50] used NMR spectroscopy to study the reaction of glycerol with benzaldehyde in DMF in the presence of PTSA*. They found that the 1,3-dioxolanes formed rapidly (kinetic product) but that isomerisation took place to give the 1,3-dioxanes [Scheme 3.19]. However, in more complex systems a variety of factors may come to bear in determining the structure of the final product. Thus benzylidenation of D-(+)-arabitol [Scheme 3.20][51] led to selective formation of only one 1,3-dioxane incorporating the C-1 and C-3 hydroxyl functions and it is noteworthy that the diastereoisomer formed contained an axial hydroxyl function. None of the corresponding diastereoisomer from the alternative C-3 and C-5 dioxane was observed. It was suggested that the outcome of this thermodynamically controlled reaction was governed by hydrogen bonding between the axial hydroxyl group and the dioxane acetal oxygens.

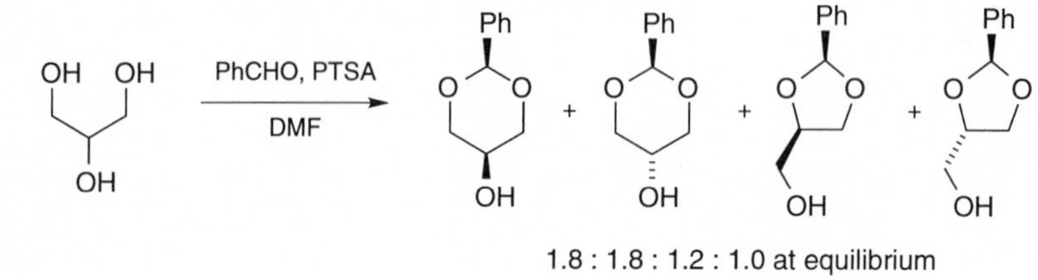

PhCHO, PTSA, DMF

1.8 : 1.8 : 1.2 : 1.0 at equilibrium

Scheme 3.19

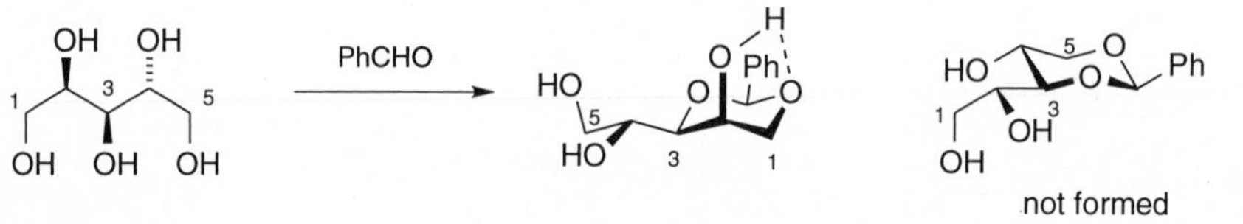

PhCHO

not formed

Scheme 3.20

The potential for mischief in benzylidenation reactions is further illustrated by the different products formed from D-gulonolactone depending on the reaction conditions [Scheme 3.21][52].

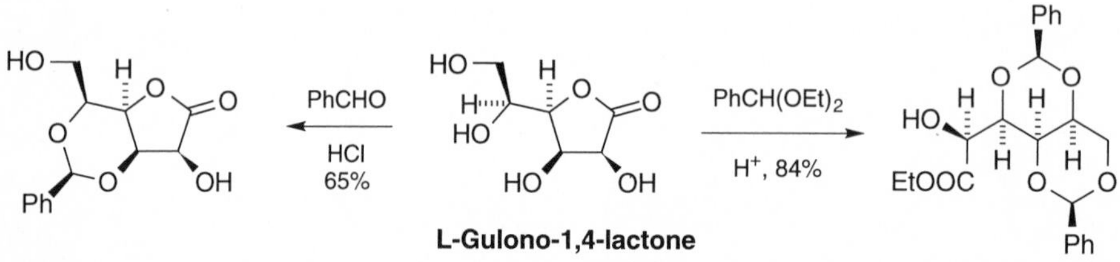

PhCHO, HCl, 65%

$PhCH(OEt)_2$, $H^+$, 84%

L-Gulono-1,4-lactone

Scheme 3.21

**(iii) NMR Data**

$^{13}C$ and $^{1}H$ NMR spectrocopy provide a valuable means for ascertaining the ring size and conformation of benzylidene acetals[53,54]. 2-Phenyl-1,3-dioxanes give signals for the acetal proton at $\delta = 5.5$ whereas the corresponding 2-phenyl-1,3-dioxolanes appear at $\delta = 5.9$–$6.3$. The acetal carbon generally appears at ca. $\delta = 102$–$106$. In a recent study of the benzylidenation of pyranose derivatives, Patroni and co-workers[46] noted only small differences in the chemical shift of the acetal carbon of 1,3-dioxolane *vs* 1,3-dioxane derivatives in some cases suggesting the need for caution in using $^{13}C$ NMR spectroscopy for determining the ring size of benzylidene acetals.

### 3.2.2 Isopropylidene Acetals

Cyclic isopropylidene acetals (also known as acetonides) have been used more frequently than any other protecting group for the protection of 1,2- and 1,3-diols. The acetals are easily prepared and they are stable to most reaction conditions except protic and Lewis acids.

**(i) Cleavage**

Acid-catalysed hydrolysis is the most common method for deprotecting isopropylidene derivatives and the acid strength and reaction time can vary widely. In most cases aqueous acetic acid, aqueous trifluoroacetic acid, dilute HCl in THF, or an ion exchange resin such as Dowex 50W [Scheme 3.22][55] will remove them rapidly. A good illustration of the wide differences in reactivity that can be observed is shown in Scheme 3.23. 1,3-Dioxanes hydrolyse faster than 1,3-dioxolanes [Scheme 3.24][56]; however, trans-fused dioxolanes hydrolyse faster than dioxanes.

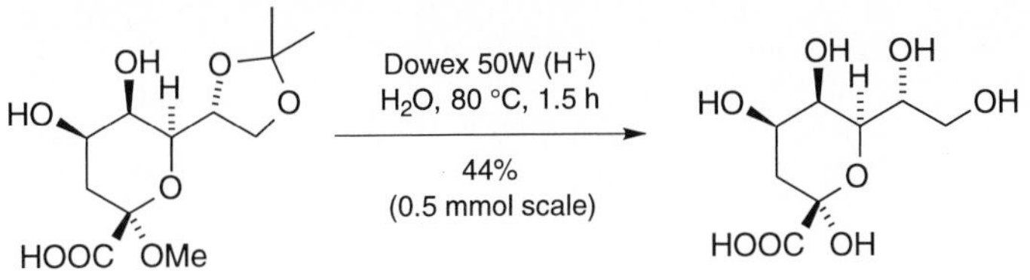

Dowex 50W ($H^+$), $H_2O$, 80 °C, 1.5 h; 44% (0.5 mmol scale)

Scheme 3.22

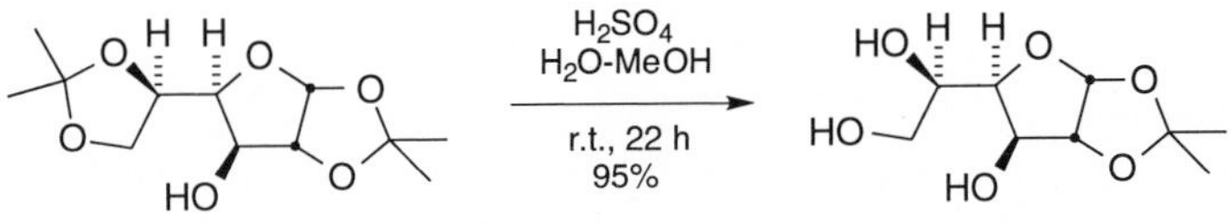

$H_2SO_4$, $H_2O$-MeOH; r.t., 22 h; 95%

Scheme 3.23

AcOH-$H_2O$

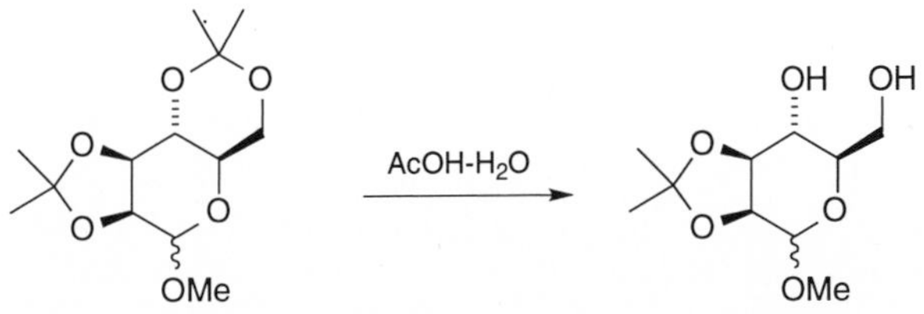

Scheme 3.24

There are many occasions in which isopropylidene acetals are remarkably stable or there are other functional groups present which cannot withstand the hydrolysis conditions. A case in point comes from Williams' synthesis of Phyllanthocin in which a dioxolane had to be hydrolysed without affecting a *tert*-butyldimethylsilyl ether group [Scheme 3.25][57]. The deprotection was accomplished by treatment of the acetal with ethanedithiol and acid. In the absence of an oxygen nucleophile, the TBS ether was stable under these conditions. Similarly, *tert*-butyldimethylsilyl ethers cannot be safely removed with HF in acetonitrile in the presence of an isopropylidene group though the transformation could easily be achieved using TBAF[58].

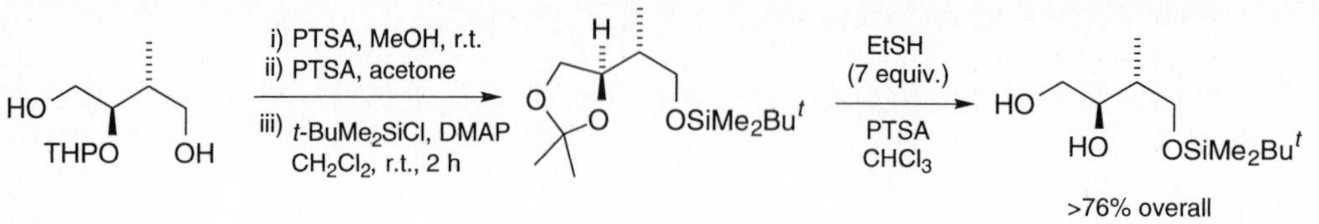

Scheme 3.25

Lewis acids can also be exploited for the cleavage of isopropylidene derivatives. For example, Ley and co-workers found that a 1,3-dioxolane resistant to the usual hydrolytic conditions deprotected successfully, albeit in only modest yield, using $Me_2BBr$ in $CH_2Cl_2$ at –78 °C [Scheme 3.26][59].

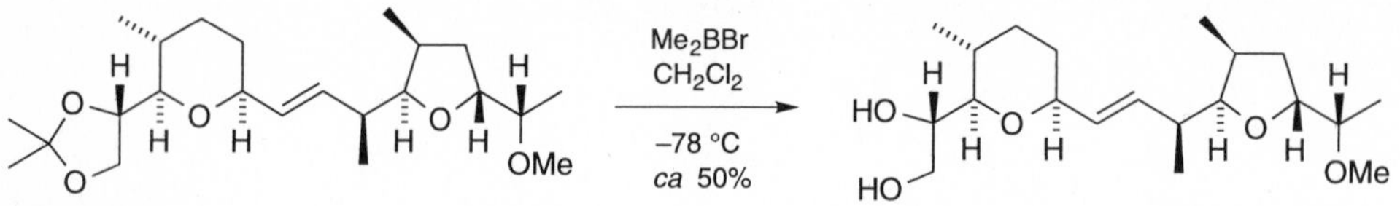

Scheme 3.26

**(ii) Formation**

Reaction of a diol with dry acetone in the presence of an acid catalyst is the oldest method used to prepare isopropylidene acetals. Typical acids are PTSA or CSA. Since a molecule of water is generated in the reaction, some means for dehydration is usually required. Owing to the low boiling point of acetone, azeoptropic removal of water is not feasible so other means are required such as molecular sieves or an inorganic dehydrating agent such as $CuSO_4$ [Scheme 3.27][60,61]. Dioxolanes are preferred to dioxanes (*vide infra*) and cis-fused dioxolanes are thermodynamically favored over trans-fused systems [Scheme 3.28][62]. In some cases reaction of a diol with acetone in the presence of $FeCl_3$[63] or $AlCl_3$[64] has been used. Formation of isopropylidene acetals and cyclohexylidene acetals is facilitated by ultrasonication[65].

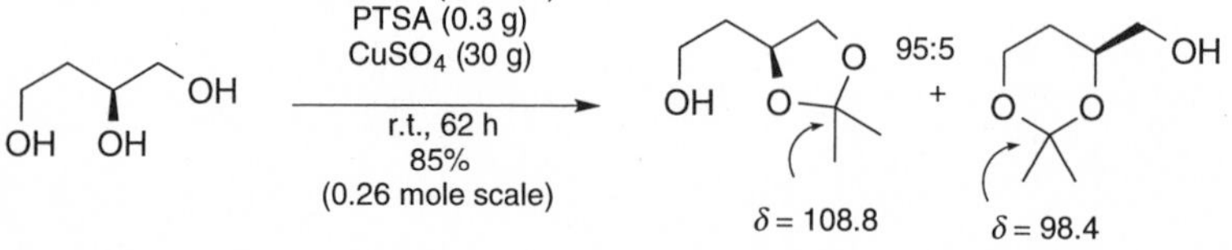

Scheme 3.27

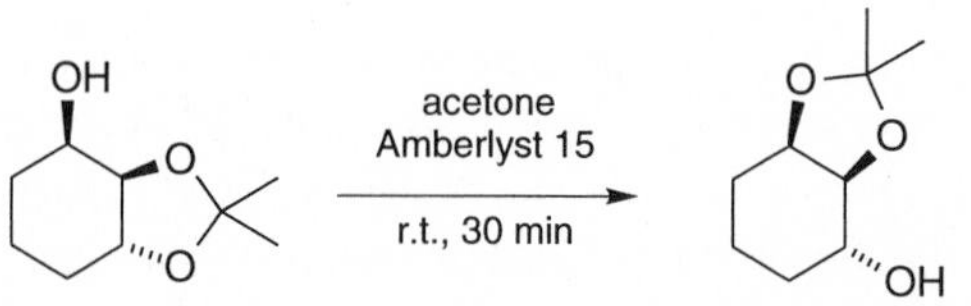

Scheme 3.28

Acetal exchange using 2,2-dimethoxypropane offers an alternative method which avoids the need for a dehydrating agent since the reaction liberates two moles of MeOH. Provided the equilibrium is favorable, as it is in most cases, the exchange will go to completion; in less favorable circumstances, 2-methoxypropene or 2-trimethylsilyloxy-propene[66] can be used [Scheme 3.29][62]. The latter reaction benefits from the formation of $(Me_3Si)_2O$ to drive the reaction to completion. Acetal exchange is the most common method for preparing isopropylidene acetals [Scheme 3.30[67], Scheme 3.31[68]] and the reaction is now used commercially to prepare 1,2:5,6-di-*O*-isopropylidene-D-mannitol (a precursor of 2,3-*O*-isopropylidene-D-glyceraldehyde) on a hundred kilogram scale using $SnCl_2$ as the catalyst [Scheme 3.32][69].

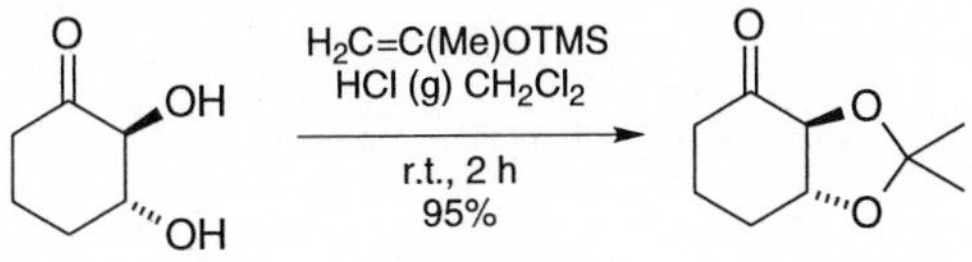
OH
OH
H2C=C(Me)OTMS
HCl (g) CH2Cl2
r.t., 2 h
95%
O
O

**Scheme 3.29**

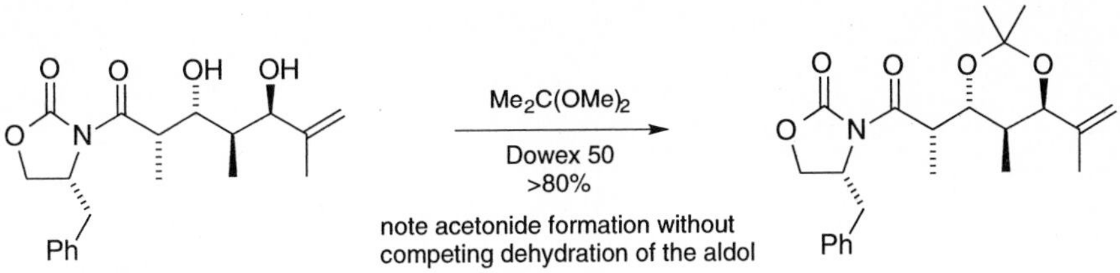
O O OH OH
N
Ph
Me2C(OMe)2
Dowex 50
>80%
note acetonide formation without
competing dehydration of the aldol

**Scheme 3.30**

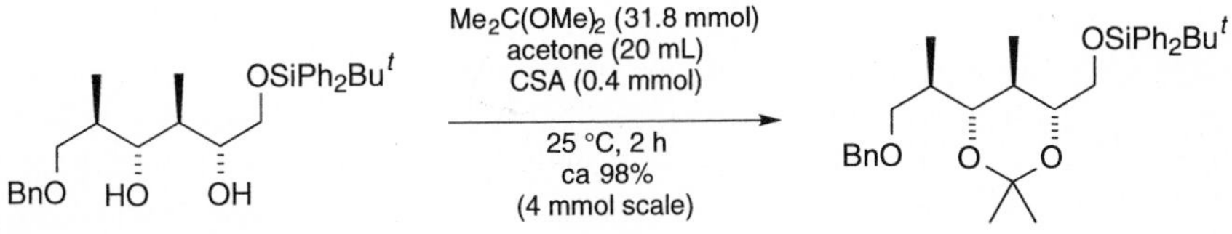
OSiPh2Bu^t
BnO HO OH
Me2C(OMe)2 (31.8 mmol)
acetone (20 mL)
CSA (0.4 mmol)
25 °C, 2 h
ca 98%
(4 mmol scale)
BnO O O

**Scheme 3.31**

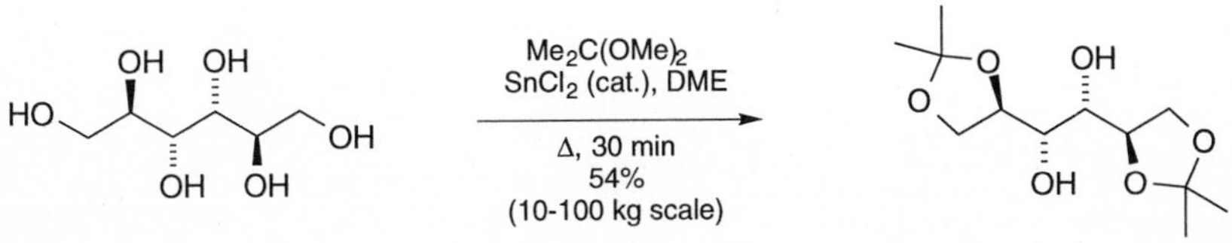
OH OH
HO
OH
OH OH
Me2C(OMe)2
SnCl2 (cat.), DME
Δ, 30 min
54%
(10-100 kg scale)

**Scheme 3.32**

Isopropylidene acetal formation is a thermodynamically driven process; hence, in cases where two (or more) acetals are possible, the more stable diastereoisomer will prevail. Two examples suffice to demonstrate the point. The preference for the formation of isopropylidene derivatives of secondary alcohols is illustrated by the isomerisation in Scheme 3.33[70,71] in which the more highly substituted *trans*-dioxolane is favored at equilibrium. Mulzer and co-workers[72] showed that a similar isomerisation occurred easily when the dioxolane **34.1** [Scheme 3.34] was treated with mineral acid to give the *trans*-dioxolane **34.2**. However, attempted isomerisation of the diastereomeric dioxolane **34.3** failed to occur under similar conditions demonstrating the penalty incurred on formation of the less stable *cis*-dioxolane **34.4**.

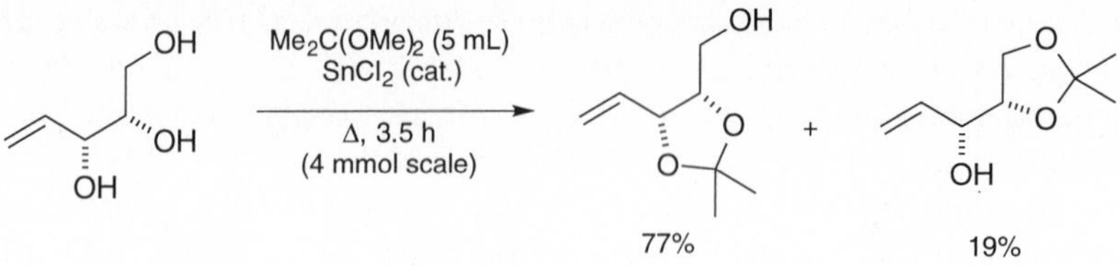

Scheme 3.33

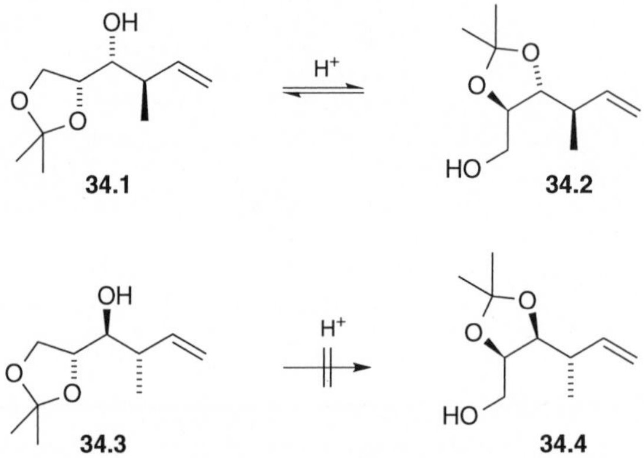

Scheme 3.34

*1,3-Dioxolane vs. 1,3-Dioxane Formation*

In acyclic 1,2,4-triols, isopropylidene acetalisation favors formation of the 5-membered 1,3-dioxolane ring even if one of the hydroxy functions is tertiary [Scheme 3.35][73] or hindered[57,74]. In the case of 1,2,4-butanetriol itself, the minor dioxane (ca. 5%) can easily be detected by $^{13}C$ NMR spectroscopy [Scheme 3.27][60,75]. The strong preference for reaction with cis-1,2-diols causes rearrangement of the stable 6-membered pyranose isomer of glucose to give 1,2:5,6-di-*O*-isopropylidene-α-D-glucofuranose in good yield [Scheme 3.36][76,77].

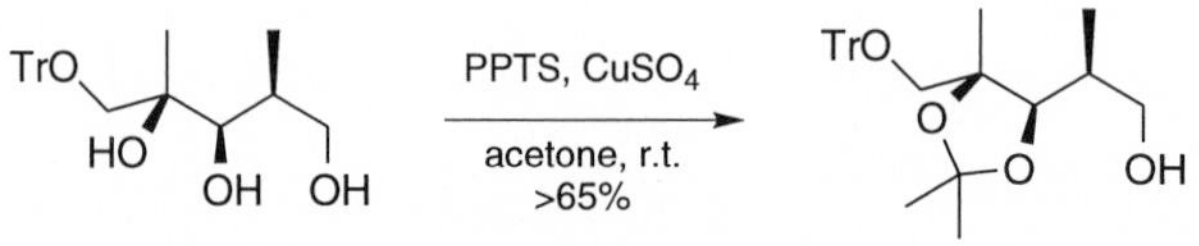

Scheme 3.35

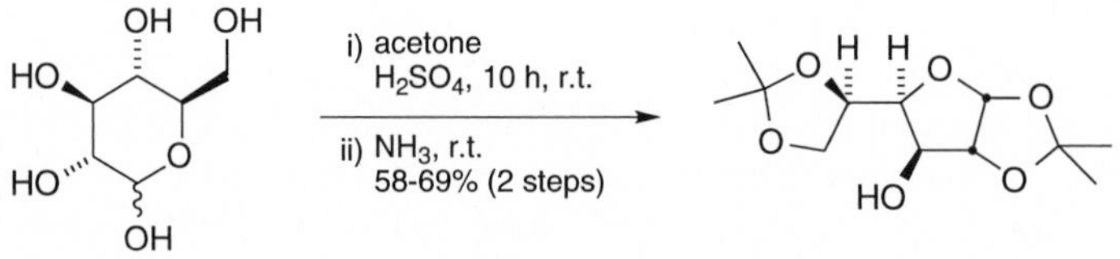

Scheme 3.36

Gelas and Horton[56] found that *kinetic* acetalisation of the 4 and 6 hydroxyl groups of pyranoses could be accomplished *without rearrangement* using 2-methoxypropene in DMF [Scheme 3.37]. Under these conditions *trans*-1,2-diols can also be converted to the highly labile isopropylidene derivative — a process that was previously not generally possible [Scheme 3.38].

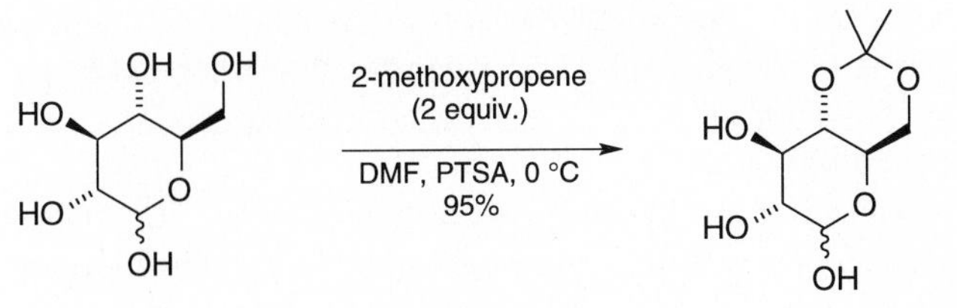

**Scheme 3.37**

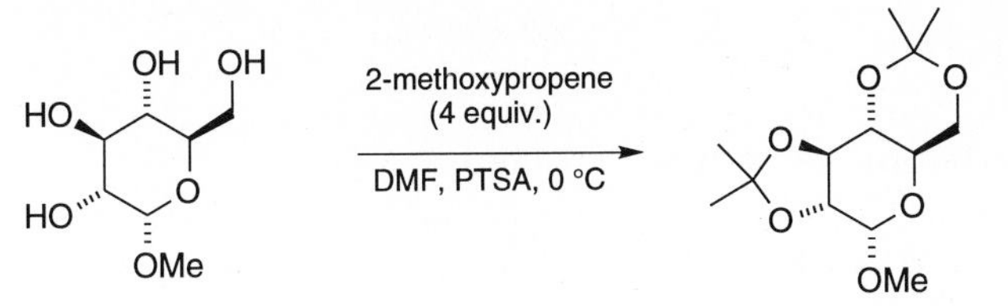

**Scheme 3.38**

Steric effects can be used to manipulate the proportion of dioxane and dioxolane isomers from a 1,2,4-triol system. For example the triol **39.1** [Scheme 3.39] gave a 9:1 mixture of dioxolane and dioxane products **39.2** and **39.3** respectively when acetone was used as the acetalisation reagent but when 3-pentanone was employed under essentially the same conditions, the dioxolane product **39.4** was formed exclusively[78]. Interestingly, the paper conveying these results also noted that boron side products derived from the reduction of a lactone with $BH_3 \cdot SMe_2$ inhibit the efficient formation of ketals.

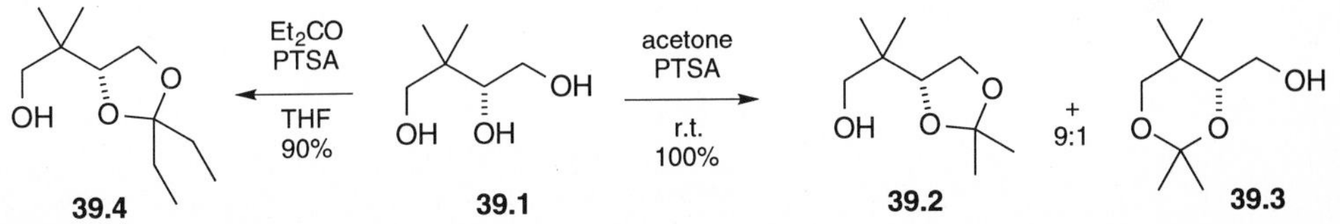

**Scheme 3.39**

Our final example of isopropylidenation involves the Lewis acid-catalysed opening of an epoxide ring using acetone as solvent as exemplified in Scheme 3.40[79]. The high stereoselectivity of the reaction is noteworthy.

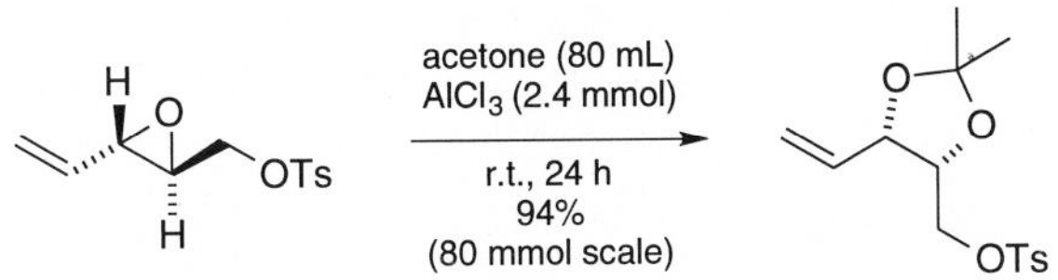

**Scheme 3.40**

**(iii) NMR Data**

$^{13}C$ NMR spectroscopy provides a sensitive method for assaying the ring size and substitution of ispropylidene acetals[80]. The chemical shifts of the acetal carbon (O–C–O) in dioxolanes appear at $\delta$ 108.1–111.4 for monocyclic or cis-fused dioxolanes to a pyranoid or cyclohexane ring and $\delta$ 111.8–112.3 for the trans-fused series; $\delta$ 111.3–115.7 when fused to a furanoid ring. The corresponding acetal carbon in the chair form of a 1,3-dioxane ring appears at $\delta$ 97.1–99.9 whereas the skew conformation appears at $\delta$ 100.6–101.1.

The relative stereochemistry of 1,3-diols can be assigned by $^{13}C$ NMR spectroscopy of the corresponding isopropylidene derivatives[81,82]. In general, the isopropylidene derivatives of *syn*-1,3-diols have methyl shifts at $\delta$ 19 and 30 and acetal carbon shifts at $\delta$ 98.5. In the isopropylidene derivatives of *anti*-1,3-diols, the methyl signals appear at $\delta$ 25 whereas the acetal carbon appears at $\delta$ 100.5. The acetal methyl chemical shifts are reliable indicators of 1,3-diol stereochemistry but the acetal carbon shift correlation should be used with caution.

### 3.2.3 Cyclohexylidene and Cyclopentylidene Acetals

The chemistry of cyclohexylidene acetals is very similar to isopropylidene acetals but the cyclic derivatives have two advantages over their acyclic counterparts. First, they diminish water solubility of low molecular weight fragments and therefore facilitate isolation. Secondly, they can show a greater bias for the formation of 1,3-dioxolanes over 1,3-dioxanes. Cyclopentylidene acetals have the advantage that they are more easily hydrolysed than isopropylidene or cyclohexylidene acetals.

**(i) Cleavage**

The relative ease of acid-catalysed hydrolysis (0.53 M sulphuric acid in 2:1 aqueous 2-propanol at r.t.) follows the order: cyclopentylidene > isopropylidene > cyclohexylidene. For example, the half life for the hydrolysis of 1,2-*O*-alkylidene-α-D-glucopyranoses followed the order cyclopentylidene (8 h), isopropylidene (20 h), cyclohexylidene (124 h)[83]. Trans-fused 1,2-*O*-cyclohexylidene derivatives cleave much faster than the corresponding cis isomers [Scheme 3.41][84].

Smith and co-workers exploited the ease of cyclopentylidene acetal cleavage in their synthesis of the disaccharide Phyllanthose: deprotection was accompanied by glycosidation with allyl alcohol

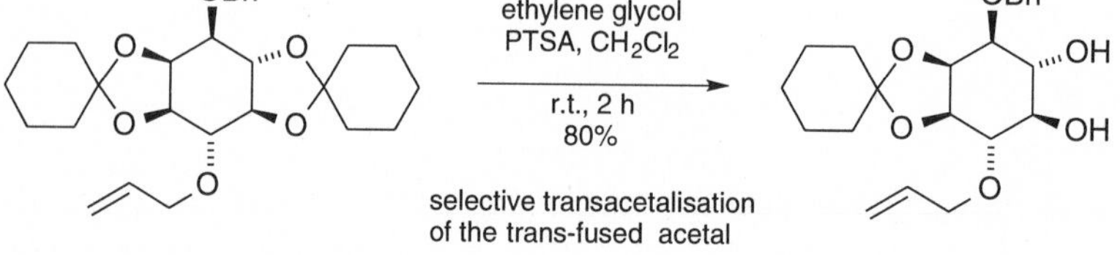

ethylene glycol
PTSA, $CH_2Cl_2$
r.t., 2 h
80%
selective transacetalisation of the trans-fused acetal

**Scheme 3.41**

allyl alcohol (6 mL)
PhH (60 mL)
CSA (cat.)
Δ, 29 h ($-H_2O$)
82–96%
(5.5 mmol scale)

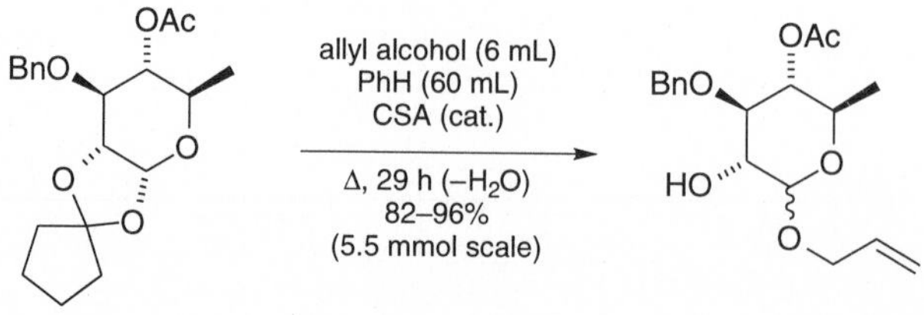

**Scheme 3.42**

under conditions mild enough to preserve an acetate function [Scheme 3.42][85]. Similar applications can be found in syntheses of Monensin[86], Erythronolide[27], and Methynolide[87]. In the latter case it was necessary to preserve a SEM protecting group whilst unmasking a 1,2-diol. Deprotection of an isopropylidene group could not be accomplished selectively but a cyclopentylidene was hydrolysed in HCl–MeOH (1:1) at r.t.

**(ii) Formation**

The cycloalkylidene acetals are formed in essentially the same way as their isopropylidene counterparts. Thus reaction of a diol with cyclohexanone [Scheme 3.43][88,89] or cyclopentanone [Scheme 3.44][86], usually with provision for the removal of water, accomplishes the transformation in good yield. In the double protection of mannitol [Scheme 3.45], triethyl orthoformate served as the dehydrating agent and $BF_3 \bullet OEt_2$ was the acid catalyst[90].

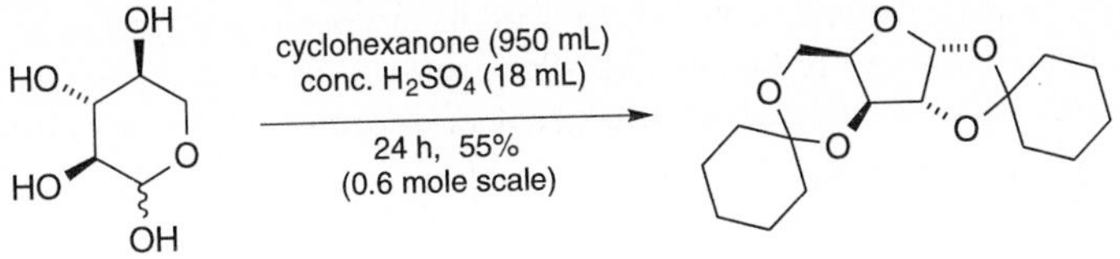

cyclohexanone (950 mL)
conc. $H_2SO_4$ (18 mL)
24 h, 55%
(0.6 mole scale)

**Scheme 3.43**

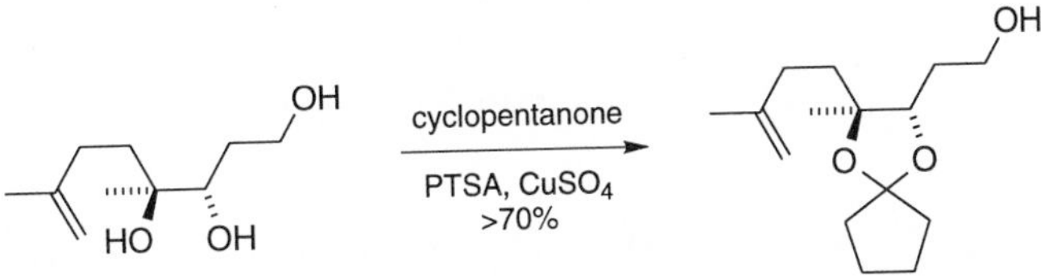

cyclopentanone
PTSA, $CuSO_4$
>70%

**Scheme 3.44**

cyclohexanone (150 mmol)
$HC(OEt)_3$ (50 mmol)
$BF_3 \bullet OEt_2$ (4 mmol)
DMSO (20 mL)
r.t., 10 h
(50 mmol scale)

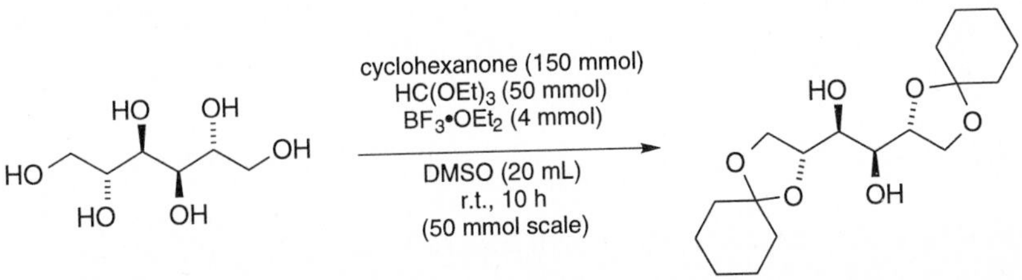

**Scheme 3.45**

Acetal exchange reactions using 1,1-dimethoxycyclohexane [Scheme 3.46][91,92] or 1,1-dimethoxycyclopentane [Scheme 3.47][93] results in the formation of the corresponding cycloalkylidene acetals.

PTSA
87%

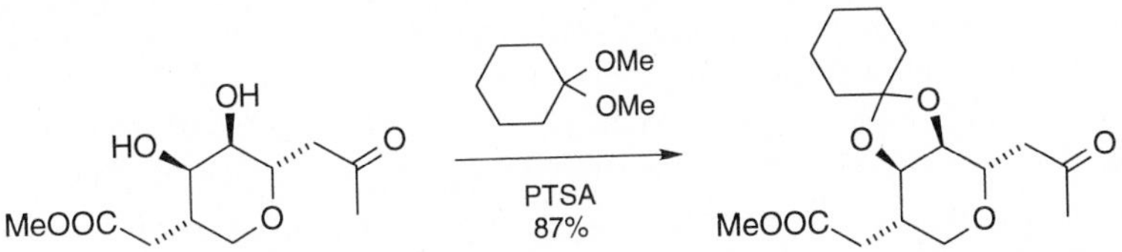

**Scheme 3.46**

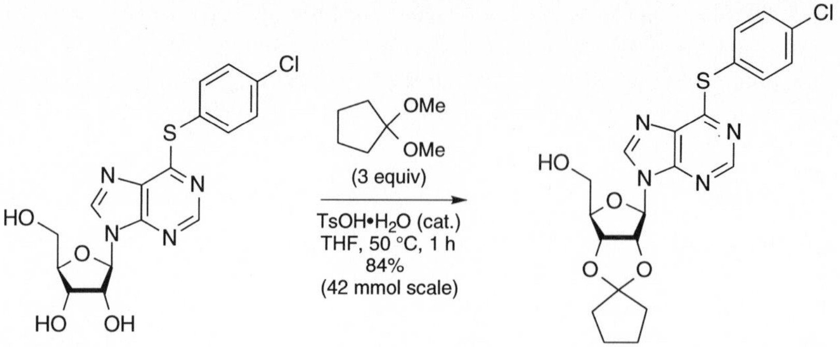

Scheme 3.47

Reaction of an epoxide in the presence of a large excess of cyclopentanone has been used to prepare a cyclopentylidene derivative on a substantial scale [Scheme 3.48][94]. The reaction requires nucleophilic opening of the epoxide ring by the oxygen atom of cyclopentanone to give an intermediate oxonium ion which closes to the cyclopentylidene acetal.

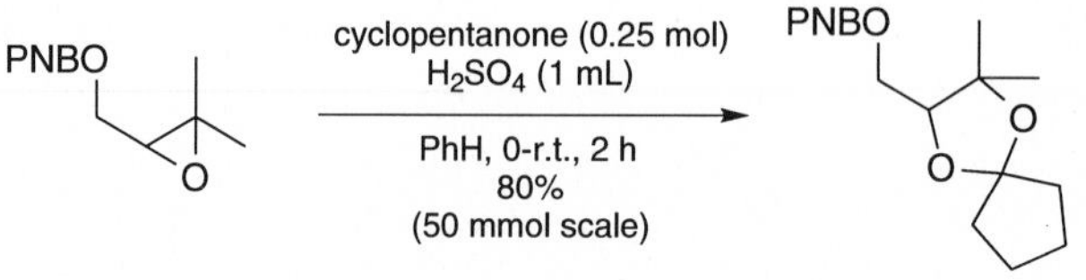

Scheme 3.48

**(iii) NMR Data**

1,1-diethoxycyclohexane: $^1$H NMR: $\delta = 1.17$ (6H, t, $J = 7$ Hz), 1.32–1.66 (10H, m), 3.46 (4H, q, $J = 7$ Hz); $^{13}$C NMR: $\delta = 15.5$ (2 x Me), 22.95, 25.7 and 33.5 (5 x $CH_2$), 54.7 (2 x $CH_2O$), 100 (acetal C). In 1,2-di-*O*-cyclohexylidene derivatives, the chemical shift of the acetal carbon shifts to ca. $\delta = 110$.

## 3.3 Silylene Derivatives

Trost first introduced the di-*tert*-butylsilylene derivative (abbreviated DTBS) as a means for protecting 1,2- and 1,3-diols during a synthesis of Pillaromycinone derivatives[95]. Di-*tert*-butylsilylene derivatives are not as robust as isopropylidene or benzylidene acetals and their use is best reserved for systems requiring deprotection under very mild conditions.

**(i) Cleavage**

1,2-Diols are more easily released from their di-*tert*-butylsilylene protectors than 1,3- or 1,4-diols. The 1,2-diol derivatives hydrolyse rapidly under basic conditions (5:1 THF–pH 10 buffer, r.t., 5 min) whereas the six- and seven-membered rings are unaffected at pH 4–10 at r.t. over several hours. Hydrolysis occurs readily with HF in MeCN or HF-Pyr complex in THF–Pyr at r.t. Scheme 3.49[95] shows that under controlled conditions a *tert*-butyldimethylsilyl group remains intact. The last step in Evans' synthesis of Cytovaricin[15] involved a comprehensive deprotection of 7 hydroxyl functions differentially protected with silicon protecting groups (3 $Et_3Si$, 1 *i*-$PrEt_2Si$, 1 *t*-$BuMe_2Si$, and a di-*tert*-butylsilylene) in the presence of a large excess of HF–Pyr [Scheme 3.50].

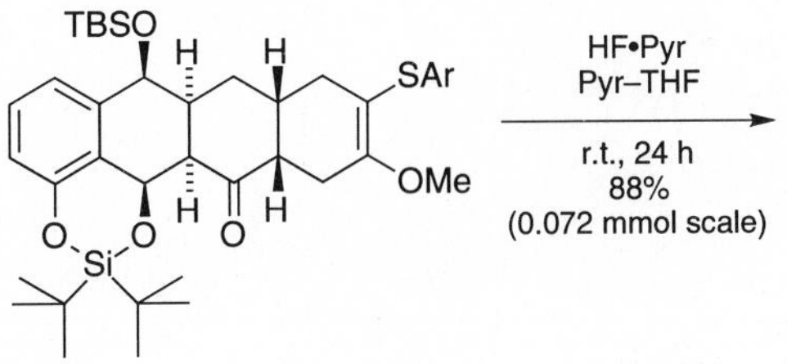
TBSO
H
H
SAr
OMe
O
Si
O
H
O
H
HF•Pyr
Pyr–THF
r.t., 24 h
88%
(0.072 mmol scale)
HO
HO

**Scheme 3.49**

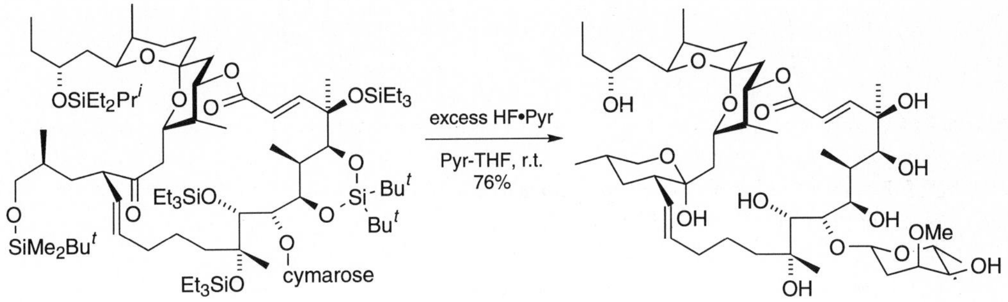
OSiEt2Pr^i
OSiEt3
Et3SiO
Si - Bu^t
Bu^t
O
SiMe2Bu^t
Et3SiO
cymarose
excess HF•Pyr
Pyr-THF, r.t.
76%
OH
OH
OH
HO
OH
OH
OMe
OH
OH

**Scheme 3.50**

**(ii) Formation**

The reaction of a diol with dichloro-di-*tert*-butylsilane [Scheme 3.51][96] in the presence of 1-hydroxybenzotriazole (HOBT) was the first method used to prepare di-*tert*-butylsilylene derivatives. Di-*tert*-butylsilyl ditriflate[97,98] effects silylene formation more rapidly and under milder conditions than the less reactive dichloride [Scheme 3.52][15].

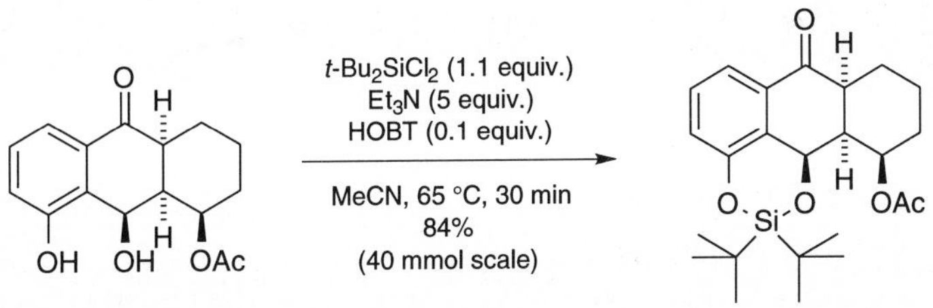
O
H
OH
OH
H
OAc
t-Bu2SiCl2 (1.1 equiv.)
Et3N (5 equiv.)
HOBT (0.1 equiv.)
MeCN, 65 °C, 30 min
84%
(40 mmol scale)
O
Si
O
OAc

**Scheme 3.51**

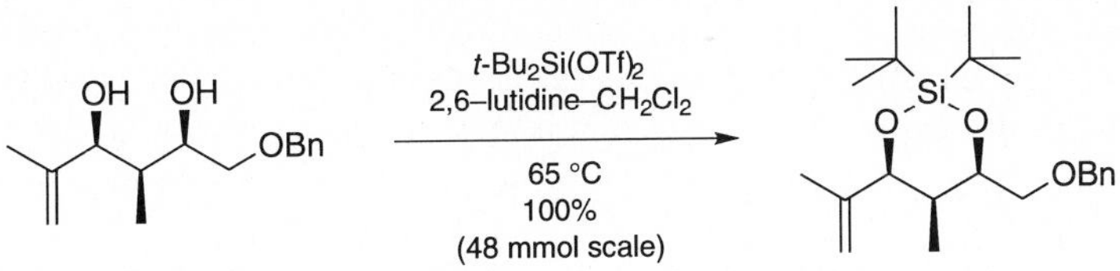
OH OH
OBn
t-Bu2Si(OTf)2
2,6–lutidine–CH2Cl2
65 °C
100%
(48 mmol scale)
Si
O
O
OBn

**Scheme 3.52**

Reaction of a *β*-hydroxy ketone with *i*-$Pr_2SiHCl$ provides a *β*-hydrosilyloxy ketone which will deliver hydride to the ketone function in the presence of $SnCl_4$. The reaction provides a highly diastereoselective method for preparing 1,3-diols via intramolecular hydrosilylation [Scheme 3.53][99].

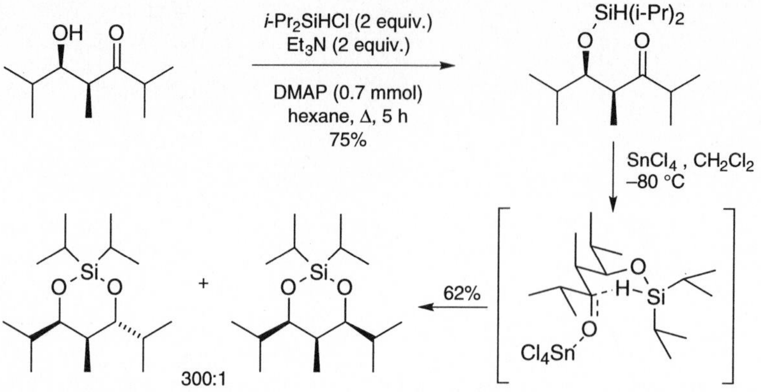

OH O

$i$-$Pr_2SiHCl$ (2 equiv.)
$Et_3N$ (2 equiv.)
DMAP (0.7 mmol)
hexane, Δ, 5 h
75%

SiH(i-Pr)2
O O

$SnCl_4$, $CH_2Cl_2$
–80 °C

$Cl_4Sn$ O H Si

62%

Si O O + Si O O

300:1

**Scheme 3.53**

**(iii) NMR Data**

$^1$H NMR: $\delta = 0.7$–$1.2$ (2s, 9H each)
$^{13}$C NMR: ca. $\delta = 27$

## 3.4 1,1,3,3-Tetraisopropyldisiloxanylidene Derivatives

The simultaneous protection of the 3′ and 5′ hydroxyl groups of nucleosides and the 4,6 or 3,4 hydroxyls of hexopyranoses is a common problem in organic synthesis. In the case of hexopyranoses, we have already seen that benzylidene acetals and, in certain circumstances, isopropylidene acetals can be used to good effect. An alternative silicon-based group would offer a wider repertoire of conditions for mild deprotection and such a group was devised by Markiewicz: the 1,1,3,3-tetraisopropyldisiloxanylidene group (abbreviated TIPDS)[100,101]. TIPDS groups are stable to water, 0.3 M PTSA in dioxane, 10% $CF_3COOH$ in $CHCl_3$, 5 M $NH_4OH$ in dioxane–$H_2O$ (4:1), and tertiary amines in pyridine.

**(i) Cleavage**

In general, conditions used to cleave TBS ethers are sufficent to remove TIPDS groups. Thus TBAF, 0.2 M HCl in aqueous dioxane, or 0.2 M NaOH in aqueous dioxane accomplishes rapid deprotection. Excess ammonium fluoride in refluxing MeOH is a cheaper alternative to TBAF which is effective in deprotecting TIPDS derivatives of nucleosides[102].

Recently, Markiewicz and co-workers[103] described the tetra-*tert*-butyloxydisiloxane-1,3-diyl group (TBDS) as a new bifunctional silyl protecting group which is more stable than the TIPDS group. The TBDS group is hydrolysed around 25 times slower than the TIPDS group in 0.2 M NaOH in dioxane and it survives 0.2 M HCl in dioxane under conditions that rapidly cleave the TIPDS group.

**(ii) Formation**

TIPDS groups are usually introduced by the reaction of the difunctional reagent 1,3-dichloro-1,1,3,3-tetraisopropyldisiloxane with the substrate in pyridine though imidazole in DMF solution can also be used. In the case of hexopyranoses, these conditions give the kinetic product, an 8-membered ring, which is formed by rapid reaction first at the least hindered hydroxyl group (primary hydroxyls react approximately 1000 times faster than secondary hydroxyls) followed by a second intramolecular silylation with the next proximate hydroxyl at C-4[104,105].

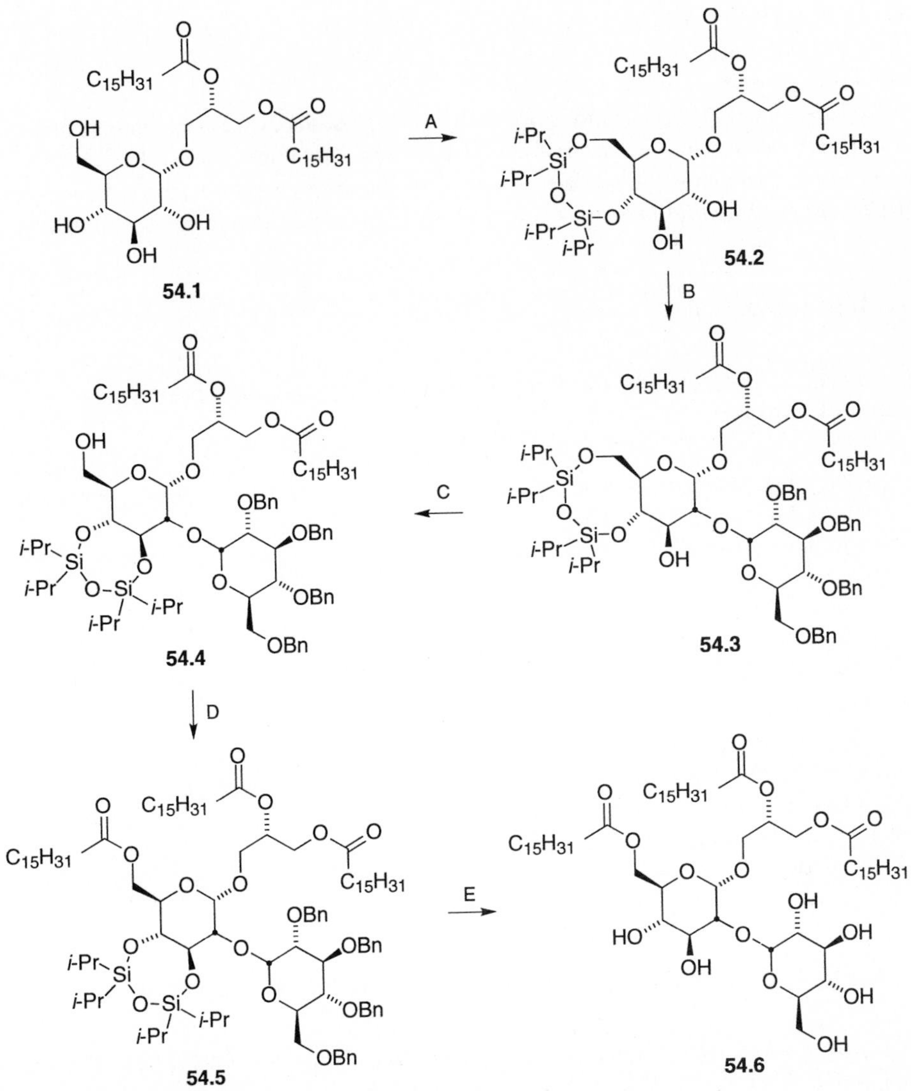

**REAGENTS AND YIELDS**

A 70% $[ClSi(i\text{-}Pr)_2]_2O$, Pyr, –15 °C, 3 h.
B 45% 1-bromo-2,3,4,6-tetra-*O*-benzyl-D-glucose, $Bu_4NBr$, $i\text{-}Pr_2NEt$, DMF, r.t., 5 d.
C 84% mesitylenesulfonic acid (cat.), DMF, r.t., 18 h.
D 96% stearoyl chloride, $CH_2Cl_2$-Pyr, r.t., 1 h.
E 74% (i) TBAF, THF, r.t., 10 min; (ii) $H_2$, Pd.

**Scheme 3.54**

The synthetic value of the TIPDS group will be illustrated here by a synthesis of a glycolipid component of *Streptococci* cell membranes [Scheme 3.54][106]. Selective protection of the C-4 and C-6 hydroxyls of the tetraol **54.1** was easily accomplished using the TIPDS group and the C-2 hydroxyl then participated in a glycosylation reaction to give the disaccharide **54.3**. The next step in the synthesis is a noteworthy feature: the 8-membered disiloxane ring rearranged under acidic conditions to give the more stable 7-membered disiloxane **54.4**. The remaining primary hydroxyl function in **54.4** was then acylated and the TIPDS group removed with TBAF and the benzyl ether functions hydrogenolysed to give the intermediate **54.6.**

The TIPDS group was also one of the *dramatis personae* in the delicate and complex methodology used to prepare other complex bacterial cell wall components implicated in immunogenic mechanisms. A superb example is the synthesis of small synthetic saccharide fragments used to prepare a vaccine against *Haemophilus influenzae* type b (Hib), a serious infection in young children[107].

# 3.5 Reviews

## 3.5.1 Reviews Concerning the Preparation of Acetals and Their Use as Protecting Groups

1 Acetals and Hemiacetals. Schmitz, E.; Eichorn, I. In *The Chemistry of the Ether Linkage*; Patai, S., Ed.; Wiley: New York, 1967; Chapter 7.
2 Protection of Alcoholic Hydroxyl Groups and Glycol Systems. Reese, C. B. In *Protective Groups in Organic Chemistry*; McOmie, J. F. W., Ed.; Plenum: London, 1973; p 95.
3 Advances in the Chemistry of Acetals, Ketals, and Ortho Esters. Bergstrom, R. G. In *The Chemistry of Ethers, Crown Ethers, Hydroxyl Groups, and their Sulphur Analogues, Part 2.*; Patai, S., Ed.; Wiley: New York, 1980; Chapter 20.
4 Methods for the Preparation of Acetals from Alcohols or Oxiranes and Carbonyl Compounds. Meskens, F. A. J. *Synthesis* **1981**, 501.
5 Protection for the Hydroxyl Group, Including 1,2- and 1,3-Diols. Greene, T. W.; Wuts, P. G. M. In *Protective Groups in Organic Synthesis*, 2nd ed.; Wiley: New York, 1990; Chapter 2.
6 O/O-Acetale. Klausener, A.; Frauenrath, H.; Lange, W.; Mikhail, G. K.; Schneider, S.; Schröder, D. In *O/O- und O/S-Acetale*; *Houben-Weyl*, 4th ed.; Vol. E14a/2; Thieme: Stuttgart, 1991.

## 3.5.2 Reviews Concerning the Mechanism for Hydolysis of Acetals

1 Mechansim and Catalysis for Hydrolysis of Acetals, Ketals, and Orthoesters. Cordes, E. H.; Bull, H. G. *Chem. Rev.* **1974**, *74*, 581.
2 General Acid Catalysis of Acetal, Ketal, and Ortho Ester Hydrolysis. Fife, T. H. *Acc. Chem. Res.* **1972**, *5*, 264.
3 *The Anomeric Effect and Related Stereoelectronic Effects at Oxygen*. Kirby, A. J.; Springer Verlag: Berlin, 1983.
4. *Stereoelectronic Effects in Organic Chemistry*. Deslongchamps, P.; Pergamon: Oxford, 1983.

## 3.5.3 Reviews Concerning Acetal Derivatives of Carbohydrates

Carbohydrate chemistry has provided a rich mine of information on the stability, selective formation, and selective cleavage of cyclic acetals and reviews devoted wholly to the synthesis and reactivity of

carbohydrate acetals should be consulted for a more detailed coverage of the subject. The Royal Society of Chemistry publishes annual reviews in its *Specialist Periodical Reports* series entitled *Carbohydrates* which gives extensive accounts of recent protecting group developments.

1 Condensation Products of Glycerol with Aldehydes and Ketones. 2-Substituted *m*-Dioxan-5-ols and 1,3-Dioxolan-4-methanols. Showler, A. J.; Darley, P. A. *Chem. Rev.* **1967**, *67*, 427.
2 Cyclic Acetal Derivatives of Sugars and Alditols. Foster, A. B. In *The Carbohydrates*; Pigman, W.; Horton, D., Eds.; Academic Press: London, 1972; Vol. IA, p 391.
3 Cyclic Acetals of Aldoses and Aldosides. De Belder, A. N. *Adv. Carbohydr. Chem. Biochem.* **1977**, *34*, 179.
4 Carbohydrate Cyclic Acetals Formation and Migration. Clode, D. M. *Chem. Rev.* **1979**, *79*, 491.
5 The Reactivity of Cyclic Acetals of Aldoses and Aldosides. Gelas, J. *Adv. Carbohydr. Chem. Biochem.* **1982**, *39*, 71.
6 Selective Removal of Protecting Groups in Carbohydrate Chemistry. Haines, A. H. *Adv. Carbohydr. Chem. Biochem.* **1982**, *39*, 13.

# References


1 Fischer, E. *Ber. Dtsch. Chem. Ges.* **1895**, *28*, 1145.
2 Horton, D.; Weckerle, W. *Carbohydr. Res.* **1988**, *174*, 305.
3 Smith, A. B.; Hale, K. J. *Tetrahedron Lett.* **1989**, *30*, 1037.
4 Ireland, R. E.; Wipf, P. *Tetrahedron Lett.* **1989**, *30*, 919.
5 Hann, R. M.; Richtmyer, N. K.; Diehl, H. W.; Hudson, C. S. *J. Am. Chem. Soc.* **1950**, *72*, 561.
6 Nicolaou, K. C.; Veale, C. A.; Hwang, C.-K.; Hutchinson, J.; Prasad, C. V. C.; Ogilvie, W. W. *Angew. Chem Int. Ed. Engl.* **1991**, *30*, 299.
7 Lipták, A.; Imre, J.; Harangi, J.; Nánási, P.; Neszmélyi, A. *Tetrahedron* **1982**, *38*, 3721.
8 Curtis, N. R.; Holmes, A. B.; Looney, M. G. *Tetrahedron Lett.* **1992**, *33*, 671.
9 Marshall, J. A.; Trometer, J. D.; Blough, B. E.; Crute, T. D. *J. Org. Chem.* **1988**, *53*, 4274.
10 Garegg, P. J.; Hultberg, H.; Wallin, S. *Carbohydr. Res.* **1982**, *108*, 97.
11 Ek, M.; Garegg, P. J.; Hultberg, H.; Oscarson, S. *J. Carbohydr. Chem.* **1983**, *2*, 305.
12 Ito, Y.; Nunomura, S.; Shibayama, S.; Ogawa, T. *J. Org. Chem.* **1992**, *57*, 1821.
13 Smith, M.; Rammler, D. H.; Goldberg, I. H.; Khorana, H. G. *J. Am. Chem. Soc.* **1962**, *84*, 430.
14 Smith, A. B.; Hale, K. J.; Laakso, L. M.; Chen, K.; Riéra, A. *Tetrahedron Lett.* **1989**, *30*, 6963.
15 Evans, D. A.; Kaldor, S. W.; Jones, T. K.; Clardy, J.; Stout, T. J. *J. Am. Chem. Soc.* **1990**, *112*, 7001.
16 Schreiber, S. L.; Wang, Z.; Schulte, G. *Tetrahedron Lett.* **1988**, *29*, 4085.
17 Kocienski, P.; Stocks, M.; Donald, D. K. In *Chirality in Drug Design and Synthesis*; C. B rown, Ed.; Academic Press: London, 1990; pp 131.
18 Joniak, D.; Kôsíková, B.; Kosáková, L. *J. Chem. Soc., Chem. Commun.* **1978**, *43*, 769.
19 Kloosterman, M.; Slaghek, T.; Hermans, J. P. G.; van Boom, J. H. *Red'l Trav. Chim. Pays-Bas* **1984**, *103*, 335.
20 Tsuri, T.; Kamata, S. *Tetrahedron Lett.* **1985**, *26*, 5195.
21 Johansson, R.; Samuelsson *J. Chem. Soc., Perkin Trans. 1* **1984**, 2371.
22 Wang, Z. *Tetrahedron Lett.* **1989**, *30*, 6611.
23 Oikawa, Y.; Yoshioka, T.; Yonemitsu, O. *Tetrahedron Lett.* **1982**, *23*, 889.
24 Nozaki, K.; Shirahama, H. *Chem. Lett.* **1988**, 1847.
25 Sviridov, A. F.; Ermolenko, M. S.; Yashunsky, D. V.; Borodkin, V. S.; Kochetkov, N. K. *Tetrahedron Lett.* **1987**, *28*, 3835.
26 Deslongchamps, P.; Moreau, C.; Fréhel, D.; Chênevert, R. *Can. J. Chem.* **1975**, *53*, 1204.
27 Stork, G.; Rychnovsky, S. *J. Am. Chem. Soc.* **1987**, *109*, 1564.
28 Sato, K.; Igarashi, T.; Yanagisawa, Y.; Kawauchi, N.; Hasimoto, H.; Yoshimura, J. *Chem. Lett.* **1991**, 1679.
29 Ziegler, F. E.; Tung, J. S. *J. Org. Chem.* **1991**, *56*, 6530.
30 Marvell, E. N.; Joncich, M. J. *J. Am. Chem. Soc.* **1951**, *73*, 973.
31 Rieche, A.; Schmitz, E.; Beyer, E. *Chem. Ber.* **1958**, *91*, 1935.

32 Hanessian, S.; Plessas, N. R. *J. Org. Chem.* **1969**, *34*, 1035.
33 Fraser-Reid, B. *Can. J. Chem.* **1976**, *54*, 2411.
34 Beau, J.-M.; Aburaki, S.; Pougny, J. R.; Sinaÿ, P. *J. Am. Chem. Soc.* **1983**, *105*, 621.
35 Hanessian, S.; Tyler, P. C.; Demailly, G.; Chapleur, Y. *J. Am. Chem. Soc.* **1981**, *103*, 6243.
36 Hanessian, S.; Desilets, D.; Rancourt, G.; Fortin, R. *Can. J. Chem.* **1982**, *60*, 2292.
37 Nakata, M.; Ikeyama, Y.; Takao, H.; Kinoshita, M. *Bull. Chem. Soc. Jpn* **1980**, *53*, 3252.
38 Collins, P. M.; Manro, A.; Opara-Mottah, E. C.; Ali, M. H. *J. Chem. Soc., Chem. Commun.* **1988**, 272.
39 Williams, D. R.; Harigaya, Y.; Moore, J. L.; D'sa, A. *J. Am. Chem. Soc.* **1984**, *106*, 2641.
40 Hines, J. N.; Peagram, M. J.; Thomas, E. J.; Whitam, G. H. *J. Chem. Soc., Perkin Trans. 1* **1973**, 2332.
41 Chapleur, Y. *J. Chem. Soc., Chem. Commun.* **1983**, 141.
42 Richtmyer, N. K. *Methods Carbohydr. Chem.* **1962**, *1*, 107.
43 Fletcher, H. G. *Methods Carbohydr. Chem.* **1962**, *2*, 307.
44 Kakinuma, K.; Otake, N.; Yonehara, H. *Tetrahedron Lett.* **1980**, *21*, 167.
45 Chittenden, G. J. F. *Rec'l. Trav. Chim. Pays-Bas* **1988**, *107*, 607.
46 Patroni, J. J.; Stick, R. V.; Skelton, B. W.; White, A. H. *Aust. J. Chem.* **1988**, *41*, 91.
47 Brockway, C.; Kocienski, P.; Pant, C. *J. Chem. Soc., Perkin Trans. 1* **1984**, 875.
48 Kocienski, P.; Street, S. D. A. *Synth. Commun,* **1984**, *14*, 1087.
49 Hikota, M.; Tone, H.; Horita, K.; Yonemitsu, O. *J. Org. Chem.* **1990**, *55*, 7.
50 Baggett, N.; Duxbury, J. M.; Foster, A. B.; Webber, J. M. *Carbohydr. Res.,* **1966**, *2*, 216.
51 Baggett, N.; Brimacombe, J. S.; Foster, A. B.; Stacey, M.; Whiffen, D. H. *J. Chem. Soc.,* **1960**, 2574.
52 Crawford, T. C.; Breitenbach, R. *J. Chem. Soc., Chem. Commun.* **1979**, 388.
53 Neszmélyi, A.; Liptá k, A.; Nánási, P. *Carbohydr. Res.* **1977**, *58*, c7.
54 Grindley, T. B.; Gulasekharem, V. *Carbohydr. Res.* **1979**, *74*, 7.
55 Martin, S. F.; Zinke, P. W. *J. Org. Chem.* **1991**, *56*, 6600.
56 Gelas, J.; Horton, D. *Heterocycles* **1981**, *16*, 1587.
57 Williams, D. R.; Sit, S.-Y. *J. Am. Chem. Soc.* **1984**, *106*, 2949.
58 Hanessian, S.; Cooke, N. G.; DeHoff, B.; Sakito, Y. *J. Am. Chem. Soc.* **1990**, *112*, 5276.
59 de Laszlo, S. E.; Ford, M. J.; Ley, S. V.; Maw, G. N. *Tetrahedron Lett.* **1990**, *31*, 5525.
60 Kocienski, P.; Yeates, C.; Street, S. D. A.; Campbell, S. F. *J. Chem. Soc., Perkin Trans. 1* **1987**, 2183.
61 Mori, K.; Iwasawa, H. *Tetrahedron* **1980**, *36*, 87.
62 Dumortier, L.; Van der Eycken, J.; Vandewalle, M. *Tetrahedron Lett.* **1989**, *30*, 3201.
63 Singh, P. P.; Gharia, M. M.; Dasgupta, F.; Srivastava, H. C. *Tetrahedron Lett.* **1977**, 439.
64 Lal, B.; Gidwani, R. M.; Rupp, R. H. *Synthesis* **1989**, 711.
65 Einhorn, C.; Luche, J. L. *Carbohydr. Res.* **1986**, *155*,
66 House, H. O.; Czuba, L. J.; Gall, M.; Olmstead, H. D. *J. Org. Chem.* **1969**, *34*, 2324.
67 Evans, D. A.; Sheppard, G. S. *J. Org. Chem.* **1990**, *55*, 5192.
68 Evans, D. A.; Dow, R. L.; Shih, T. L.; Takacs, J. M.; Zahler, R. *J. Am. Chem. Soc.* **1990**, *112*, 5290.
69 Schmid, C. R.; Bryant, J. D.; Dowlatzedah, M.; Phillips, J. L.; Prather, D. E.; Schantz, R. D.; Sear, N. L.; Vianco, C. S. *J. Org. Chem.* **1991**, *54*, 4058.
70 Jäger, V.; Häfele, B. *Synthesis* **1987**, 801.
71 Coe, J. W.; Roush, W. R. *J. Org. Chem.* **1989**, *54*, 915.
72 Mulzer, J.; de Lasalle, P.; Friessler, A. *Liebigs Ann. Chem.* **1986**, 1152.
73 Nakata, T.; Fukui, M.; Oishi, T. *Tetrahedron Lett.* **1988**, *29*, 2219.
74 Rollin, P.; Pougny, J.-R. *Tetrahedron* **1986**, *42*, 3479.
75 Meyers, A. I.; Lawson, J. P. *Tetrahedron Lett.* **1982**, *23*, 4883.
76 Schmidt, O. T. *Methods Carbohydr. Chem.* **1963**, *2*, 318.
77 Stevens, J. D. *Methods Carbohydr. Chem.* **1972**, *6*, 124.
78 Lavalée, P.; Ruel, R.; Grenier, L.; Bissonnette, M. *Tetrahedron Lett.* **1986**, *27*, 679.
79 Wershofen, S.; Scharf, H.-D. *Synthesis* **1988**, 854.
80 Buchanan, J. G.; Edgar, A. R.; Rawson, D. I.; Shahidi, P.; Wightman, R. H. *Carbohydr. Res.* **1982**, *100*, 75.
81 Rychnovsky, S. D.; Rogers, D.; Yang, G. *J. Org. Chem.* **1993**, *58*, 3511.
82 Evans, D. A.; Rieger, D. L.; Gage, J. R. *Tetrahedron Lett.* **1990**, *31*, 7099.
83 van Heeswijk, W. A. R.; Goedhart, J. B.; Vliegenthart, J. F. G. *Carbohydr. Res.* **1977**, *58*, 337.
84 Dreef, C. E.; Tuinman, R. J.; Lefeber, A. W. M.; Elie, C. J. J.; van der Marel, G. A.; van Boom, J. H. *Tetrahedron* **1991**, *47*, 4709.
85 Smith, A. B.; Rivero, R. A.; Hale, K. J.; Vaccaro, H. A. *J. Am. Chem. Soc.* **1991**, *113*, 2092.
86 Collum, D. B.; McDonald, J. H.; Still, W. C. *J. Am. Chem. Soc.* **1980**, *102*, 2118.
87 Ditrich, K. *Liebigs Ann. Chem.* **1990**, 789.
88 Heyns, K.; Lenz, J. *Chem. Ber.* **1961**, *94*, 348.

89 Liu, D.; Caperelli, C. A. *Synthesis* **1991**, 933.
90 Yin, H.; Franck, R. W.; Chen, S.-L.; Quigley, G. J.; Todaro, L. *J. Org. Chem.* **1992**, *57*, 644.
91 White, J. D.; Theramongkol, P.; Kuroda, C.; Engebrecht, J. R. *J. Org. Chem.* **1988**, *53*, 5909.
92 Curran, D. P.; Suh, Y.-G. *Tetrahedron Lett.* **1984**, *25*, 4179.
93 Buck, I. M.; Reese, C. B. *J. Chem. Soc., Perkin Trans. 1* **1990**, 2937.
94 Smith, A. B.; Kingery-Wood, J.; Leenay, T. L.; Nolen, E. G.; Sunazuka, T. *J. Am. Chem. Soc.* **1992**, *114*, 1438.
95 Trost, B. M.; Caldwell, C. G.; Murayama, E.; Heissler, D. *J. Org. Chem.* **1983**, *48*, 3252.
96 Trost, B. M.; Caldwell, C. G. *Tetrahedron Lett.* **1981**, *22*, 4999.
97 Corey, E. J.; Hopkins, P. B. *Tetrahedron Lett.* **1982**, *23*, 4871.
98 Van Speybroeck, R.; Guo, H.; Van der Eycken, J.; Vandewalle, M. *Tetrahedron* **1991**, *47*, 4675.
99 Anwar, S.; Bradley, G.; Davis, A. P. *J. Chem. Soc., Perkin Trans. 1* **1991**, 1383.
100 Markiewicz, W. T.; Wiewiórowski, M. *Nucl. Acids Res. Spec. Pub. #4* **1978**, s185.
101 Markiewicz, W. T. *J. Chem. Res.* **1979**, 24.
102 Zhang, W.; Robins, M. J. *Tetrahedron Lett.* **1992**, *33*, 1177.
103 Markiewicz, W. T.; Nowakowska, B.; Adrych, K. *Nucl. Acids Res. Sym. Ser. 18* **1987**, 149.
104 Theim, J.; Duckstein, V.; Prahst, A.; Matzke, M. *Liebigs Ann. Chem.* **1987**, 289.
105 Paulsen, H.; Krogmann, C. *Liebigs Ann. Chem.* **1989**, 1203.
106 van Boeckel, C. A. A.; van Boom, J. H. *Tetrahedron* **1985**, *41*, 4567.
107 van Boeckel, C. A. A. In *Modern Synthetic Methods 1992*; R. Scheffold, Ed.; VCH: Basel, 1992; Vol. 6, p 467.

# Chapter 4 Carboxyl Protecting Groups

# 4.1 Introduction

In Chapter 5 of their recent monograph on *Protecting Groups in Organic Synthesis*, Greene and Wuts list 83 different protecting groups for carboxylic acids of which 72 are esters in one form or another*. Only the more common groups which have found general favor will be covered here but there are many more esoteric groups which have been developed to satisfy the stringent demands of, for example, peptide and nucleoside synthesis. The vast majority of carboxyl protecting groups merely serve to prevent reaction of the acidic carboxyl hydrogen with basic reagents; consequently many of the protecting groups used are identical to hydroxyl protecting groups in terms of introduction or cleavage. However, protection of the carbonyl function against nucleophilic attack or metal hydride reduction has only rarely been achieved and Corey's 2,6,7-trioxabicyclo[2.2.2]octane (OBO) protecting group represents a significant advance.

# 4.2 Esters

## 4.2.1 General Comments on the Esterification of Carboxylic Acids

Stability, selectivity, and expense will usually dictate how an ester is to be prepared and there are a large number of methods to choose from. The traditional methods, listed below, will be covered in greater detail when we examine some of the specific ester groups

(i) *Direct preparation from an acid and alcohol.* Since the alcohol is used in large excess, it must be cheap and volatile. Hence the method is good for methyl, ethyl, isopropyl esters, etc. Since the reaction requires a strong acid catalyst, the substrate must be stable towards acid.

(ii) *Reaction of acid chlorides and anhydrides with alcohols.* This is probably the most widely used method since it is amenable to a wide range of substrates and conditions are sufficiently mild to tolerate sensitive functionality. The reaction is usually done in the presence of a mild base such as pyridine or triethylamine. With sterically hindered alcohols, the reaction may be slow. In this case DMAP can be added, which accelerates acylation reactions by a factor of $10^4$ compared with pyridine*.

(iii) *Reaction of carboxylate salts with alkyl halides.* This method is growing in popularity as a means for preparing methyl, ethyl, allyl, and benzyl esters. Since the reaction proceeds by an $S_N2$ mechanism, it is obviously limited to primary halides. Use of cesium salts or tetra-alkylammonium salts* in dipolar aprotic solvents* is particularly efficacious.

(iv) *Reaction of carboxylic acids with olefins.* The preparation of *tert*-butyl esters from isobutylene and carboxylic acids catalysed by mineral acid best exemplifies this method. It is obviously restricted in scope.

(v) *Reaction of carboxylic acid with diazoalkanes.* Provided the requisite diazoalkanes are readily available, this method offers one of the mildest and most efficient methods of esterification. It is most frequently used to make methyl and benzhydryl esters.

### 4.2.2 Recent Methods for the Esterification of Carboxylic Acids

We are now about to make a slight detour because the problem of esterification extends far beyond the bounds of protecting group manipulation. Since there is a persistent need to develop methods for protecting carboxyl groups under highly specific and mild conditions in polyfunctional molecules, some of the more recent methods for achieving esterification will be included below. All of these methods essentially involve the formation of a mixed anhydride which then undergoes selective reaction with the requisite alcohol and they have been selected to display the virtues of the methods in demanding circumstances. In the ensuing discussion some of these methods will reappear in simpler cases when we discuss the individual protecting groups.

(i) Dicyclohexylcarbodiimide (DCC) activation via *O*-acylisoureas. Esters of primary, secondary, and tertiary alcohols react with carboxylic acids in the presence of DCC and DMAP (Steglich esterification) to give the esters in good yield at room temperature[1-3]. The method has been applied to the esterification of very base-sensitive alcohols such as the one shown in [Scheme 4.1] which is prone to dehydration[4]. A similar method based on reaction of carboxylic acids with *O*-alkylisoureas* is also effective (see *tert*-butyl esters). The latter reaction can be used to prepare Me, Et, *t*-Bu, benzyl esters.

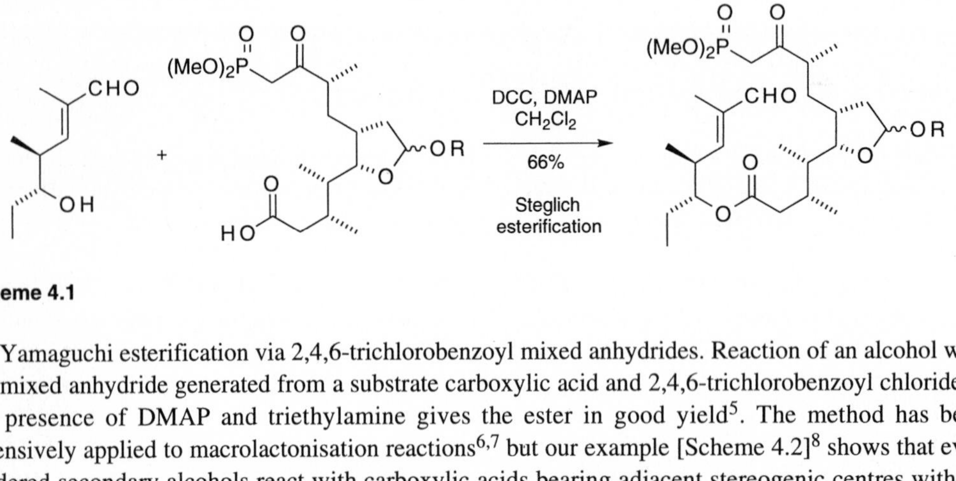

DCC, DMAP
$CH_2Cl_2$
66%
Steglich esterification

**Scheme 4.1**

(ii) Yamaguchi esterification via 2,4,6-trichlorobenzoyl mixed anhydrides. Reaction of an alcohol with the mixed anhydride generated from a substrate carboxylic acid and 2,4,6-trichlorobenzoyl chloride in the presence of DMAP and triethylamine gives the ester in good yield[5]. The method has been extensively applied to macrolactonisation reactions[6,7] but our example [Scheme 4.2][8] shows that even hindered secondary alcohols react with carboxylic acids bearing adjacent stereogenic centres without racemisation. Recently a study of some of the various methods currently available for macrolactonisation reactions has appeared in which 10-membered ring closure was used as a basis for comparison and the Yamaguchi method was shown to be one of the better methods[9].

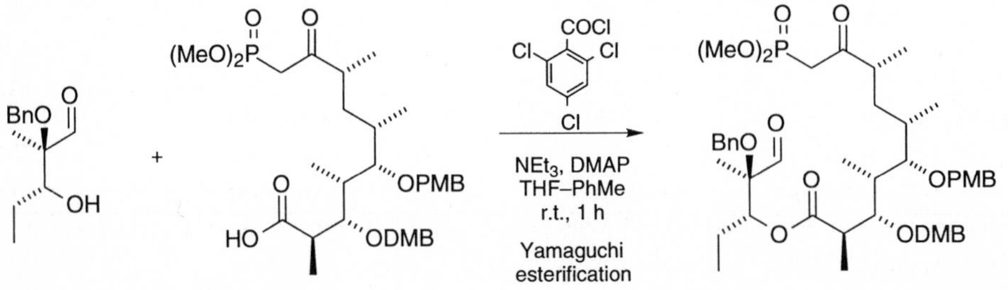

$NEt_3$, DMAP
THF–PhMe
r.t., 1 h
Yamaguchi esterification

**Scheme 4.2**

(iii) Esterification, including macrolactonisation[10], can be achieved via activation with *N*-alkyl-2-halopyridinium salts*. Scheme 4.3[11] shows an application of the method to the synthesis of a 9-membered ring which is very difficult to accomplish.

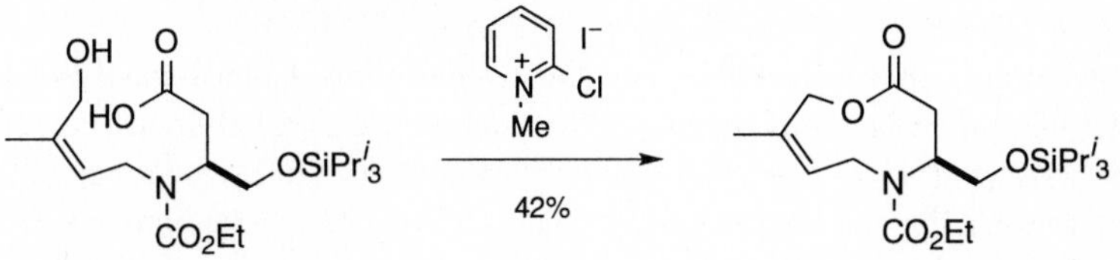

Scheme 4.3

(iv) Mitsunobu esterification* proceeds in neutral conditions and in high yield. This method differs from those cited above in that esterification proceeds by *hydroxyl activation* rather than carboxyl activation[12]; consequently, esterification of secondary alcohols proceeds with inversion of configuration. Seebach and co-workers used the Mitsunobu reaction to achieve a macrolactonisation with concomitant inversion of configuration in their synthesis of the seed germination inhibitor Gloeosporone [Scheme 4.4][13]. In a recent detailed study of various macrolactonisation methods prompted by the need for a practical and efficient synthesis of various grain beetle pheromones, Stevens and co-workers verified the superiority of the Mitsunobu reaction in the modification of Justus and Steglich[14]. For another significant application of Mitsunobu macrolactonisation in the synthesis of complex natural products, see the synthesis of Verrucarin A by Still and Ohmizu[15].

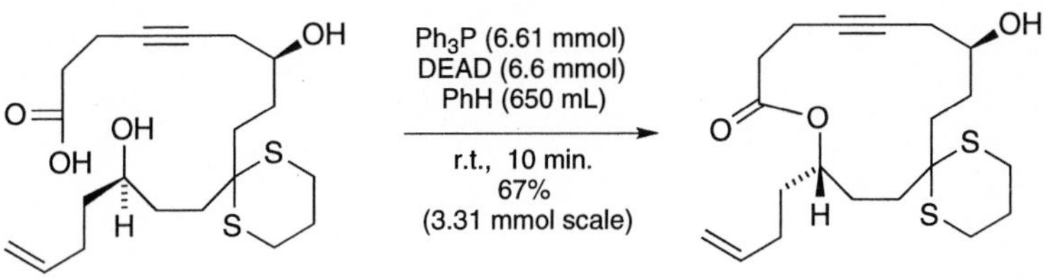

Scheme 4.4

## 4.2.3 Methyl Esters and Derivatives

### 4.2.3.1 Methyl Esters

Simplicity, low steric bulk, simple $^1$H NMR spectra, and ease of preparation are prized assets making methyl esters the most common of the carboxylic protecting groups.

**(i) Cleavage**

Base-catalysed hydrolysis using alkali metal hydroxides or carbonates in aqueous MeOH or THF remains the commonest method for cleaving simple esters limited mainly by the stability of the substrate to the basic conditions. Scheme 4.5[16] illustrates an example of selective hydrolysis using the traditional hydrolytic method in which participation of the hydroxyl group is the likely source of selectivity.

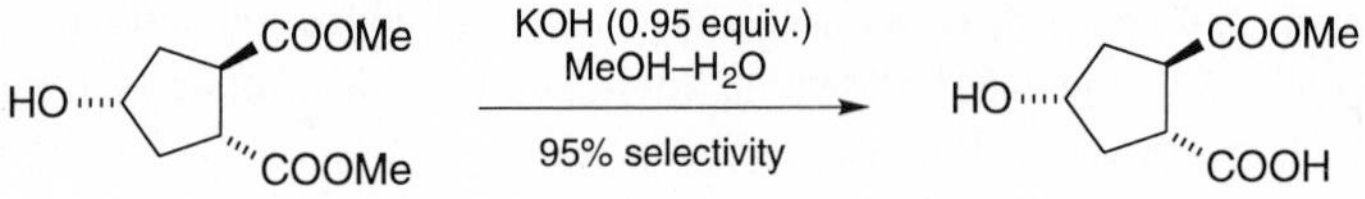
COOMe
HO
COOMe
KOH (0.95 equiv.)
MeOH–H2O
95% selectivity
COOMe
HO
COOH

**Scheme 4.5**

There are many circumstances in which the harsh conditions of base-catalysed hydrolysis are precluded. One alternative involves nucleophilic attack at the *O*-alkyl bond (i.e. an $S_N2$ reaction) rather than the carbonyl. Carboxylate anion is not a particularly good leaving group but halide, thiolate or cyanide ions, whose nucleophilicity has been enhanced by using dipolar aprotic solvents, will attack methyl, ethyl, and benzyl esters at elevated temperature. The synthetic significance of this reaction was first reported by Polish workers in 1956 (and they are seldom given due credit for their work)[17]. Methyl esters react about 70 times faster than the corresponding ethyl esters and therefore they can be selectively cleaved[18]. The procedure has been well reviewed*.

Nucleophilic cleavage is especially useful for initiating the decarboxylation of *β*-keto esters without the retro-Claisen competition which frequently attends traditional base-catalysed hydrolysis. In the case of malonates, only one of the ester functions is cleaved and the reaction is specific for esters activated by carbonyl, cyano, or sulfone groups. Scheme 4.6[19], Scheme 4.7[20], and Scheme 4.8[21] illustrate the range of substrates and reaction conditions. Although NaCl and LiI are used most frequently, other nucleophiles have been employed such as cyanide, thiocyanate, thiols, various amines, DBN, and DBU.

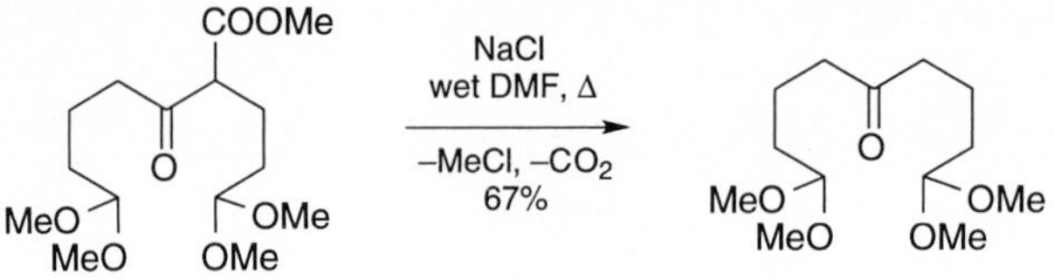
COOMe
O
MeO
MeO
OMe
OMe
NaCl
wet DMF, Δ
–MeCl, –CO2
67%
O
MeO
MeO
OMe
OMe

**Scheme 4.6**

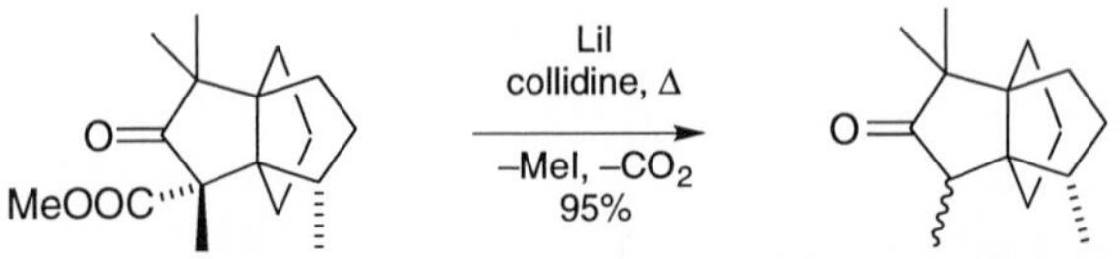
O
MeOOC
LiI
collidine, Δ
–MeI, –CO2
95%
O

**Scheme 4.7**

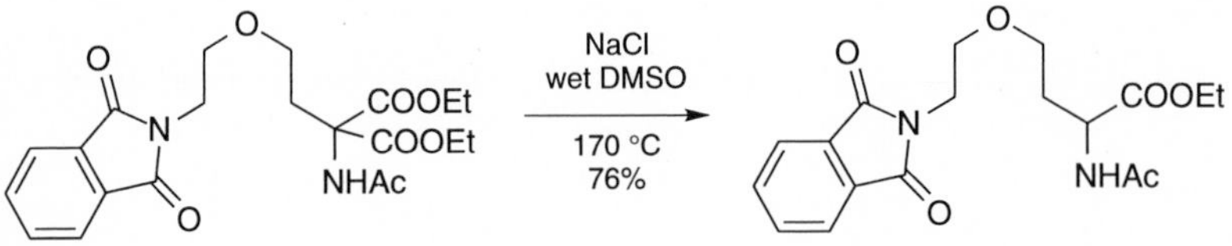
O
O
N
O
COOEt
COOEt
NHAc
NaCl
wet DMSO
170 °C
76%
O
O
N
O
COOEt
NHAc

**Scheme 4.8**

An obvious disadvantage of the methods outlined above is the narrow scope imposed by the requirement for additional activation and the high temperatures at which dealkylation takes place. Simple esters can also be cleaved, and at much lower temperature, by exploiting the high nucleophilicity of thiolate anions in dipolar aprotic solvents [Scheme 4.9][22,23] or by using Lewis acid assisted cleavage in the presence of thioether nucleophiles [Scheme 4.10][24].

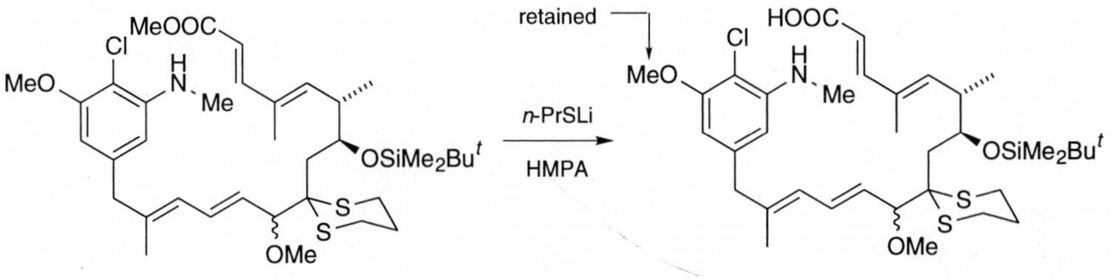

Scheme 4.9

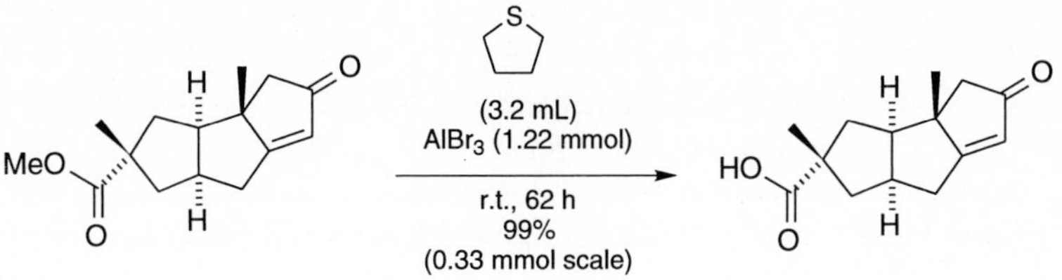

Scheme 4.10

We have already noted the value of enzymes in the cleavage and formation of esters in our discussion of hydroxyl protecting groups (section 2.2) and as promised, we now return to the subject to show how the superb selectivity and mild conditions attending enzymatic methods can be exploited to accomplish carboxyl deprotection in synthetically demanding circumstances. Thus, in another example of the "meso trick", regioselective hydrolysis of a *meso*-diester [Scheme 4.11] was used to launch the synthesis of the *ansa* side chain of the antibiotic Rifamycin S without the complication of retroaldolisation or dehydration[25,26].

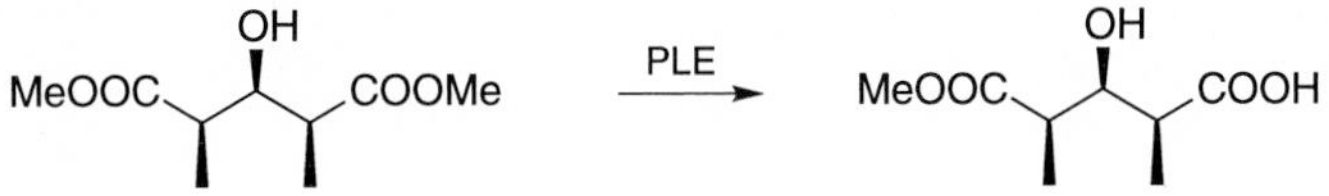

Scheme 4.11

A Merck group found a practical application of the "meso trick" during a synthesis of a receptor antagonist of Leukotriene $B_4$ having potential for the treatment of asthma [Scheme 4.12][27]. A noteworthy feature of this synthesis is the selectivity achieved given the distance separating the ester functions.

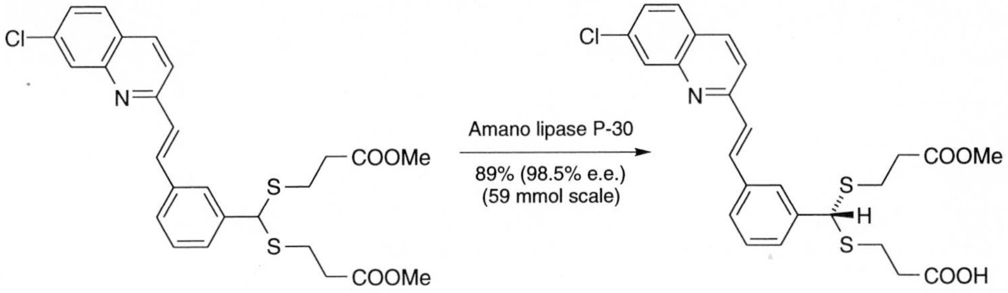

Scheme 4.12

Attempts to hydrolyse the terminal methyl ester in the Pseudomonic acid derivative shown in Scheme 4.13 using traditional chemical means failed owing to competing attack at the $\alpha,\beta$-unsaturated ester. However, the desired transformation was accomplished by using the protease Subtilisin[28].

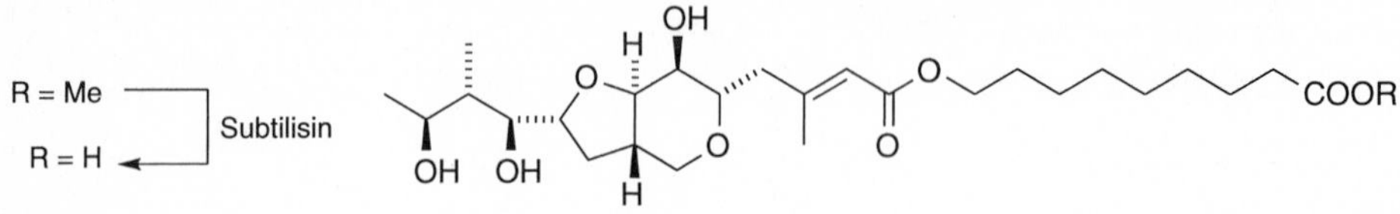

R = Me → R = H (Subtilisin)

**Scheme 4.13**

**(ii) Formation**

The reaction of a carboxylic acid with diazomethane is much prized for its mildness and efficiency. Diazomethane is usually prepared by reaction of KOH with *N*-methyl-*N*-nitroso-*p*-toluenesulfonamide (**HAZARD**: carcinogenic) and used in ether solution since it is volatile, toxic, and explosive[29]. Therefore, the method is most suitable for small scale reactions. A useful feature of the reaction is that diazomethane is intensely yellow and consumption of the reagent is easily detected by the disappearance of the colour. It may be convenient to prepare the diazomethane *in situ* [30]. Trimethylsilyldiazomethane has been recommended as a safer alternative to diazomethane for the preparation of methyl esters[31,32].

Carboxylic acids are esterified in MeOH in the presence of 2 equivalents of TMSCl[33]. It is a useful method for esterifying amino acids such as glucosaminic acid [Scheme 4.14][34]. The reaction first generates TMS esters which are converted to the corresponding methyl esters *in situ*. An even older method used for the esterification of amino acids generates HCl by the reaction of thionyl chloride in MeOH at –10 °C[35].

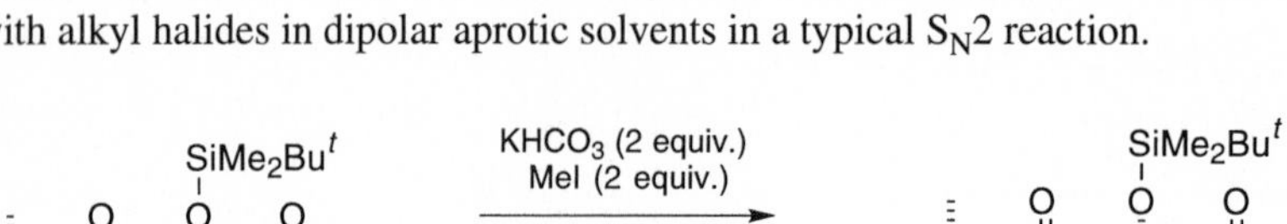

$Me_3SiCl$ (14 mmol), MeOH (35 mL); r.t., 20 h; 76% (7.6 mmol scale)

**Scheme 4.14**

Sodium, potassium [Scheme 4.15][36], and tetraalkylammonium salts [Scheme 4.16][37] of carboxylic acids react with alkyl halides in dipolar aprotic solvents in a typical $S_N2$ reaction.

$KHCO_3$ (2 equiv.), MeI (2 equiv.); DMF, r.t., 42 h; 96% (78 mmol scale)

**Scheme 4.15**

R = H → R = Me ($Na_2CO_3$, $Bu_4NBr$, $(MeO)_2SO_2$)

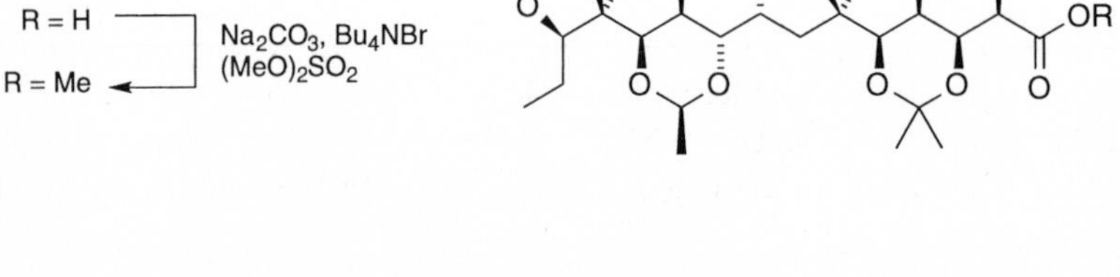

**Scheme 4.16**

Cesium salts prepared from cesium carbonate are particularly reactive and now enjoy widespread use as illustrated in Scheme 4.17 involving a synthesis of KDO, an inhibitor of lipopolysaccharide biosynthesis in Gram-negative bacteria[38]. The cesium salts are not hygroscopic and since the cesium ion is completely solvated, the carboxylates are very reactive. A recent compilation of references concerning the "cesium effect" offers some discussion as to the origin of the effect[39]. A comparison of the factors affecting the reactivity of the alkali metal cations has been reported[40].

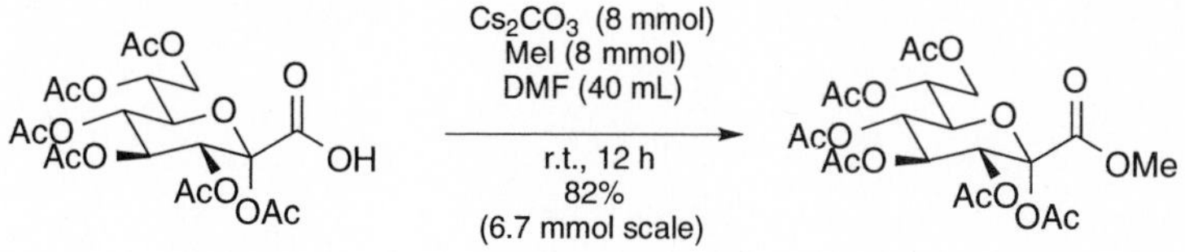

$Cs_2CO_3$ (8 mmol)
MeI (8 mmol)
DMF (40 mL)
r.t., 12 h
82%
(6.7 mmol scale)

**Scheme 4.17**

Alkylation with *N,N′*-dialkyl-*O*-methylisoureas offers a mild and efficient method which deserves wider acquaintance[41]. The *O*-methylisourea starting materials are easy to prepare by addition of MeOH to the corresponding carbodiimide and they can be stored for protracted periods in the absence of moisture. *The esterification can be performed in the presence of unprotected hydroxyl groups.*

**(iii) NMR Data**
$^1$H NMR: $\delta = 3.7$ (3H, s)
$^{13}$C NMR: $\delta = 50$ ($COOCH_3$)

### 4.2.3.2 *tert*-Butyl Esters

A feature which has been much prized in organic synthesis is the acid lability of *tert*-butyl esters which allows them to be selectively removed in the presence or primary alkyl esters[42-44]. On the other hand, primary alkyl esters hydrolyse in aqueous base much faster than *tert*-butyl esters. Compared with primary alkyl esters, *tert*-butyl esters offer a degree of steric shielding which makes them resistant to attack by a wide range of nucleophiles. An indication of the degree of protection is exemplified by the selective attack on a methyl ester in the presence of a *tert*-butyl ester [Scheme 4.18][45]. Furthermore, the enolates from *tert*-butyl esters are far more stable than enolates from simpler esters and therefore more synthetically useful[46-48]. The instability of unhindered ester enolates has been well-documented[49-51].

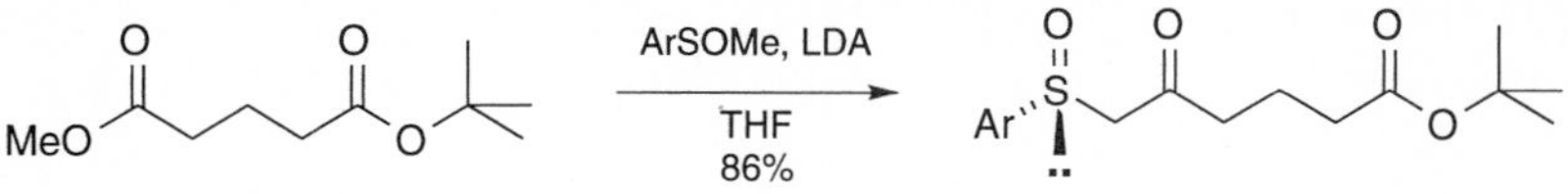

ArSOMe, LDA
THF
86%

**Scheme 4.18**

**(i) Cleavage**
*tert*-Butyl esters rapidly decompose in $CF_3COOH$ (neat or $CH_2Cl_2$) at r.t. [Scheme 4.19][48] with loss of isobutylene by a mechanism which is effectively the reverse of the classical method of preparation. Other strong acids such as formic acid or PTSA (catalytic amount) in refluxing benzene can also be used[47]. If milder acidic conditions are required, acetic acid in refluxing 2-propanol is effective and

these conditions leave methyl esters intact as shown in Scheme 4.20 which is taken from a recent synthesis of the indole alkaloid Vincamine[52].

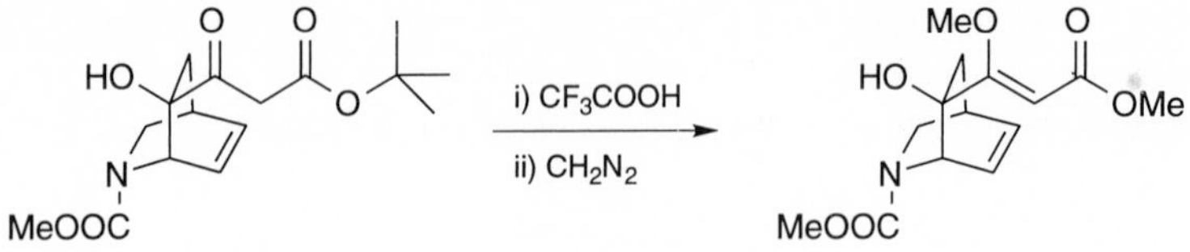

i) $CF_3COOH$
ii) $CH_2N_2$

**Scheme 4.19**

HOAc, *i*-PrOH, $H_2O$ (4:4:1)
100 °C, 15 h
100%
(2.9 mmol scale)

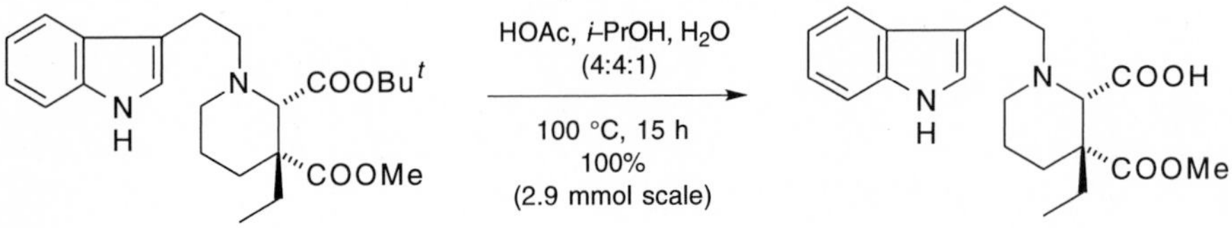

**Scheme 4.20**

The *tert*-butyl cation released in the decomposition of a *tert*-butyl ester is a powerful electrophile which may react with a functional group in the substrate. In such cases it is beneficial to add a nucleophilic scavenger such as 1,3-dimethoxybenzene or thioanisole[53-55]. In the presence of an excess of a scavenger, even highly reactive phenolic and indole substrates can be freed of a *tert*-butyl ester [Scheme 4.21][56]. Unfortunately, the reaction conditions do not allow the selective removal of a *tert*-butyl ester in the presence of a *tert*-butoxycarbonyl (Boc) moiety [Scheme 4.22] which is likewise acid-labile[57].

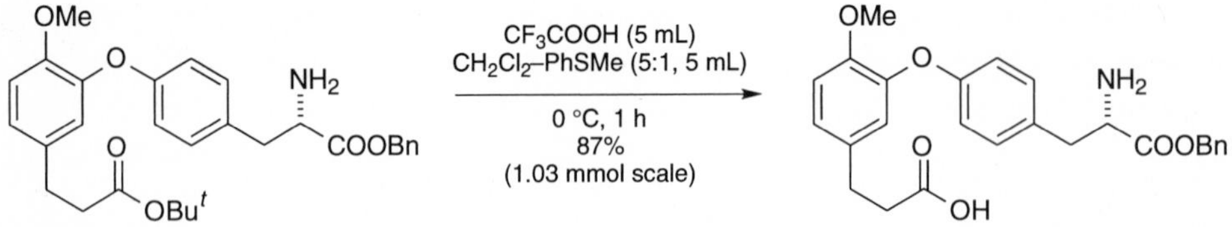

$CF_3COOH$ (5 mL)
$CH_2Cl_2$–PhSMe (5:1, 5 mL)
0 °C, 1 h
87%
(1.03 mmol scale)

**Scheme 4.21**

$CF_3COOH$ (30 mL, neat)
1,3-dimethoxybenzene (1 g)
0 °C, 3 h
95%
(5.3 mmol scale)

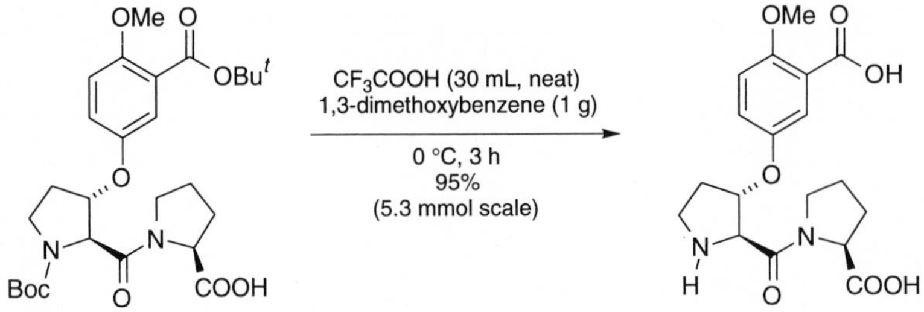

**Scheme 4.22**

*tert*-Butyldimethylsilyl triflate is sufficiently Lewis acidic to induce decomposition of a *tert*-butyl ester with formation of a *tert*-butyldimethylsilyl ester which can then be hydrolysed under very mild conditions ($K_2CO_3$ in MeOH). *tert*-Butyldimethylsilyl triflate also offers a mild method for removing Boc groups from amines[58]. Trimethylsilyl triflate is more reactive than *tert*-butyldimethylsilyl triflate and it will cleave *tert*-butyl esters rapidly [Scheme 4.23][59].

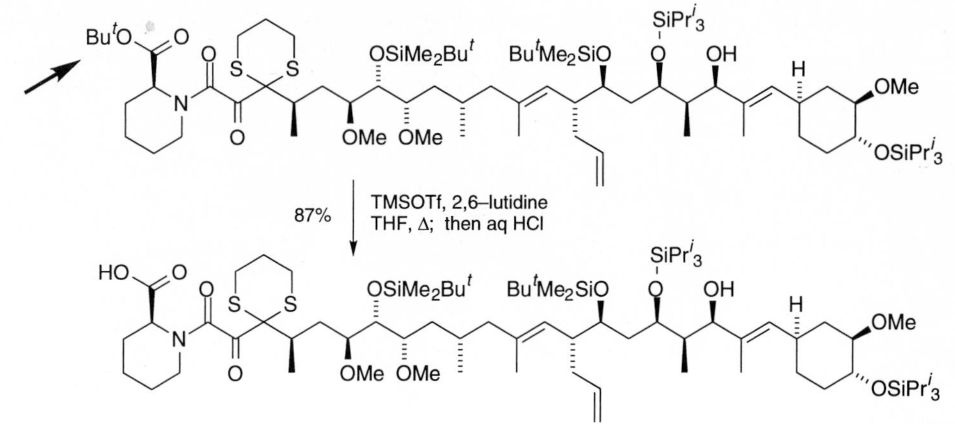

**Scheme 4.23**

*tert*-Butyl, PMB, and benzhydryl esters deprotect in the presence of $CF_3COOH$ in a phenol matrix at 45 °C. At 60 °C, the $CF_3COOH$ can be omitted. The method is mild enough for use in $\beta$-lactam chemistry [Scheme 4.24][60].

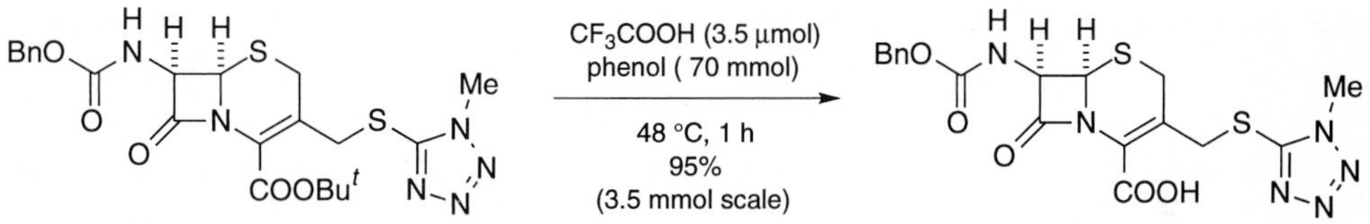

**Scheme 4.24**

*tert*-Butyl esters might be regarded as unlikely substrates for enzymatic hydrolysis. Nevertheless, Kunz and associates[61] have recently reported a remarkable alkaline serine protease (Thermitase) from the thermophilic microorganism *Thermactinomyces vulgaris* which is capable of hydrolysing sterically hindered esters ordinarily unaffected by other hydrolases. Because of the high esterase/protease activity, the enzyme is especially relevant to deprotection reactions in glycopeptides [Scheme 4.25]. Sterically less demanding methyl esters have previously been hydrolysed in peptide substrates using Chymotrypsin[62].

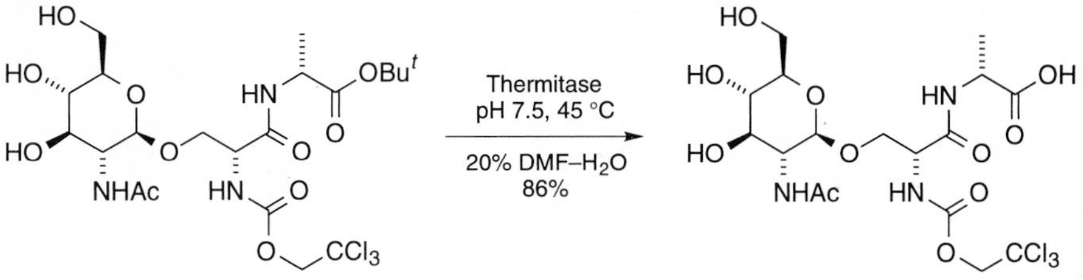

**Scheme 4.25**

Pivaloyloxymethyl (Pom) esters are useful as prodrugs of penicillin and other $\beta$-lactam antibiotics owing to their easy hydrolysis *in vivo* by ubiquitous non-specific esterases. Recently Mascaretti and co-workers showed that Pom esters can also by cleaved under mild conditions by simply treating with 2 equivalents of $(Bu_3Sn)_2O$ as shown in Scheme 4.26[63]. The intermediate tributylstannyl esters are readily hydrolysed on treatment with water to release the carboxylic acid. These conditions are also useful for the cleavage of methyl esters of amino acids without racemisation and it is noteworthy that functional groups such as aldehydes, thioacetals, amides, vinyl bromides, and nitro compounds are compatible[64].

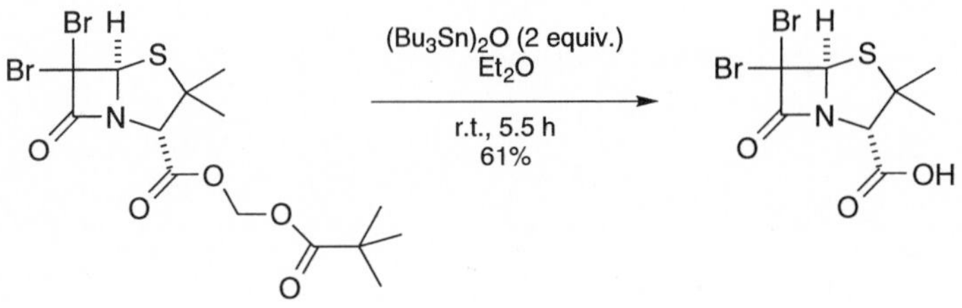

$(Bu_3Sn)_2O$ (2 equiv.), $Et_2O$, r.t., 5.5 h, 61%

**Scheme 4.26**

### (ii) Formation

*O-tert*-Butyl trichloroacetimidate, prepared in 70% yield by reacting *t*-BuOK with trichloroacetonitrile, reacts with carboxylic acids and alcohols in the presence of a catalytic amount of $BF_3$ at r.t. for 16–21 h in cyclohexane–$CH_2Cl_2$[65]. The method also converts alcohols to *tert*-butyl ethers (see section 2.4.5). A very similar reaction which allows *tert*-butylation under essentially neutral conditions involves reaction of a carboxylic acid with 3-4 equivalents of *N,N′*-diisopropyl-*O-tert*-butylisourea[41] [Scheme 4.27] — a reaction we have already described in regard to the preparation of methyl esters (*vide supra*). The reaction proceeds via a tertiary carbonium ion intermediate and since capture of the cation is inefficient, 3-4 equivalents of the isourea are required. The presence of alcohols is tolerated but not thiols or unhindered amines.

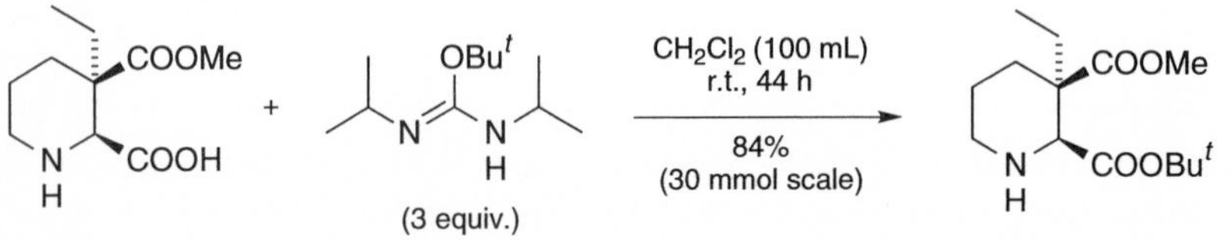

COOMe, COOH + OBu$^t$ (3 equiv.); $CH_2Cl_2$ (100 mL), r.t., 44 h, 84% (30 mmol scale); COOMe, COOBu$^t$

**Scheme 4.27**

The classical method for making *t*-Bu esters involves mineral acid-catalysed addition of the carboxylic acid to isobutene but it is a rather brutal procedure for use in any but the most insensitive of substrates [Scheme 4.28][66-68].

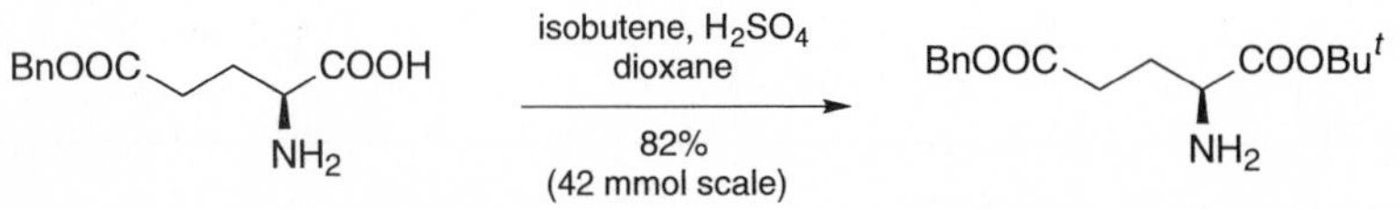

BnOOC, COOH, $NH_2$; isobutene, $H_2SO_4$, dioxane, 82% (42 mmol scale); BnOOC, COOBu$^t$, $NH_2$

**Scheme 4.28**

A milder method for forming *t*-Bu esters is the Yamaguchi esterification which involves reaction of a carboxyl group with 2,4,6-trichlorobenzoyl chloride and triethylamine in THF. The resultant mixed anhydride intermediate reacts with *t*-BuOH in the presence of DMAP to give the *tert*-butyl ester in good yield[5].

**(iii) NMR Data**

$^{1}$H NMR: $\delta = 1.5$ (9H, s)
$^{13}$C NMR: $\delta = 80$ (3C), 28 (1C)

#### 4.2.3.3 MOM, MEM, BOM, MTM, SEM Esters

These esters are seldom used which is surprising since they are very easy to make in good yield and they can be removed under mild conditions. Their inclusion here is based more on promise rather than performance.

**(i) Cleavage**

Acetal esters are sensitive to mild aqueous acid in THF; and mild Lewis acids such as $MgBr_2$ (2 equiv., in $Et_2O$ at r.t. for 1–24 h) cleave all the esters including MTM esters. MEM is cleaved the slowest[69]. A SEM ester was recently used in a synthesis of the cyclodepsipeptide Didemnin by Joullié and co-workers[70]. In one step [Scheme 4.29], mild acid deprotection was accomplished by reaction with aqueous HF in acetonitrile at –10 °C without detriment to Boc (*tert*-butoxycarbonyl) and Cbz (benzyloxycarbonyl) groups. In addition, BOM esters can be removed by catalytic hydrogenation[71].

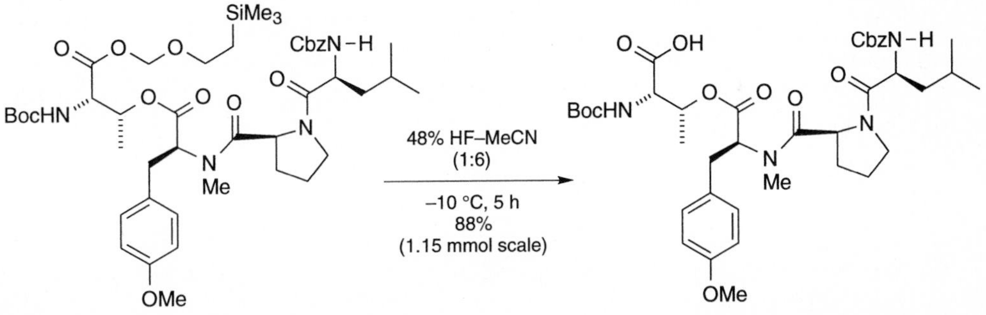

SiMe3 O O O O BocHN O O O CbzN–H N N Me OMe
48% HF–MeCN (1:6)
–10 °C, 5 h
88%
(1.15 mmol scale)
O OH BocHN O O O O CbzN–H N N Me OMe

**Scheme 4.29**

Methylthiomethyl (MTM) are useful in relay deprotection. They have the stability of most typical unhindered esters towards base but their lability can be greatly increased by oxidation with ammonium molybdate and $H_2O_2$ to the corresponding methylsulfonylmethyl esters[72]. A striking example of the value of MTM esters in activating a carboxyl group towards nucleophilic attack can be found in the synthesis of the pyrollizidine alkaloid Integerrimine [Scheme 4.30] reported by Narasaka and co-workers[73]. The MTM ester group was deployed twice: in the first instance a methylsulfonylmethyl ester (**30.2**) was selectively hydrolysed in the presence of a neighboring MTM ester in order to form the carboxylic acid precursor to symmetrical anhydride **30.3**. After acylation of the protected Retronecine derivative **30.7**, selective activation of the terminal carboxyl group was accomplished via oxidation of the MTM ester to the corresponding methylsulfonylmethyl ester **30.5** which underwent macrolactonisation to give the 12-membered bislactone bridge. Finally removal of the MOM ether afforded Integerrimine. MTM esters can also be hydrolysed by heating with MeI in aqueous acetone[74].

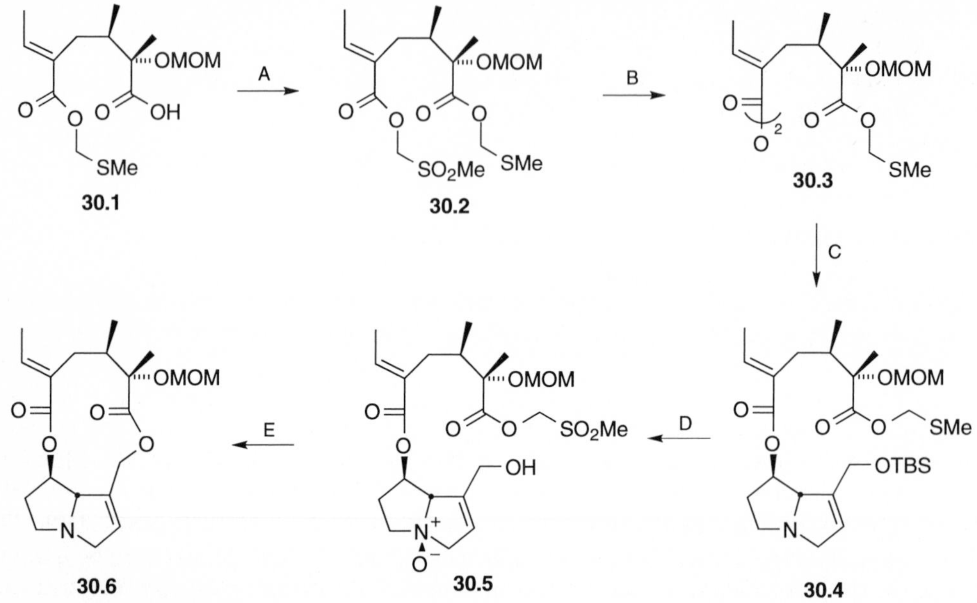

30.1 30.2 30.3 30.4 30.5 30.6 30.7

REAGENTS AND YIELDS

A 71% (i) Mo(VI), $H_2O_2$; (ii) $MeSCH_2Cl$, *i*-$Pr_2NEt$, DME, r.t., 6 h;
B 91% (i) NaOH, aq. DME, r.t., 5 h;
(ii) 1-methyl-2-chloropyridinium iodide, $Et_3N$, $CH_2Cl_2$;
C 81% alkoxide **30.7,** THF, r.t., 4.5 h;
D 60% (i) $NH_4F$ (10 equiv.), MeOH-$H_2O$ (2:1), 60 °C, 6 h; (ii) Mo(VI), $H_2O_2$;
E 24% (i) 1 equiv. $Ph_3CLi$, THF, –78 °C (41%); (ii) Zn, $H_2SO_4$, DME, r.t., 1 h (59%).

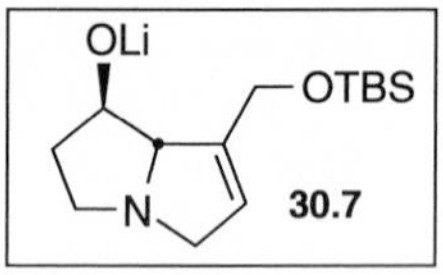

**Scheme 4.30**

## (ii) Formation

Reaction of carboxylate salts with RO-$CH_2$-Cl takes place under conditions similarly used to protect hydroxyl functions [Scheme 4.31][70]. In the absence of acid, the resultant acetal esters display reactivity typical of alkyl esters.

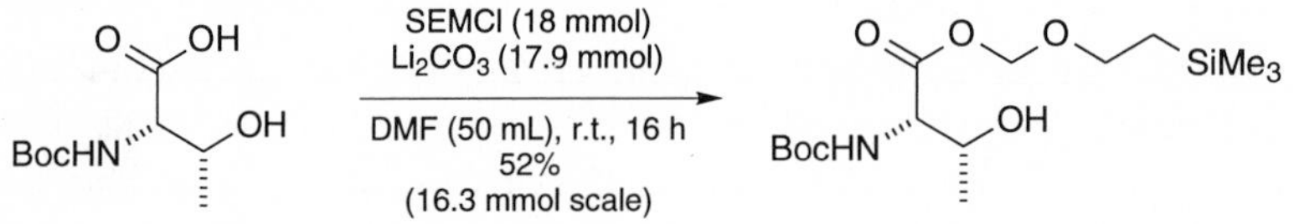

SEMCl (18 mmol)
$Li_2CO_3$ (17.9 mmol)
DMF (50 mL), r.t., 16 h
52%
(16.3 mmol scale)

**Scheme 4.31**

MTM esters have been prepared by reaction of triethylammonium salts of carboxylic acids with chlorodimethylsulfonium chloride[75] or by reaction of the potassium salts of carboxylic acids with $MeSCH_2Cl$ in the presence of NaI and 18-crown-6[76].

**(iii) NMR Data**

$^{1}$H NMR:

BOM $\delta = 4.7$ (2H, s, $PhCH_2$), 5.7 (2H, s, $OCH_2O$), 7.3 (5H)
MOM $\delta = 5.0$ (2H, s, $OCH_2O$), 3.3 (3H, s, OMe)
MTM $\delta = 5.0$ (2H, s, $OCH_2S$), 2.15 (3H, s, SMe)
MEM $\delta = 5.0$ (2H, s, $OCH_2O$), 3.4 (4H, m, $OCH_2CH_2O$), 3.3 (3H, s, OMe)
SEM $\delta = 4.3$ (2H, m or s, $OCH_2O$), 3.7 (2H, m, $TMSCH_2\mathbf{CH_2}$), 0.9–1.0 (2H, m, $TMSCH_2$).

$^{13}$C NMR:

MOM ($MeCOOCH_2OMe$) $\delta = 170.6$ (C=O), 90.4 ($OCH_2O$), 57.6 (OMe), 21.0 (**Me**C=O)
MEM $\delta = 172$ (C=O), 88 ($OCH_2O$), 73 ($OCH_2CH_2O$), 71 ($CH_2OMe$), 58 (OMe)

## 4.2.4 β-Substituted Ethyl Ester Derivatives

### 4.2.4.1 2,2,2-Trichloroethyl Esters

**(i) Cleavage**

For a synthesis of Cephalosporin C, Woodward and co-workers required a protecting group which could be removed under mild and highly specific conditions. From this work, the 2,2,2-trichloroethyl (TCE) group was born and its specific removal on treatment with Zn was the concluding step to their pioneering synthesis[77]. The group is still used extensively in peptide synthesis as shown in an early step of a recent synthesis of the antibiotic Lavendomycin [Scheme 4.32][78] and more recently it has been extended to the protection of phosphate esters in the formation of internucleotide links in oligonucleotide synthesis[79-81]. Other reductive methods which have been recommended include electrolysis[82] and treatment with catalytic amounts of Se metal in the presence of sodium borohydride in DMF at 40–50 °C[83].

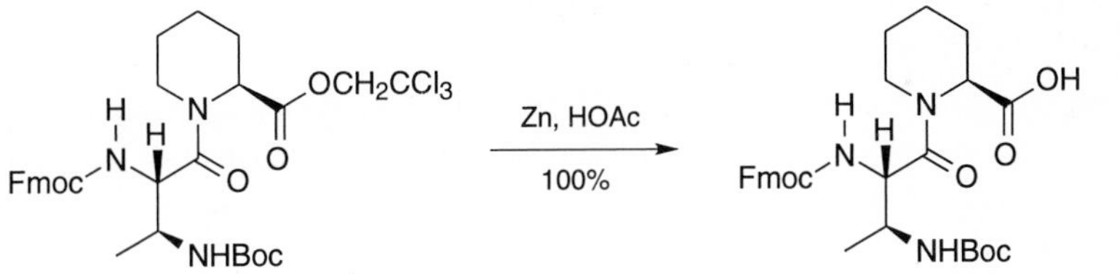

Scheme 4.32

Another denizen of the small reductive–elimination orthogonal set is the phenacyl group which, like the TCE ester, is useful for deprotecting the C-terminus of a peptide under conditions that retain such stalwart N-protecting groups as Fmoc, Boc, and Cbz (see Chapter 6). Its utility is exemplified in a the C-terminal deprotection [Scheme 4.33] of a dipeptide in Schmidt's recent synthesis of the immunosuppressive and antiviral cyclodepsipeptide Didemnin[84].

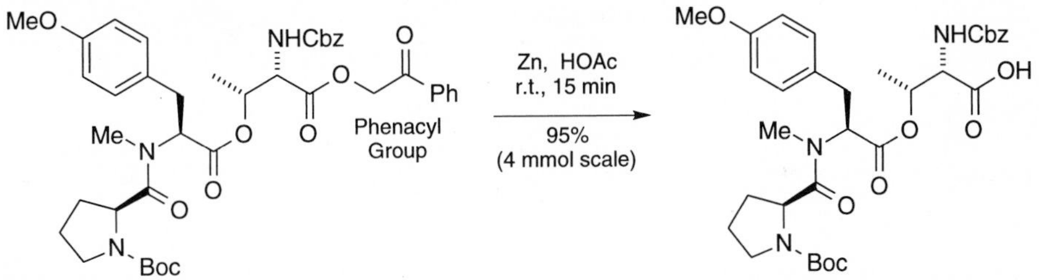

Scheme 4.33

**(ii) Formation**

The Steglich esterification of a carboxylic acid with 2,2,2-trichloroethanol using DCC–DMAP affords a mild and convenient method for preparing TCE esters [Scheme 4.34][85]. More traditional methods involving the reaction of an acid chloride with trichloroethanol in the presence of pyridine have also been used.

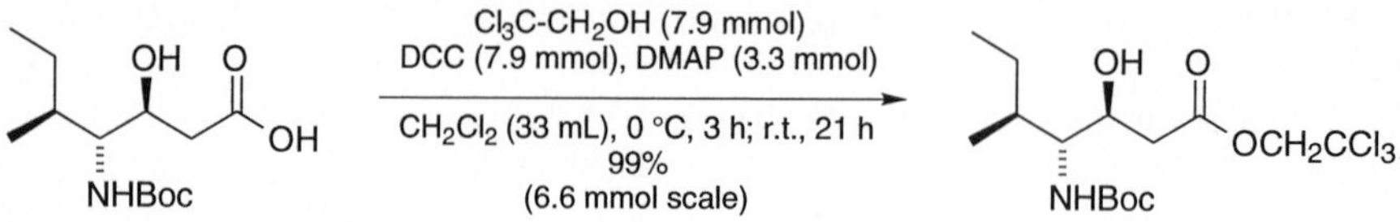

$Cl_3C\text{-}CH_2OH$ (7.9 mmol)
DCC (7.9 mmol), DMAP (3.3 mmol)
$CH_2Cl_2$ (33 mL), 0 °C, 3 h; r.t., 21 h
99%
(6.6 mmol scale)

OH O, OH, NHBoc → OH O, $OCH_2CCl_3$, NHBoc

**Scheme 4.34**

**(iii) NMR Data**

$^1H$ NMR: $\delta = 4.8$ (2H)

$^{13}C$ NMR: $\delta = 153$ (C=O), 95 ($Cl_3C$), 75 ($CH_2$)

#### 4.2.4.2 2-(Trimethylsilyl)ethyl Esters

The 2-(trimethylsilyl)ethyl (TMSE) ester function is prized for the ease and selectivity of its deprotection. Like the SEM protecting group, 2-(trimethylsilyl)ethyl esters undergo a fragmentation reaction on treatment with TBAF to give ethylene, TMSF, and the tetra-*n*-butylammonium salt of the carboxylic acid which is then recovered on acidification. Gerlach and Sieber independently developed this group for the unmasking of specific ester functions in the presence of other esters[86,87].

**(i) Cleavage**

During a synthesis of the potent insect toxin Pederin, a Southampton group was faced with the problem of unleashing a carboxylic acid in a fragment which was both acid and base sensitive. After much experimentation TMSE ester protection provided the sole practical solution [Scheme 4.35][88]. Thus, treatment of TMSE ester **35.1** with TBAF in THF resulted in the expected fragmentation reaction but attempts to recover the carboxylic acid by acidification of the tetra-*n*-butylammonium salt **35.2** only led to rapid decomposition. However, if the salt was simply partitioned between ether and water, the desired carboxylic acid **35.3** extracted into the ether layer *without acidification* and could be isolated in quantitative yield. In polypeptide systems, similar fragmentation reactions have been secured by using a mixture of tetra-*n*-butylammonium chloride and $KF \cdot 2H_2O$ in DMF[89].

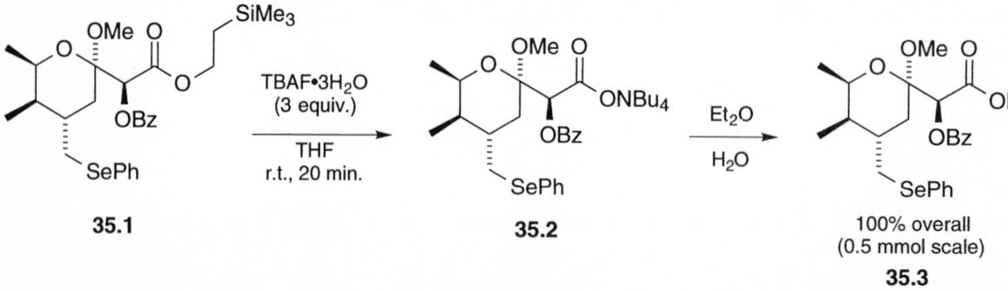

OMe O, O, $SiMe_3$, OBz, SePh — **35.1**

$TBAF \cdot 3H_2O$ (3 equiv.), THF, r.t., 20 min.

OMe O, $ONBu_4$, OBz, SePh — **35.2**

$Et_2O$, $H_2O$

OMe O, OH, OBz, SePh — 100% overall (0.5 mmol scale) **35.3**

**Scheme 4.35**

The high susceptibility of silicon to attack by fluoride has been used extensively to accomplish the simultaneous deprotection of hydroxyls (protected as silyl ethers) and carboxyls (protected as TMSE esters) as a prelude to lactonisation [Scheme 4.36][90]. The power of the method is nicely illustrated [Scheme 4.37] by the penultimate step of a synthesis of the tumour growth inhibitor Verrucarin A[15]. The challenge here was the selective deprotection of only one of three ester functions whilst laying bare a hydroxyl function (protected as a *tert*-butyldiphenylsilyl ether) at the same time. The desired transformation was accomplished by treating the $\beta$-(trimethylsilyl)ethyl ester **37.1** with TBAF to produce an intermediate hydroxy acid **37.2** which underwent Mitsunobu macrolactonistion to afford Verrucarin A in 52% yield.

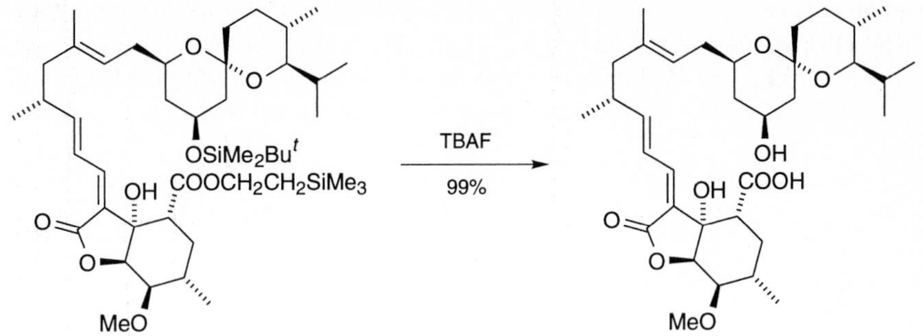

$OSiMe_2Bu^t$ OH $COOCH_2CH_2SiMe_3$ MeO — TBAF, 99% → OH, OH COOH, MeO

**Scheme 4.36**

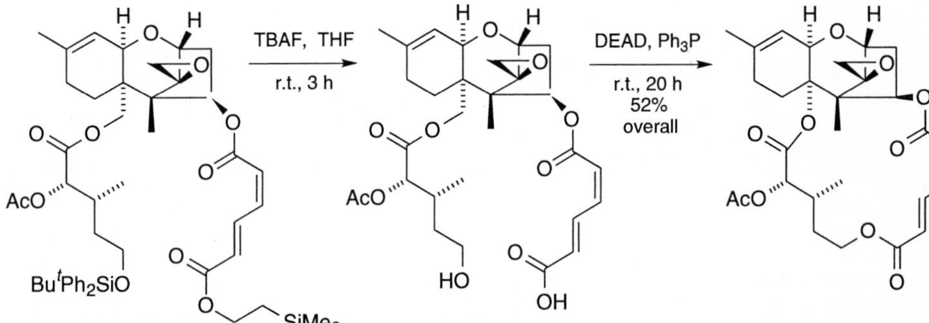

**37.1** ($Bu^tPh_2SiO$, AcO, $SiMe_3$) — TBAF, THF, r.t., 3 h → **37.2** (HO, AcO, OH) — DEAD, $Ph_3P$, r.t., 20 h, 52% overall → **37.3** (AcO)

**Scheme 4.37**

The synthesis of Fulvine by Vedejs and Larsen[91] [Scheme 4.38] also unleashed a carboxyl group as a prelude to macrolactonisation but in this case the nucleophilicity of the nascent tetra-*n*-butyl

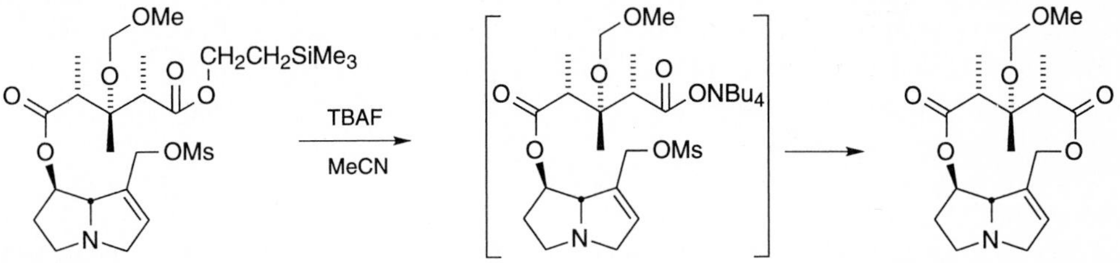

OMe, $CH_2CH_2SiMe_3$, OMs — TBAF, MeCN → [OMe, $ONBu_4$, OMs] → OMe

**Scheme 4.38**

ammonium carboxylate was exploited in an intramolecular displacement to close the 12-membered bis-lactone ring. It is noteworthy that the requisite deprotection took place in the midst of a sensitive allylic mesylate moiety.

**(ii) Formation**

Esterification of a carboxylic acid with 2-(trimethylsilyl)ethanol (commercially available) using DCC–DMAP (Steglich esterification) [Scheme 4.39][92] has the advantage of wide application and mild conditions. Carboxyl activation using 2-chloro-1-methylpyridinium chloride (Mukaiyama's reagent) has also been used[93]. Traditional methods based on esterification using an acid chloride or anhydride and 2-(trimethylsilyl)ethanol are equally effective. However, with less reactive acylating agents, esterification in the presence of acid or base poses a potential problem in the decomposition of the 2-(trimethylsilyl)ethanol via Peterson olefination. One solution to the problem of enhancing acylation rates exploits the oxyphilicity of aluminium. Thus, reaction of the anhydride in Scheme 4.40 with [2-(trimethylsilyl)ethoxy]dimethylalane afforded the desired ester in >85% yield[91].

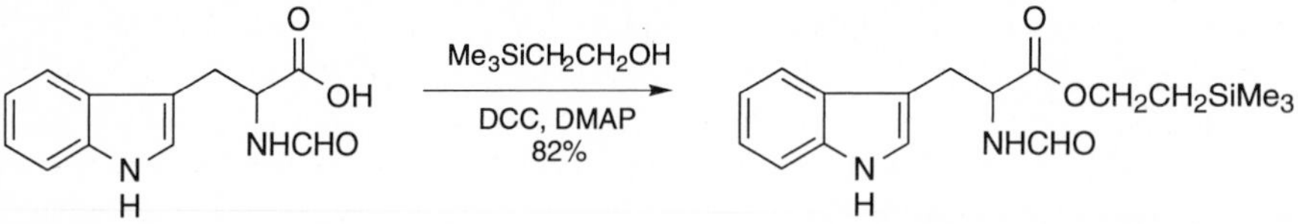

**Scheme 4.39**

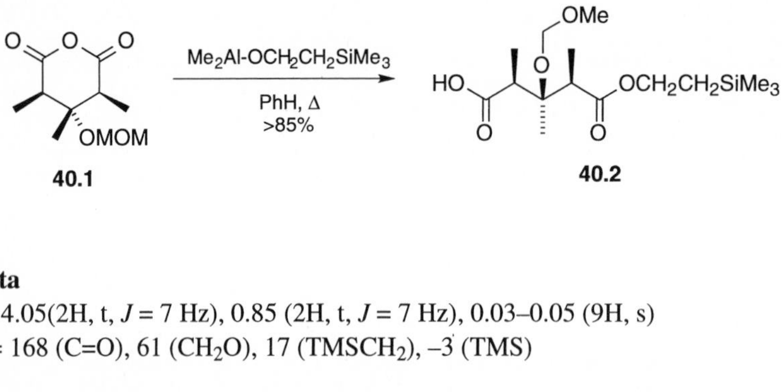

**Scheme 4.40**

**(iii) NMR Data**

$^1H$ NMR: $\delta = 4.05$(2H, t, $J = 7$ Hz), 0.85 (2H, t, $J = 7$ Hz), 0.03–0.05 (9H, s)
$^{13}C$ NMR: $\delta = 168$ (C=O), 61 ($CH_2O$), 17 ($TMSCH_2$), –3 (TMS)

### 4.2.4.3 2-(*p*-Toluenesulfonyl)ethyl Esters and Related Base-Labile Groups

In 1968 Miller and Stirling showed that the 2-(*p*-toluenesulfonyl)ethyl ester function (abbreviated TSE) underwent easy base-catalysed elimination in the presence of M NaOH or $Na_2CO_3$ (but not $NaHCO_3$) in aqueous dioxane at r.t. to give *p*-toluenesulfonylethene and a carboxylate[94]. Thus, this ester function complements the methylsulfonylmethyl function (see section 4.2.3.3) derived from MTM esters in its base-sensitivity. Electron withdrawing groups (e.g., *p*-$NO_2$) on the aryl ring increase base lability — a feature which has been exploited for the protection of the 2′-hydroxyl function in oligoribonucleotide synthesis[95].

**(i) Cleavage**

In our earlier discussion of the use of fluoride in the deprotection of silyl ethers, the potential for mischief caused by the basicity of various fluoride reagents was noted. However, the basicity of fluoride reagents can be harnessed to good use[96]. A case in point is the easy elimination of TSE esters induced by TBAF in non-aqueous media. The reaction is so fast that TBS groups remain intact [Scheme 4.41][97].

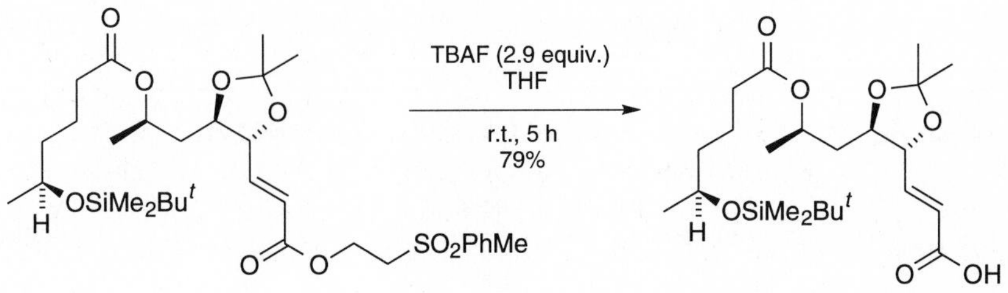

O
O
O
O
OSiMe2Bu^t
H
O
O
SO2PhMe
TBAF (2.9 equiv.)
THF
r.t., 5 h
79%
OH

**Scheme 4.41**

Deprotection of a 2-(*p*-toluenesulfonyl)ethyl ester using a catalytic amount of DBU in benzene was an important step in the finale of Raphael's synthesis of Pyrenophorin[98] which has recently been modified[7] [Scheme 4.42]. Nowadays this transformation would probably be accomplished using $\beta$-(trimethylsilyl)ethyl esters. 2-(Methylsulfonyl)ethyl esters have also been cleaved under basic conditions to release a protected carboxylic acid. Scheme 4.43 shows an example from the closing stages of a synthesis of Surugatoxin[99].

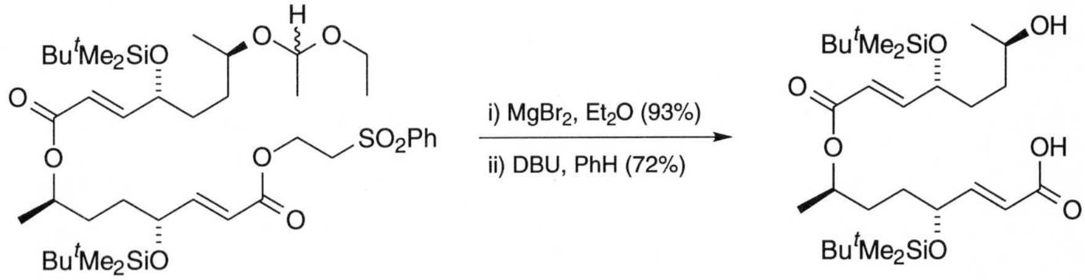

Bu^tMe2SiO
SO2Ph
i) MgBr2, Et2O (93%)
ii) DBU, PhH (72%)
OH
OH

**Scheme 4.42**

MeSO2CH2CH2OOC
COMe
C=O OBn
Br
Ac
OBn
NaHCO3–Na2CO3
acetone, pH 10.2
r.t., 2 h
60%
HOOC

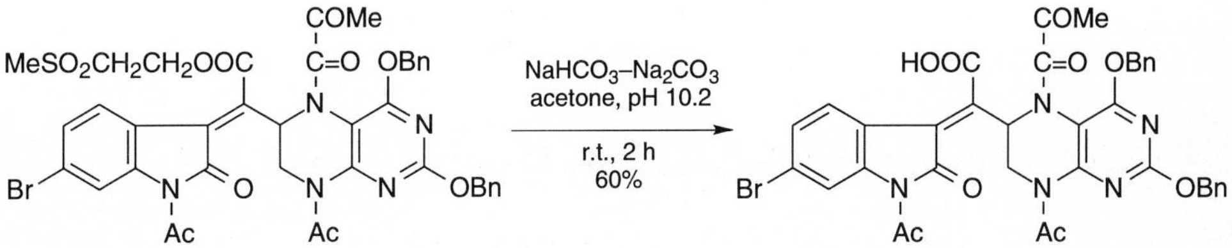

**Scheme 4.43**

9-Fluorenylmethoxycarbonyl (Fmoc) groups enjoy widespread use in the protection of amines during solid phase peptide synthesis (see Chapter 6) and recently their useful base-lability has been adapted to the protection of carboxylic acids as 9-fluorenylmethyl esters[100-102].

Many of the carboxylate protecting groups we have discussed so far can also be used to protect phosphates. A protecting group from the $\beta$-elimination orthogonal set which has been particularly effective for the protection of phosphates in oligonucleotide synthesis is the $\beta$-cyanoethyl ester function[103]. However, the transformation we have chosen to exemplify the method [Scheme 4.44] comes from recent model studies directed towards a synthesis of the tumour promoter Calyculin[104]. In this example, attempts to accomplish the desired $\beta$-elimination with DBU alone led to loss of only one of the cyanoethyl groups whereas use of DBU in the presence of TMSCl resulted in complete elimination.

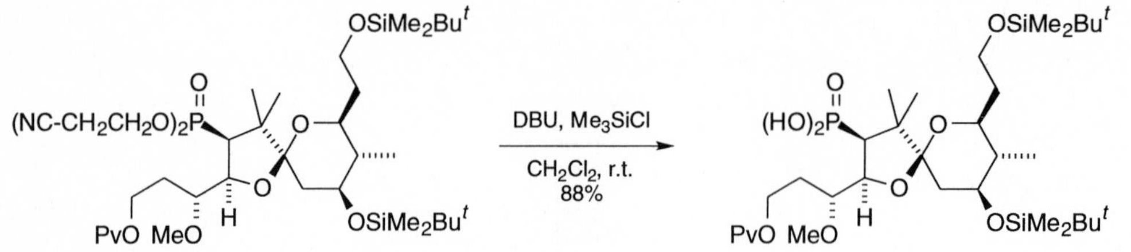

Scheme 4.44

(ii) **Formation**

The method used by Miller and Stirling for preparing TSE esters in their pioneering study involved esterification with 2-(*p*-toluenesulfonyl)ethanol in the presence of dicyclohexylcarbodiimide [Scheme 4.45]. The resultant TSE esters are stable to triethylamine and ammonia and they survive conditions for peptide bond formation using the *p*-nitrophenyl ester method or the mixed carbonic anhydride method of carboxyl activation. If the base sensitivity of the TSE ester poses a problem in a multistage synthesis, the sulfonyl group can be introduced by oxidation of the corresponding thioether using ammonium molybdate [$(NH_4)_6Mo_7O_{24}$] and $H_2O_2$ — conditions we have already seen used in the oxidation of MTM esters (section 4.2.3.3).

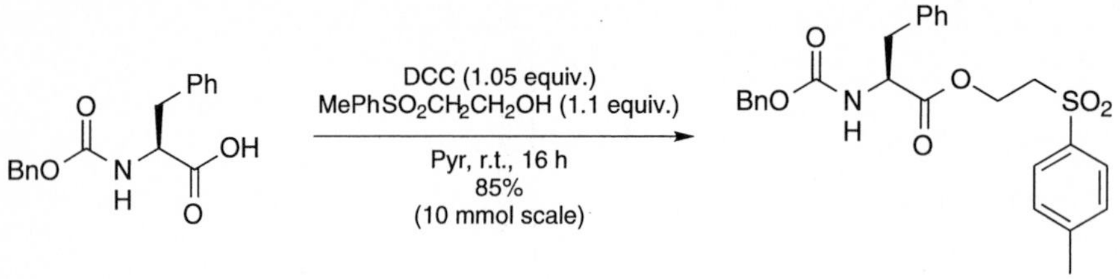

Scheme 4.45

(iii) **NMR Data**

$^1$H NMR: $\delta = 4.4$ (2H, m, $CH_2$-O), 3.4 (2H, m, $ArSO_2$-$CH_2$)

### 4.2.5 Benzyl, Benzhydryl, and *p*-Nitrobenzyl Esters

The benzyl ester group, like the benzyl ether, is an early and versatile addition to the armamentarium of carboxyl protecting groups, which has long been prized for its ease of introduction and removal[105]. There are many methods that can be used to achieve both.

(i) **Cleavage**

Being comparatively unhindered, benzyl esters are susceptible to acid or base hydrolysis. However, the mildest and most attractive method involves catalytic hydrogenolysis which is particularly easy in the case of benzhydryl esters. For example, simultaneous hydrogenolysis of a benzyloxycarbamate and benzhydryl ester (Pd–C) was used to unleash the amino acid residue in Clavalamine [Scheme 4.46][106]. Further examples gleaned from Merck's extensive *β*-lactam programme underscore the value of *p*-nitrobenzyl esters in the synthesis of demanding targets. Hydrogenolysis of *p*-nitrobenzyl ester **47.1** using Pt as the catalyst was a terminal step in the synthesis of Thienamycin **47.2** [Scheme 4.47][107]. In an earlier synthesis of the same target, the Merck group[108] used LiI in hot collidine to cleave a bis-*p*-nitrobenzyl malonate **47.3**. The remaining three *p*-nitrobenzyl ester groups in the product **47.4** were then removed simultaneously from the resultant product to give Thienamycin. Benzyl esters, benzyl phosphates, benzyl carbonates and benzyl carbamates are selectively and efficiently deprotected in the presence of benzyl ethers, benzyloxymethyl ethers, and *N*-Bn groups by catalytic transfer hydrogenation using 10% Pd-C in EtOH and cyclohexadiene as the hydrogen donor[109].

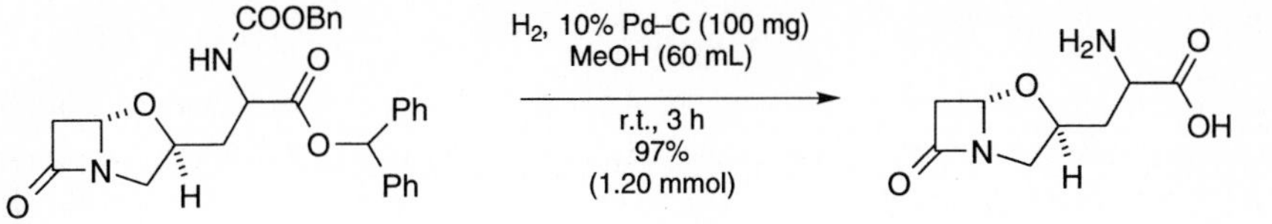

COOBn HN O O N O H O Ph Ph

$H_2$, 10% Pd–C (100 mg) MeOH (60 mL) r.t., 3 h 97% (1.20 mmol)

$H_2N$ O O N O H OH

**Scheme 4.46**

OH H H N S H N PNB O O O O PNB **47.1**

$H_2$, Pt

OH H H N S $NH_2$ O OH O **47.2**

$H_2$, cat.

O O PNB O H H N S H N PNB O O O O O O PNB PNB **47.3**

LiI, collidine 120 °C, 30 min 47%

O O PNB O H H N S H N O PNB O O O O PNB **47.4**

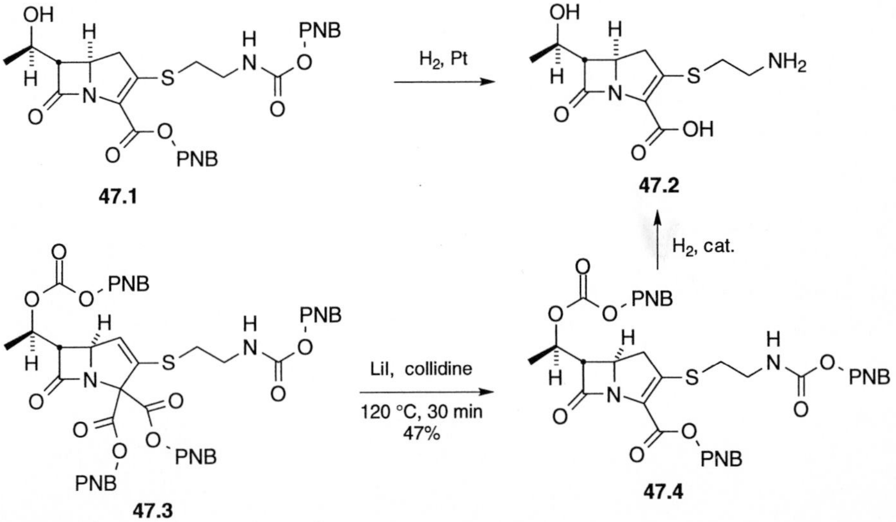

**Scheme 4.47**

Catalytic hydrogenolysis of benzyl ethers, esters and carbamates can be prevented by traces of poisons such as S or Se. For example, a Swern oxidation in a previous step can introduce sufficient sulfur by-products to thwart even the most active catalyst. In such cases, Raney nickel can be used to remove sulfur contaminants and the sample then subjected to the usual hydrogenolysis conditions and it is

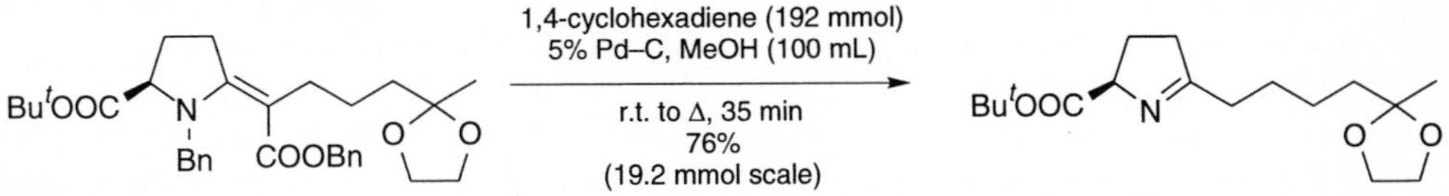

$Bu^tOOC$ N Bn COOBn O O

1,4-cyclohexadiene (192 mmol) 5% Pd–C, MeOH (100 mL) r.t. to Δ, 35 min 76% (19.2 mmol scale)

$Bu^tOOC$ N O O

**Scheme 4.48**

noteworthy that Raney nickel will tolerate the presence of a disubstituted double bond[110]. Alternatively, catalytic transfer hydrogenation can be used [Scheme 4.48] as exemplified in Rapoport's synthesis of Anatoxin[111].

Benzyl esters can be cleaved with $t$-$BuMe_2SiH$ in the presence of $Pd(OAc)_2$ without affecting a double bond [Scheme 4.49][58]. An advantage of this route is the generation of a TBS ester as the product; furthermore, cleavage of benzyl esters can be accomplished in the presence of benzyl ethers.

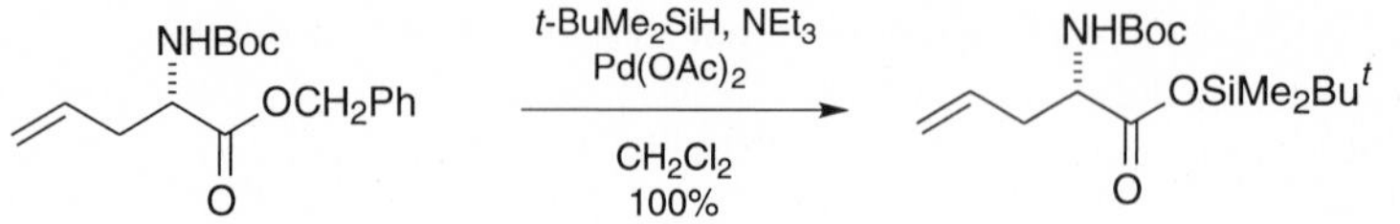

NHBoc, $OCH_2Ph$, O — $t$-$BuMe_2SiH$, $NEt_3$, $Pd(OAc)_2$, $CH_2Cl_2$, 100% → NHBoc, $OSiMe_2Bu^t$, O

**Scheme 4.49**

Benzhydryl esters are cleaved in the presence of strong acids such as neat formic acid at 40 °C[112] or HF in a mixture of nitromethane and acetic acid (12:2:1)[113]. However, best yields are obtained using $CF_3COOH$ in thioanisole [Scheme 4.50][114] or anisole to scavenge the highly electrophilic benzhydryl carbocation released in the reaction. The method has been applied to $\beta$-lactams[115] and sensitive peptides bearing unusual and reactive functionality[116].

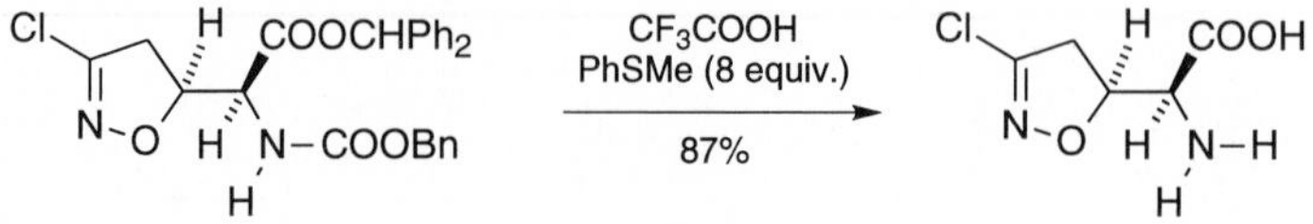

Cl, H, $COOCHPh_2$, N–O, H, N–COOBn, H — $CF_3COOH$, PhSMe (8 equiv.), 87% → Cl, H, COOH, N–O, H, N–H, H

**Scheme 4.50**

By essentially the same mechanism as proton-mediated cleavage, Lewis acids [e.g. $AlCl_3$] coordinate to an ester carbonyl resulting in loss of a relatively stable carbocation such as benzhydryl[117]. The reaction is valuable when hydrogenolysis is precluded by the presence of incompatible functionality as in the $BCl_3$-mediated cleavage of a benzyl malonate in the presence of a labile $\alpha$-chloro ester [Scheme 4.51][84].

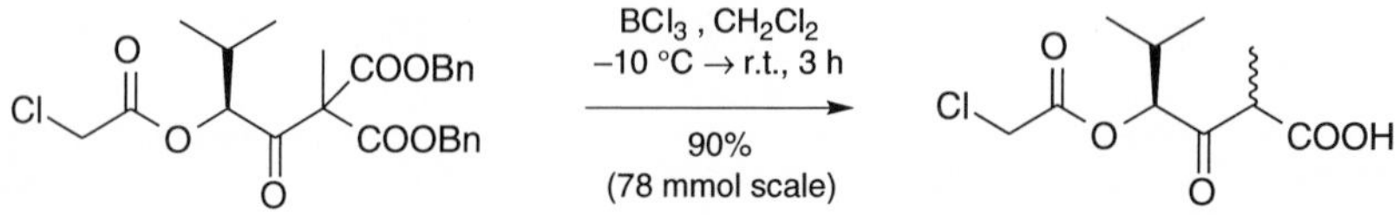

Cl, O, O, O, COOBn, COOBn — $BCl_3$, $CH_2Cl_2$, −10 °C → r.t., 3 h, 90% (78 mmol scale) → Cl, O, O, O, COOH

**Scheme 4.51**

A $p$-nitrobenzyl ester has better stability than a Bn ester against the acidic conditions used for removal of amino acid and peptide protecting groups and has been recommended for the protection of glutamic acid and aspartic acid side chains in solid phase peptide synthesis. Previous methods used to cleave PNB esters include $Na_2S$, zinc, or $Na_2S_2O_4$ reduction in addition to the more usual catalytic hydrogenolysis and Birch reduction. TBAF will also cleanly remove PNB esters in amino acids. The reaction is complete (10–30 min) at r.t. in THF, DMF, or DMSO using 2–9 equivalents of commercial TBAF•$3H_2O$ in THF. Obviously silicon-based protecting groups would be labile as are Fmoc groups.

Phosphoric acids can also be protected as their benzyl esters. In addition to catalytic hydrogenolysis, phosphate $O$-benzyl groups have been cleaved by TMSBr in $CH_2Cl_2$ or $CHCl_3$ to give, initially, the phosphate silyl esters which hydrolyse on aqueous workup to give the phosphoric acid [Scheme 4.52][118,119].

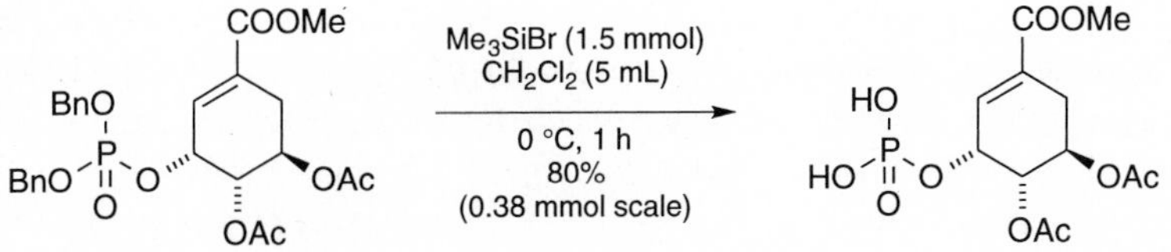

Scheme 4.52

**(ii) Formation**

Benzyl and *p*-nitrobenzyl esters are usually made by classical methods as well as recent procedures such as those developed by Yamaguchi and Mukaiyama (see general esterification methods). Benzhydryl esters are usually synthesised by esterification with diphenyldiazomethane which is a red crystalline solid (m.p. 29–32 °C) first reported by Staudinger in 1916. It is easily prepared by oxidation of benzophenone hydrazone with yellow HgO in the presence of base and it can be stored in the cold with protection from light[120]. Since benzhydryl esters are colourless, the course of the esterification can be easily monitored by the disappearance of the intense red colour. The yield is usually excellent [Scheme 4.53][106,121]. Similarly, 9-fluorenyl esters are easily prepared by the reaction of 9-diazofluorene with a carboxylic acid and they provide a useful alternative to the benzhydryl esters in the protection of amino acids. Such esters are cleaved by mild acidolysis or hydrogenolysis[122].

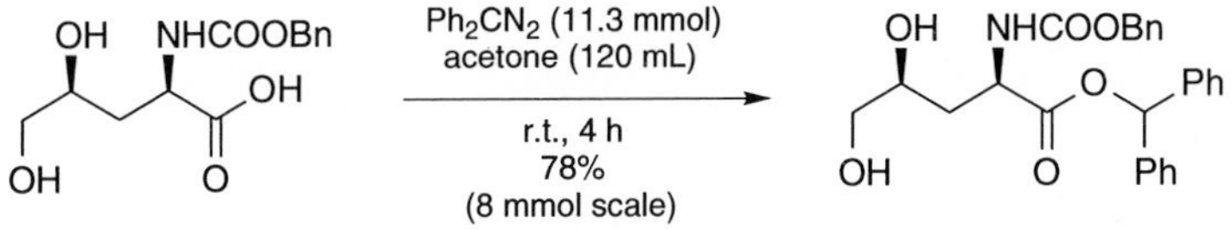

Scheme 4.53

**(iii) NMR Data**

$^1$H NMR:

benzyl $\delta = 7.2$ (5H, s), 5.1 (2H, s)
benzhydryl $\delta = 6.8$ (1H, s, RCOO-**C**H$Ph_2$)
*p*-nitrobenzyl $\delta = 8.2$ (2H, d, $J = 8.7$ Hz), 7.5 (2H, d, $J = 8.7$ Hz), 5.3 (2H, s)

$^{13}$C NMR:

benzyl $\delta = 170$ (C=O), 136 (s), ca 128 (3C, d) 66 (Ar**C**$H_2$)
benzhydryl $\delta = 140.3$ (2C, **C**$_{Ar}$-CH-O), 128.7 (4C), 128.1 (4C), 127.3 (2C), 75.3 (1C, CH-O)
*p*-nitrobenzyl $\delta = 170$ (C=O), 147 ($C_{Ar}$-$NO_2$), 143 (**C**$_{Ar}$-$CH_2$O), 128 (2C), 123 (2C), 65 ($CH_2$O)

## 4.2.6 Allyl Esters

We have already seen in Chapter 2 that the robust nature of the allyl ether moiety and its easy transition-metal catalysed isomerisation to an acid-sensitive enol ether can be usefully exploited in the selective unmasking of alcohols in complex substrates such as carbohydrates. These same advantages have also been exploited for the protection of carboxylic acids as the allyl esters and amines as the alloxycarbonyl function (see Chapter 6). Although rearrangement of an allyl ester to an enol ester using RhCl[$Ph_3P$]$_3$ followed by hydrolysis has been used[123], even milder methods were demanded for the synthesis of glycopeptides, in which the problems of peptide and carbohydrate chemistry merge. To that end, Kunz and co-workers[124-126] have developed a number of methods based on Pd(0)-catalysed allyl transfer to a suitable nucleophile which, for its efficiency, specificity, and gentleness is eminently suited to the synthesis of polyfunctional targets such as glycopeptides.

### (i) Cleavage

Early indications of the potential of Pd–mediated cleavage of allylic carboxylates were provided by Hey and Arpe[127] who demonstrated that Pd(0) catalysed the extrusion of $CO_2$ from allyl formate. Later Tsuji and co-workers accomplished the destruction of allyl esters under reducing conditions [$PdCl_2$, $Ph_3P$, and $HCO_2NH_4$ in refluxing dioxane] and a more recent variant accomplished the deprotection-decarboxylation sequence [Scheme 4.54] of an $\alpha$-cyano ester with $Pd(OAc)_2$ in the presence of HCOOH as part of a synthesis of the macrolide Brefeldin[128]. The formic acid reduces the $\pi$-allyl Pd intermediate to a PdH species which undergoes reductive elimination to give propene and carbon dioxide; the Pd(0) then re-enters the catalytic cycle. The Pd-catalysed decarboxylation of allylic $\beta$-keto esters has also been described[129] but a potential problem associated with the reductive cleavage method is the saturation of alkene bonds. Selective deprotection of allyloxycarbonyl groups has also been accomplished by the Pd–catalysed hydrostannolysis with $Bu_3SnH$ — a method we discussed previously with regard to allyl ether deprotection[130].

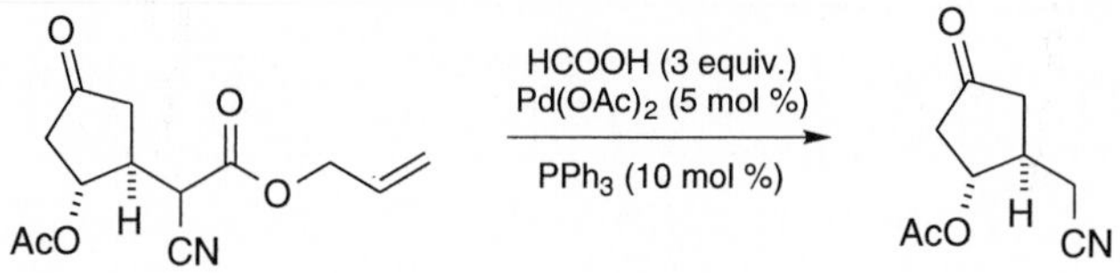

HCOOH (3 equiv.)
$Pd(OAc)_2$ (5 mol %)
$PPh_3$ (10 mol %)

**Scheme 4.54**

The full impact of Pd-catalysed deprotection of allyl esters was first appreciated by Jeffrey and McCombie[131] who showed that an allyl ester could be deprotected to *the potassium salt of the corresponding carboxylic acid* under Pd catalysis — a transformation that is particularly valuable for acid-sensitive substrates. The method was recently exploited by Ruediger and Solomon[132] to deprotect a $\beta$-lactam [Scheme 4.55]. In this case, the displacement of the exchange equilibrium over to the desired product is assured by employing an alkali metal·salt of 2–ethylhexanoate which is soluble in most common solvents except hydrocarbons. The sodium salt can also be generated[133].

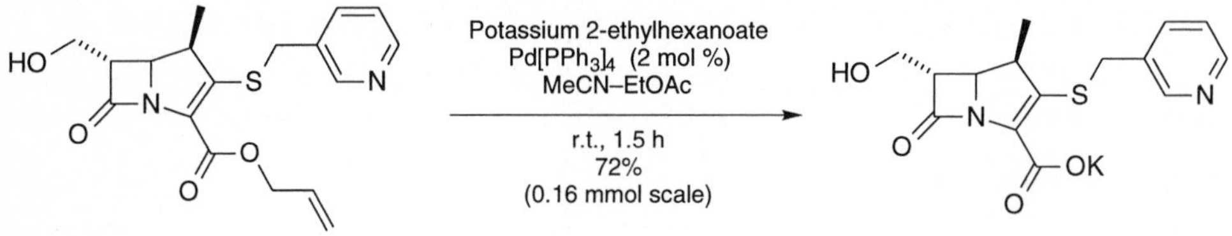

Potassium 2-ethylhexanoate
$Pd[PPh_3]_4$ (2 mol %)
MeCN–EtOAc
r.t., 1.5 h
72%
(0.16 mmol scale)

**Scheme 4.55**

The scope of the Jeffrey–McCombie method is necessarily rather narrow since success depends on the precipitation of the potassium salt of the desired product from the reaction mixture. A far more general method was subsequently designed by Kunz and Waldmann[124] for use in glycopeptide synthesis. Their procedure involves an *irreversible* Pd(0)-catalysed transfer of allyl to weakly basic morpholine and the conditions are sufficiently mild for use in deprotecting allyl esters of *O*-glycosyl-serine and -threonine derivatives which are both acid-labile as well as base-labile [Scheme 4.56]. 1,3-Dimethylbarbituric acid may be used as the nucleophile in place of morpholine[134,135]. The method has also been extended to the deprotection of allyloxycarbonyl derivatives of amines (see Chapter 6) with equal facility. A mechanism which accounts for the catalytic role of Pd is given in Scheme 4.57.

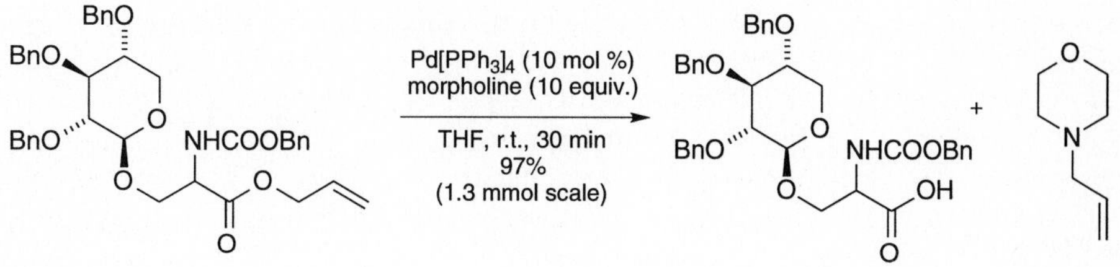

BnO BnO BnO NHCOOBn O O O O Pd[PPh$_3$]$_4$ (10 mol %) morpholine (10 equiv.) THF, r.t., 30 min 97% (1.3 mmol scale) BnO BnO BnO NHCOOBn OH O + O N

**Scheme 4.56**

R O O Pd L OCOR Pd(0)L + NuH RCOOH Nu Pd L Nu

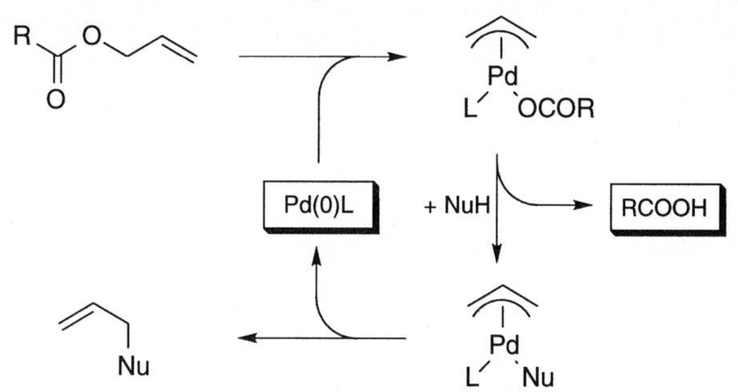

**Scheme 4.57**

The Pd-catalysed allyl transfer method has recently been used to unmask a carboxyl terminus in the tetrapeptide antibiotic Lavendomycin by Schmidt and co-workers[78] [Scheme 4.58] and further testimony to its synthetic value is underscored in syntheses of the macrolides Pyrenophorol[7] and Mycinolide [Scheme 4.59][136].

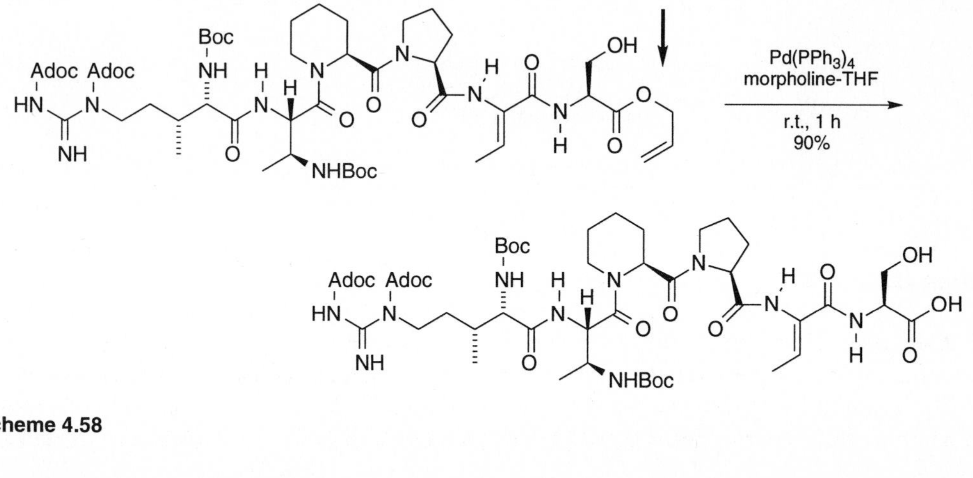

Adoc Adoc HN N NH Boc NH H N H N O O NHBoc N O O H N N H O OH O Pd(PPh$_3$)$_4$ morpholine-THF r.t., 1 h 90%

Adoc Adoc HN N NH Boc NH H N H N O O NHBoc N O O H N N H O OH OH O

**Scheme 4.58**

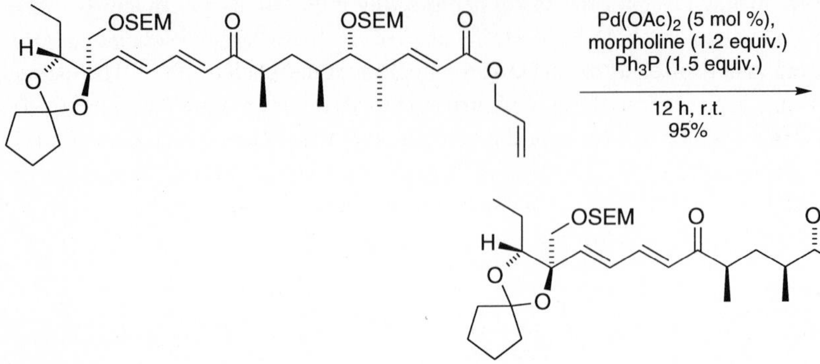

$Pd(OAc)_2$ (5 mol %), morpholine (1.2 equiv.) $Ph_3P$ (1.5 equiv.)

12 h, r.t.
95%

Scheme 4.59

4-(Trimethylsilyl)but-2-enyl esters undergo Pd-catalysed extrusion of butadiene to give the very labile but neutral trimethylsilyl ester of the carboxylic acid[137] [Scheme 4.60]. Simple workup with MeOH is sufficient to destroy the TMS ester and give the carboxylic acid. 4-(Trimethylsilyl)but-2-enyl esters survive treatment with M HCl–MeOH (1:5) for 16 h at 5 °C under conditions which hydrolyse TBS ethers.

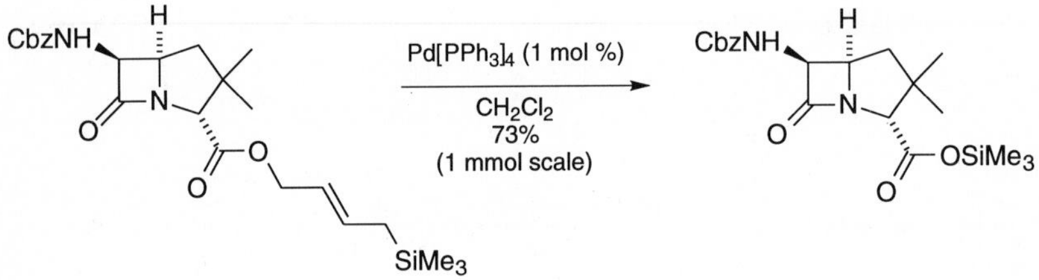

$Pd[PPh_3]_4$ (1 mol %)

$CH_2Cl_2$
73%
(1 mmol scale)

Scheme 4.60

### (ii) Formation

Esterification of an activated carboxylic acid [Scheme 4.61][138] using allyl alcohol is typically used to prepare allyl esters. Alternatively, alkylation of cesium carboxylates with allyl bromide[139] or phase transfer catalysed reactions are gentle and efficient methods[126]. The traditional method involving acid-catalysed esterification of carboxylic acids in the presence of a strong acid such as TsOH has been employed in the esterification of amino acids[123].

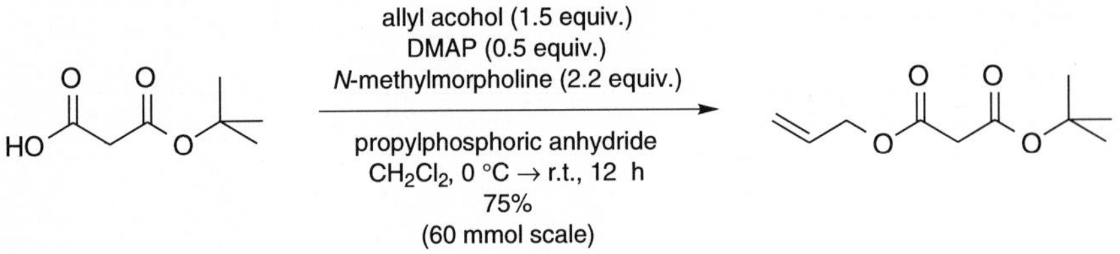

allyl acohol (1.5 equiv.)
DMAP (0.5 equiv.)
*N*-methylmorpholine (2.2 equiv.)

propylphosphoric anhydride
$CH_2Cl_2$, 0 °C → r.t., 12 h
75%
(60 mmol scale)

Scheme 4.61

### (iii) NMR Data

$^1$H NMR: $\delta$ = 5.95 (1H, m, **CH**=$CH_2$), 5.35 (1H, dd, $J$ = 16, 2 Hz, CH=$\mathbf{CH_A}H_B$), 5.23 (1H, dd, $J$ = 10, 2 Hz, CH=$CH_A\mathbf{H_B}$), 4.55 (2H, dt, $J$ = 6, 1 Hz)

$^{13}$C NMR: $\delta$ = 132 (d, **CH**=$CH_2$), 118 (t, CH=$\mathbf{CH_2}$), 65 (t, O-$CH_2$)

## 4.2.7 Silyl Esters

Rigid application of our rule including only those protecting groups which have found wide favor with synthetic chemists would have excluded silyl esters from consideration because they are too labile to mild acid or base to survive even simple manipulation. Their inclusion here is intended to underline their lability in comparison to the corresponding silyl ethers. Like the silyl ethers, the stability of the silyl esters parallels the steric bulk of the substituents on the silicon atom. Silyl esters protect acids against reduction with hydroborating agents[140].

### (i) Cleavage

Potassium carbonate in MeOH or HOAc–$H_2O$–THF (3:1:1) at r.t. is sufficient to cleave even *tert*-butyldimethylsilyl esters. The esters cleave faster than the corresponding ethers [Scheme 4.62][141]. Their high base lability allows removal of *tert*-butyldimethylsilyl esters in the presence of THP groups and *tert*-butyldimethylsilyl ethers [Scheme 4.63][142].

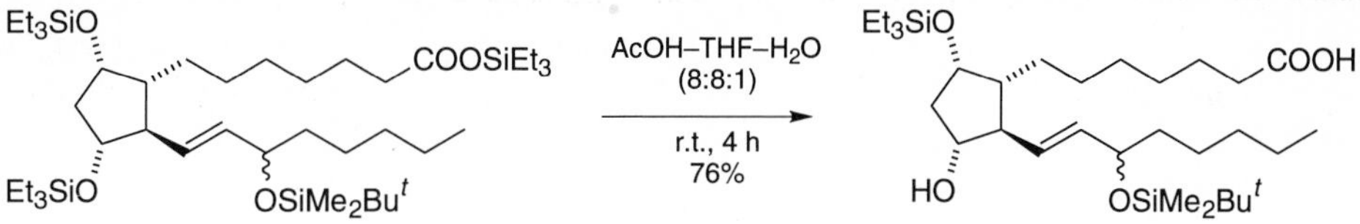

$Et_3SiO$, $COOSiEt_3$, $Et_3SiO$, $OSiMe_2Bu^t$ → AcOH–THF–$H_2O$ (8:8:1), r.t., 4 h, 76% → $Et_3SiO$, COOH, HO, $OSiMe_2Bu^t$

**Scheme 4.62**

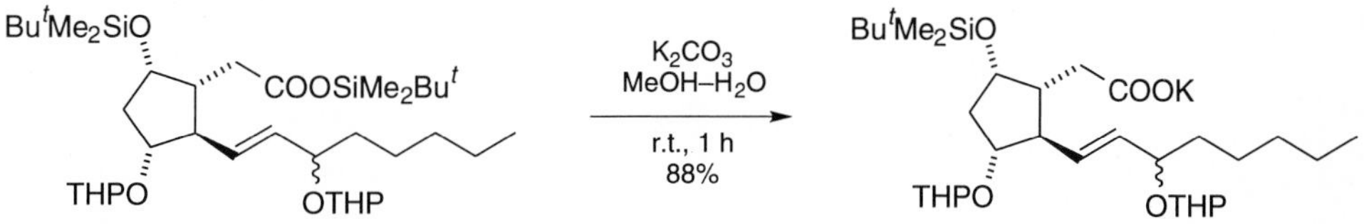

$Bu^tMe_2SiO$, $COOSiMe_2Bu^t$, THPO, OTHP → $K_2CO_3$, MeOH–$H_2O$, r.t., 1 h, 88% → $Bu^tMe_2SiO$, COOK, THPO, OTHP

**Scheme 4.63**

Like the silyl ethers, the stability of silyl esters is strongly dependent on steric effects. However, the silyl esters are much more labile as shown by the fact that *tert*-butyldiphenylsilyl esters cleave in the presence of 1% HF–MeCN at r.t. for 15 min [Scheme 4.64][143] whereas *tert*-butyldiphenylsilyl ethers generally require much longer reaction times (see section 1.2.4). The di-*tert*-butylmethylsilyl group affords the most stable of the silyl esters prepared to date and it is sufficiently stable to allow selective removal of a THP group using PPTS in warm ethanol[144].

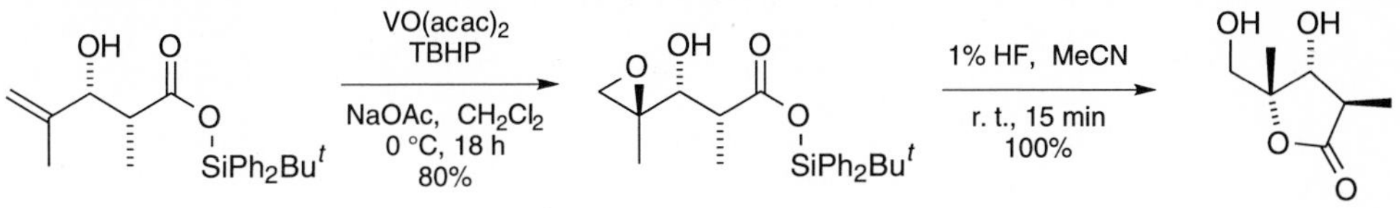

OH, O, O, $SiPh_2Bu^t$ → $VO(acac)_2$, TBHP, NaOAc, $CH_2Cl_2$, 0 °C, 18 h, 80% → O, OH, O, O, $SiPh_2Bu^t$ → 1% HF, MeCN, r. t., 15 min, 100% → OH, OH, O, O

**Scheme 4.64**

The high lability of silyl esters is occasionally beneficial. A group at Smith Kline Beecham[145] were able to form an *N*-formyl aminal of a 7$\alpha$–aminocephalosporanic acid on a kilogram scale in which the lability of a TMS ester was crucial to the success of the last step. The elegance of the route is best appreciated by seeing it in its entirety [Scheme 4.65].

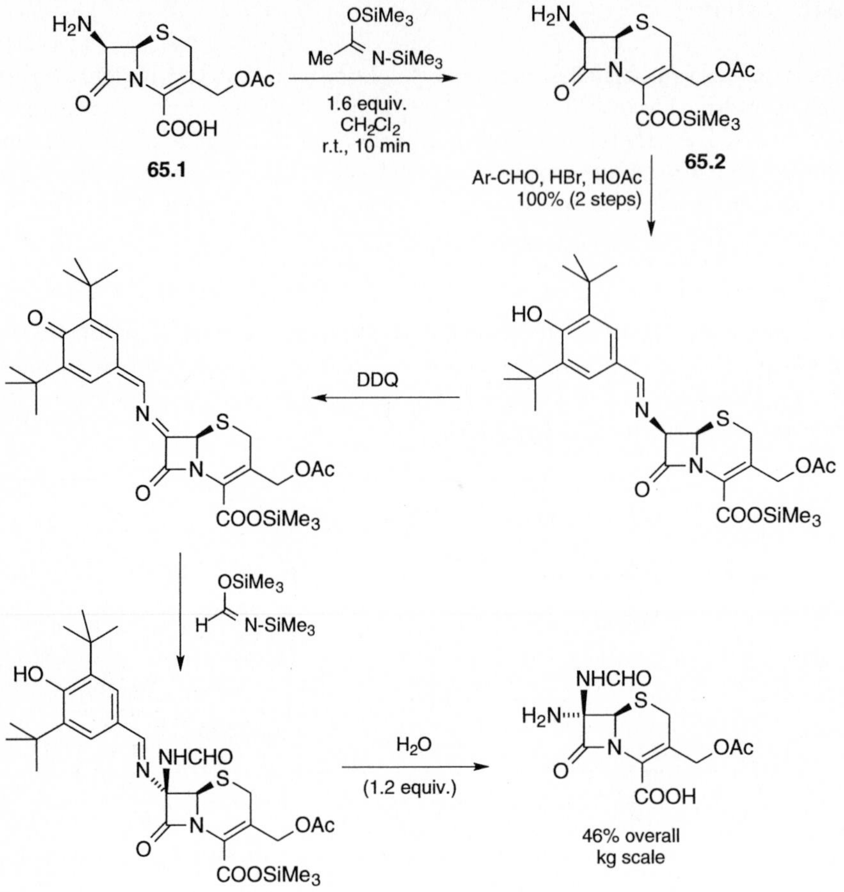

H2N
S
N
O
OAc
COOH
65.1
OSiMe3
Me
N-SiMe3
1.6 equiv.
CH2Cl2
r.t., 10 min
COOSiMe3
65.2
Ar-CHO, HBr, HOAc
100% (2 steps)
HO
DDQ
NHCHO
H2O
(1.2 equiv.)
46% overall
kg scale


**Scheme 4.65**

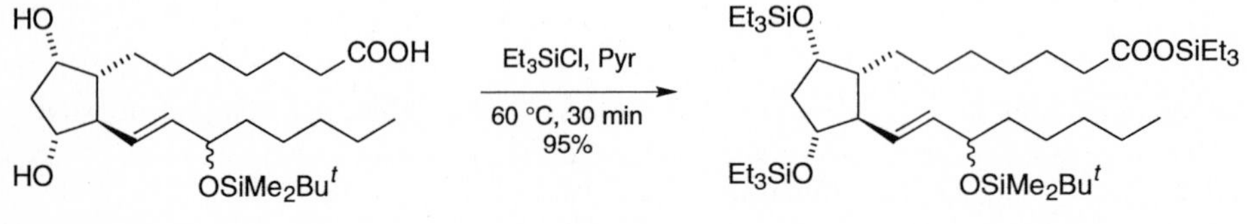

HO
COOH
OSiMe2Bu^t
Et3SiCl, Pyr
60 °C, 30 min
95%
Et3SiO
COOSiEt3


**Scheme 4.66**

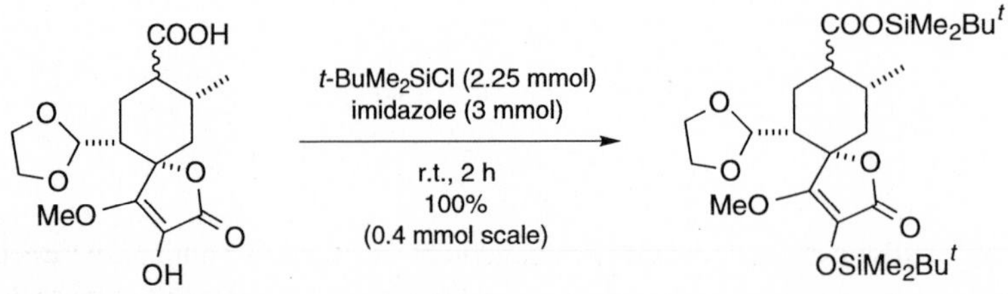

COOH
MeO
OH
t-BuMe2SiCl (2.25 mmol)
imidazole (3 mmol)
r.t., 2 h
100%
(0.4 mmol scale)
COOSiMe2Bu^t
OSiMe2Bu^t


**Scheme 4.67**

**(ii) Formation**

Silyl esters are generally prepared from the carboxylic acid and the corresponding silyl chloride or silyl triflate in the presence of base (pyridine, triethylamine, imidazole, etc) in DMF or THF as shown in Scheme 4.66[141] and Scheme 4.67[146].

Trimethylsilyl esters are very moisture sensitive and so their preparation is best achieved by methods that avoid aqueous workup such as reaction of the carboxylic acid **65.1** with *N,O*-bis(trimethylsilyl)acetamide to give the trimethylsilyl ester **65.2** [Scheme 4.65]. Another mild method for preparing TMS esters which avoids aqueous workup involves Pd-catalysed extrusion of butadiene from 4-(trimethylsilyl)but-2-enyl esters as discussed in section 4.2.4.2[137].

# 4.3 2,6,7-Trioxabicyclo[2.2.2]octanes [OBO] and Other Ortho Esters

All of the protecting groups for carboxylic acids covered so far are designed to prevent the *hydroxy* group from doing any mischief in the course of a synthetic transformation. In order to shield the *carbonyl* group of carboxylic acids and esters from attack by strongly nucleophilic reagents such as metal hydrides, Grignard reagents, organolithiums, or alkoxides, Corey[147] developed the ortho ester 2,6,7-trioxabicyclo[2.2.2]octane [abbreviated OBO] protecting group. Ortho esters are generally resistant to attack by strong nucleophiles provided they are not associated with a Lewis acid. For example, Grignard reagents do react with ortho esters at elevated temperature because of the presence of magnesium halides but the reaction is very slow at low temperature[148,149]. They are obviously sensitive to mild acid hydrolysis. Reviews on the chemistry of ortho esters have recently appeared*.

**(i) Cleavage**

Ortho oxygen esters are usually cleaved by acid hydrolysis using conditions similar to those normally employed for acetals; however, OBO groups can be hydrolysed in the presence of acetals[147]. As with acetals, cyclic ortho esters are more stable than their acyclic counterparts. The product of mild acid hydrolysis is itself an ester which must then be further treated with base (pH 10–11) to release the acid. The acid catalyst for the first step can be very mild. For example pyridinium tosylate[150] or $NaHSO_4$[151] have proved effective. The two-step deprotection sequence is illustrated by steps [Scheme 4.68][152] in a synthesis of the aggregation pheromone (*S*)-(+)-3-dodecen-11-olide of the male flat grain beetle. The OBO group has been exploited in a number of other natural product syntheses including Retigeranic acid[153], Ginkolide[147], Prostaglandin $E_1$[151], Prostaglandin $D_2$[154], Hydridalactone[155] (hydrolysis of an OBO group in the presence of a 1,2-disubstituted epoxide using $NaHSO_4$ in aq. DME), 12-HETE[156], Cephalotaxine[157], and the cotton boll weevil antifeedant $\alpha$-Eleostearic acid[158].

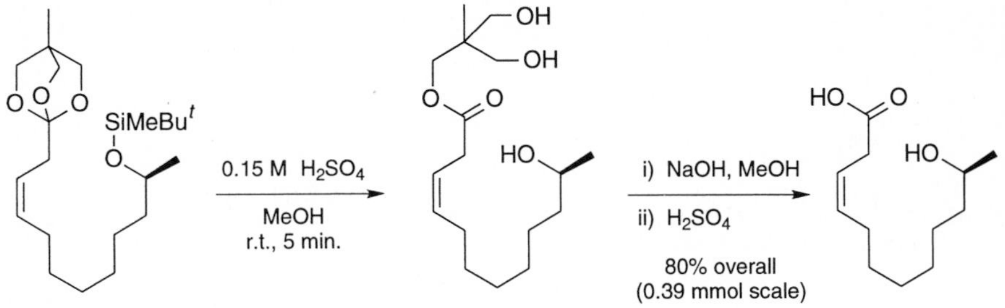

SiMeBu$^t$

0.15 M $H_2SO_4$ / MeOH / r.t., 5 min.

i) NaOH, MeOH / ii) $H_2SO_4$ / 80% overall (0.39 mmol scale)

**Scheme 4.68**

Ortho dithio esters are much more stable to acid hydrolysis than their oxygen analogues. However, they can be unleashed under essentially neutral conditions using Hg(II) catalysis [Scheme 4.69][159]. A lactone was similarly released from its ortho dithio ester precursor under mild conditions using a Ag(I)-catalysed hydrolysis reaction [Scheme 4.70][160].

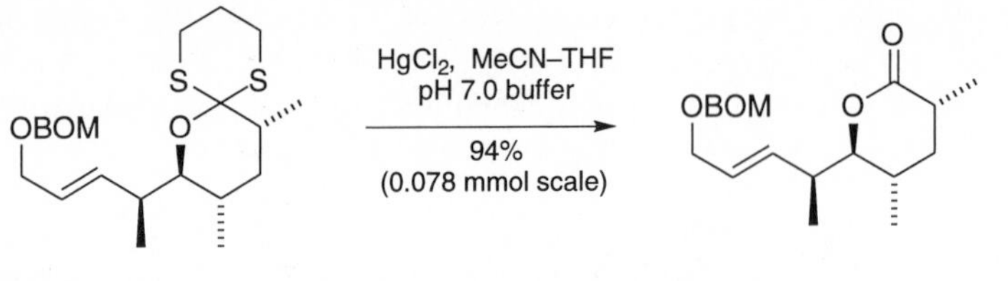
OBOM
HgCl2, MeCN–THF
pH 7.0 buffer
94%
(0.078 mmol scale)
OBOM

**Scheme 4.69**

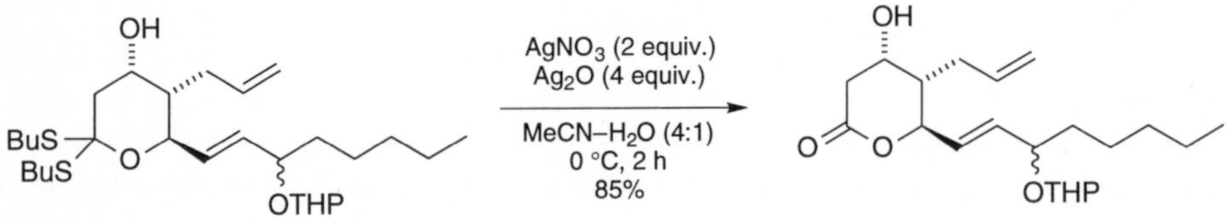
OH
BuS
BuS
OTHP
AgNO3 (2 equiv.)
Ag2O (4 equiv.)
MeCN–H2O (4:1)
0 °C, 2 h
85%
OH
OTHP

**Scheme 4.70**

Ortho trithio esters can also be used to introduce a protected carboxyl function. In 1967 Seebach[161] showed that metallation of trimethylthiomethane afforded an anion which was sufficiently stable and nucleophilic to react with a variety of carbon electrophiles. The concept was used by Damon and Schlessinger[162] in a synthesis of Protolichesterenic Acid [Scheme 4.71].

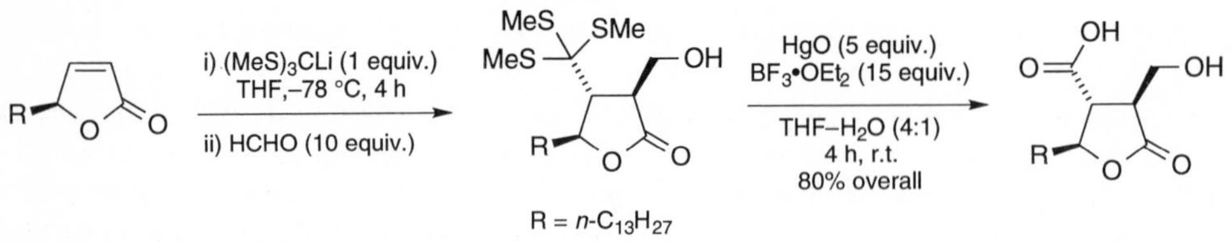
i) (MeS)3CLi (1 equiv.)
THF,–78 °C, 4 h
ii) HCHO (10 equiv.)
MeS
SMe
MeS
OH
HgO (5 equiv.)
BF3•OEt2 (15 equiv.)
THF–H2O (4:1)
4 h, r.t.
80% overall
R = n-C13H27

**Scheme 4.71**

2-Trimethylsilyl-1,3-dithianes can also serve as masked carboxyl groups. The principal is exemplified by the transformations shown in Scheme 4.72 which form part of an elegant large-scale synthesis of the *β*-lactam antibiotic Thienamycin[163].

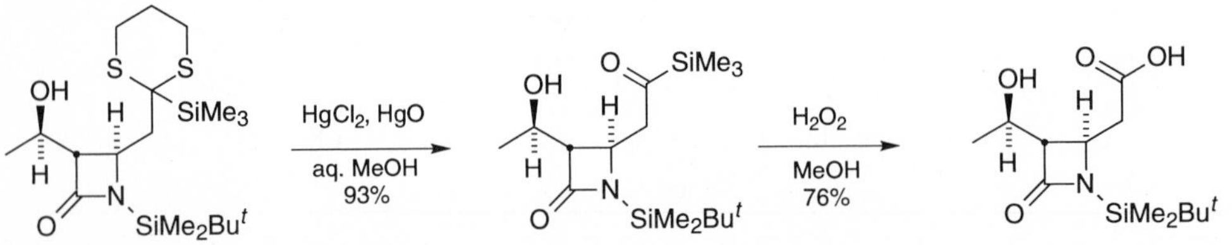
OH
SiMe3
SiMe2Bu^t
HgCl2, HgO
aq. MeOH
93%
SiMe3
SiMe2Bu^t
H2O2
MeOH
76%
OH
SiMe2Bu^t

**Scheme 4.72**

**(ii) Formation**

Ortho esters cannot usually be prepared efficiently from esters by reaction with alcohols under acid conditions analogous to the preparation of acetals from aldehydes and ketones[164,165]. One of the few protocols for converting a carboxylic acid to an ortho ester was perfected by Corey and Raju[166]. The method involves reaction of an acid chloride with 3-methyl-3-(hydroxymethyl)oxetane[167] followed by treatment of the resultant oxetane ester with $BF_3$•$OEt_2$ as illustrated in early transformations [Scheme 4.73] in Keinan's flat grain beetle pheromone synthesis[152].

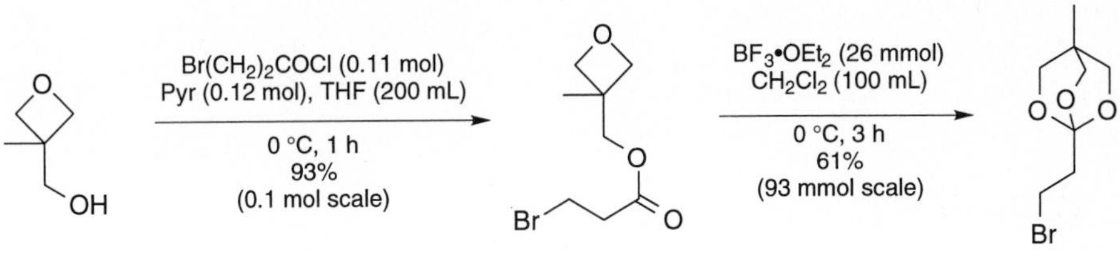

$Br(CH_2)_2COCl$ (0.11 mol), Pyr (0.12 mol), THF (200 mL); 0 °C, 1 h; 93%; (0.1 mol scale)

$BF_3$•$OEt_2$ (26 mmol), $CH_2Cl_2$ (100 mL); 0 °C, 3 h; 61%; (93 mmol scale)

**Scheme 4.73**

The Pinner reaction is a standard method for preparing ortho esters which involves treatment of a nitrile with an alcohol in the presence of anhydrous HCl. Voss and Gerlach[168] used a Pinner reaction to prepare a robust ortho ester from *cis*-cyclohexane-1,3,5-triol [Scheme 4.74].

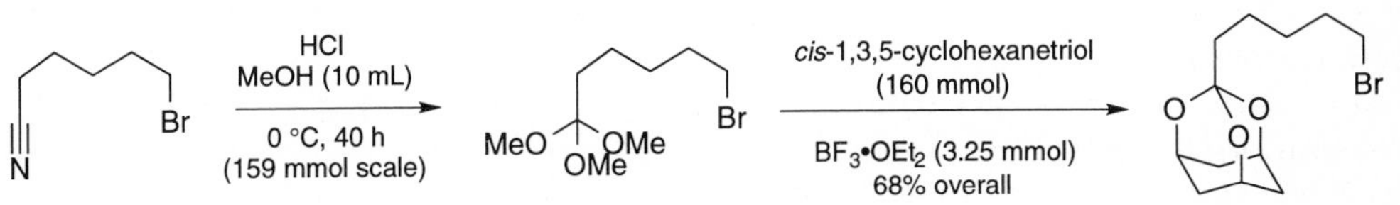

HCl, MeOH (10 mL); 0 °C, 40 h; (159 mmol scale)

*cis*-1,3,5-cyclohexanetriol (160 mmol); $BF_3$•$OEt_2$ (3.25 mmol); 68% overall

**Scheme 4.74**

Unlike ortho oxygen esters, ortho dithio esters can be formed directly from esters and lactones. However, pains must be taken to provide an effective method for dehydration. Corey and Beames[169] used the oxyphilic properties of aluminium to achieve formation of an ortho dithio lactone from a butyrolactone and an aluminium reagent prepared from ethane-1,2-dithiol [Scheme 4.75]. A spirocyclic ortho dithio ester can also be prepared by intramolecular cyclisation of an alcohol onto a ketene dithioacetal under acid catalysis. Thus, the diastereoselective spirocyclisation reaction shown in Scheme 4.76 was used to control the stereochemistry of a methyl substituent in a synthesis of the macrolide Protomycinolide[159]. In this case the problem of dehydration was solved at the stage of ketene dithioacetal formation.

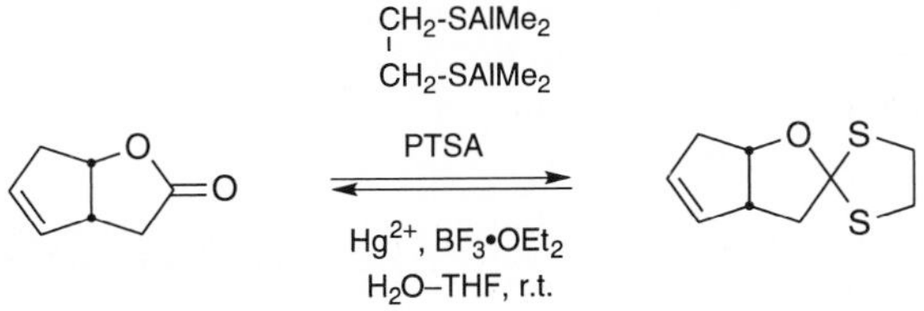

$CH_2$-$SAlMe_2$ / $CH_2$-$SAlMe_2$, PTSA

$Hg^{2+}$, $BF_3$•$OEt_2$, $H_2O$–THF, r.t.

**Scheme 4.75**

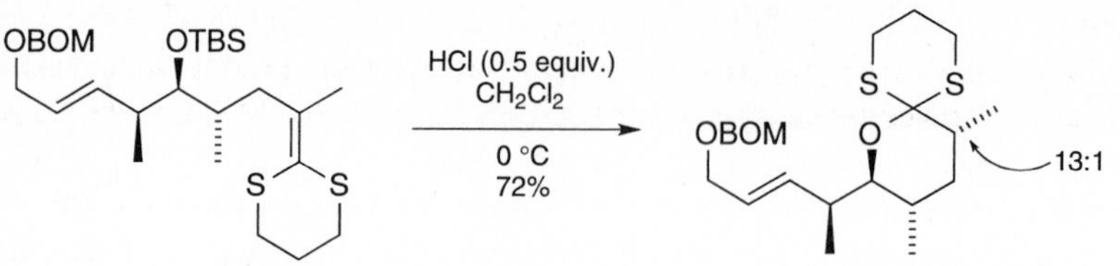

OBOM OTBS HCl (0.5 equiv.) $CH_2Cl_2$ 0 °C 72% OBOM 13:1

Scheme 4.76

**(iii) NMR Data [OBO]**
$^1H$ NMR: $\delta = 3.9$ (6H, s), 0.8 (3H, s)
$^{13}C$ NMR: $\delta = 118, 72, 30.2, 14.4$

# 4.4 Oxazolines

2-Substituted-1,3-oxazolines may be considered as masked carboxylic esters in which the carbonyl oxygen has been replaced by nitrogen. Homochiral oxazolines have been extensively employed in asymmetric synthesis and much of their chemistry parallels the chemistry of esters. Within the context of carboxyl protecting groups, they are useful in situations where powerful nucleophilic reagents such as lithium aluminium hydride, Grignard or organolithium reagents would be incompatible with the presence of an ester function. The use of oxazolines in asymmetric synthesis has been reviewed*.

**(i) Cleavage**
Oxazolines are not easily hydrolysed and forcing conditions may be required which are incompatible with sensitive functional groups. Typical hydrolytic methods involve strong mineral acids (3 M HCl–EtOH, reflux) or *N*-methylation with MeI followed by hydrolysis of the resultant iminium salt with aqueous NaOH[170]. The resistance of oxazolines to attack by strong nucleophiles is the principal feature which has commanded the attention of synthetic chemists. Carboxyl groups masked as oxazolines survive the formation of Grignard reagents[171] or formation of an aryllithium from the corresponding bromide using lithium–halogen exchange. Oxazoline groups also activate aromatic rings to *o*-metallation[172]. Two examples suffice to demonstrate the synthetic utility of oxazolines. The first example is taken from an asymmetric synthesis of the AB ring system of Aklavinone. The key step in the sequence used the homochiral oxazoline **77.3** to activate a naphthalene ring toward diastereoselective nucleophilic addition by an alkenyllithium [Scheme 4.77][173]. The resultant adduct underwent acidolysis to give the dihydronaphthalene **77.4** which was converted to the target **77.5** in a further 10 steps.

The second example [Scheme 4.78] shows how $\alpha$-metallation of an oxazoline followed by [2,3]-sigmatropic rearrangement can be used to achieve a highly stereoselective synthesis of trisubstituted alkenes[174]. Interestingly, only one out of a possible 8 diastereoisomers was formed. The latent carboxyl function was released from the $\alpha$-hydroxyoxazoline to give an advanced intermediate in a synthesis of the macrolide antibiotic Nargenicin.

**(ii) Formation**
4,4-Dimethyloxazolines are formed directly from the reaction of carboxylic acids with $H_2N$-$C(Me)_2CH_2OH$ in refluxing toluene[175] but a two-step procedure involving reaction of $H_2N$-$C(Me)_2CH_2OH$ with an acid chloride followed by treatment of the resultant amide with excess thionyl chloride as a dehydrating agent is generally preferred [Scheme 4.79][176,177]. Alternatives include

reaction of dimethylaziridine with a carboxylic acid in the presence of DCC to form the *N*-acylaziridine followed by acid-catalysed rearrangement[178]; or reaction of an ortho ester, or an imidate ester, with an amino alcohol as illustrated by the conversion of **77.1** to **77.3** [Scheme 4.77][179,180].

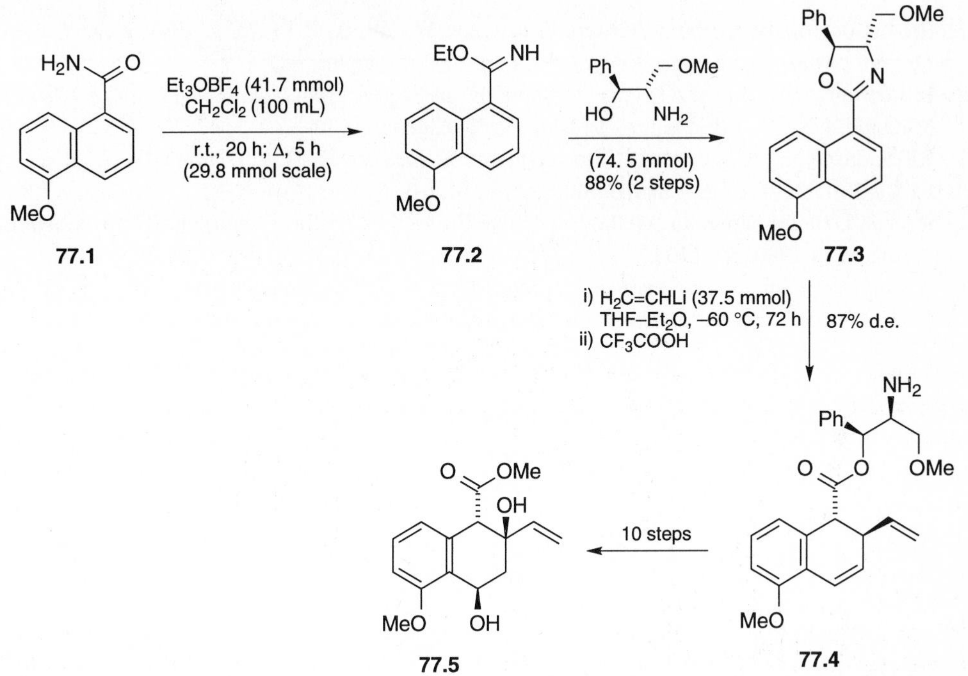

**Scheme 4.77**

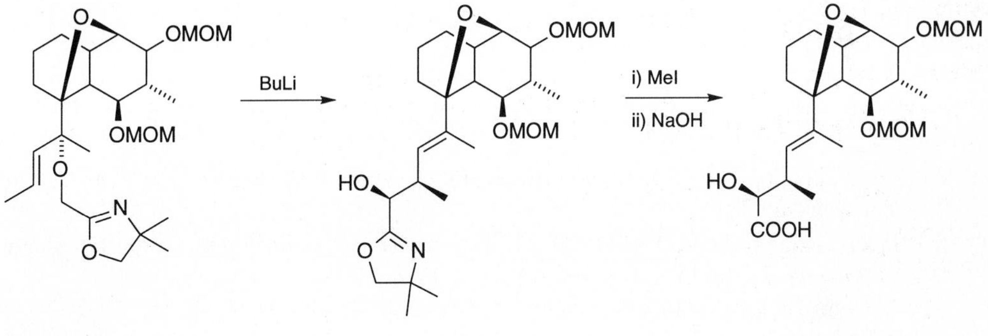

**Scheme 4.78**

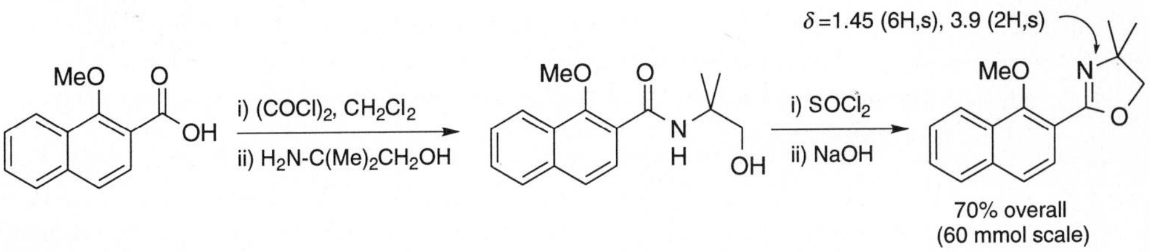

**Scheme 4.79**

# 4.5 Reviews

## 4.5.1 Reviews Concerning the Specific Use of Esters as Protecting Groups

1 Protection for the Carboxyl Group. Greene, T. W.; Wuts, P. G. M. In *Protective Groups in Organic Synthesis*, 2nd ed.; Wiley: New York, 1990; Chapter 5.
2 Protection of Carboxyl Groups. Haslam, E. in *Protective Groups in Organic Chemistry*, McOmie, J. F. W., Ed.; Plenum: London, 1973; Chapter 5.
3 Blockierung und Schutz der $\alpha$-Carboxy-Funktion. Deimer, K.-H. In *Houben–Weyl*, 4th ed., Vol. E15/1; Wünsch, E., Ed.; Thieme: Stuttgart, 1974; p 315.
4 Recent Developments in Methods for the Protection of the Carboxyl Group. Haslam, E. *Tetrahedron* **1980**, *36*, 2409.
5 Recent Developments in Chemical Deprotection of Ester Functional Groups. Salomon, C.; Mata, E. G.; Mescaretti, O. A. *Tetrahedron* **1993**, *49*, 3691.

## 4.5.2 Reviews Concerning the Preparation of Esters

1 Esterification and Alkylation Reactions Employing Isoureas. Mathias, L.J. *Synthesis* **1979**, 561.
2 *The Chemistry of Carboxylic Acid Derivatives. Supplement B;* Patai, S., Ed.; Wiley: New York, 1979.
3 Synthesis of Esters, Activated Esters and Lactones. Mulzer, J. In *Comp. Org. Synth.*, Vol. 6; Pergamon: Oxford, 1991; p 323.
4 New Synthetic Methods Based on the Onium Salts of Aza–Arenes. Mukaiyama, T. *Angew. Chem. Int. Ed. Engl.* **1979**, *18*, 707.
5 The Use of Diethylazodicarboxylate and Triphenylphosphine in Synthesis and Transformations of Natural Products. Mitsunobu, O. *Synthesis* **1981**, 1.
6 The Mitsunobu Reaction. Hughes, D. L. *Org. React.* **1992**, *40*, 335.

## 4.5.3 Reviews Concerning Techniques for Facilitating Esterification

1 4-Dimethylaminopyridines: Super Acylation and Alkylation Catalysts. Scriven, E. F. V. *Chem. Soc. Rev.* **1983**, *12*, 129.
2 4-Dimethylaminopyridine (DMAP) as a Highly Active Acylation Agent. Höfle, G.; Steglich, W.; Vorbrüggen, H. *Angew. Chem. Int. Ed. Engl.*, **1978**, *17*, 569.
3 The Solvent Dimethyl Sulfoxide. Martin, D.; Weise, A.; Niclas, H.–J. *Angew. Chem. Int. Ed. Engl.* **1967**, *6*, 318.
4 Hexamethylphosphoramide. Normant, H. *Angew. Chem. Int. Ed. Engl.* **1967**, *6*, 1046.
5 Solvents and Solvent Effects in Organic Chemistry. Reichardt, C. VCH: Weinheim, 1988.
6 Quaternary Ammonium Compounds in Organic Synthesis. Dockx, J. *Synthesis* **1973**, 441.
7 Ring Closure Methods in the Synthesis of Macrocyclic Natural Products. Meng, Q.; Hesse, M. *Top. Curr. Chem.* **1991**, *161*, 107.

### 4.5.4 Reviews Concerning the Use of Allyl Esters in Carboxyl Protection

1 New Synthetic Reactions of Allyl Alkyl Carbonates, Allyl $\beta$–Keto Carboxylates, and Allyl Vinylic Carbonates Catalysed by Pd Complexes. Tsuji, J.; Minami, I. *Acc. Chem. Res.* **1987**, *20*, 140.

### 4.5.5 Reviews Concerning the Nucleophilic Cleavage of Esters

1 Synthetic Applications of Dealkoxycarbonylation of Malonate Esters, $\beta$–Keto Esters, $\alpha$–Cyano Esters and Related Compounds in Dipolar Aprotic Solvents. Krapcho, A. *Synthesis* **1982,** 805, 893.
2 Ester Cleavages via $S_N2$-Type Dealkylation. McMurry, J. *Org. React.* **1976**, *24*, 187.

### 4.5.6 Reviews Concerning the Chemistry of Ortho Esters and Oxazolines

1 Ortho Esters and Dialkoxycarbenium Ions: Reactivity, Stability, Structure, and New Synthetic Applications. Pindur, U.; Müller, J.; Flo, C.; Witzel, H. *Chem. Soc. Rev.* **1987**, 75.
2 Synthesis of Carboxylic and Carbonic Ortho Esters. DeWolfe, R. H. *Synthesis* **1974**, 153.
3 *Carboxylic Ortho Ester Derivatives. Preparation and Synthetic Applications.* DeWolfe, R. H.; Academic Press: New York, 1970.
4 Asymmetric Synthesis via Chiral Oxazolines. Lutomski, K. A.; Meyers, A. I. In *Asymmetric Synthesis*, Vol. 3; Morrison, J. D., Ed.; Academic Press: New York, 1984; p 213.

## References

1 Neises, B.; Steglich, W. *Angew. Chem Int. Ed. Engl.* **1978**, *17*, 522.
2 Hassner, A.; Alexanian, V. *Tetrahedron Lett.* **1978**, 4475.
3 Boden, E. P.; Keck, G. E. *J. Org. Chem.* **1985**, *50*, 2394.
4 Schlessinger, R. H.; Poss, M. A.; Richardson, S. *J. Am. Chem. Soc.* **1986**, *108*, 3112.
5 Inanaga, J.; Hirata, K.; Saeki, H.; Katsuki, T.; Yamaguchi, M. *Bull. Chem. Soc. Jpn.* **1979**, *52*, 1989.
6 Ditrich, K. *Liebigs Ann. Chem.* **1990**, 789.
7 Dommerholt, F. J.; Thijs, L.; Zwanenburg, B. *Tetrahedron Lett.* **1991**, *32*, 1499.
8 Nakajima, N.; Hamada, T.; Tanaka, T.; Oikawa, Y.; Yonemitsu, O. *J. Am. Chem. Soc.* **1986**, *108*, 4645.
9 Bartra, M.; Vilarrasa, J. *J. Org. Chem.* **1991**, *56*, 5132.
10 Millar, J. G.; Oehlschlager, A. C. *J. Org. Chem.* **1984**, *49*, 2332.
11 Cooper, J.; Knight, D. W.; Gallagher, P. T. *J. Chem. Soc., Perkin Trans. 1* **1991**, 705.
12 Kurihara, T.; Nakajima, Y.; Mitsunobu, O. *Tetrahedron Lett.* **1976**, 2455.
13 Seebach, D.; Adam, G.; Zibuck, R.; Simon, W.; Rouilly, M.; Meyer, W. L.; Hinton, J. F.; Privett, T. A.; Templeton, G. E.; Heiny, D. K.; Gisi, U.; Binder, H. *Liebigs Ann. Chem.* **1989**, 1233.
14 Justus, K.; Steglich, W. *Tetrahedron Lett.* **1991**, *32*, 5781.
15 Still, W. C.; Ohmizu, H. *J. Org. Chem.* **1981**, *46*, 5242.
16 Honda, M.; Hirata, K.; Sueoka, H.; Katsuki, T.; Yamaguchi, M. *Tetrahedron Lett.* **1981**, *22*, 2679.
17 Taschner, E.; Liberek, B. *Rocz. Chem.* **1956**, *30*, 323. C.A. **1957**, *51*, 1039d.
18 Müller, P.; Siegfried, B. *Helv. Chim. Acta.* **1974**, *57*, 987.
19 Stevens, R. V.; Lee, A. W. M. *J. Am. Chem. Soc.* **1979**, *101*, 7032.
20 Wrobel, J.; Takahashi, K.; Honkan, V.; Lannoye, G.; Cook, J. M.; Bertz, S. H. *J. Org. Chem.* **1983**, *48*, 139.
21 Keith, D. D.; Tortora, J. A.; Yang, R. *J. Org. Chem.* **1978**, *43*, 3711.
22 Corey, E. J.; Weigel, L. O.; Floyd, D.; Bock, M. G. *J. Am. Chem. Soc.* **1978**, *100*, 2916.

23 Stevens, K. E.; Yates, P. *J. Chem. Soc., Chem. Commun.* **1980**, 990.
24 Greene, A. E.; Luche, M.-J.; Deprés, J.-P. *J. Am. Chem. Soc.* **1983**, *105*, 2435.
25 Mohr, P.; Waespe-Sarceivic, N.; Tamm, C.; Gawronska, K.; Gawronski, J. K. *Helv. Chim. Acta.* **1983**, *66*, 2501.
26 Tschamber, T.; Waespe-Sarceivic, N.; Tamm, C. *Helv. Chim. Acta.* **1986**, *69*, 621.
27 Hughes, D. L.; Bergan, J. J.; Amato, J. S.; Bhupathy, M.; Leazer, J. L.; McNamara, J. M.; Sidler, D. R.; Reider, P. J.; Grabowski, E. J. J. *J. Org. Chem.* **1990**, *55*, 6252.
28 Sime, J. T.; Pool, C. R.; Tyler, J. W. *Tetrahedron Lett.* **1987**, *28*, 5169.
29 Hudlicky, M. *J. Org. Chem.* **1980**, *45*, 5377.
30 Hecht, S. M.; Kozarich, J. W. *Tetrahedron Lett.* **1973**, 1397.
31 Hirai, Y.; Aida, T.; Inoue, S. *J. Am. Chem. Soc.* **1989**, *111*, 3062.
32 Shioiri, T.; Aoyama, T. *Org. Synth.* **1989**, *68*, 1.
33 Brook, M. A.; Chan, T. H. *Synthesis* **1983**, 201.
34 Gerspracher, M.; Rapoport, H. *J. Org. Chem.* **1991**, *56*, 3700.
35 Brenner, M.; Huber, W. *Helv. Chim. Acta* **1953**, *36*, 1109.
36 Karanewsky, D. S.; Malley, M. F.; Gougoutas, J. Z. *J. Org. Chem.* **1991**, *56*, 3744.
37 Stork, G.; Rychnovsky, S. *J. Am. Chem. Soc.* **1987**, *109*, 1565.
38 Luthman, K.; Orbe, M.; Wagland, T.; Claesson, A. *J. Org. Chem.* **1987**, *52*, 3777.
39 Dijkstra, G.; Kruizinga, W. H.; Kellogg, R. M. *J. Org. Chem.* **1987**, *52*, 4230.
40 Pfeiffer, P. E.; Silbert, L. S. *J. Org. Chem.* **1976**, *41*, 1373.
41 Mathias, L. J. *Synthesis* **1979**, 561.
42 Roeske, R. W. *Chem. Ind.* **1959**, 1121.
43 Taschner, E.; Chimiak, A.; Bator, B.; Sokowlowska, T. *Liebigs Ann. Chem.* **1961**, *646*, 134.
44 Schröder, E.; Lübke, K. *Liebigs Ann. Chem.* **1962**, *655*, 211.
45 Solladié, G.; Hamdouchi, C.; Ziani-Chérif, C. *Tetrahedron Asymmetry* **1991**, *2*, 457.
46 Heathcock, C. H.; Mahaim, C.; Schlecht, M. F.; Utawanit, T. *J. Org. Chem.* **1984**, *49*, 3264.
47 Hermann, J. L.; Berger, M. H.; Schlessinger, R. H. *J. Am. Chem. Soc.* **1973**, *95*, 7923.
48 Wender, P. A.; Schaus, J. M.; White, A. W. *J. Am. Chem. Soc.* **1980**, *102*, 6157.
49 Seebach, D.; Amstutz, R.; Laube, T.; Schweizer, W. B.; Dunitz, J. D. *J. Am. Chem. Soc.* **1985**, *107*, 5403.
50 Siekaly, H. R.; Tidwell, T. T. *Tetrahedron* **1986**, *42*, 2587.
51 Fehr, C.; Galindo, J. *J. Org. Chem.* **1988**, *53*, 1828.
52 Gmeiner, P.; Feldman, P. L.; Chu-Moyer, M. Y.; Rapoport, H. *J. Org. Chem.* **1990**, *55*, 3068.
53 Inouye, K.; Sumitomo, Y.; Shin, M. *Bull. Chem. Soc. Jpn.* **1976**, *49*, 3620.
54 Wünsch, E.; Jaeger, E.; Kisfaludy, L.; Löw, M. *Angew. Chem. Int. Ed. Engl.* **1977**, *16*, 317.
55 Fujii, N.; Otaka, A.; Ikemura, O.; Akaji, K.; Funakoshi, S.; Hayashi, Y.; Kuroda, Y.; Yajima, H. *J. Chem. Soc., Chem. Commun.* **1987**, 274.
56 Evans, D. A.; Ellman, J. A. *J. Am. Chem. Soc.* **1989**, *111*, 1063.
57 Schmidt, U.; Lieberknecht, A.; Bökens, H.; Griesser, H. *J. Org. Chem.* **1983**, *48*, 2680.
58 Sakaitani, M.; Ohfune, Y. *Tetrahedron Lett.* **1985**, *26*, 5543.
59 Jones, A. B.; Villalobos, A.; Linde, R. G.; Danishhefsky, S. J. *J. Org. Chem.* **1990**, *55*, 2786.
60 Torii, S.; Tanaka, H.; Taniguchi, M.; Kameyama, Y.; Sasaoka, M.; Shioiri, T.; Kikuchi, R.; Kawahara, I.; Shimabayashi, A.; Nagao, S. *J. Org. Chem.* **1991**, *56*, 3633.
61 Schultz, M.; Hermann, P.; Kunz, H. *Synlett* **1992**, 37.
62 Walton, E.; Rodin, J. O.; Stammer, C. H.; Holly, F. W. *J. Org. Chem.* **1962**, *27*, 2255.
63 Mata, E. G.; Mascaretti, O. A. *Tetrahedron Lett.* **1988**, *29*, 6893.
64 Salomon, C. J.; Mata, E. G.; Mescaretti, O. A. *Tetrahedron Lett.* **1991**, *32*, 4239.
65 Armstrong, A.; Brackenridge, I.; Jackson, R. F. W.; Kirk, J. M. *Tetrahedron Lett.* **1988**, *24*, 2483.
66 Olsen, R. K.; Ramaswamy, K.; Emery, T. *J. Org. Chem.* **1984**, *49*, 3527.
67 Valerio, R. M.; Alewood, P. F.; Johns, R. B. *Synthesis* **1988**, 786.
68 Schwyzer, R.; Kappeler, H. *Helv. Chim. Acta.* **1961**, *44*, 1991.
69 Kim, S.; Park, Y. H.; Kee, I. S. *Tetrahedron Lett.* **1991**, *32*, 3099.
70 Li, W.-R.; Ewing, W. R.; Harris, B. D.; Joullié, M. M. *J. Am. Chem. Soc.* **1990**, *112*, 7659.
71 Zoretic, P. A.; Soja, P.; Conrad, W. E. *J. Org. Chem.* **1975**, *40*, 2962.
72 Gerdes, J. M.; Wade, L. G. *Tetrahedron Lett.* **1979**, 689.
73 Narasaka, K.; Sakakura, T.; Uchimaru, T.; Guédin-Vuong, D. *J. Am. Chem. Soc.* **1984**, *106*, 2954.
74 Ho, T.-S.; Wong, C. M. *J. Chem. Soc., Chem. Commun.* **1973**, 224.
75 Ho, T.-S. *Synth. Commun.* **1979**, *9*, 267.
76 Wade, L. G.; Gerdes, J. M.; Wirth, R. P. *Tetrahedron Lett.* **1978**, 731.
77 Woodward, R. B.; Heusler, K.; Gosteli, J.; Naegeli, R.; Oppolzer, W.; Ramage, R.; Ranganathan, S.; Vorbrüggen, H. *J. Am. Chem. Soc.* **1966**, *88*, 852.

78 Schmidt, U.; Mundinger, K.; Mangold, R.; Lieberknecht, A. *J. Chem. Soc., Chem. Commun.* **1990**, 1216.
79 Eckstein, F.; Rizk, I. *Angew. Chem. Int. Ed. Engl.* **1967**, *6*, 949.
80 Adamiak, R. W.; Biala, E.; Grzeskowiak, K.; Kierzek, R.; Kraszewski, A.; Markiewicz, W. T.; Stawinski, J.; Wiewiorowski, M. *Nucl. Acids Res.* **1977**, *4*, 2321.
81 Ogilvie, K. K.; Theriault, N. Y.; Seifert, J.-M.; Pon, R. T.; Nemer, M. J. *Can. J. Chem.* **1980**, *58*, 2686.
82 Semmelhack, M. F.; Heinsohn, G. E. *J. Am. Chem. Soc.* **1972**, *94*, 5139.
83 Huang, Z.-Z.; Zhou, X.-J. *Synthesis* **1989**, 693.
84 Schmidt, U.; Kroner, M.; Griesser, H. *Synthesis* **1991**, 294.
85 Hamada, Y.; Kondo, Y.; Shibata, M.; Shioiri, T. *J. Am. Chem. Soc.* **1989**, *111*, 669.
86 Gerlach, H. *Helv. Chim. Acta.* **1977**, *60*, 3039.
87 Sieber, P. *Helv. Chim. Acta* **1977**, *60*, 2711.
88 Willson, T. M.; Kocienski, P.; Jarowicki, K.; Isaac, K.; Faller, A.; Campbell, S. F.; Bordner, J. *Tetrahedron* **1990**, *46*, 1757.
89 Forsch, R. A.; Rosowsky, A. *J. Org. Chem.* **1984**, *49*, 1305.
90 Karim, S.; Parmee, E. R.; Thomas, E. J. *Tetrahedron Lett.* **1991**, *32*, 2269.
91 Vedejs, E.; Larsen, S. D. *J. Am. Chem. Soc.* **1984**, *106*, 3030.
92 Bourne, G. T.; Howell, D. C.; Pritchard, M. C. *Tetrahedron.* **1991**, *47*, 4763.
93 White, J. D.; Jayasinghe, L. R. *Tetrahedron Lett.* **1988**, *29*, 2139.
94 Miller, A. W.; Stirling, C. J. M. *J. Chem. Soc. C* **1968**, 2612.
95 Pfister, M.; Farkas, S.; Charubala, R.; Pfleiderer, W. *Nucleosides Nucleotides* **1988**, *7*, 595.
96 Clark, J. H. *Chem. Rev.* **1980**, *80*, 429.
97 Tsutsui, H.; Muto, M.; Motoyoshi, K.; Mitsunobu, O. *Chemistry Lett.* **1987**, 1595.
98 Colvin, W.; Purcell, T. A.; Raphael, R. A. *J. Chem. Soc., Perkin Trans.1* **1976**, 1718.
99 Inoue, S.; Okada, K.; Tanino, H.; Hashizume, K.; Kakoi, H. *Tetrahedron Lett.* **1984**, *25*, 4407.
100 Bednarek, M. A.; Bodanszky, M. *Int. J. Pept. Protein Res.* **1983**, *21*, 196.
101 Kessler, H.; Siegmeier, R. *Tetrahedron Lett.* **1983**, 281.
102 Chong, J. M.; Lajoie, G.; Tjepkema, M. W. *Synthesis* **1992**, 819.
103 Letsinger, R. L.; Ogilvie, K. K. *J. Am. Chem. Soc.* **1967**, *89*, 4801.
104 Evans, D. A.; Gage, J. R.; Leighton, J. L. *J. Org. Chem.* **1992**, *57*, 1964.
105 Bergmann, M.; Zervas, L. *Ber.* **1933**, *66*, 1288.
106 De Barnardo, S.; Tengi, J. P.; Sasso, G. J.; Weigele, M. *J. Org. Chem.* **1985**, *50*, 3457.
107 Melillo, D. G.; Shinkai, I.; Liu, T.; Ryan, K.; Sletzinger, M. *Tetrahedron Lett.* **1980**, *21*, 2783.
108 Johnston, D. B. R.; Schmitt, S. M.; Bouffard, F. A.; Christensen, B. G. *J. Am. Chem. Soc.* **1978**, *100*, 313.
109 Bajwa, J. S. *Tetrahedron Lett.* **1992**, *33*, 2299.
110 Hashimoto, S.; Miyazaki, Y.; Shinoda, T.; Ikegani, S. *Tetrahedron Lett.* **1989**, *30*, 7195.
111 Petersen, J. S.; Fels, G.; Rapoport, H. *J. Am. Chem. Soc.* **1984**, *106*, 4539.
112 Kametani, T.; Sekine, H.; Honda, T. *Chem. Pharm. Bull.* **1982**, *30*, 4545.
113 Hillis, L. R.; Ronald, R. C. *J. Org. Chem.* **1985**, *50*, 470.
114 Silverman, R. B.; Holladay, M. W. *J. Am. Chem. Soc.* **1981**, *103*, 7357.
115 Alexander, R. P.; Beeley, N. R. A.; O'Driscoll, M.; O'Neill, F. P.; Millican, T. A.; Pratt, A. J.; Willenbrock, F. W. *Tetrahedron Lett.* **1991**, *32*, 3269.
116 Baldwin, J. E.; Jesudason, C. D.; Maloney, M. G.; Rhys Morgan, D.; Pratt, A. J. *Tetrahedron* **1991**, *47*, 5603.
117 Ohtani, M.; Watanabe, F.; Narisada, M. *J. Org. Chem.* **1984**, *49*, 5271.
118 Chahoua, L.; Baltas, M.; Gorrichon, L.; Tisnès, P.; Zedde, C. *J. Org. Chem.* **1992**, *57*, 5798.
119 Kozikowski, A. P.; Fauq, A. H.; Powis, G.; Kurian, P.; Crews, F. T. *J. Chem. Soc., Chem. Commun.* **1992**, 362.
120 Miller, J. B. *J. Org. Chem.* **1959**, *24*, 560.
121 Mortimore, M.; Cockerill, G. S.; Kocienski, P.; Treadgold, R. *Tetrahedron Lett.* **1987**, *28*, 3747.
122 Froussios, C.; Kolovos, M. *Tetrahedron Lett.* **1989**, *30*, 3413.
123 Waldmann, H.; Kunz, H. *Liebigs Ann. Chem.* **1983**, 1712.
124 Kunz, H.; Waldmann, H. *Angew. Chem. Int. Ed. Engl.* **1984**, *23*, 71.
125 Kunz, H. *Angew. Chem. Int. Ed. Engl.* **1987**, *26*, 294.
126 Friedrich-Bochnitschek, S.; Waldmann, H.; Kunz, H. *J. Org. Chem.* **1989**, *54*, 751.
127 Hey, H.; Arpe, H.-J. *Angew. Chem. Int. Ed. Engl.* **1973**, *12*, 928.
128 Nokami, J.; Ohkura, M.; Dan-oh, Y.; Sakamoto, Y. *Tetrahedron Lett.* **1991**, *32*, 2409.
129 Tsuji, J.; Nisar, M.; Shimizu, I. *J. Org. Chem.* **1985**, *50*, 3416.
130 Dangles, O.; Guibé, F.; Balavoine, G.; Lavielle, S.; Marquet, A. *J. Org. Chem.* **1987**, *52*, 4984.
131 Jeffrey, P. D.; McCombie, S. W. *J. Org. Chem.* **1982**, *47*, 587.
132 Ruediger, E. H.; Solomon, C. *J. Org. Chem.* **1991**, *56*, 3183.
133 Junghaim, L. N. *Tetrahedron Lett.* **1989**, *30*, 1889.

134 Kunz, H.; März, J. *Angew. Chem. Int. Ed. Engl.* **1988**, *27*, 1375.
135 Paulsen, H.; Merz, G.; Brockhausen, I. *Liebigs Ann. Chem.* **1990**, 719.
136 Hoffmann, R. W.; Ditrich, K. *Liebigs Ann. Chem.* **1990**, 23.
137 Mastalerz, H. *J. Org. Chem.* **1984**, *49*, 4092.
138 Montforts, F.-P.; Schwartz, U. M. *Liebigs Ann. Chem.* **1991**, 709.
139 Kunz, H.; Waldmann, H.; Unverzagt, C. *Int. J. Pept. Protein Res.* **1985**, *26*, 493.
140 Kabalka, G. W.; Bierer, D. E. *Synth. Commun.* **1989**, *19*, 2783.
141 Hart, T. W.; Metcalfe, D. A.; Scheinmann, F. *J. Chem. Soc., Chem. Commun.* **1979**, 156.
142 Morton, D. R.; Thompson, J. L. *J. Org. Chem.* **1978**, *43*, 2102.
143 McCarthy, P. A. *Tetrahedron Lett.* **1982**, *23*, 4199.
144 Bhide, R. S.; Levison, B. S.; Sharma, R. B.; Ghosh, S.; Salomon, R. G. *Tetrahedron Lett.* **1986**, *27*, 671.
145 Berry, P. D.; Brown, A. C.; Hanson, J. C.; Kaura, A. C.; Milner, P. H.; Moores, C. J.; Quick, J. K.; Saunders, R. N.; Southgate, R.; Whittall, N. *Tetrahedron Lett.* **1991**, *32*, 2683.
146 Hirsenkorn, R.; Schmidt, R. R. *Liebigs Ann. Chem.* **1990**, 883.
147 Corey, E. J.; Kang, M.-c.; Desai, M. C.; Ghosh, A. K.; Houpis, I. N. *J. Am. Chem. Soc.* **1988**, *110*, 649.
148 Bailey, W. F.; Rivera, A. D. *J. Org. Chem.* **1987**, *52*, 1559.
149 Eliel, E. L.; Nader, F. W. *J. Am. Chem. Soc.* **1970**, *92*, 584.
150 Just, G.; Luthe, C.; Viet, M. T. P. *Can. J. Chem.* **1983**, *61*, 712.
151 Corey, E. J.; Niimura, K.; Konishi, Y.; Hashimoto, S.; Hamada, Y. *Tetrahedron Lett.* **1986**, *27*, 2199.
152 Keinan, E.; Sinha, S. C.; Singh, S. P. *Tetrahedron* **1991**, *47*, 4631.
153 Corey, E. J.; Desai, M. C.; Engler, T. A. *J. Am. Chem. Soc.* **1985**, *107*, 4339.
154 Corey, E. J.; Shimoji, K. *J. Am. Chem. Soc.* **1983**, *105*, 1662.
155 Corey, E. J.; De, B. *J. Am. Chem. Soc.* **1984**, *106*, 2735.
156 Corey, E. J.; Kyler, K.; Raju, N. *Tetrahedron Lett.* **1984**, *25*, 5115.
157 Burkholder, T. P.; Fuchs, P. L. *J. Am. Chem. Soc.* **1988**, *110*, 2341.
158 Trost, B. M.; Tometzki, G. B. *Synthesis* **1991**, 1235.
159 Suzuki, K.; Tomooka, K.; Katayama, E.; Matsumoto, T.; Tsuchihashi, G. *J. Am. Chem. Soc.* **1986**, *108*, 5221.
160 Corey, E. J.; Shibasaki, M.; Knolle, J.; Sugahara, T. *Tetrahedron Lett.* **1977**, 785.
161 Seebach, D. *Angew. Chem. Int. Ed. Engl.* **1967**, *6*, 442.
162 Damon, R. E.; Schlessinger, R. H. *Tetrahedron Lett.* **1976**, 1561.
163 Salzmann, T. N.; Ratcliffe, R. W.; Christensen, B. G.; Bouffard, F. A. *J. Am. Chem. Soc.* **1980**, *102*, 6161.
164 White, J. D.; Kuo, S.-c.; Vedananda, T. R. *Tetrahedron Lett.* **1987**, *28*, 3061.
165 Wakamatsu, T.; Hara, H.; Ban, Y. *J. Org. Chem.* **1985**, *50*, 108.
166 Corey, E. J.; Raju, N. *Tetrahedron Lett.* **1983**, *24*, 5571.
167 Pattison, D. B. *J. Am. Chem. Soc.* **1957**, *79*, 3455.
168 Voss, G.; Gerlach, H. *Helv. Chim. Acta.* **1983**, *66*, 2294.
169 Corey, E. J.; Beames, D. J. *J. Am. Chem. Soc.* **1973**, *95*, 5829.
170 Meyers, A. I.; Temple, D. L.; Nolen, R. L.; Mihelich, E. D. *J. Org. Chem.* **1974**, *39*, 2778.
171 Meyers, A. I.; Temple, D. L. *J. Am. Chem. Soc.* **1972**, *92*, 6646.
172 Mori, S.; Ohno, T.; Harada, H.; Aoyama, T.; Shioiri, T. *Tetrahedron* **1991**, *47*, 5051.
173 Meyers, A. I.; Higashiyama, K. *J. Org. Chem.* **1987**, *52*, 4592.
174 Rossano, L. T.; Plata, D. J.; Kallmerten, J. *J. Org. Chem.* **1988**, *53*, 5189.
175 Wehrmeister, H. L. *J. Org. Chem.* **1961**, *26*, 3821.
176 Meyers, A. I.; Lutomski, K. A. *Synthesis* **1983**, 105.
177 Schow, S. R.; Bloom, J. D.; Thompson, A. S.; Winzenberg, K. N.; Smith, A. B. *J. Am. Chem. Soc.* **1986**, *108*, 2662.
178 Haidukewych, D.; Meyers, A. I. *Tetrahedron Lett.* **1972**, 3031.
179 Meyers, A. I.; Knaus, G.; Kamata, K. *J. Am. Chem. Soc.* **1974**, *96*, 268.
180 Meyers, A. I.; Lutomski, K. A. *J. Am. Chem. Soc.* **1982**, *104*, 879.

# Chapter 5 Carbonyl Protecting Groups

## 5.1 Introduction

The electrophilicity of the carbonyl group is a dominant feature of its extensive chemistry and a major problem in a synthesis of any length is to shield a carbonyl from nucleophilic attack until such time as its electrophilic properties must be exploited. If the carbonyl is a structural component of a target rather than a reactive functional group in an intermediate, we are faced with the even more daunting problem of preserving the carbonyl throughout the entire chain of synthetic reactions only to lay it bare at the very end. The protection of aldehydes and ketones has been served by a relatively small repertoire of protecting groups and of these, acetals and thioacetals have proven the most serviceable. One of the reasons why there are comparatively few protecting groups for carbonyls is because they reveal a useful range of reactivity in both the protection and deprotection steps which allows a modicum of selectivity.

Although we will be viewing acetals through the prism of protecting group strategy, and therefore extolling their virtues of inertness, it should be remembered that there is a corresponding range of fascinating and valuable reactivity* to be explored, that will provide us with a few welcome digressions.

A word about nomenclature. Once upon a time, chemists made a useful distinction between acetals (derived from aldehydes) and ketals (derived from ketones) which has since perished. The International Union of Pure and Applied Chemistry (IUPAC) has decided (Rule C–331.1) that the term ketal is redundant and that the term acetal should now apply to all 1,1-*bis*-ethers whether derived from aldehyde or ketones.

## 5.2 *O,O*-Acetals

Before discussing the hydrolytic cleavage of acetals as a deprotection reaction, we should recall some of the problems associated with acetal stability. Acetals are cleaved by Lewis acids at varying rates depending on the metal. Most cyclic acetals will tolerate magnesium and zinc halides at low temperature but they are imperilled at elevated temperature. Stronger oxyphilic Lewis acids such as aluminium, titanium, boron and trialkylsilyl halides will induce reactions with acetals some of which have been harnessed for the good. For example Grignard reagents[1], cuprates[2], dialkylzincs[3], alanes[4], and allyl silanes (see section 5.2.2) will displace one of the oxygen atoms of an acetal with the assistance of titanium or boron Lewis acids. In the case of acetal derivatives of aromatic carbonyls, hydrogenolysis and reductive cleavage under dissolving metal conditions can be a problem. However, acetals are stable to metal hydride reduction and organolithium reagents[5-7] and they are inert toward strong aqueous or alcoholic bases, catalytic hydrogenation (unless benzylic), and Na or Li in liquid

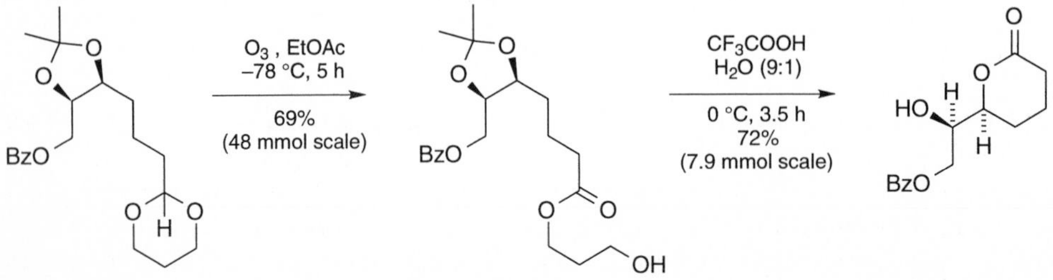

$O_3$, EtOAc, –78 °C, 5 h; 69% (48 mmol scale)

$CF_3COOH$ $H_2O$ (9:1), 0 °C, 3.5 h; 72% (7.9 mmol scale)

**Scheme 5.1**

ammonia. Most non-acidic oxidising conditions leave them intact but the C–H bond of dioxolane derivatives of aldehydes is rapidly attacked by ozone. Indeed, ozonolytic cleavage of 1,3-dioxanes has been deliberately used in synthesis to convert an aldehyde acetal to an ester whilst leaving a ketonic acetal unscathed [Scheme 5.1][8]. In a similar vein, acetals can be oxidatively cleaved to esters with *tert*-butyl hydroperoxide catalysed by Ruthenium[9].

## 5.2.1 Cyclic Acetals

### (i) Cleavage

Acid-catalysed hydrolysis is the most common method for the deprotection of acetals and as a rough guide, the ease of hydrolysis parallels the ease of formation. Thus, in general, 1,3-dioxane derivatives of *ketones* hydrolyse faster than the corresponding 1,3-dioxolane derivatives, whereas the 1,3-dioxolane derivatives of *aldehydes* hydrolyse faster than the corresponding 1,3-dioxane derivatives. The ease of hydrolysis can vary enormously: in some cases hydrolysis is so easy that inadvertent loss of the protecting group during chromatography becomes a nuisance (such as acetals of enones). On the other hand acetals may be so robust that forcing conditions (mineral acid and heat) are required. For example, substrates which contain a basic amino function — even if it is remote — are quite resistant to hydrolysis because protonation first takes place at the more basic amino function. The resultant positive charge repels the second *O*-protonation required to set in motion the hydrolysis. An acid-catalysed hydrolysis of a basic acetal that required refluxing with 6 M HCl in acetone for 6–10 h[10] is illustrated in Scheme 5.2. Perhaps the mildest conditions that have been generally employed to date involve heating the acetal in aqueous acetone in the presence of PPTS as exemplified by a deprotection step taken from a recent synthesis of 1,25-Dihydroxycholecalciferol [Scheme 5.3][11]. Our survey of the reagents and conditions for the hydrolysis of acetals can be abbreviated because the subject has already been considered in some depth in chapter 3 under the guise of 1,2- and 1,3-diol protection.

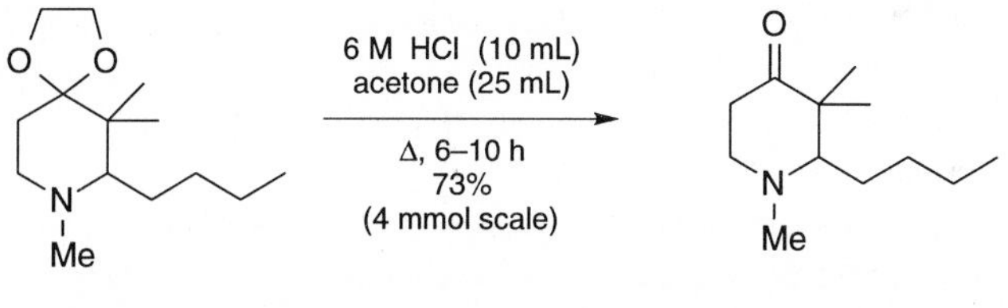

6 M HCl (10 mL)
acetone (25 mL)
Δ, 6–10 h
73%
(4 mmol scale)

**Scheme 5.2**

PPTS
$H_2O$–acetone
Δ
>60%

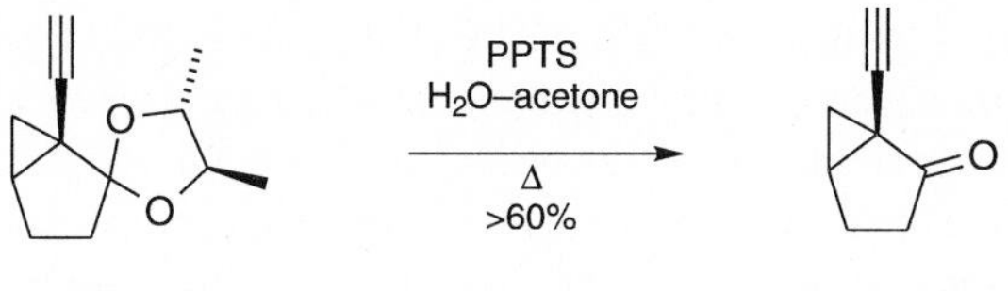

**Scheme 5.3**

Lewis acids can also be used to effect deacetalisation under mild conditions. A stringent example comes from a synthesis of Milbemycin by Ley and co-workers [Scheme 5.4][12] in which deacetalisation released a *β*-hydroxy ketone which was prone to elimination. However, by simply treating the acetal with $PdCl_2(MeCN)_2$ in acetone at r.t.[13], the requisite product was obtained in 94% yield. Under these conditions, acetals can be released in the presence of *tert*-butyldiphenylsilyl ethers or oxiranes[14] without incident.

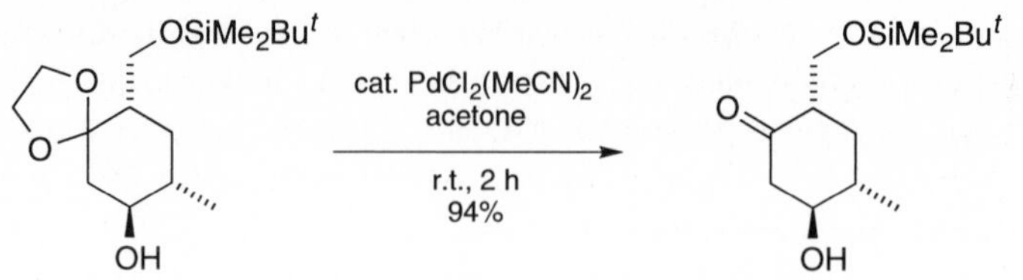

OSiMe$_2$Bu$^t$ cat. $PdCl_2(MeCN)_2$ acetone r.t., 2 h 94% OSiMe$_2$Bu$^t$ OH OH

**Scheme 5.4**

Another mild method for deacetalisation requires treatment of the substrate with $FeCl_3$ adsorbed on silica[15]. The conditions are sufficiently mild to release a ketone in the presence of an easily epimerisable centre [Scheme 5.5][16]. For more rugged substrates a greater latitude of Lewis acidity can be employed using reagents such as $Me_2BBr$ or TMSI. For conditions and references see chapter 3.

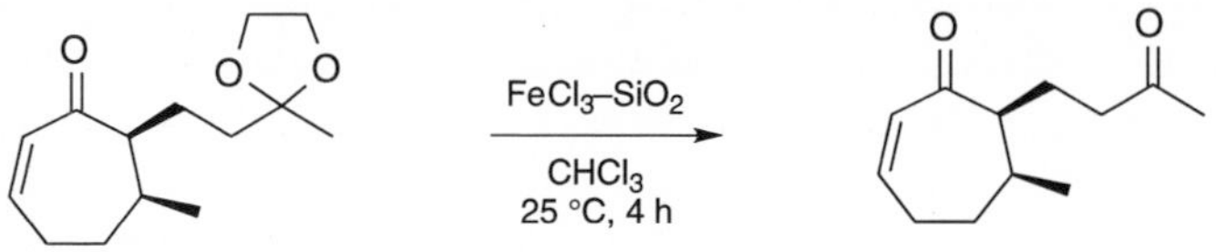

$FeCl_3$–$SiO_2$ $CHCl_3$ 25 °C, 4 h

**Scheme 5.5**

Among the less common reagents that have recently been recommended for the cleavage of dioxolanes of aldehydes and ketones can be included a concoction of $SmCl_3$ and TMSCl in THF[17] and DDQ in aqueous acetonitrile at r.t.[18]. In the case of DDQ, the hydrolysis is probably an acid catalysed process since the $pK_a$ of DDQ in water is 3.42[19].

**(ii) Formation**

Before elaborating some of the specific methods for acetal formation, we might first consider a few general principles which apply to acyclic as well as cyclic acetal formation:

1. Acetals are more easily prepared from aldehydes than from ketones;
2. Cyclic acetals are easier to form than acyclic acetals;
3. Conjugation deactivates the carbonyl group towards acetalisation;
4. Sterically hindered carbonyls react more slowly (if at all);
5. In aromatic aldehydes and ketones, electron donating substituents on the arene ring retard acetal formation whereas electron withdrawing substituents facilitate it.

1,3-Dioxolanes and 1,3-dioxanes are the most commonly used cyclic acetals and they are usually prepared by reaction of the carbonyl with ethane-1,2-diol or propane-1,3-diol in the presence of an acid-catalyst. A study of the relative ease of acetal formation with ketones showed the reactivity order 2,2-dimethylpropane-1,3-diol > ethane-1,2-diol > propane-1,3-diol[20]. A molar equivalent of water is liberated in the reaction and its efficient removal is necessary to drive the reaction to completion. Dehydration can be accomplished by physical means or chemical means. The favorite physical method entails continuous azeotropic distillation using benzene or toluene and a Dean–Stark apparatus; alternatively water can be removed by 4A molecular sieves in cases where small scale reactions are involved. If the elevated temperature required by the azeotropic method is undesirable, a chemical dehydrating agent can be employed such as magnesium sulfate, calcium sulfate, copper(II) sulfate, or alumina. The most common acid-catalysts are PTSA, CSA, PPTS, or acid ion exchange resins

(Amberlyst, Dowex). The higher reactivity of aldehydes to acetalisation than that of ketones is illustrated in Scheme 5.6[21] and a recent example of the use of an acid exchange resin as the catalyst to form a 5,5-dimethyl-1,3-dioxane[22] is shown in Scheme 5.7. The advantages to this method are (a) the 2,2-dimethyl-1,3-propanediol is a crystalline solid (cheap and commercially available), (b) NMR analysis of the product is simpler than that for ordinary 1,3-dioxanes, (c) the 5,5-dimethyl-1,3-dioxane is more stable towards acid than the corresponding 1,3-dioxane, and (d) aqueous workup is not necessary since the insoluble catalyst can be simply recovered by filtration and then recycled.

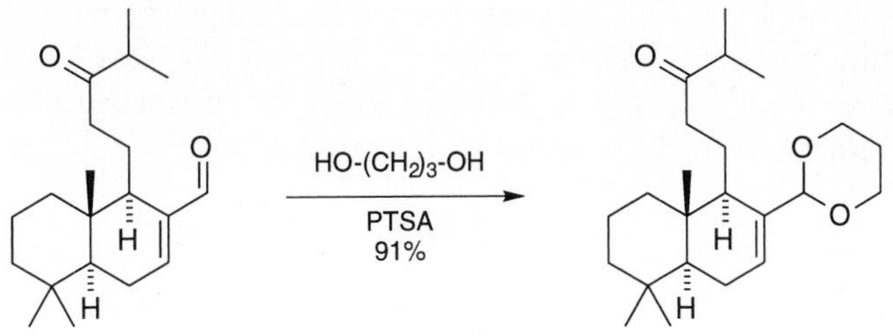

$HO-(CH_2)_3-OH$
PTSA
91%

**Scheme 5.6**

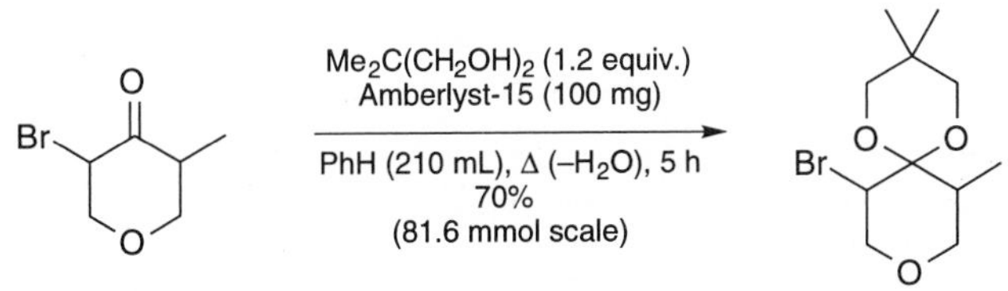

$Me_2C(CH_2OH)_2$ (1.2 equiv.)
Amberlyst-15 (100 mg)
PhH (210 mL), Δ ($-H_2O$), 5 h
70%
(81.6 mmol scale)

**Scheme 5.7**

Using the traditional acid-catalysed dioxolanation methods (PTSA, benzene, reflux), selective protection of the less hindered of a pair of ketone functions can be achieved [Scheme 5.8][23].

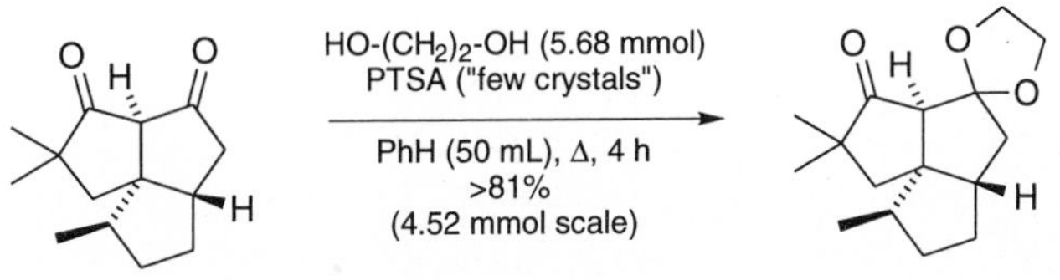

$HO-(CH_2)_2-OH$ (5.68 mmol)
PTSA ("few crystals")
PhH (50 mL), Δ, 4 h
>81%
(4.52 mmol scale)

**Scheme 5.8**

Chan and co-workers[24] showed that TMSCl was both an effective catalyst and dehydrating agent in acetalisation reactions. The example is taken from Schmidt's synthesis of Chlorothricolide [Scheme 5.9][25].

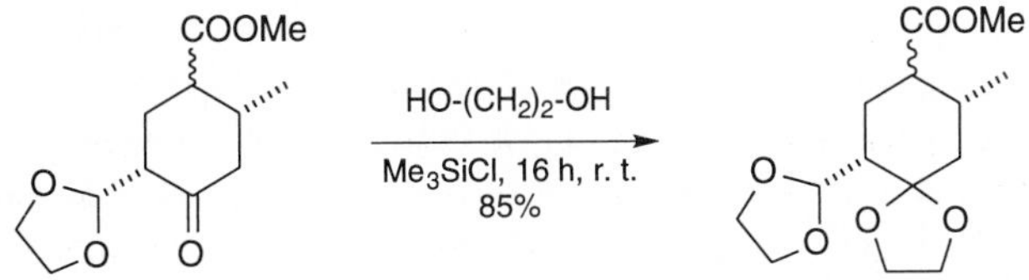

$HO-(CH_2)_2-OH$
$Me_3SiCl$, 16 h, r. t.
85%

**Scheme 5.9**

Noyori and co-workers[26,27] showed that ketones could be converted to the corresponding dioxolanes by reaction with the bis-trimethylsilyl ether of ethane-1,2-diol in the presence of TMSOTf at low temperature. These conditions permit acetalisation reactions of acid-sensitive substrates and since $(TMS)_2O$ is the by-product of the reaction rather than water, the Noyori method is especially suited to systems resistant to the usual methods of dehydration. We have selected three examples to illustrate the value of Noyori's method. The first [Scheme 5.10] is taken from a synthesis of Hygromycin A — a compound which inhibits hemagglutination and shows antitreponemal activity[28] — in which a ketone function was protected in the presence of a phenolic glycoside, where other standard methods failed. Our second example [Scheme 5.11] was complicated by competing retroaldolisation reactions following BOM ether cleavage when the standard conditions (ethane-1,2-diol in benzene in the presence of PTSA) were used[29]. Finally the double protection of the cyclobutenedione in Scheme 5.12 shows that sensitive substrates can be protected on a practical scale[30].

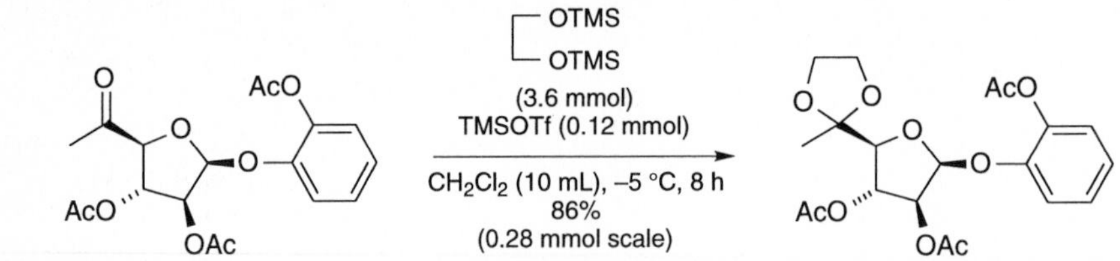

OTMS
OTMS
(3.6 mmol)
TMSOTf (0.12 mmol)
$CH_2Cl_2$ (10 mL), –5 °C, 8 h
86%
(0.28 mmol scale)

**Scheme 5.10**

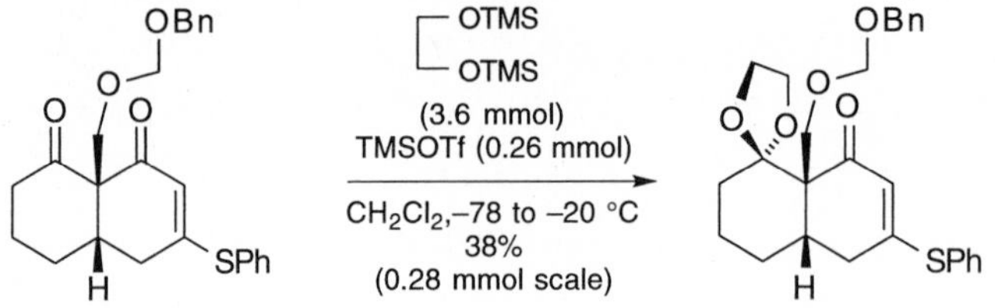

OTMS
OTMS
(3.6 mmol)
TMSOTf (0.26 mmol)
$CH_2Cl_2$,–78 to –20 °C
38%
(0.28 mmol scale)

**Scheme 5.11**

OTMS
OTMS
(150 mmol)
TMSOTf (31 mmol)
PhH, Δ, 6 h
76%
(58 mmol scale)

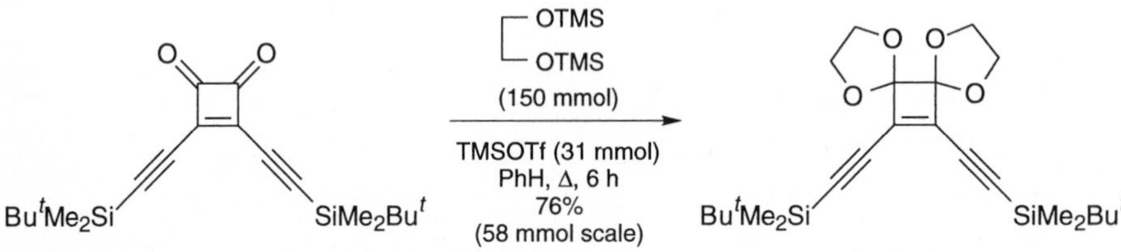

**Scheme 5.12**

The Noyori conditions were recently adapted to provide a method for the selective protection of a ketone in the presence of an aldehyde. The method [Scheme 5.13][31] requires the addition of dimethyl sulfide to the reaction mixture in order to provide temporary protection of the more reactive aldehyde function whilst leaving the ketone free to react in the desired dioxolanation reaction.

An earlier but effective method for accomplishing unfavorable acetalisation reactions uses acetal exchange which avoids the liberation of water. Scheme 5.14 illustrates the method as well as the higher reactivity of a saturated ketone in the presence of an unsaturated ketone[32-35]. In this case, acetal exchange accomplished a clean transfer of the dioxolane ring from the commercially available

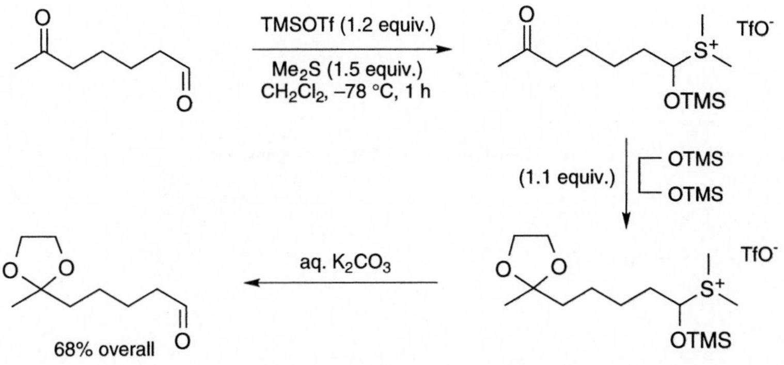
O
O
TMSOTf (1.2 equiv.)
$Me_2S$ (1.5 equiv.)
$CH_2Cl_2$, –78 °C, 1 h
O
S+
TfO-
OTMS
(1.1 equiv.)
OTMS
OTMS
aq. $K_2CO_3$
O O
O
68% overall
O O
S+
TfO-
OTMS

**Scheme 5.13**

2-methyl-2-ethyl-1,3-dioxolane to the Wieland–Miescher ketone liberating thereby 2-butanone. This same transformation was achieved (10 g scale) by McMurry and Isser[36] in the early stages of their Longifolene synthesis, but a large excess of ethane-1,2-diol and PTSA was used in refluxing benzene. Under these conditions, the diketal (3%) along with recovered starting diketone (7%) was obtained. Interestingly, the Noyori conditions (see above) reverse the selectivity of acetalisation; i.e., ketones react in preference to aldehydes[37]. In the example shown [Scheme 5.15] there is a further bonus: the double bond does not rearrange out of conjugation as it is wont to do under certain traditional acid-catalysed conditions (see below).

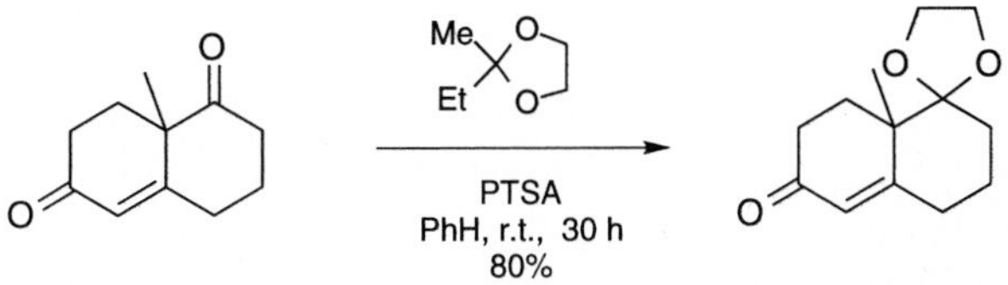
O
O
Me
Et
O
O
PTSA
PhH, r.t., 30 h
80%
O O
O

**Scheme 5.14**

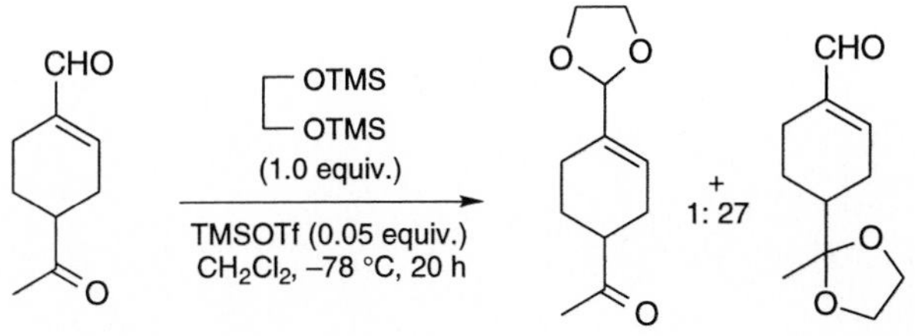
CHO
O
OTMS
OTMS
(1.0 equiv.)
TMSOTf (0.05 equiv.)
$CH_2Cl_2$, –78 °C, 20 h
O O
O
+
1: 27
CHO
O
O

**Scheme 5.15**

The extent of migration of the double bond of an enone which takes place on dioxolane formation depends on the strength of the acid catalyst. As can be seen from Scheme 5.16, weaker acids ($pK_a \geq 3$) leave the double bond intact whereas stronger acids cause partial or complete isomerisation[38]. The isomerisation of the double bond is not necessarily a disaster; indeed, in some cases it can prove a useful synthetic tool. For example, Heathcock[39] exploited the isomerisation in a concise synthesis of Confertin, the initial steps of which are shown in Scheme 5.17, and there are many other examples of a similar nature[40-44].

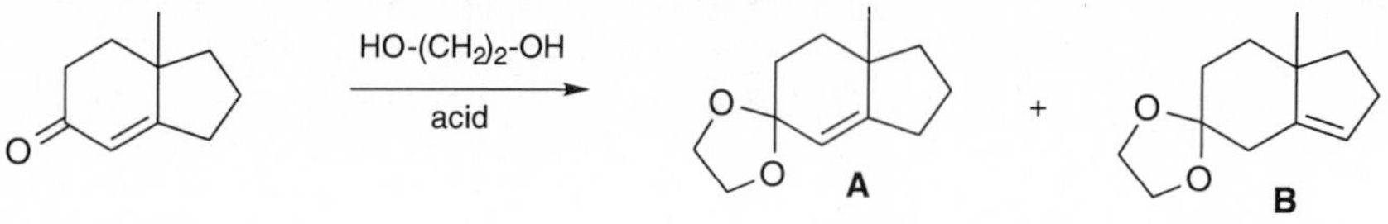

HO-(CH$_2$)$_2$-OH, acid → A + B

| acid | p$K_a$ | A:B |
|---|---|---|
| fumaric acid | 3.03 | 10:0 |
| phthalic acid | 2.89 | 7:3 |
| oxalic acid | 1.23 | 8:2 |
| PTSA | <1.0 | 0:10 |

**Scheme 5.16**

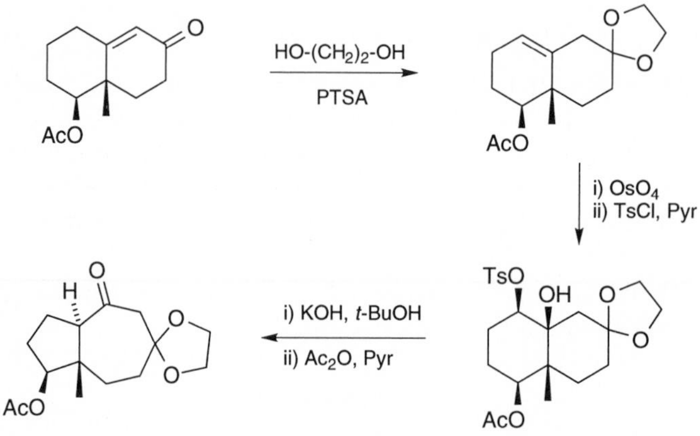

HO-(CH$_2$)$_2$-OH, PTSA; i) $OsO_4$ ii) TsCl, Pyr; i) KOH, *t*-BuOH ii) $Ac_2O$, Pyr

**Scheme 5.17**

Happily, the rearrangement of the double bond in the protection of a conjugated enone is reversible since re-conjugation takes place on hydrolysis to give the more thermodynamically stable product as illustrated in the closing stages of Sarett's classic synthesis of Cortisone acetate [Scheme 5.18][45]. By a judicious choice of conditions, a dioxolane protecting group can also be hydrolysed without reconjugation of the double bond[46].

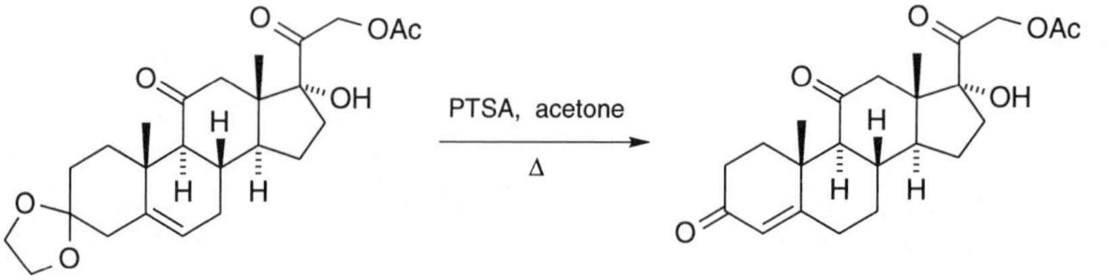

PTSA, acetone, Δ

**Scheme 5.18**

Protection of an unsubstituted vinyl ketone is problematic because of the ease with which such moieties undergo Michael addition and polymerisation. Corey and Magriotis[47] surmounted the problem indirectly during their synthesis of 7,20-diisocyanoadociane [Scheme 5.19] by first performing a conjugate addition of benzeneselenol to the enone followed by acetalisation. The double bond was then recovered by an easy selenoxide elimination reaction without affecting the acetal function. A similar problem was confronted by Schmidt and co-workers in their synthesis of the cyclic tetrapeptide Chlamydocin [Scheme 5.20][48]. Addition of HCl to the starting enone **20.1** produced a $\beta$-chloro ketone, which was subsequently transformed to the dioxolane **20.2**. The product was then used as a precursor to the tetrapeptide fragment, from which the enone intermediate **20.4** was recovered later by hydrolysis of the dioxolane, followed by base-catalysed elimination of HCl.

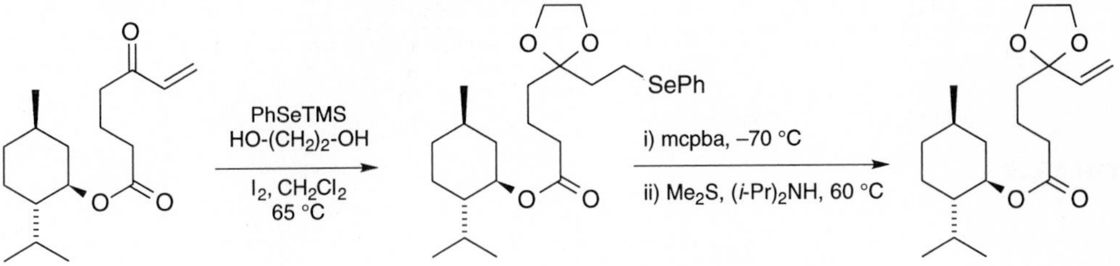

PhSeTMS, HO-$(CH_2)_2$-OH, $I_2$, $CH_2Cl_2$, 65 °C; SePh; i) mcpba, –70 °C; ii) $Me_2S$, $(i\text{-}Pr)_2NH$, 60 °C

**Scheme 5.19**

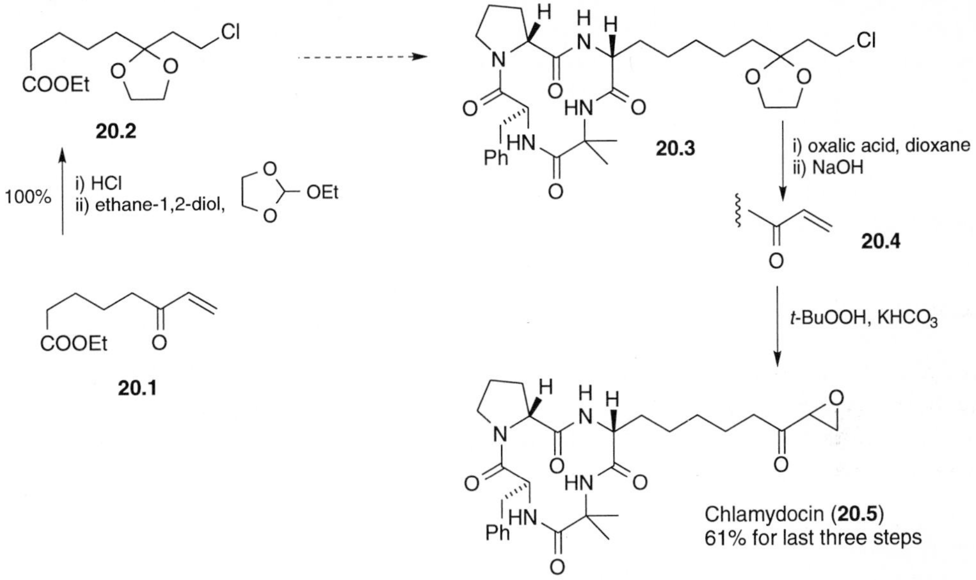

COOEt; Cl; 20.2; 100%; i) HCl; ii) ethane-1,2-diol, OEt; COOEt; 20.1; H; N; HN; Ph; 20.3; i) oxalic acid, dioxane; ii) NaOH; 20.4; *t*-BuOOH, $KHCO_3$; Chlamydocin (**20.5**) 61% for last three steps

**Scheme 5.20**

The use of transition metal catalysts in acetalisation reactions is ripe for exploitation, though comparatively little work has appeared to date despite encouraging preliminary results. Rhodium catalysts bearing the tridentate triphos ligand catalyse acetalisation under mild conditions (refluxing benzene) and with a substrate:catalyst ratio = 1: 2,000–10,000 [Scheme 5.21][49]. An obvious advantage to transition metal complexes as Lewis acids is the possibility of simultaneous coordination of both the alcohol and the carbonyl to the metal thus enhancing the potential for reaction.

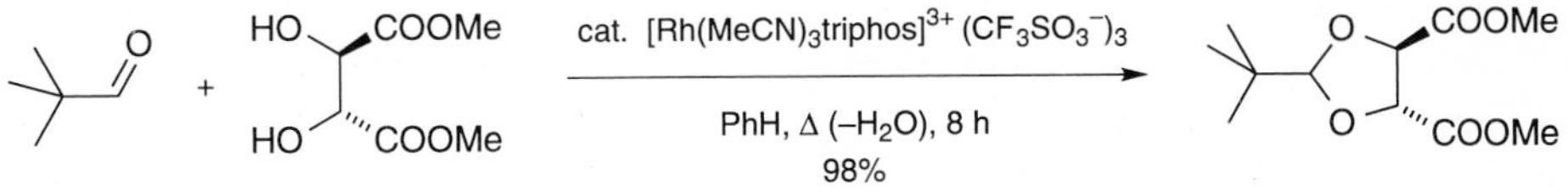

HO, COOMe; HO, COOMe; cat. $[Rh(MeCN)_3triphos]^{3+}(CF_3SO_3^-)_3$; PhH, Δ ($-H_2O$), 8 h; 98%; COOMe; COOMe

**Scheme 5.21**

Acid catalysis is the common feature which links all of the methods of acetal formation discussed so far. Dioxolanation can also be accomplished under basic conditions using 2-bromoethanol[50] as shown in the $\alpha$-oxo amide protection step [Scheme 5.22] taken from a recent synthesis of Strychnine[51].

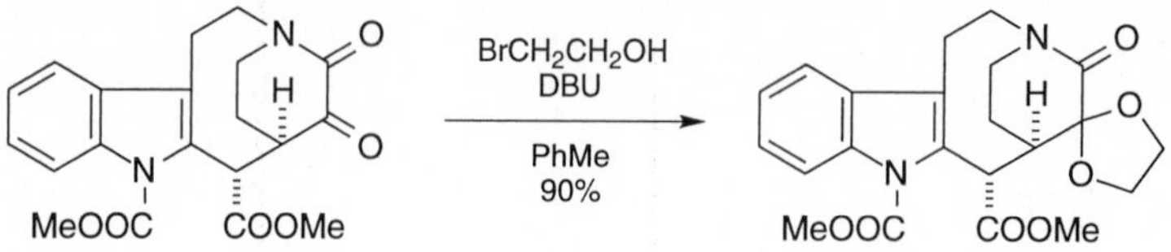

**Scheme 5.22**

## 5.2.2 *O,O*-Acetals in Asymmetric Synthesis

Our attention has so far been focused on the *passive* role of acetals as protecting groups and we have not considered the implications of stereochemistry in acetal formation or cleavage. One way in which the more traditional role of acetals as protecting groups has been exploited in asymmetric synthesis is in the use of acetals derived from homochiral diols to create an asymmetric environment in which further diastereoselective reactions can take place. However, acetals are also valuable *reactive* functional groups in asymmetric synthesis — a subject pioneered by the Kishi[52] and Johnson groups[53]. Since this section is yet another diversion from our narrow path we will have to be brief but a fuller treatment can be found in recent reviews*.

The fundamental principle at the heart of the use of acetals in asymmetric synthesis is the diastereoselective polarisation of only one of the two C–O bonds in an asymmetric acetal leading to an intimate ion pair which is capable of highly selective substitution by carbon nucleophiles with clean inversion[54-56]. The power of the reaction is illustrated by the asymmetric directed aldol reaction deployed in an elegant synthesis of Aklavinone [Scheme 5.23][52]. Yamamoto's synthesis of the side chain of vitamin E [Scheme 5.24] shows that remarkable diastereoselectivity is also possible in the $S_N2'$-like reaction of unsaturated $C_2$-symmetric acetals mediated by Lewis acidic reagents[57].

Diastereoselective elimination reactions have also been observed as shown by the Al-mediated cleavage of a 1,3-dioxane, which provided a key enol ether intermediate in a synthesis of Lardolure [Scheme 5.25][58].

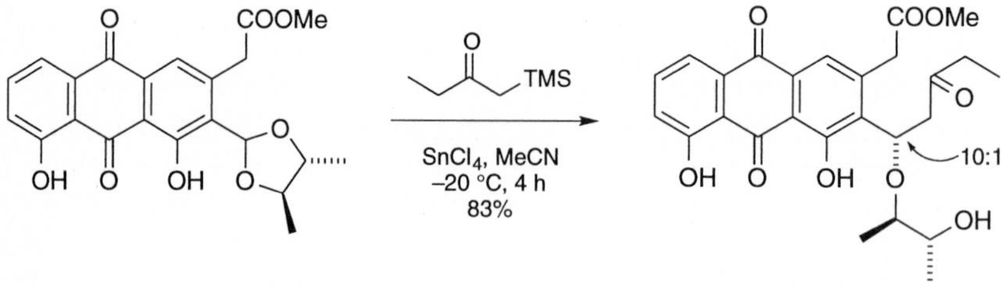

**Scheme 5.23**

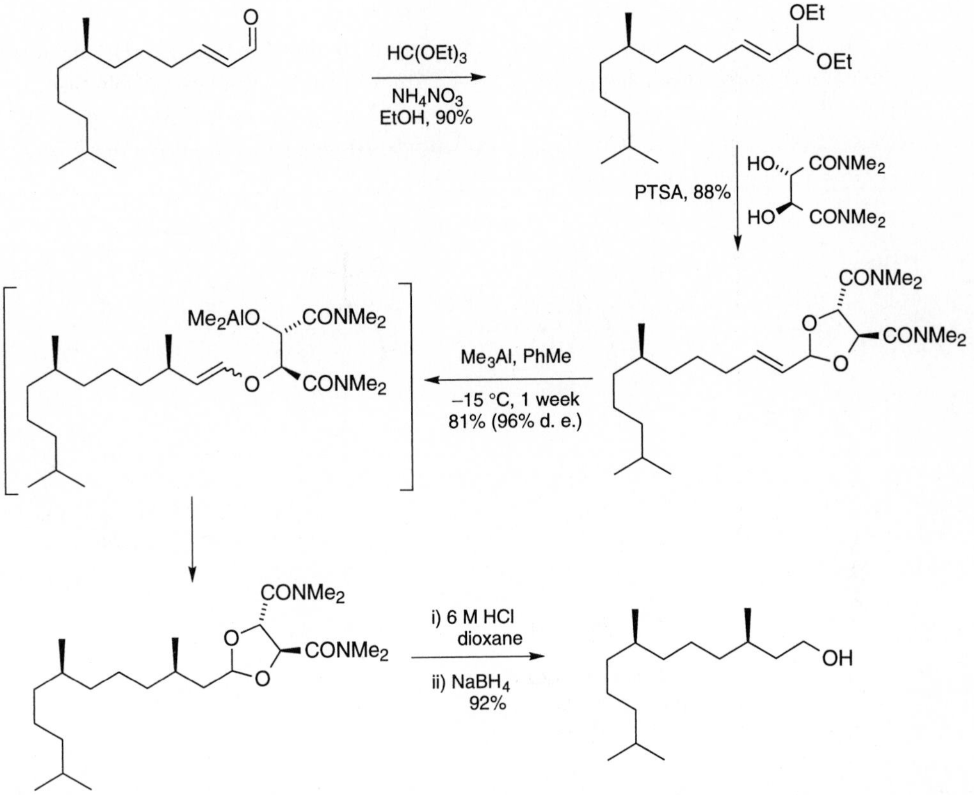

$HC(OEt)_3$, $NH_4NO_3$, EtOH, 90%

PTSA, 88%

Me$_3$Al, PhMe, −15 °C, 1 week, 81% (96% d. e.)

i) 6 M HCl dioxane
ii) $NaBH_4$ 92%

**Scheme 5.24**

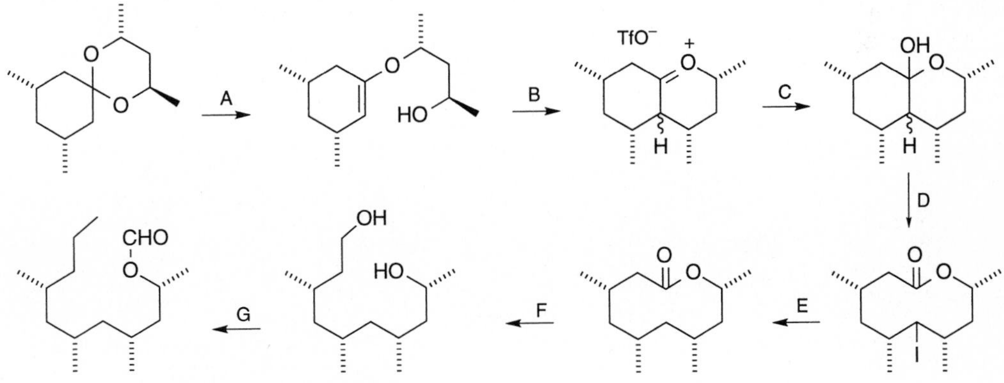

**REAGENTS AND YIELDS**

A - *i*-$Bu_3Al$, −78 °C to 0 °C;
B - $Tf_2O$, *i*-$Pr_2NEt$, $CH_2Cl_2$, −78 °C;
C 83% aq HOAc;
D 88% $PhI(OAc)_2$, $I_2$, *hv*;
E 100% $Bu_3SnH$, AIBN, THF, Δ;
F 92% DIBALH, −78 °C;
G 53% (i) TsCl (1.2 equiv.), DMAP, Pyr, −20 °C; (ii) $Me_2CuLi$, $Et_2O$; (iii) HCOOH, 65 °C.

**Scheme 5.25**

Meyers and Romo[59] showed that oxazolidines prepared from homochiral *t*-leucinol afford unique opportunities for the simultaneous protection of a carbonyl and the inducement of further diastereoselective reactions*. The example [Scheme 5.26] taken from a synthesis of the insecticide Deltamethrin, illustrates the $\pi$-facial differentiation of bicyclic enone **26.4** resulting in diastereoselective cyclopropanation. Note that the cyclopropanation reaction has taken place from the more hindered concave or *endo* face of the ring system.

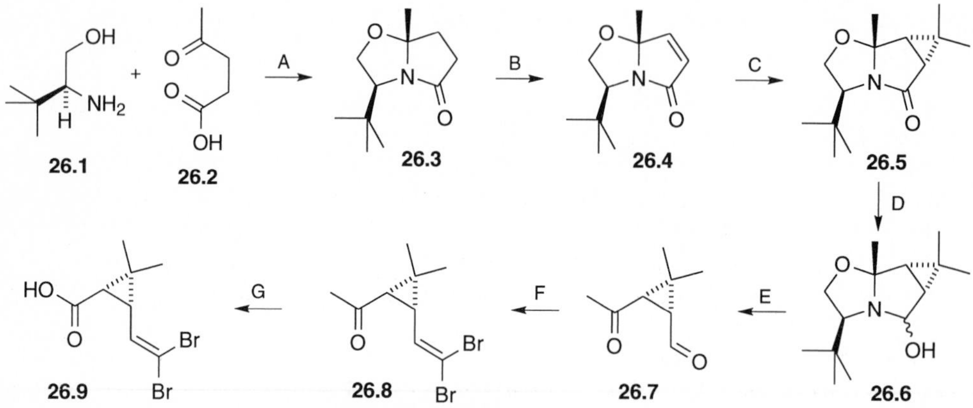

**REAGENTS AND YIELDS**

A 92% PhMe, $\Delta$ ($-H_2O$);
B 80% (i) LDA; (ii) PhSeBr; (iii) $H_2O_2$;
C 94% $Me_2C{=}SPh_2$;
D - $[(MeOCH_2CH_2O)_2AlH_2]Na$, THF, r.t.;
E - $Bu_4NH_2PO_4$, $CH_2Cl_2$–$H_2O$ (1:1), 96 h, r.t.;
F 42% $CBr_4$, $Ph_3P$;
G 81% $Br_2$, NaOH.

**Scheme 5.26**

An asymmetric synthesis of a key fragment of the anthracyclinone antibiotic $\beta$-Rhodomycinone[60] [Scheme 5.27] is our last example of the use of acetals in asymmetric synthesis and it illustrates the

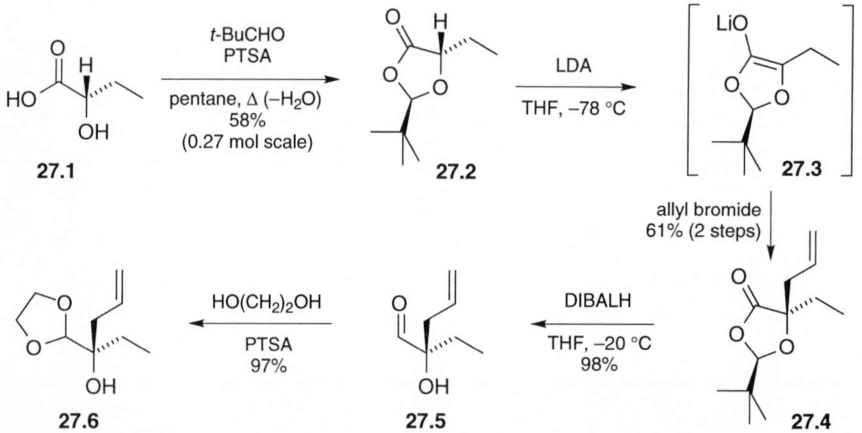

**Scheme 5.27**

"principle of self-reproduction of chirality centres" first propounded by Seebach and Naef in 1981[61,62]. In essence, the stereochemical information embedded in the starting $\alpha$-hydroxy acid **27.1** was used to create a second *temporary* stereogenic centre *via* acetal formation. The original stereochemistry $\alpha$ to the carbonyl can then be destroyed via enolate formation, and the stereogenicity now carried by the acetal centre in **27.3** is used to control the reintroduction of stereochemistry during alkylation $\alpha$ to the carbonyl to give **27.4**. The stereogenic acetal centre, having served its role as a caretaker of stereochemical information, can then be destroyed leaving behind the $\alpha$-substituted $\alpha$-hydroxy aldehyde **27.5**. As in the Deltamethrin synthesis [Scheme 5.26], a stereogenic acetal centre was used to control the conformation of an intermediate resulting in $\pi$-facial differentiation.

### 5.2.3 Acyclic *O,O*-Acetals

Acyclic acetals are usually only used when selective or very mild hydrolysis is required. Otherwise their more difficult preparation and chromatographic instability is best avoided by using the more robust cyclic acetals. The ease of hydrolysis depends on substitution; thus the relative rates of hydrolysis follow the order $CH_2(OEt)_2$ [1] : $MeCH(OEt)_2$ [6,000] : $Me_2C(OEt)_2$ [1.8 x $10^7$][63].

**(i) Cleavage**

Because of the relative fragility of acyclic acetals, the principal challenge in their use lies in defining the mildest conditions for their hydrolytic removal. All of our examples are taken from syntheses in which the liberated carbonyl is prone to further acid-catalysed reactions. However, we begin with an example of selective hydrolysis of only one of three differentially protected carbonyls spanning a mere 6-carbon chain [Scheme 5.28][64]. The acyclic acetal was easily removed in the presence of a dioxolane and a dithiane.

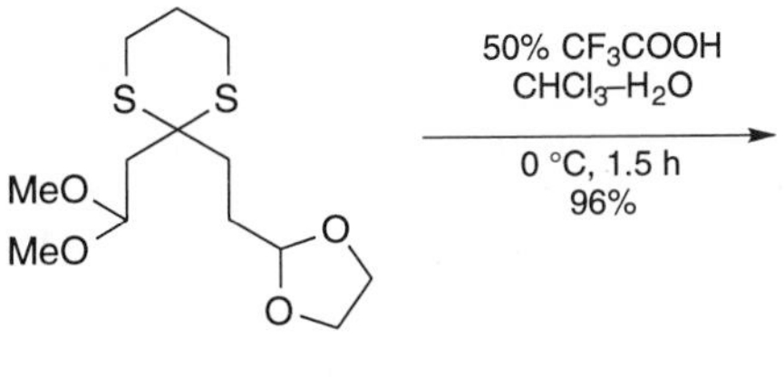

50% $CF_3COOH$
$CHCl_3$–$H_2O$
0 °C, 1.5 h
96%

**Scheme 5.28**

Acyclic acetal functions were used to protect both an aldehyde and an alcohol function in a synthesis of the labile anti-inflammatory agent Manoalide. Even so, the mild conditions used (70% aqueous acetic acid), only returned a yield of 55% [Scheme 5.29][65].

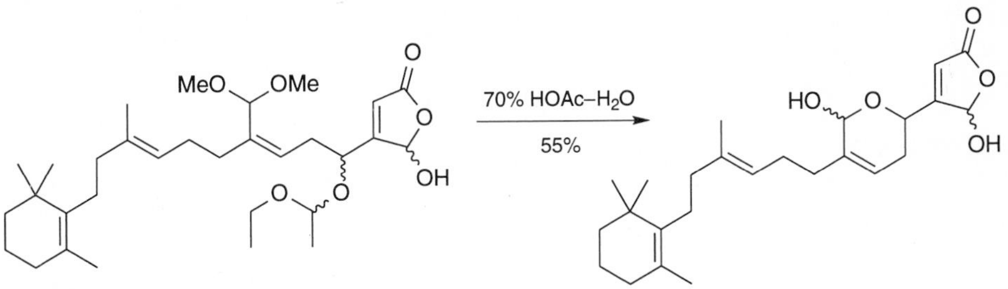

70% HOAc–$H_2O$
55%

**Scheme 5.29**

Corey's synthesis of the macrodiolide Vermiculine reveals mild hydrolytic conditions that preserved the product, which was susceptible to further hydrolysis as well as acid-catalysed $\beta$-elimination [Scheme 5.30][66].

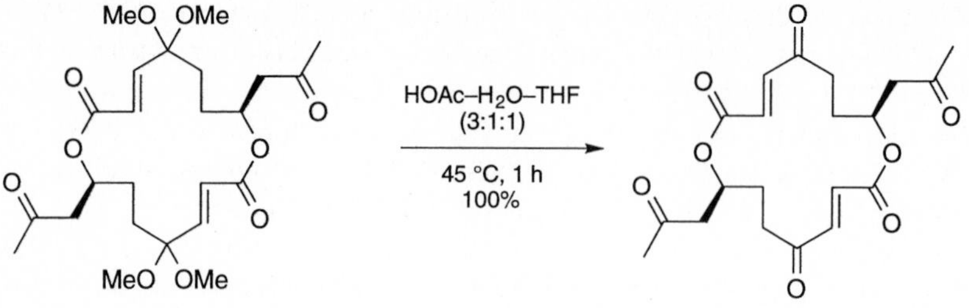

HOAc–$H_2O$–THF (3:1:1), 45 °C, 1 h, 100%

**Scheme 5.30**

A synthesis of the fragrant terpenoid Sinensal [Scheme 5.31] is noteworthy for two reasons[67]. First, the method used to prepare the diethyl acetal is a rare example of an acetal synthesis which takes place under *basic* conditions; secondly, the hydrolysis of the acetal product was only achieved after considerable effort by reaction with aqueous oxalic acid adsorbed on silica gel[68].

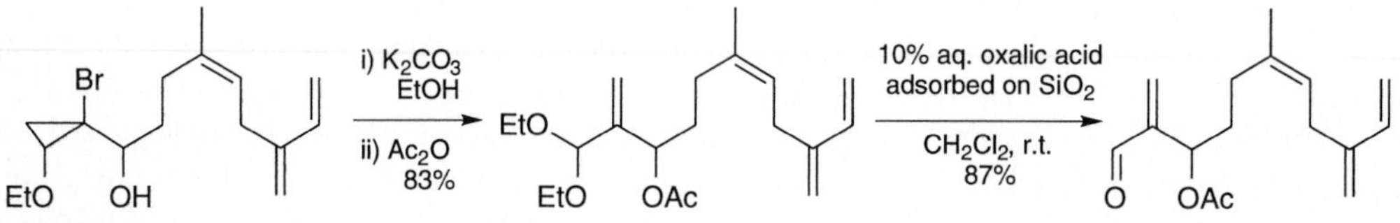

i) $K_2CO_3$, EtOH; ii) $Ac_2O$, 83%; 10% aq. oxalic acid adsorbed on $SiO_2$, $CH_2Cl_2$, r.t., 87%

**Scheme 5.31**

Two examples from the synthesis of the Neocarzinostatin core structure illustrate the problems of manipulating acetal protecting groups in substrates harbouring sensitive oxirane rings. Magnus and

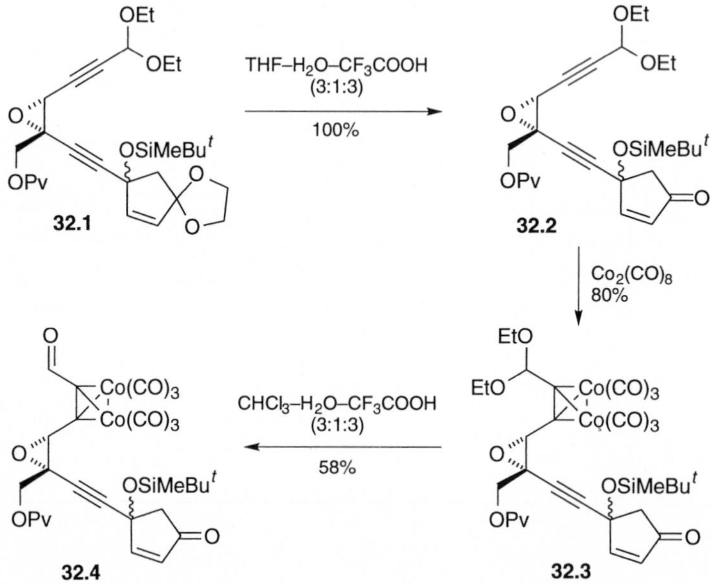

**32.1** → THF–$H_2O$–$CF_3COOH$ (3:1:3), 100% → **32.2** → $Co_2(CO)_8$, 80% → **32.3** → $CHCl_3$–$H_2O$–$CF_3COOH$ (3:1:3), 58% → **32.4**

**Scheme 5.32**

Davies[69] expected a simultaneous hydrolysis of both the dioxolane ring and the diethyl acetal function in the substrate **32.1** [Scheme 5.32]; instead they observed that selective hydrolysis of the dioxolane and prolonged reaction times only lead to decomposition. Successful hydrolysis of the acyclic acetal was only achieved after expulsion of EtOH (i.e. oxonium ion formation) was facilitated by conversion of the proximate alkyne function in **32.2** to the cobalt complex **32.3,** but even then the yield was rather modest (58%).

The usual acid-sensitivity of acyclic acetals was similarly of no avail in the related system studied by Meyers and co-workers [Scheme 5.33][70]. In this case, treatment of the acetal **33.1** with trichloroacetic acid generated an oxonium ion intermediate, that was rapidly and efficiently intercepted with hydrogen peroxide — a reagent which is much more nucleophilic than water. The resultant $\alpha$-methoxy hydroperoxide **33.2** was then reductively cleaved *under neutral conditions* to produce a hemiacetal, which lost methanol to give the desired aldehyde **33.3.**

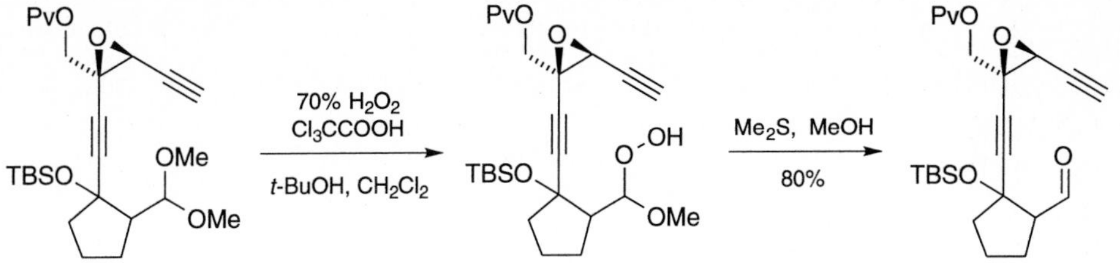

PvO, TBSO, OMe, OMe — 33.1 → (70% $H_2O_2$, $Cl_3CCOOH$, *t*-BuOH, $CH_2Cl_2$) → 33.2 (OOH, OMe) → ($Me_2S$, MeOH, 80%) → 33.3

**Scheme 5.33**

Husson's asymmetric synthesis of the frog toxin Pumiliotoxin[71,72] demanded the release of a protected ketone [Scheme 5.34] in the presence of an acid-sensitive aminal. In this case, success was achieved by exploiting the mild Lewis acidic properties of the lithium cation under conditions first reported by Lipshutz and Harvey[73]. Thus treatment of the dimethyl acetal with $LiBF_4$ in acetonitrile containing 2% water returned the desired ketone in >95% yield.

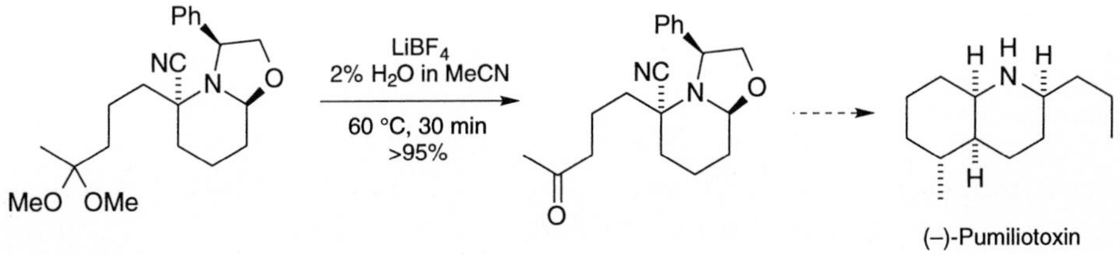

Ph, NC, N, O, MeO OMe → ($LiBF_4$, 2% $H_2O$ in MeCN, 60 °C, 30 min, >95%) → Ph, NC, N, O, O ⇢ (–)-Pumiliotoxin

**Scheme 5.34**

Viala and Santelli[74] devised a simple and efficient synthesis of 1,4-dienes which is practical and amenable to large scale. The method has been applied to a number of natural products including the methyl ester of Arachidonic Acid. A key step in the sequence [Scheme 5.35] is the mild hydrolysis of a diisopropyl acetal[75] under conditions which do not provoke rearrangement of the double bond into conjugation with the aldehyde function.

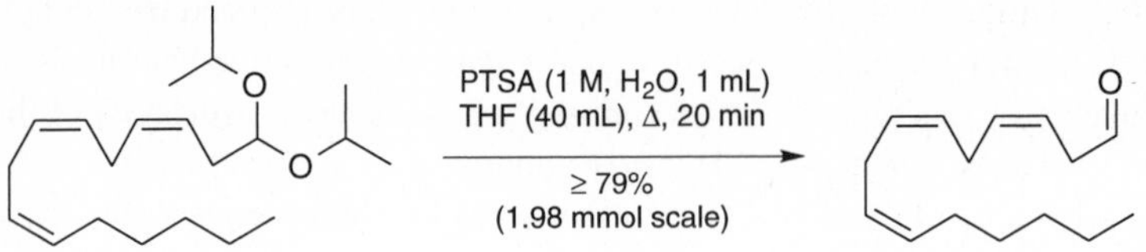

PTSA (1 M, $H_2O$, 1 mL)
THF (40 mL), Δ, 20 min

≥ 79%
(1.98 mmol scale)

**Scheme 5.35**

**(ii) Formation**

The formation of acyclic dialkyl acetals is generally restricted to aldehydes and more reactive ketones. The most common methods involve acid-catalysed acetalisation using an alcohol solvent (usually MeOH) or, more usually, by transacetalisation, which is preferred because it provides the most efficient means for removing water liberated in the reaction. Scheme 5.36 illustrates acetal formation catalysed by $CeCl_3$[76]. $LaCl_3$ has also been used to similar effect[77]. Protic acid catalysts such as PTSA can also be used and our example shows the formation of a dimethyl acetal from a ketone [Scheme 5.37][78].

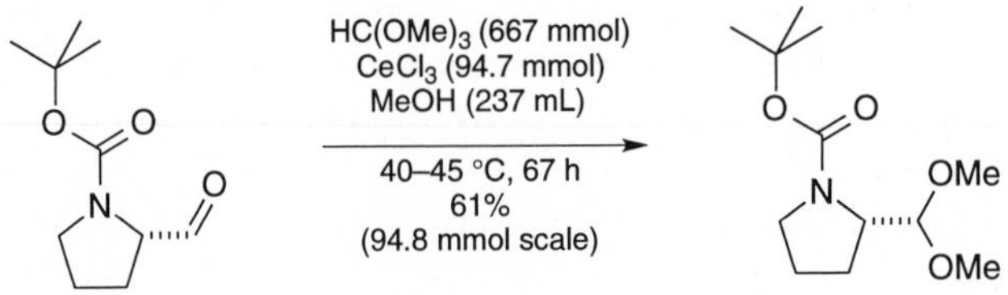

$HC(OMe)_3$ (667 mmol)
$CeCl_3$ (94.7 mmol)
MeOH (237 mL)

40–45 °C, 67 h
61%
(94.8 mmol scale)

**Scheme 5.36**

$Me_2C(OMe)_2$ (15 mL)
PTSA (0.25 mmol)

Δ, 18 h
76%
(10 mmol scale)

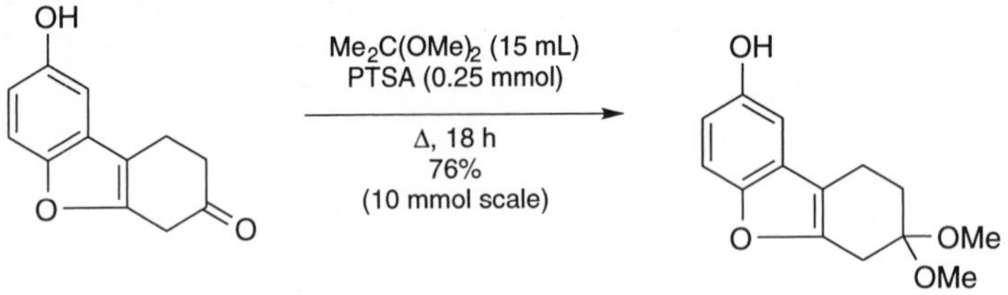

**Scheme 5.37**

The Noyori acetalisation reaction conditions (see section 5.2.1) can also be used to prepare acyclic acetals. During their asymmetric synthesis of Loganin, Vandewalle and co-workers[79], protected the aldehyde [Scheme 5.38] in good yield by reaction with TMSOMe in the presence of TMSOTf.

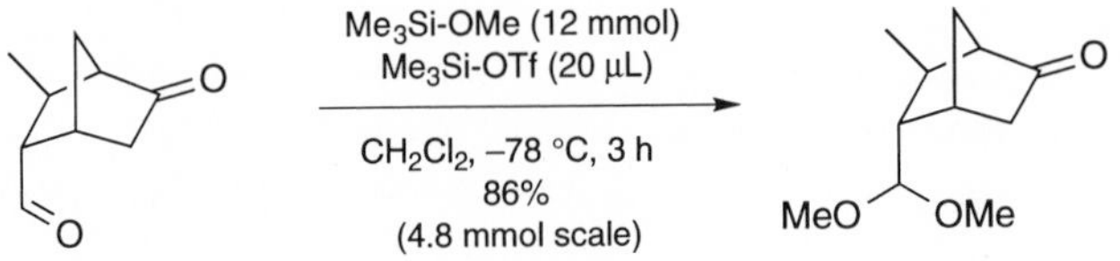

$Me_3Si$-OMe (12 mmol)
$Me_3Si$-OTf (20 μL)

$CH_2Cl_2$, −78 °C, 3 h
86%
(4.8 mmol scale)

**Scheme 5.38**

## 5.3 *S,S*-Acetals

There are two objections to the use of *S,S*-acetals both of which are rooted in their environmental impact. Firstly, most thiols and dithiols have obnoxious odours which require extraordinary feats of experimental skill and planning to hide and otherwise docile neighbours can be provoked to spirited rebellion by the merest whiff of a part per million of a fetid thiol. Secondly, hydrolysis frequently requires heavy metal catalysis, which then creates a toxic waste disposal problem. A third objection to their use is that traces of sulfur-containing by-products can subsequently poison Pd and Pt catalysts which may be later required to assist the catalytic reduction of an alkene, or the hydrogenolysis of a benzyl group. In certain cases reduction reactions can be accomplished in the presence of sulfur using Pd but high pressures and catalyst loading are often required[80,81]; alternatively, more reactive catalysts such as $[Ir(cod)pyPCy_3]PF_6$ may be employed[82]. Despite these serious detractions, the hydrolytic stability of thioacetals and the highly specific and mild conditions used for their removal offer compensations, which can be invaluable in a complex synthesis.

Unlike the corresponding oxygen analogues, there is little distinction in the ease of formation or hydrolytic stability of cyclic *S,S*-acetals (e. g. 1,3-dithiolanes and 1,3-dithianes) compared with acyclic congeners. The higher boiling points of ethane-1,2-dithiol (bp 146–148 °C) and propane-1,3-dithiol (bp 58–60°C/11 mm Hg) make them marginally less obnoxious to handle than ethanethiol (bp 34 °C) and so cyclic *S,S*-acetals are more popular. However, practical considerations aside, there are important strategic reasons why cyclic *S,S*-acetals in general and 1,3-dithianes in particular have commanded the attention of synthetic chemists. 1,3-Dithianes can be metallated with BuLi [$pK_a$ 1,3-dithiane = 31; $pK_a$ 2-methyl-1,3-dithiane = 38] and the resultant anions are sufficiently stable to serve as effective nucleophiles in C–C bond-forming reactions. The role of dithianes in the development of the useful concept of synthons (Corey) and umpolung (Seebach) is well beyond the scope of this survey, and although the subject will not be covered explicitly here, several of the examples cited below exploit the use of metallated dithianes as acyl anion equivalents. For a more detailed treatment of this vast subject the reader should consult the reviews* cited in section 5.5.3.

**(i) Cleavage**

We have already stressed that *O,O*-acetals are far more susceptible to hydrolysis than *S,S*-acetals and Schemes 5.39 and 5.40 attest to the assertion[83,84]. The high acid stability of *S,S*-acetals compared with *O,O*-acetals may be attributed to the lower Brönsted basicity of sulfur compared with oxygen and the barrier to formation of a thionium ion with its 2π–3π bond. The large number of methods which address the problem of dethioacetalisation may be taken as evidence that deprotection often proves refractory — especially in sensitive substrates. We will concentrate here on the methods which have found greatest favor with synthetic chemists but a comprehensive coverage of the problem can be found in an excellent review* by Gröbel and Seebach.

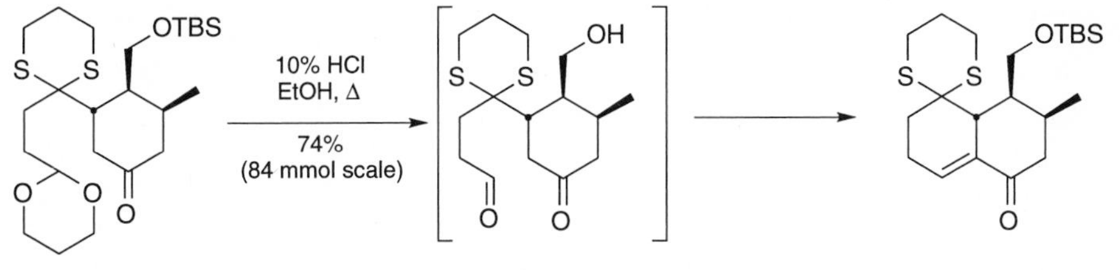

10% HCl
EtOH, Δ
74%
(84 mmol scale)

**Scheme 5.39**

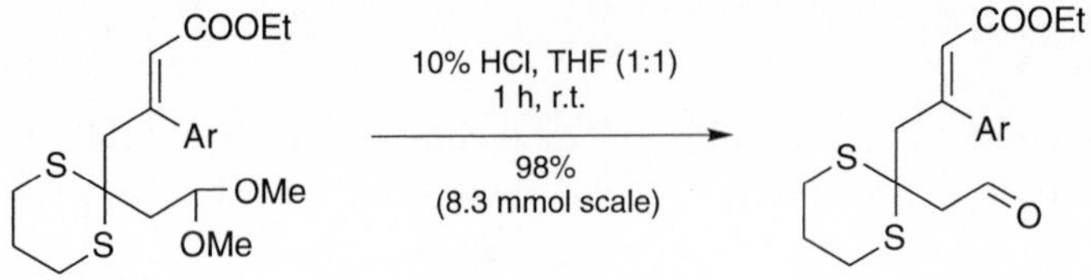
COOEt
Ar
S
S
OMe
OMe
10% HCl, THF (1:1)
1 h, r.t.
98%
(8.3 mmol scale)
COOEt
Ar
S
S
O

**Scheme 5.40**

Three general methods have been used to enhance the nucleofugacity (i.e., leaving group ability) of sulfur: metal coordination, alkylation, and oxidation. The most popular method for deprotection of *S,S*-acetals exploits the high thermodynamic affinity of heavy metals such as Hg(II), Ag(I), Ag(II), Cu(II),and Tl(III) for sulfur. Neither the concept nor the method is new: Fischer first used Hg(II) in 1894 to deprotect sugar *S,S*-acetals[85] and it is this method which continues to find greatest favor amongst synthetic chemists. Provided a suitable base is added to destroy the acid generated during hydrolysis (HgO, $CaCO_3$, $CdCO_3$, $BaCO_3$), the method will tolerate a wide range of sensitive functionality as illustrated by the double deprotection which unveiled the macrocyclic dialdehyde during Kinoshita's synthesis of Elaiophylin [Scheme 5.41][86].

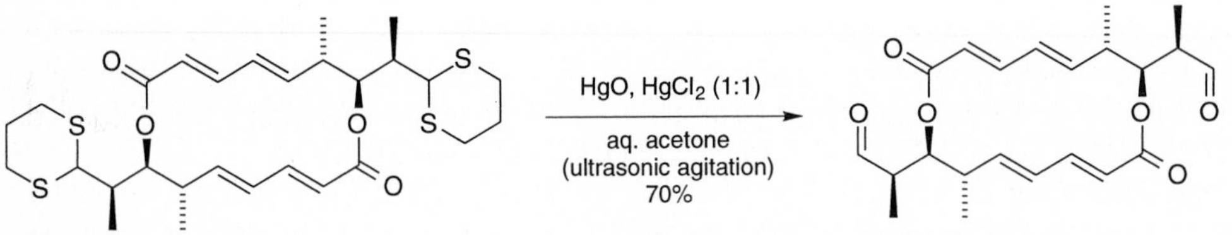
HgO, $HgCl_2$ (1:1)
aq. acetone
(ultrasonic agitation)
70%

**Scheme 5.41**

In order to avoid complications arising from solvomercuration of sensitive alkene substrates, care must be taken to ensure that one equivalent of Hg(II) be used for each sulfur atom. Alternatively, Ag salts may be used [Scheme 5.42][87-89].

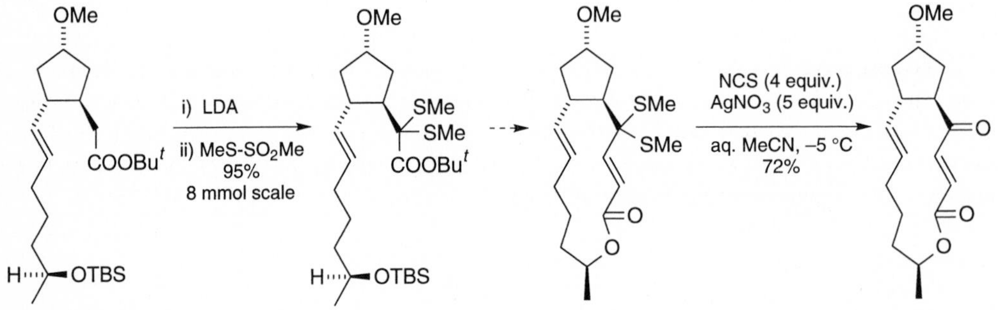
OMe
COOBu$^t$
H OTBS
i) LDA
ii) MeS-$SO_2$Me
95%
8 mmol scale
SMe
SMe
COOBu$^t$
NCS (4 equiv.)
$AgNO_3$ (5 equiv.)
aq. MeCN, –5 °C
72%

**Scheme 5.42**

The combination of HgO and $BF_3$•$OEt_2$ provides a mild method for deprotecting some dithianes [Scheme 5.43][90,91]. However, with substrates which are unusually susceptible to acid-catalysed side reactions, further transformations may be observed under these conditions [Scheme 5.44][92].

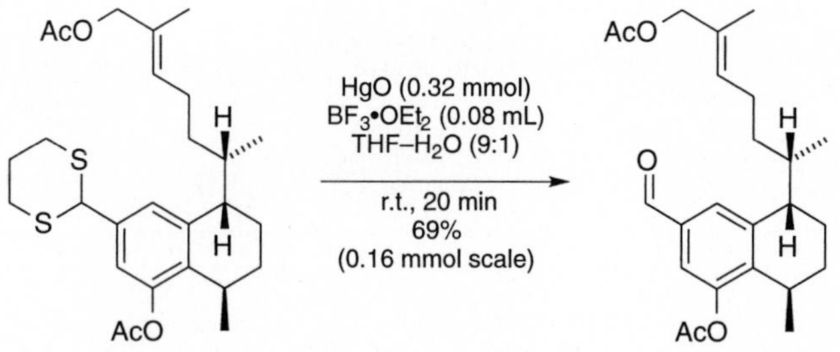

AcO AcO H H S S AcO HgO (0.32 mmol) $BF_3$•$OEt_2$ (0.08 mL) THF–$H_2O$ (9:1) r.t., 20 min 69% (0.16 mmol scale)

**Scheme 5.43**

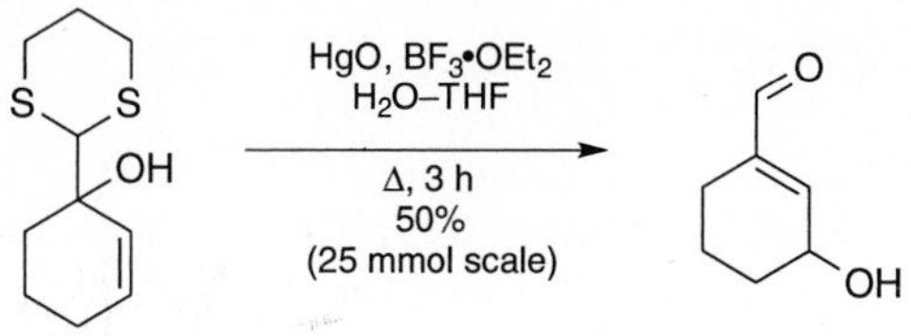

S S OH HgO, $BF_3$•$OEt_2$ $H_2O$–THF Δ, 3 h 50% (25 mmol scale) O OH

**Scheme 5.44**

The second mild method for enhancing the nucleofugacity of sulfur exploits the ease with which sulfur is alkylated with reactive alkylating agents such as MeI, $Me_3OBF_4$, $Et_3OBF_4$, and $MeOSO_2CF_3$ to form the corresponding trialkylsulfonium salts. A single example, taken from Ley's synthesis of Azadirachtin [Scheme 5.45][93-95], will suffice to illustrate the procedure.

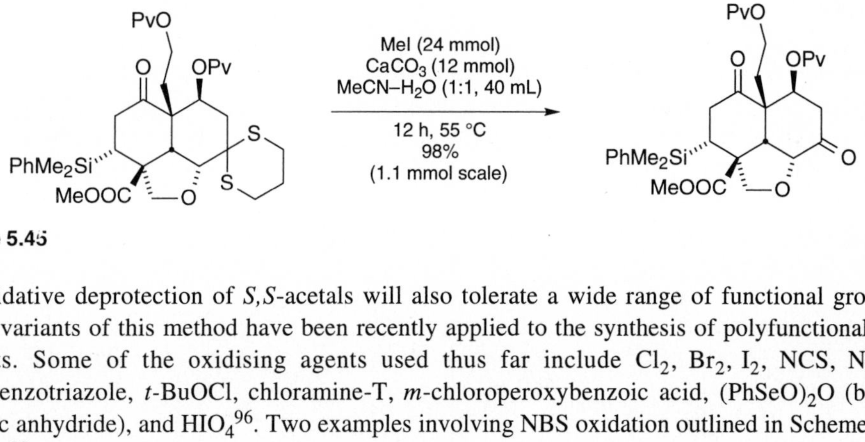

PvO O OPv $PhMe_2Si$ MeOOC O S S MeI (24 mmol) $CaCO_3$ (12 mmol) MeCN–$H_2O$ (1:1, 40 mL) 12 h, 55 °C 98% (1.1 mmol scale)

**Scheme 5.45**

The oxidative deprotection of *S,S*-acetals will also tolerate a wide range of functional groups and several variants of this method have been recently applied to the synthesis of polyfunctional natural products. Some of the oxidising agents used thus far include $Cl_2$, $Br_2$, $I_2$, NCS, NBS, *N*-chlorobenzotriazole, *t*-BuOCl, chloramine-T, *m*-chloroperoxybenzoic acid, $(PhSeO)_2O$ (benzeneseleninic anhydride), and $HIO_4$[96]. Two examples involving NBS oxidation outlined in Schemes 5.46[97] and 5.47[98] show that free hydroxyl groups, disubstituted alkenes, MEM ethers, TBDPS ethers, 1,3-dioxanes, and phenolic benzyl ethers survive intact.

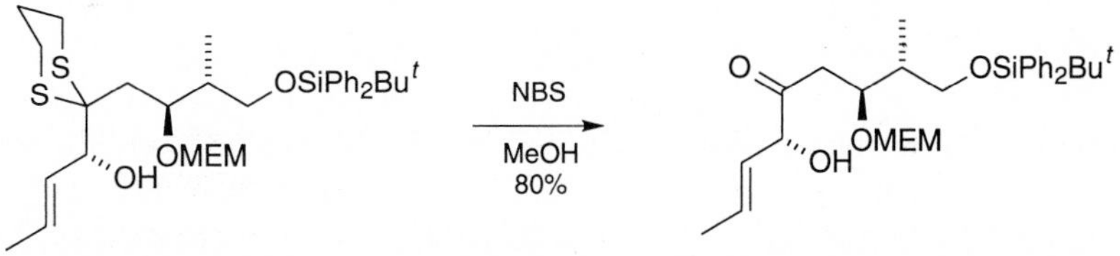

S S OH OMEM $OSiPh_2Bu^t$ NBS MeOH 80% O OH OMEM $OSiPh_2Bu^t$

**Scheme 5.46**

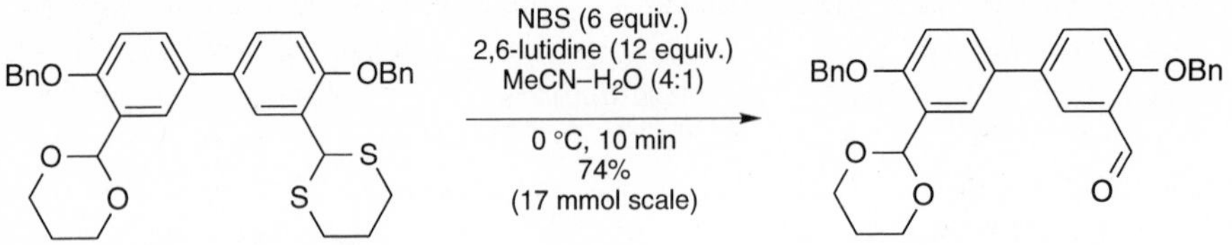

BnO OBn
NBS (6 equiv.)
2,6-lutidine (12 equiv.)
MeCN–$H_2O$ (4:1)
0 °C, 10 min
74%
(17 mmol scale)

**Scheme 5.47**

Several heavy metal oxidants have also been exploited for cleaving *S,S*-acetals including Pb(IV), Tl(III) [Scheme 5.48][99], and Ce(IV) [Scheme 5.49][100].

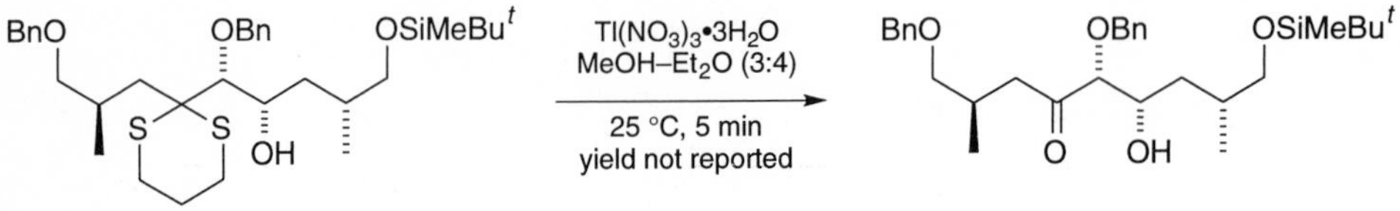

BnO OBn $OSiMeBu^t$ OH
$Tl(NO_3)_3$•$3H_2O$
MeOH–$Et_2O$ (3:4)
25 °C, 5 min
yield not reported

**Scheme 5.48**

$P(O)Ph_2$
$Ce(NH_4)_2(NO_3)_6$ (2 equiv.)
acetone–$H_2O$
r.t., 12 h
99%
(13 mmol scale)

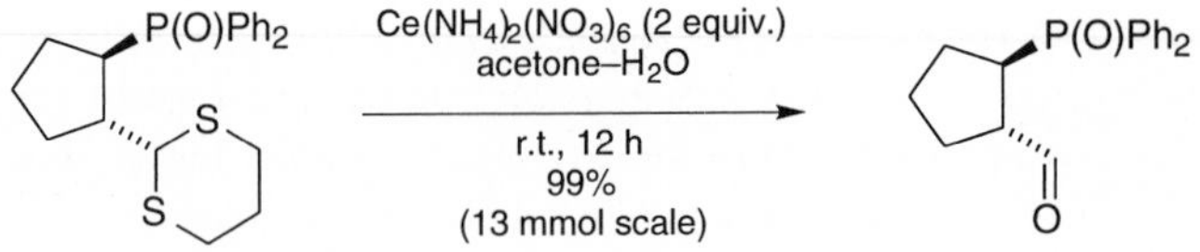

**Scheme 5.49**

During a synthesis of Bertyadionol, the Smith group invested several months of effort in the cleavage of the dithiane ring shown in Scheme 5.50 but most of the known methods failed dismally[101]. The problem related to the generation of a carbocation at C14 which was vinylogously $\alpha$ to the cyclopropane ring and therefore poised to react in multifarious ways. Eventually their beleaguered efforts succeeded by oxidising the dithiane ring with *m*-chloroperoxybenzoic acid to the dithianemonosulfoxide which decomposed by a "Pummerer-like" reaction on heating with acetic anhydride.

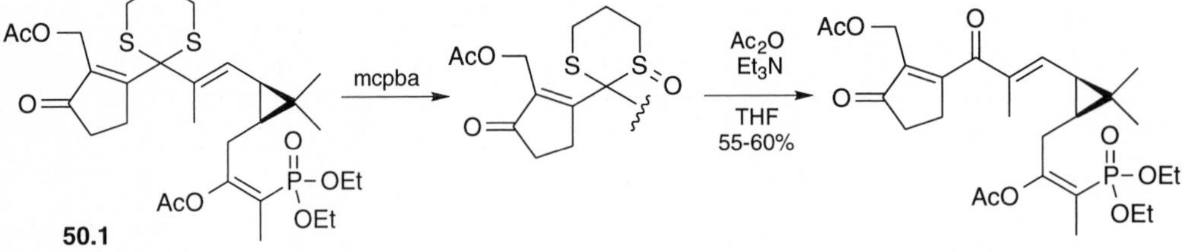

AcO
P–OEt
OEt
**50.1**
mcpba
$Ac_2O$
$Et_3N$
THF
55-60%

**Scheme 5.50**

Two recent syntheses of the immunosuppressant FK-506 offer excellent testimony to the mildness of oxidative dethioacetalisation procedures. In both cases it was found preferable to effect a transacetalisation reaction first to generate a dimethyl acetal, which was subsequently hydrolysed to the desired aldehyde. The Merck synthesis[102] effected the transacetalisation [Scheme 5.51] in 75%

yield using NCS and $AgNO_3$ in the presence of 2,6-lutidine to scavenge acid. The resultant sensitive dimethyl acetal was then hydrolysed with glyoxylic acid. Schreiber and co-workers[103] remarked on problems associated with hydrolysis of a closely related dithiane. After several standard methods failed to return the desired aldehyde in yields exceeding 25-30%, a two-step procedure was likewise employed in which the oxidant was [bis(trifluoroacetoxy)iodo]benzene. Hydrolysis of the intermediate dimethyl acetal afforded the desired aldehyde in 59% overall yield. The oxidant [bis(trifluoroacetoxy)iodo]benzene tolerates a wide range of functionality including thioesters, amines, esters, nitriles, secondary amides, alcohols, halides, alkenes, and alkynes[104]. The method has been again employed by Schreiber in an approach to Rapamycin [Scheme 5.52][105].

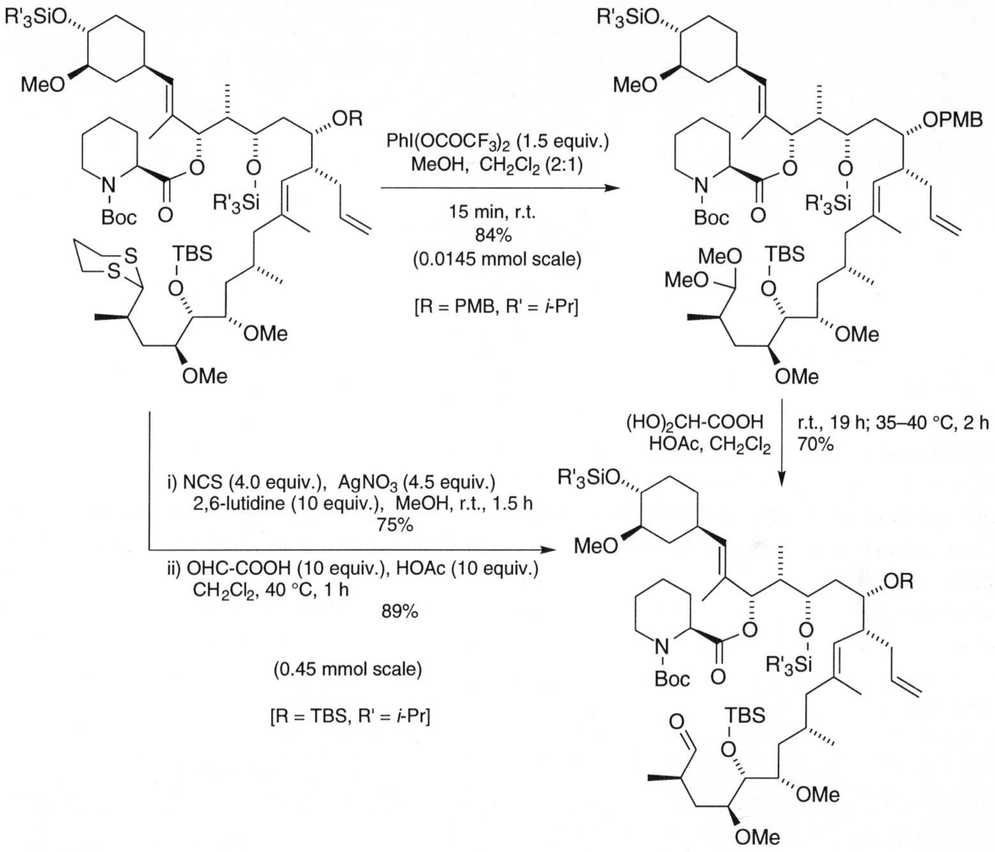

$PhI(OCOCF_3)_2$ (1.5 equiv.)
MeOH, $CH_2Cl_2$ (2:1)
15 min, r.t.
84%
(0.0145 mmol scale)
[R = PMB, R' = *i*-Pr]

$(HO)_2CH$-COOH
HOAc, $CH_2Cl_2$
r.t., 19 h; 35–40 °C, 2 h
70%

i) NCS (4.0 equiv.), $AgNO_3$ (4.5 equiv.)
2,6-lutidine (10 equiv.), MeOH, r.t., 1.5 h
75%
ii) OHC-COOH (10 equiv.), HOAc (10 equiv.)
$CH_2Cl_2$, 40 °C, 1 h
89%
(0.45 mmol scale)
[R = TBS, R' = *i*-Pr]

**Scheme 5.51**

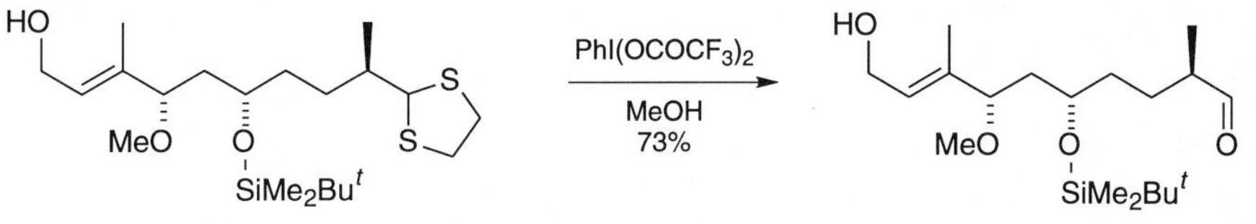

$PhI(OCOCF_3)_2$
MeOH
73%

**Scheme 5.52**

**(ii) Formation**

The methods for making *S,S*-acetals largely parallel the methods used for preparing the corresponding oxygen analogues, i.e., the carbonyl compound is simply treated with the dithiol in the presence of an acid catalyst. However, such is the hydrolytic stability of the *S,S*-acetal moiety that no special provision need be made for the removal of water during the course of the reaction. Both protic acids and Lewis acids can be used to effect reaction, with $BF_3•OEt_2$ being the most popular of the Lewis acids [Scheme 5.53][102]. Barring extenuating steric factors, aldehydes react in preference to ketones and enones react in preference to saturated ketones *without rearrangement of the alkene* [Scheme 5.54][34].

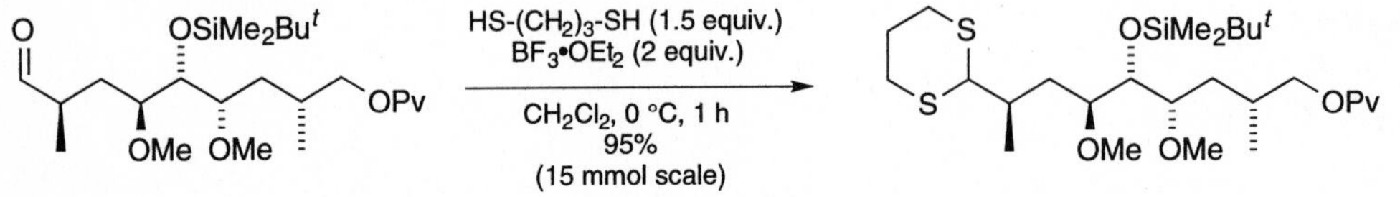
O
OSiMe2But
OPv
OMe OMe
HS-(CH2)3-SH (1.5 equiv.)
BF3•OEt2 (2 equiv.)
CH2Cl2, 0 °C, 1 h
95%
(15 mmol scale)
S
S
OSiMe2But
OPv
OMe OMe

**Scheme 5.53**

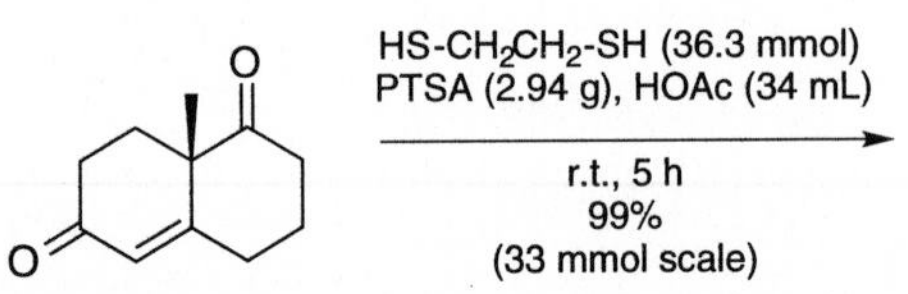
O
O
HS-CH2CH2-SH (36.3 mmol)
PTSA (2.94 g), HOAc (34 mL)
r.t., 5 h
99%
(33 mmol scale)
O
S
S

**Scheme 5.54**

For sensitive molecules milder methods are required. A synthesis of a urinary metabolite of $PGD_2$[106] included the protection of a cyclopentanone [Scheme 5.55] which was prone to easy *β*-elimination reactions under the usual conditions (e. g. $BF_3•OEt_2$ and $MgSO_4$). In this case ethane-1,2-dithiol in the presence of $Zn(OTf)_2$ achieved the formation of the dithiolane in excellent yield.

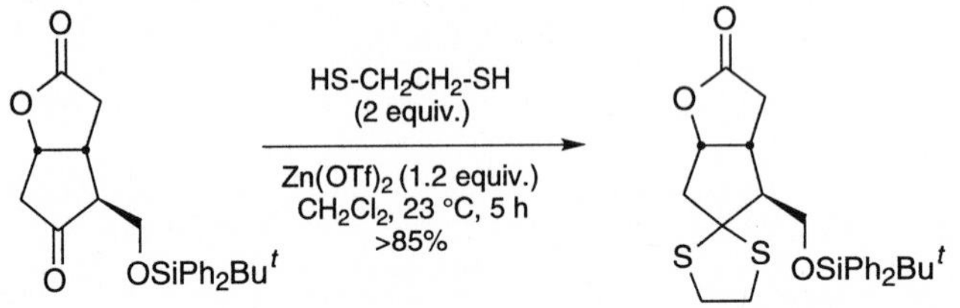
O
O
O
OSiPh2But
HS-CH2CH2-SH
(2 equiv.)
Zn(OTf)2 (1.2 equiv.)
CH2Cl2, 23 °C, 5 h
>85%
O
O
S
S
OSiPh2But

**Scheme 5.55**

Protection of an enone as the dithiane derivative without destruction of an accompanying dioxane was accomplished during a synthesis of Aphidicolin[107] by using the bis-trimethylsilyl ether of propane-1,3-dithiol in the presence of $ZnI_2$ as the catalyst [Scheme 5.56][108].

Acetals and hemiacetals imbedded in 5- and 6-membered rings are remarkably stable and their deliberate conversion to the acyclic chain tautomers can be a problem. The high thermodynamic stability of the corresponding *S,S*-acetals can be used to good effect as shown in Scheme 5.57[109] and Scheme 5.58[110]. $TiCl_4$ has also been recommended as the catalyst for the preparation of *S,S*-acetals from lactols[105,111].

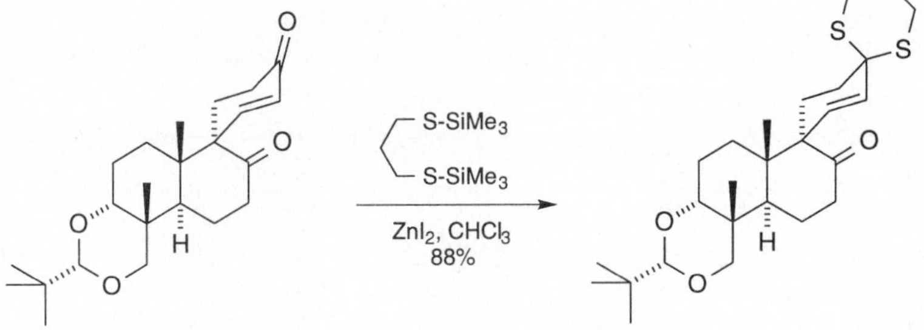

S-SiMe3 S-SiMe3 ZnI2, CHCl3 88%

Scheme 5.56

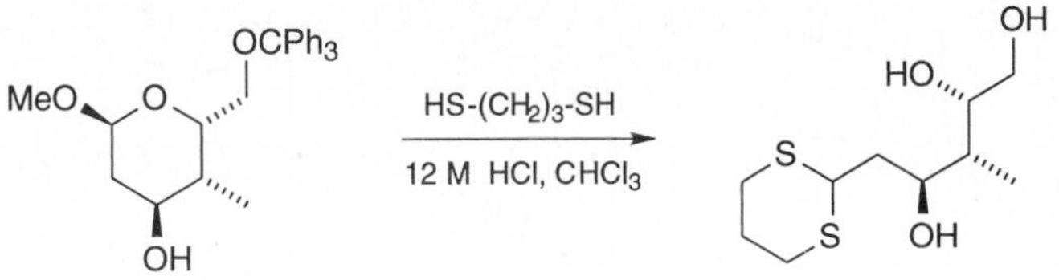

HS-(CH2)3-SH 12 M HCl, CHCl3

Scheme 5.57

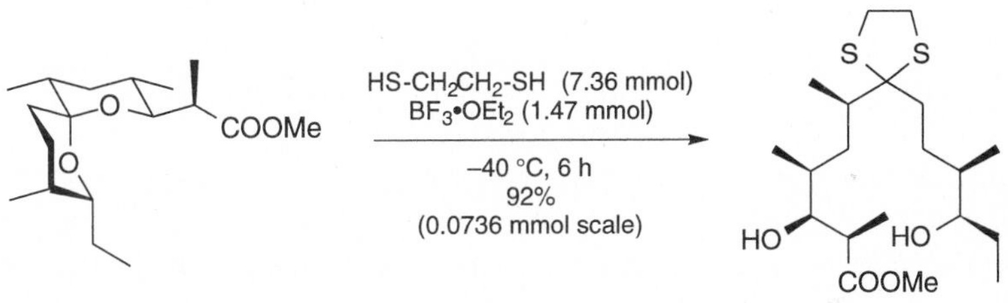

HS-CH2CH2-SH (7.36 mmol) BF3•OEt2 (1.47 mmol) –40 °C, 6 h 92% (0.0736 mmol scale)

Scheme 5.58

Esters and lactones react with thiols in the presence of $Me_3Al$ to give the corresponding ketene *S,S*-acetal — a reaction which was exploited in a synthesis of Aplasmomycin [Scheme 5.59][112]. A related conversion of carboxylic acids to 1,3-dithianes using 1,3-dithia-2-borinane and stannous chloride has also been described[113].

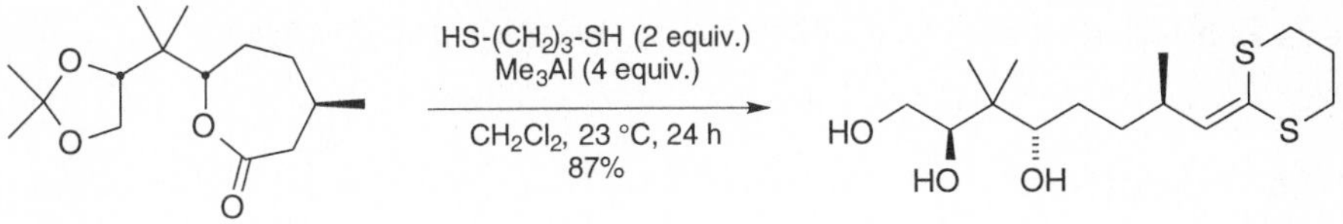

HS-(CH2)3-SH (2 equiv.) Me3Al (4 equiv.) CH2Cl2, 23 °C, 24 h 87%

Scheme 5.59

A spiroannulation process can be used to append a dithiolane or a dithiane to the $\alpha$-position of a cycloalkanone under basic conditions. The procedure, first developed by Woodward and co-workers, involves the reaction of an $\alpha$-hydroxymethylene ketone[114] or an enamine with ethylene dithiotosylate (for dithiolanes) or trimethylene dithiotosylate (1,3-dithianes). The resultant spirocyclic dithianes can be used to accomplish a regiospecific ring cleavage [Scheme 5.60][115] or a 1,2-carbonyl transposition [Scheme 5.61][116].

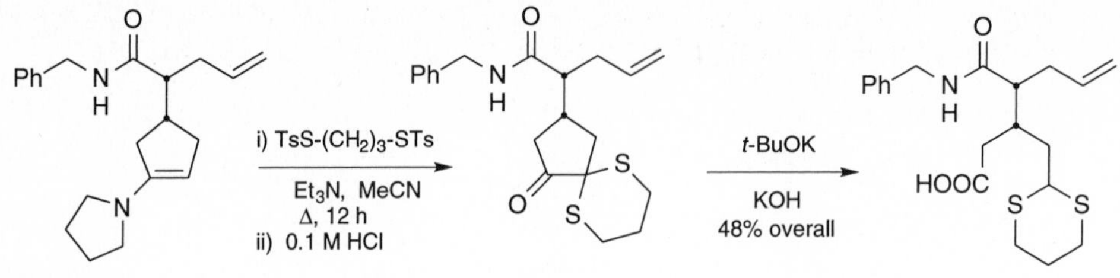

i) TsS-(CH$_2$)$_3$-STs
Et$_3$N, MeCN
Δ, 12 h
ii) 0.1 M HCl

*t*-BuOK
KOH
48% overall

**Scheme 5.60**

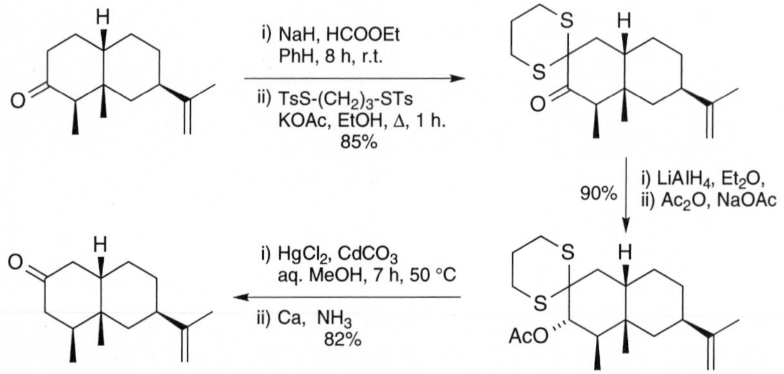

i) NaH, HCOOEt
PhH, 8 h, r.t.
ii) TsS-(CH$_2$)$_3$-STs
KOAc, EtOH, Δ, 1 h.
85%

90%
i) LiAlH$_4$, Et$_2$O,
ii) Ac$_2$O, NaOAc

i) HgCl$_2$, CdCO$_3$
aq. MeOH, 7 h, 50 °C
ii) Ca, NH$_3$
82%

**Scheme 5.61**

**(iii) NMR Data**

$^1$H NMR:

2-methyl-1,3-dithiane $\delta = 4.2$ (1H, q, $J = 7$ Hz), 2.8 (4H, m), 2.1 (2H, m), 1.5 (3H, d, $J = 7$ Hz)
2,2-dimethyl-1,3-dithiane (CCl$_4$) $\delta = 2.74$ (4H, t), 1.90 (2H, m), 1.62 (6H, s)
2-methyl-1,3-dithiolane $\delta = 4.55$ (1H, q, $J = 6.5$ Hz), 3.09 (4H, s), 1.65 (3H, d, $J = 6.5$ Hz)
2,2-dimethyldithiolane (CCl$_4$) $\delta = 3.30$ (4H, s), 1.75 (6H, s)

$^{13}$C NMR (C$_6$D$_6$):

2-methyl-1,3-dithiane $\delta = 41.9$ (S-C-S), 30.6 (2C, S-CH$_2$), 25.2 (C-**C**-C), 21.1 (Me)
2,2-dimethyl-1,3-dithiane $\delta = 45.1$ (S-C-S), 30.8 (2C, S-CH$_2$), 27.0 (2C, Me), 25.2 (C-**C**-C)
2-methyl-1,3-dithiolane $\delta = 48.1$ (S-C-S), 39.1 (2C), 24.9 (Me)
2,2-dimethyl-1,3-dithiolane $\delta = 62.0$ (S-C-S), 40.2 (2C), 34.2 (2C, Me)

## 5.4 *O,S*-Acetals

1,3-Oxathianes have been considered as potential replacements for 1,3-dithianes because they can be metallated under similar conditions[117,118] and they hydrolyse about 10,000 times faster. However, their value is diminished by the limited stability of the lithio derivatives and their inherent lack of symmetry which introduces the complications of diastereoisomerism. However, Eliel and co-workers have exploited the inherent diastereoisomerism of 1,3-oxathianes imbedded in a homochiral framework [Scheme 5.62]. By relying on the differential Lewis basicities of the oxygen and sulfur atoms in the 2-acyl-1,3-oxathiane intermediate **62.2**, a chelation-controlled addition of Grignard reagents occurred leading, after hydrolysis, to an asymmetric synthesis of $\alpha$-hydroxy aldehydes[119-121].The procedure has been applied to a synthesis of (*R*)-(+)-Mevalonolactone[122].

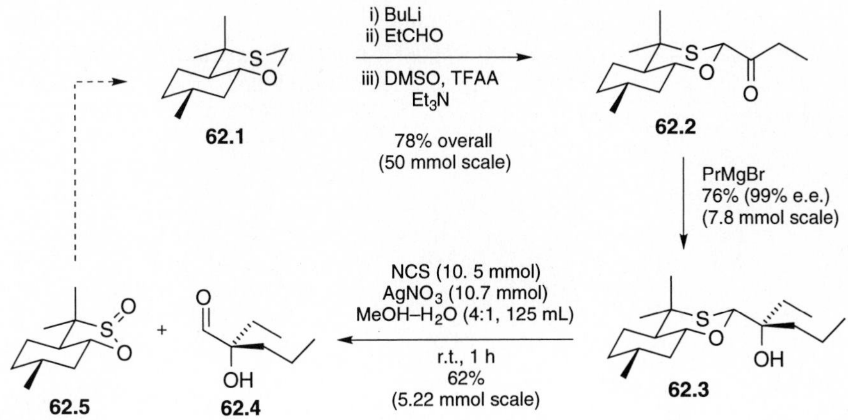

**Scheme 5.62**

Like their *S,S*-acetal counterparts, the *O,S*-acetals can be hydrolysed after *S*-alkylation [Scheme 5.63][123] or Hg(II) catalysis [Scheme 5.64][124]. In the latter case, the selective destruction of the methylthiomethyl ether in the presence of a dithiane bears witness to the enhanced lability of *O,S*-acetals. Under similar conditions an *O,S*-acetal can be cleaved in the presence of a dioxolane[125]. For a comprehensive list of reagents and conditions for cleaving *O,S*-acetals, the interested reader should consult the review by Wimmer*.

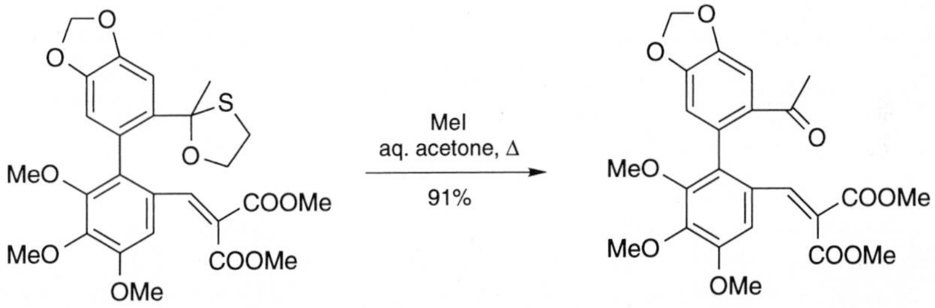

**Scheme 5.63**

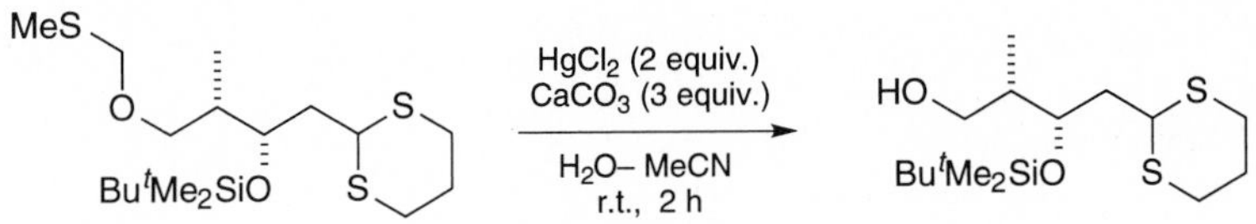

**Scheme 5.64**

There is an old adage which says that you cannot have your cake and eat it too. Conventional wisdom has never really put a break on human ingenuity and our final carbonyl protecting group displays an attempt to retain the carbanion-stabilising properties of the dithiane *S,S*-acetal whilst increasing its hydrolytic lability. The result is the symmetrical dihydrodithiazine ring system incorporating a strongly Lewis basic tertiary amine subunit to assist hydrolysis[126]. A recent synthesis of Didemnone [Scheme 5.65][127] illustrates the use of 4,5-dihydro-2-lithio-5-methyl-1,3,5-dithiazine (**65.2**) as a formyl anion equivalent. As expected, hydrolysis was easier though it still required Hg(II) catalysis; however, unlike the dithianes, which can be metallated and alkylated twice at the 2-position, the dihydrodithiazine ring system bearing a substituent at C2 cannot be metallated a second time. Another example [Scheme 5.66] demonstrates a further use of metallated dihydrodithiazines in chain appendage chemistry[128].

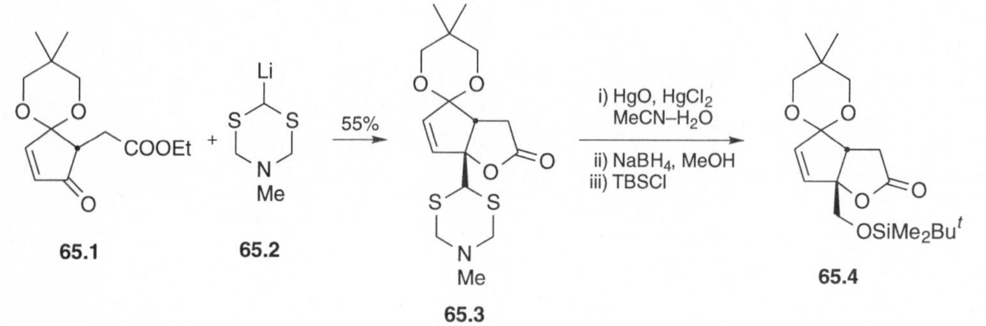

**Scheme 5.65**

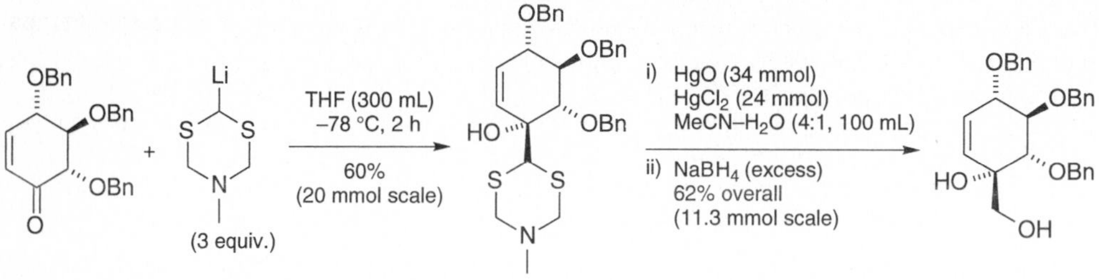

**Scheme 5.66**

**(iii) NMR Data** (oxathiane **64.1**)
$^1$H NMR ($CDCl_3$): $\delta = 5.03$ (1H, d, $J = 10$ Hz), 4.65 (1H, d, $J = 10$ Hz), 3.34 (1H, dt, $J = 4, 10$ Hz), 1.42 and 1.25 (3H each, s), 0.91 (3H, d, $J = 7$ Hz)
$^{13}$C NMR ($CDCl_3$): $\delta = 76.4$ (O-C-S), 67.1, 51.5, 41.8, 41.5, 34.7, 31.3, 29.4, 24.4, 22.1, 21.8

# 5.5 Reviews

## 5.5.1 Reviews Concerning the Protection of Carbonyl Groups

1 Protection of Aldehydes and Ketones. Lowenthal, H. J. E. In *Protective Groups in Organic Chemistry*; McOmie, J. F. W., Ed.; Plenum: London, 1973; Chapter 9.
2 Protection for the Carbonyl Group. Greene, T. W.; Wuts, P. G. M. In *Protective Groups in Organic Synthesis*, 2nd ed.; Wiley: New York, 1990; Chapter 4.

### 5.5.2 Reviews Concerning the Use of *O,O*-Acetals as Reactive Functionality

For further reviews on the preparation and reactions of acetals including aspects of acetal hydrolysis see section 3.5

1 Chiral Acetals in Asymmetric Synthesis. Alexakis, A.; Mangeney, P. *Tetrahedron Asymmetry* **1990**, *1*, 477.
2 Enantiomerically Pure Compound Syntheses with C–C Bond Formation *via* Acetals and Enamines. Seebach, D.; Imwinkelried, R.; Weber, T. In *Modern Synthetic Methods*, Vol. 4; Scheffold, R., Ed.; Wiley: New York, 1986; p 128.
3 Chiral Non-Racemic Bicyclic Lactams. Vehicles for the Construction of Natural and Unnatural Products Containing Quaternary Carbon Centres. Romo D.; Meyers, A. I. *Tetrahedron* **1991**, *47*, 9503.
4 Chemistry of Spiroacetals. Perron, F.; Albizati, K. F. *Chem. Rev.* **1989**, *89*, 1617.

### 5.5.3 Reviews Concerning the Preparation, Chemistry, and Hydrolysis of *O,S*- and *S,S*-Acetals

1 Methods and Possibilities of Nucleophilic Acylation. Seebach, D. *Angew. Chem. Int. Ed. Engl.* **1969**, *8*, 639.
2 Umpolung of the Reactivity of Carbonyl Compounds Through Sulphur-Containing Reagents. Gröbel, B. T.; Seebach, D. *Synthesis* **1977**, 357.
3 Methods of Reactivity Umpolung. Seebach, D. *Angew. Chem. Int. Ed. Engl.* **1979**, *18*, 239.
4 Formyl and Acyl Anions: $^{-}$CH=O and $^{-}$CR=O. Ager, D. In *Umpoled Synthons* Hase, T.; Ed.; Wiley: New York, 1987; Chapter 2.
5 Synthetic Uses of the 1,3-Dithiane Group from 1977-1988. Page, P. C. B.; van Niel, M. B.; Prodger, J. C. *Tetrahedron* **1989**, *45*, 7643.
6 Mechanisms of Hydrolysis of Thioacetals. Satchell, D. P. N.; Satchell, R. S. *Chem. Soc. Rev.* **1990**, *19*, 55.
7 New Synthetic Applications of the Dithiane Functionality. Luh, T.-Y. *Acc. Chem. Res.* **1991**, *24*, 257.
8 O/S-Acetale. Wimmer, P. In *O/O- und O/S-Acetale*, *Houben-Weyl*, Vol E14a/1; Hagemann, H.; Klamann, D., Eds.; Thieme: Stuttgart, 1991; p 785.

# References

1 Ishikawa, H.; Mukaiyama, T.; Ikeda, S. *Bull. Chem. Soc. Jpn.* **1981**, *54*, 776.
2 Ghribi, A.; Alexakis, A.; Normant, J. F. *Tetrahedron Lett.* **1984**, *25*, 3075.
3 Reetz, M. T.; Westermann, J.; Steinbach, R. *Angew. Chem. Int. Ed. Engl.* **1980**, *19*, 900.
4 Barbot, F.; Miginiac, P. *J. Organomet. Chem.* **1979**, *170,* 1.
5 Cherkauskas, J. P.; Cohen, T. *J. Org. Chem.* **1992**, *57*, 6.
6 Hulce, M.; Mallomo, J. P.; Frye, L. L.; Kogan, T. P.; Posner, G. H. *Org. Synth. Coll. Vol. VII* **1990**, 495.
7 Smith, A. B.; Branca, S. J.; Guaciaro, M. A.; Wovkulich, P. M.; Korn, A. *Org. Synth. Coll. Vol. VII* **1990**, 271.
8 Cohen, N.; Banner, B. L.; Lopresti, R. J.; Wong, F.; Rosenberger, M.; Liu, Y.-Y.; Thom, E.; Liebman, A. A. *J. Am. Chem. Soc.* **1983**, *105*, 3661.
9 Murahashi, S.; Oda, Y.; Naoka, T. *Chem. Lett.* **1992**, 2237.

10 Dodd, D. S.; Oehlschlager, A. C.; Georgopapadakou, N. H.; Polak, A.-M.; Hartman, P. G. *J. Org. Chem.* **1992**, *57*, 7227.
11 Kabat, M.; Kiegiel, J.; Toth, K.; Wovkulich, P. M.; Uskokovic, M. R. *Tetrahedron Lett.* **1991**, *32*, 2343.
12 Anthony, N. J.; Clarke, T.; Jones, A. B.; Ley, S. V. *Tetrahedron Lett.* **1987**, *28*, 5755.
13 Lipshutz, B. H.; Pollart, D.; Monforte, J.; Kotsuki, H. *Tetrahedron Lett.* **1985**, *26*, 705.
14 McKillop, A.; Taylor, R. J. K.; Watson, R. J.; Lewis, N. *Synlett* **1992**, 1005.
15 Kim, K. S.; Song, Y. H.; Lee, B. H.; Hahn, C. S. *J. Org. Chem.* **1986**, *51*, 404.
16 Still, I. W. J.; Shi, Y. *Tetrahedron Lett.* **1987**, *28*, 2489.
17 Ukagi, Y.; Koumoto, N.; Fujisawa, T. *Chem. Lett.* **1989**, 1623.
18 Tanemura, K.; Suzuki, T.; Horaguchi, T. *J. Chem. Soc., Chem. Commun.* **1992**, 979.
19 Oku, A.; Kinogasa, M.; Kamada, T. *Chem. Lett.* **1993**, 163.
20 Smith, S. W.; Newman, M. S. *J. Am. Chem. Soc.* **1968**, *90*, 1249.
21 Okawara, H.; Nakai, H.; Ohno, M. *Tetrahedron Lett.* **1982**, *23*, 1087.
22 Smith, A. B.; Fukui, M.; Vaccaro, H. A.; Empfield, J. R. *J. Am. Chem. Soc.* **1991**, *113*, 2071.
23 Crimmins, M. T.; DeLoach, J. A. *J. Am. Chem. Soc.* **1986**, *108*, 800.
24 Chan, T. H.; Brook, M. A.; Chaly, T. *Synthesis* **1983**, 203.
25 Hirsenkorn, R.; Schmidt, R. R. *Liebigs Ann. Chem.* **1990**, 883.
26 Tsunoda, T.; Suzuki, M.; Noyori, R. *Tetrahedron Lett.* **1980**, *21*, 1357.
27 Yoshinera, J.; Horita, S.; Hashimoto, H. *Chem. Lett.* **1981**, 375.
28 Chida, N.; Ohtsuka, M.; Nakazawa, K.; Ogawa, S. *J. Org. Chem.* **1991**, *56*, 2976.
29 Chan, T. H.; Schwerdtfeger, A. E. *J. Org. Chem.* **1991**, *56*, 3294.
30 Rubin, Y.; Knobler, C. B.; Diederich, F. *J. Am. Chem. Soc.* **1990**, *112*, 1607.
31 Kim, S.; Kim, Y. G.; Kim, D.-i. *Tetrahedron Lett.* **1992**, *33*, 2565.
32 Bauduin, G.; Bondon, D.; Pietrasanta, Y.; Pucci, B. *Tetrahedron* **1978**, *34*, 3269.
33 Smith, A. B.; Mewshaw, R. *J. Org. Chem.* **1984**, *49*, 3685.
34 Bosch, M. P.; Camps, F.; Coll, J.; Guerrero, A.; Tatsuoka, T.; Meinwald, J. *J. Org. Chem.* **1986**, *51*, 773.
35 Paquette, L. A.; Sauer, D. R.; Cleary, D. G.; Kinsella, M. A.; Blackwell, C. M.; Anderson, L. G. *J. Am. Chem. Soc.* **1992**, *114*, 7375.
36 McMurry, J. E.; Isser, S. J. *J. Am. Chem. Soc.* **1972**, *94*, 7132.
37 Hwu, J. R.; Leu, L.-C.; Robl, J. A.; Anderson, D. A.; Wetzel, J. M. *J. Org. Chem.* **1987**, *52*, 188.
38 De Leeuw, J. W.; De Waard, E. R.; Beetz, T.; Huisman, O. *Recl Trav. Chim. Pays-Bas* **1973**, *92*, 1047.
39 Heathcock, C. H.; DelMar, E. G.; Graham, S. L. *J. Am. Chem. Soc.* **1982**, *104*, 1907.
40 Marshall, J. A.; Pike, M. T.; Carroll, R. D. *J. Org. Chem.* **1966**, *31*, 2933.
41 Grieco, P. A.; Nishizawa, M. *J. Chem. Soc., Chem. Commun.* **1976**, 582.
42 Grieco, P. A.; Oguri, T.; Gilman, S.; De Titta, G. T. *J. Am. Chem. Soc.* **1978**, *100*, 1616.
43 Grieco, P. A.; Nishizawa, M.; Oguri, T.; Burke, S. D.; Marinovic, N. *J. Am. Chem. Soc.* **1977**, *99*, 5773.
44 Kato, M.; Kurihara, H.; Yoshikoshi, A. *J. Chem. Soc., Perkin Trans. 1* **1979**, 2740.
45 Sarett, L. H.; Arth, G. E.; Lukes, R. M.; Beyler, R. E.; Poos, G. I.; Johns, W. F.; Constantin, J. M. *J. Am. Chem. Soc.* **1952**, *74*, 4974.
46 Babler, J. H.; Casey Malek, N.; Coghlan, M. J. *J. Org. Chem.* **1978**, *43*, 1821.
47 Corey, E. J.; Magriotis, P. A. *J. Am. Chem. Soc.* **1987**, *109*, 287.
48 Schmidt, U.; Beuttler, T.; Lieberknecht, A.; Griesser, A. *Tetrahedron Lett.* **1983**, *24*, 3573.
49 Ott, J.; Ramos Tombo, G. M.; Schmid, B.; Venanzi, L. M.; Wang, G.; Ward, T. R. *Tetrahedron Lett.* **1989**, *30*, 6151.
50 Newkome, G. R.; Sauer, J. D.; McClure, G. L. *Tetrahedron Lett.* **1973**, 1599.
51 Magnus, P.; Giles, M.; Bonnert, R.; Johnson, G.; McQuire, L.; Deluca, M.; Merritt, A.; Kim, C. S.; Vicker, N. *J. Am. Chem. Soc.* **1993**, *115*, 8116.
52 McNamara, J. M.; Kishi, Y. *J. Am. Chem. Soc.* **1982**, *104*, 7371.
53 Johnson, W. S.; Kelson, A. B.; Elliott, J. D. *Tetrahedron Lett.* **1988**, *29*, 3757.
54 Denmark, S. E.; Willson, T. M. *J. Am. Chem. Soc.* **1989**, *111*, 3475.
55 Denmark, S. E.; Almstead, N. G. *J. Am. Chem. Soc.* **1991**, *113*, 8089.
56 Sammakia, T.; Smith, R. S. *J. Org. Chem.* **1992**, *57*, 2997.
57 Fujiwara, J.; Fukutani, Y.; Hasegawa, M.; Maruoka, K.; Yamamoto, H. *J. Am. Chem. Soc.* **1984**, *106*, 5004.
58 Kaino, M.; Naruse, Y.; Ishihara, K.; Yamamoto, H. *J. Org. Chem.* **1990**, *55*, 5814.
59 Meyers, A. I.; Romo, D. *Tetrahedron Lett.* **1989**, *30*, 1745.
60 Krohn, K.; Hamann, I. *Liebigs Ann. Chem.* **1988**, 949.
61 Seebach, D.; Naef, R. *Helv. Chim. Acta* **1981**, *64*, 2704.
62 Fráter, G.; Müller, U.; Günther, W. *Tetrahedron Lett.* **1981**, *22*, 4221.
63 Reese, C. B.; Saffhill, R.; Sulston, J. E. *J. Am. Chem. Soc.* **1967**, *89*, 3366.

64 Ellison, R. A.; Lukenbach, E. R.; Chiu, C.-W. *Tetrahedron Lett.* **1975**, 499.
65 Katsumura, S.; Fujiwara, S.; Isoe, S. *Tetrahedron Lett.* **1988**, *29*, 1173.
66 Corey, E. J.; Nicolaou, K. C.; Toru, T. *J. Am. Chem. Soc.* **1975**, *97*, 2287.
67 Hiyama, T.; Kanakura, A.; Yamamoto, H.; Nozaki, H. *Tetrahedron Lett.* **1978**, 3051.
68 Huet, F.; Lechevallier, A.; Pellet, M.; Conia, J. M. *Synthesis* **1978**, 63.
69 Magnus, P.; Davies, M. *J. Chem. Soc., Chem. Commun.* **1991**, 1522.
70 Meyers, A. G.; Fundy, M. A. M.; Lindstrom, P. A. *Tetrahedron Lett.* **1988**, *29*, 5609.
71 Bonin, M.; Royer, J.; Grierson, D. S.; Husson, H.-P. *Tetrahedron Lett.* **1986**, *27*, 1569.
72 Bonin, M.; Grierson, D. S.; Rojer, J.; Husson, H.-P. *Org. Synth.* **1991**, *70*, 54.
73 Lipshutz, B. H.; Harvey, D. F. *Synth. Commun.* **1982**, *12*, 267.
74 Viala, J.; Santelli, M. *J. Org. Chem.* **1988**, *53*, 6121.
75 Battersby, A. R.; Buckley, D. G.; Staunton, J.; Williams, P. J. *J. Chem. Soc., Perkin Trans. 1* **1979**, 2550.
76 Mori, S.; Ohno, T.; Harada, H.; Aoyama, T.; Shioiri, T. *Tetrahedron* **1991**, *47*, 5051.
77 Gemal, A. L.; Luche, J.-L. *J. Org. Chem.* **1979**, *44*, 4187.
78 Rupprecht, K. M.; Boger, J.; Hoogsteen, K.; Nachbar, R. B.; Springer, J. P. *J. Org. Chem.* **1991**, *56*, 6180.
79 Vandewalle, M.; Van der Eycken, J.; Oppolzer, W.; Vullioud, C. *Tetrahedron* **1986**, *42*, 4035.
80 Vasilevskis, J.; Gualtieri, J. A.; Hutchings, S. D.; West, R. C.; Scott, J. W.; Parrish, D. R.; Bizzarro, F. T.; Field, G. F. *J. Am. Chem. Soc.* **1978**, *100*, 7423.
81 Confalone, P. N. *Helv. Chim. Acta* **1976**, *59*, 1005.
82 Schreiber, S. L.; Sommer, T. J. *Tetrahedron Lett.* **1983**, *24*, 4781.
83 Rosen, T.; Taschner, M. J.; Thomas, J. A.; Heathcock, C. H. *J. Org. Chem.* **1985**, *50*, 1190.
84 Baldwin, S. W.; Aubé, J.; McPhail, A. T. *J. Org. Chem.* **1991**, *56*, 6546.
85 Fischer, E. *Ber. Dtsch. Chem. Ges.* **1894**, *27*, 673.
86 Toshima, T.; Tatsuta, K.; Kinoshita, M. *Bull. Chem. Soc. Jpn.* **1988**, *61*, 1281.
87 Le Drian, C.; Greene, A. E. *J. Am. Chem. Soc.* **1982**, *104*, 5473.
88 Trost, B. M.; Grese, T. A. *J. Org. Chem.* **1992**, *57*, 686.
89 Reece, C. A.; Rodin, J. O.; Brownlee, R. G.; Duncan, W. G.; Silverstein, R. M. *Tetrahedron* **1968**, *24*, 4249.
90 Vedejs, E.; Fuchs, P. L. *J. Org. Chem.* **1971**, *36*, 366.
91 Uemura, M.; Nishimura, H.; Minami, T.; Hayashi, Y. *J. Am. Chem. Soc.* **1991**, *113*, 5402.
92 Rigby, H. L.; Neveu, M.; Pailly, D.; Ranu, B. C.; Hudlicky, T. *Org. Synth.* **1988**, *67*, 205.
93 Kolb, H. C.; Ley, S. V.; Slawin, A. M. Z.; Williams, D. J. *J. Chem. Soc., Perkin Trans. 1* **1992**, 2735.
94 Ley, S. V.; Maw, G. N.; Trudell, M. L. *Tetrahedron Lett.* **1990**, *31*, 5521.
95 Cardani, S.; Bernardi, A.; Colombo, L.; Gennari, C.; Scolastico, C.; Venturini, I. *Tetrahedron* **1988**, *44*, 5563.
96 Corey, E. J.; Kang, M.-c.; Desai, M. C.; Ghosh, A. K.; Houpis, I. N. *J. Am. Chem. Soc.* **1988**, *110*, 649.
97 Williams, D. R.; Jass, P. A.; Allan Tse, H.-L.; Gaston, R. D. *J. Am. Chem. Soc.* **1990**, *112*, 4552.
98 Schmidt, U.; Meyer, R.; Leitemberger, V.; Griesser, H.; Lieberknecht, A. *Synthesis* **1992**, 1025.
99 Gu, R.-L.; Sih, C. J. *Tetrahedron Lett.* **1990**, *31*, 3283.
100 Okada, A.; Minami, T.; Umezu, Y.; Nishikawa, S.; Mori, R.; Nakayama, Y. *Tetrahedron Asymmetry* **1991**, *2*, 667.
101 Smith, A. B.; Dorsey, B. D.; Visnick, M.; Maeda, T.; Malamas, M. S. *J. Am. Chem. Soc.* **1986**, *108*, 3110.
102 Jones, T. K.; Reamer, R. A.; Desmond, R.; Mills, S. G. *J. Am. Chem. Soc.* **1990**, *112*, 2998.
103 Nakatsuka, M.; Ragan, J. A.; Sammakia, T.; Smith, D. B.; Uehling, D. E.; Schreiber, S. L. *J. Am. Chem. Soc.* **1990**, *112*, 5583.
104 Stork, G.; Zhao, K. *Tetrahedron Lett.* **1989**, *30*, 287.
105 Meyer, S. D.; Miwa, T.; Nakatsuka, M.; Schreiber, S. L. *J. Org. Chem.* **1992**, *57*, 5058.
106 Corey, E. J.; Shimoji, K. *J. Am. Chem. Soc.* **1983**, *105*, 1662.
107 Corey, E. J.; Tius, M. A.; Das, J. *J. Am. Chem. Soc.* **1980**, *102*, 1742.
108 Evans, D. A.; Truesdale, L. K.; Grimm, K. G.; Nesbitt, S. L. *J. Am. Chem. Soc.* **1977**, *99*, 5009.
109 Corey, E. J.; Weigel, L. O.; Chamberlin, A. R.; Lipshutz, B. *J. Am. Chem. Soc.* **1980**, *102*, 1439.
110 Ireland, R. E.; Daub, J. P.; Mandel, G. S.; Mandel, N. S. *J. Org. Chem.* **1983**, *48*, 1312.
111 Bulman-Page, P. C.; Roberts, R. A.; Paquette, L. A. *Tetrahedron Lett.* **1983**, *24*, 3555.
112 Corey, E. J.; Pan, B.-C.; Hua, D. H.; Deardorff, D. R. *J. Am. Chem. Soc.* **1982**, *104*, 6816.
113 Kim, S.; Kim, S. S.; Lim, S. T.; Shim, S. C. *J. Org. Chem.* **1987**, *52*, 2114.
114 Woodward, R. B.; Pachter, I. J.; Scheinbaum, M. L. *Org. Synth. Coll. Vol. VI* **1988**, 590.
115 Takano, S.; Takahashi, M.; Hatakeyama, S.; Ogasawara, K. *J. Chem. Soc., Chem. Commun.* **1979**, 556.
116 Marshall, J. A.; Roebke, H. *J. Org. Chem.* **1969**, *34*, 4188.
117 Fuji, K.; Ueda, M.; Sumi, K.; Kajiwara, K.; Fujita, E.; Iwashita, T.; Miura, I. *J. Org. Chem.* **1985**, *50*, 657.
118 Juaristi, E.; Gordillo, B.; Aparicio, D. M. *Tetrahedron Lett.* **1985**, 1927.
119 Eliel, E. L.; Morris-Natschke, S. *J. Am. Chem. Soc.* **1984**, *106*, 2937.
120 Lynch, J. E.; Eliel, E. L. *J. Am. Chem. Soc.* **1984**, *106*, 2943.

121 Eliel, E. L.; Lynch, J. E.; Kume, F.; Frye, S. V. *Org. Synth.* **1987**, *65*, 215.
122 Eliel, E. L.; Soai, K. *Tetrahedron Lett.* **1981**, *22*, 2859.
123 Ziegler, F. E.; Fowler, K. W.; Sinha, N. D. *Tetrahedron Lett.* **1978**, 2767.
124 Corey, E. J.; Bock, M. G. *Tetrahedron Lett.* **1975**, 2643.
125 Jansen, B. J. M.; Peperzak, R. M.; de Groot, A. *Recl Trav. Chim. Pays-Bas* **1987**, *106*, 505.
126 Balanson, R. D.; Kobal, V. M.; Schumaker, R. R. *J. Org. Chem.* **1977**, *42*, 393.
127 Bauermeister, H.; Riechers, H.; Schomburg, D.; Washausen, P.; Winterfeldt, E. *Angew. Chem. Int. Ed. Engl.* **1991**, *30*, 191.
128 Paulsen, H.; Mielke, B.; von Deyn, W. *Liebigs Ann. Chem.* **1987**, 439.

# Chapter 6 Amino Protecting Groups

## 6.1 Introduction

Nitrogen is the element of faction neatly dividing organic chemists into camps. Those who have mastered its vagaries, are undaunted by the prospect of contending with its reactivity. Then there are those in the minority antipodal camp (like the author) who suspect that nitrogen is best confined to a cylinder and used to shield precious reactions and reagents from the ravages of air and moisture! This, the penultimate chapter of our survey, concerns the amino function whose nucleophilicity and basicity is a problem of signal importance in the synthesis of a diverse array of biological molecules such as amino acids, peptides, glycopeptides, aminoglycosides, $\beta$-lactams, nucleosides, sphingosines, and alkaloids to name but a few. Since we must retard what we cannot repel, and palliate what we cannot cure, our success will depend on deft use of protecting groups of which there are a bewildering array. However, in keeping with our resolve to include only the more generally useful protecting groups, we will focus on only about 20 of the 250 protecting groups reported to date.

## 6.2 *N*-Acyl Derivatives

We will begin by considering 10 *N*-acyl protecting groups the bulk of which are carbamates. All of these groups are easy to introduce and offer a wide range of deprotection conditions. The reader will be excused a sense of *déjà vu* if our survey is reminiscent of Chapter 4 since all of the carbamate protecting groups are simply carboxyl protection groups cleverly adapted to the needs of amino protection. The adaptation works because *O*-alkyl cleavage of carbamates releases carbamic acids which are unstable and decompose with loss of $CO_2$ to give the free amine.

### 6.2.1 Phthalimides

Phthalimides (sometimes abbreviated Phth or Pht) are stable to $Pb(OAc)_4$, ozone (at –78 °C), 30% $H_2O_2$, $SOCl_2$, HBr in HOAc, most methods for oxidising alcohols including the Jones oxidation, $OsO_4$, and most methods for forming acetals from carbonyls or diols. Phthalimides will also survive transesterification of acetate groups using NaOMe in MeOH[1]; however, they are labile towards $Na_2S{\bullet}9H_2O$[2] and many metal hydride reducing agents including $NaBH_4$. It has also been claimed[3] that they are sensitive to piperidine under conditions used to cleave Fmoc groups (see section 6.2.7). Phthalimides are especially useful in the protection of amino functions in aminoglycoside syntheses[4,5] because they are stable to Koenigs–Knorr glycosidation reactions [Scheme 6.1][6] in which they help control the stereochemistry by neighboring group participation .

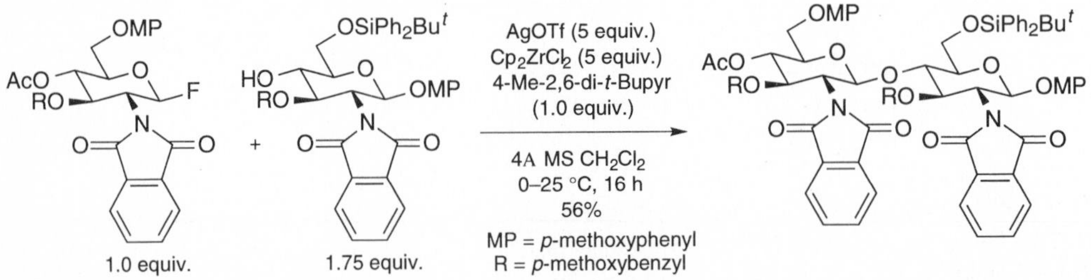

AgOTf (5 equiv.)
$Cp_2ZrCl_2$ (5 equiv.)
4-Me-2,6-di-*t*-Bupyr
(1.0 equiv.)
4A MS $CH_2Cl_2$
0–25 °C, 16 h
56%

1.0 equiv. 1.75 equiv.

MP = *p*-methoxyphenyl
R = *p*-methoxybenzyl

**Scheme 6.1**

### (i) Cleavage

The classical Gabriel reaction* evolved over a century ago[7] as the first generally useful method for preparing primary amines free of contamination by secondary and tertiary amines. The method involved the *N*-alkylation of the sodium or potassium salt of phthalimide with a primary alkyl halide followed by hydrolysis of the *N*-alkyl phthalimide. A significant advance in the utility of the phthalimide function as a *protecting group* for primary amines was the discovery[8] that hydrazinolysis offered a milder and more efficient method for deprotection than base hydrolysis, and hydrazinolysis remains the method of choice. For example, Wasserman and co-workers[9] used hydrazinolysis to selectively cleave a phthalimide in the presence of a Boc function in order to free a primary amine as a prelude to ring expansion in a synthesis of the macrocyclic polyamine alkaloid Chaenorhine [Scheme 6.2].

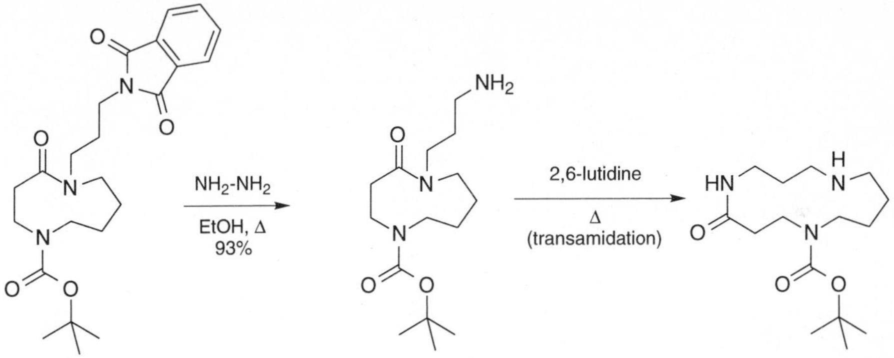

NH2-NH2, EtOH, Δ, 93% — 2,6-lutidine, Δ (transamidation)

**Scheme 6.2**

An unusual problem surfaced during a recent synthesis of (–)-Calicheamicinone[10]. Attempts to accomplish deprotection of the phthalimide function in **3.1** [Scheme 6.3] using hydrazine led to partial reduction of the acetylene function owing to the formation of diimide (presumably by aerial oxidation of the hydrazine) as a side reaction. In this case, cleavage to the desired enediyne **3.2** was better accomplished using methylhydrazine. Attempts to remove a phthalimide function from the monobactam antibiotic Nocardicinic acid failed with a variety of reagents including hydrazine, but success was eventually achieved[11] with $Me_2N{-}(CH_2)_3{-}NH_2$. In other cases where standard hydrazinolysis has been inefficient, reductive cleavage using $NaBH_4$ has been recommended[12].

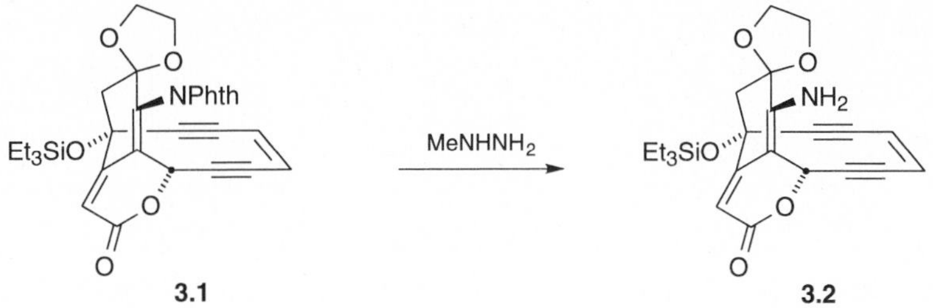

3.1 (NPhth, $Et_3SiO$) — MeNHNH2 → 3.2 ($NH_2$, $Et_3SiO$)

**Scheme 6.3**

### (ii) Formation

The phthalimide function is typically introduced by reaction of a primary amine with phthalic anhydride in $CHCl_3$ at 70 °C for 4 h (85–93% yield)[13], *o*-$(MeOOC)C_6H_4COCl$ in the presence of base

[Scheme 6.4][14,15], or *N*-ethoxycarbonylphthalimide[16]. The latter reagent is especially useful for the *N*-protection of $\alpha$-amino acids [Scheme 6.5][17].

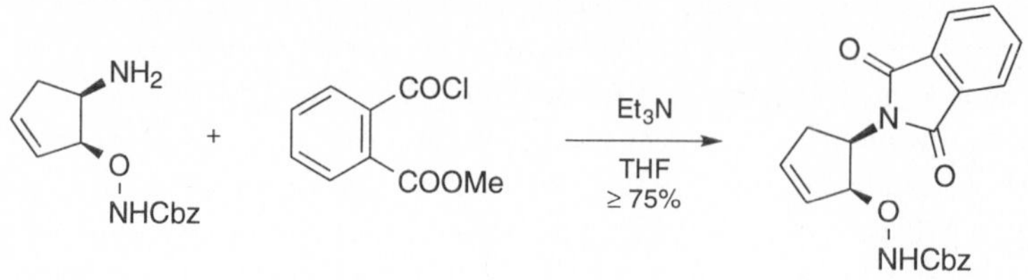

$Et_3N$
THF
≥ 75%

**Scheme 6.4**

$Na_2CO_3$ (0.56 g)
DMSO (6 mL)
16 h, r.t.
87%
(1.52 mmol scale)

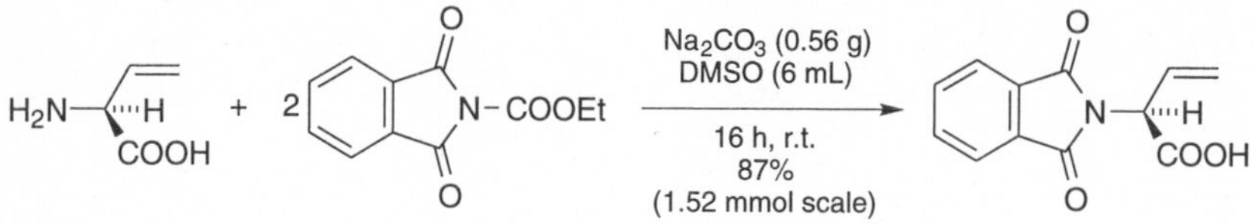

**Scheme 6.5**

*N*-Phthaloylation of the vinylogous amide **6.1** was accomplished by a rather circuitous route [Scheme 6.6][10]. Reaction of **6.1** with phthaloyl chloride in the presence of pyridine occurred preferentially at oxygen resulting in **6.2**. However, hydrolysis of the enol ester and activation of the intermediate phthalamic acid **6.3** with acetic anhydride gave the required phthalimide **6.4** in 78% overall yield.

A milder alternative to the classical Gabriel synthesis which allows the appendage of a protected primary amine to a *secondary* alkyl group exploits the Mitsunobu inversion[18] as illustrated in Scheme 6.7[19]. The method can also be adapted to the synthesis of *N*-protected hydroxylamines [Scheme 6.8][15].

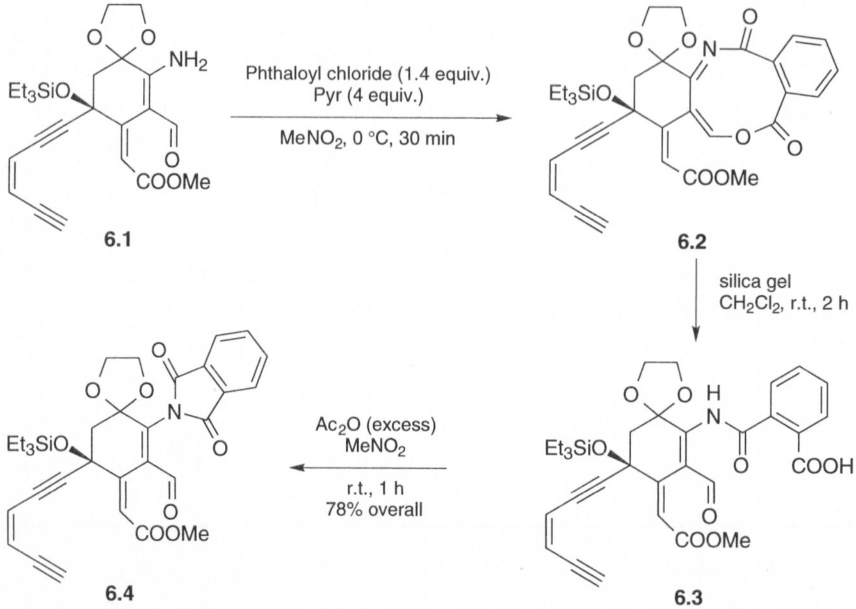

Phthaloyl chloride (1.4 equiv.)
Pyr (4 equiv.)
$MeNO_2$, 0 °C, 30 min

silica gel
$CH_2Cl_2$, r.t., 2 h

$Ac_2O$ (excess)
$MeNO_2$
r.t., 1 h
78% overall

**6.1** **6.2** **6.3** **6.4**

**Scheme 6.6**

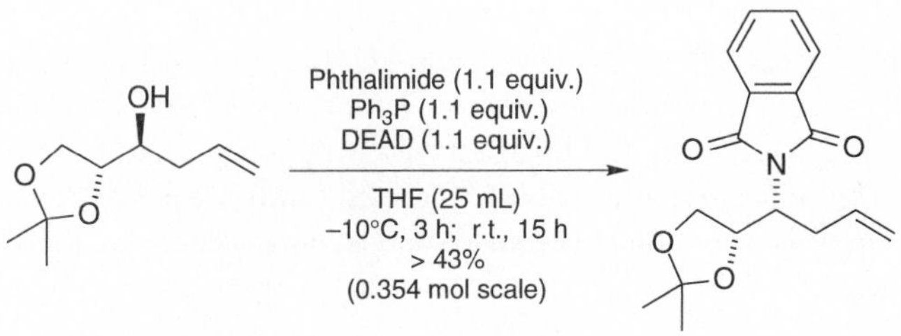

Phthalimide (1.1 equiv.)
$Ph_3P$ (1.1 equiv.)
DEAD (1.1 equiv.)
THF (25 mL)
−10°C, 3 h; r.t., 15 h
> 43%
(0.354 mol scale)

**Scheme 6.7**

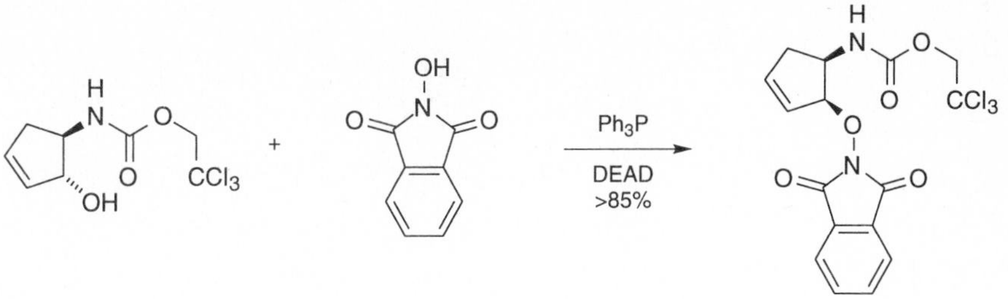

$Ph_3P$
DEAD
>85%

**Scheme 6.8**

**(iii) NMR Data** (recorded for *N*-(5-hexenyl)phthalimide)
$^1$H NMR: $\delta = 7.79$ and 7.67 (2H each, dd, $J = 5.2, 3.1$ Hz), 3.65 (2H, t, $J = 7.2$ Hz, $CH_2$–N)
$^{13}$C NMR: $\delta = 168.4$ (2C, C=O), 133.9 (2C, CH), 132.2 (2C), 123.2 (2CH), 37.9 (C–N)

## 6.2.2 Trifluoroacetamides

Simple amide derivatives are usually worthless as protecting groups because the conditions required to remove them are harsh. A proximate nucleophile may accelerate the cleavage of an amide to the point where the reaction becomes synthetically useful and no doubt this is one of the factors contributing to the mild conditions by which proteases cleave amide bonds. Chymotrypsin, for example, cleaves 3-phenylpropionamides and *N*-benzoylphenylalanine amides at pH 7 at 37 °C[20], and enzymes with broad substrate specificity hold great promise for routine amide deprotections hitherto deemed impractical[21,22]. The advantages of an intramolecular nucleophile in amide cleavage were appreciated in 1903[23] — long before the advent of enzymatic reagents — as illustrated by the deprotection of the chloroacetamide shown in Scheme 6.9[24] which proceeds by an initial *S*-alkylation of piperidinethiocarboxamide[25]. Thiourea will accomplish the same task[26,27].

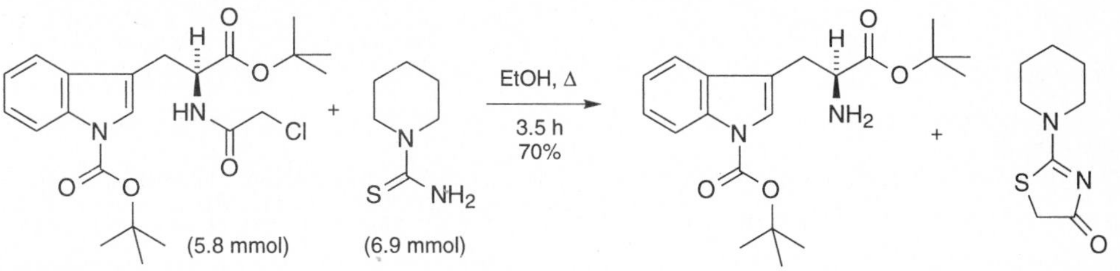

(5.8 mmol) (6.9 mmol)
EtOH, Δ
3.5 h
70%

**Scheme 6.9**

### (i) Cleavage

The trifluoroacetamide group is exceptionally labile to basic hydrolysis and therefore useful in the protection of primary and secondary amines — a fact appreciated for some time by peptide chemists[28,29]. It is readily cleaved by $K_2CO_3$ in aqueous MeOH under conditions that preserve simple methyl esters[30,31]. In an impressive synthesis of Macbecin I, Baker and Castro[32] easily accomplished the simultaneous hydrolysis of a trifluoroacetamide and an ethyl ester using LiOH without cleaving a TBS ether [Scheme 6.10].

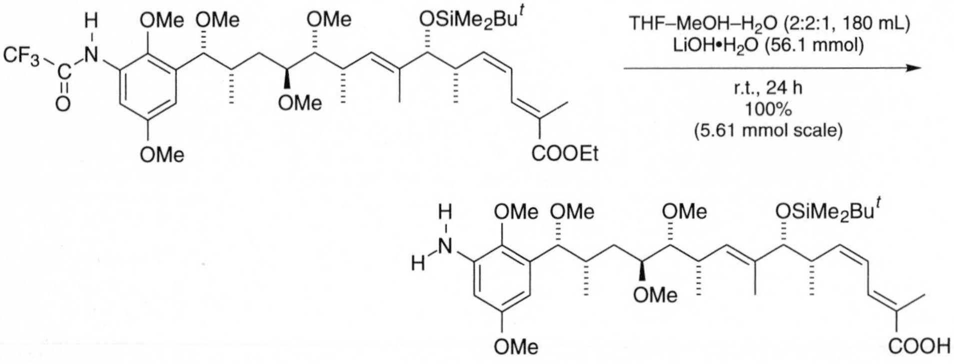

THF–MeOH–$H_2O$ (2:2:1, 180 mL)
LiOH•$H_2O$ (56.1 mmol)
r.t., 24 h
100%
(5.61 mmol scale)

**Scheme 6.10**

### (ii) Formation

Trifluoroacetamides are usually prepared by acylation of the amine with $(CF_3CO)_2O$ in the presence of a suitable base such as $NEt_3$ or pyridine in $CH_2Cl_2$ [Scheme 6.11][32]. For the selective *N*-trifluoroacetylation of a primary amine in the presence of a secondary amine, *N*-trifluoroacetoxysuccinimide has been used[31].

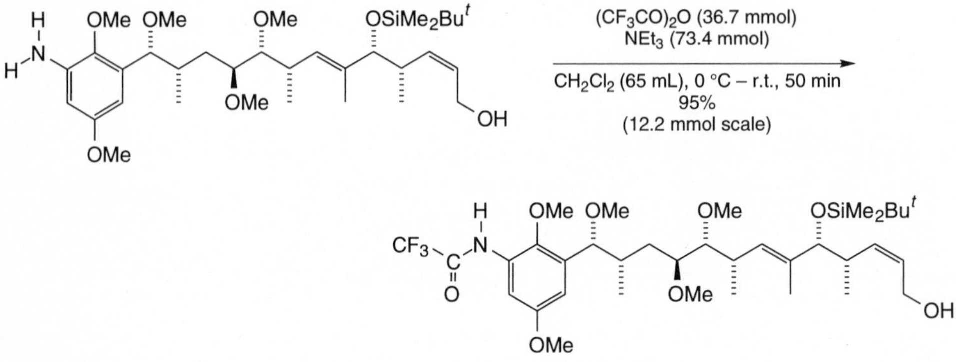

$(CF_3CO)_2O$ (36.7 mmol)
$NEt_3$ (73.4 mmol)
$CH_2Cl_2$ (65 mL), 0 °C – r.t., 50 min
95%
(12.2 mmol scale)

**Scheme 6.11**

### (iii) NMR Data

The signals in the $^{13}C$ NMR spectra of trifluoroacetamides are coupled to $^{19}F$. The trifluoroacetamide function in Scheme 6.10 gave signals at $\delta = 154.65$ (q, $J = 37.3$ Hz, CO) and 118.01 (q, $J = 288$ Hz, $CF_3$).

### 6.2.3 Methoxy- and Ethoxycarbonyl

**(i) Cleavage**

Methoxy- and ethoxycarbonyl derivatives are the simplest of the carbamate-type protecting groups in common use since 1903[33]. They are stable to most oxidising agents but metal hydrides may react. Thus N–COOEt groups can be reduced to N-Me using $LiAlH_4$ in THF[34] but they can also be removed using $NaAlH_2(OCH_2CH_2OMe)_2$ in PhH at r.t.[35] They are stable to aqueous base under conditions that typically hydrolyse methyl esters or acetates. However, hydrolysis can be achieved under vigorous conditions such as aqueous KOH in ethylene glycol at 100°C for 12 h[36] or with $Ba(OH)_2$ in aqueous MeOH at 110 °C for 12 h[37]. Hydrolysis of carbamates of indoles, pyrroles, and imidazoles is much easier[38]. Like the related ester functions, N–COOMe and N–COOEt groups undergo *O*-alkyl cleavage — albeit more sluggishly — with powerful nucleophiles such as thiolates though this method is seldom used. If the deactivating effect of the nitrogen lone pair can be reduced by delocalisation into an aromatic framework (aniline derivatives, indoles etc.), the method becomes practicable as illustrated in Scheme 6.12[39]. A more fruitful tack involves deliberate exploitation of the enhanced Lewis basicity of the carbamate carbonyl group relative to an ester function leading to easy *O*-alkyl cleavage with Lewis acids. One of the most popular reagents is TMSI[40,41] which accomplished the simultaneous cleavage of an N–COOMe group and a methoxypyridine in a synthesis of Huperzine A [Scheme 6.13], an anticholinesterase inhibitor isolated from a Chinese club moss, which is a useful lead for the treatment of Alzheimer's disease[42]. For comparison, a similar deprotection of an N–COOMe group in an analogue of Huperzine required 20% KOH in ethylene glycol at 100 °C for 48 h. Note that cleavage of an N–COOEt group in the presence of a methyl ester[43] or an N–COOMe group in the presence of a methyl ether or acetate function[44] has been successful. A practical bonus of the method is the fact that the TMSI can be generated *in situ* by the reaction of iodine with hexamethyldisilazane [Scheme 6.14][45].

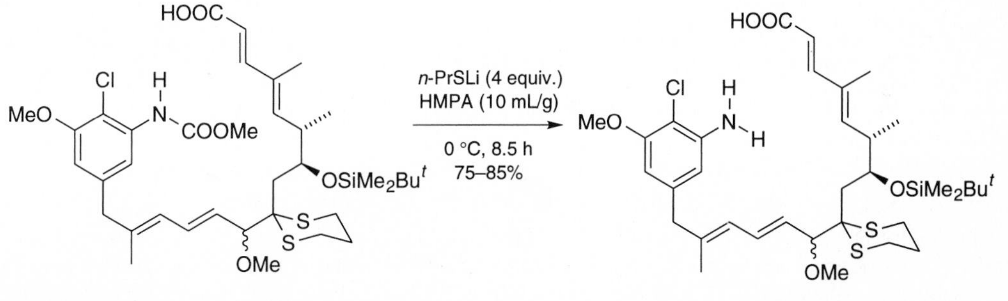

HOOC, Cl, H, N, COOMe, MeO, OSiMe$_2$Bu$^t$, S, S, OMe → *n*-PrSLi (4 equiv.), HMPA (10 mL/g), 0 °C, 8.5 h, 75–85% → HOOC, Cl, H, N, H, MeO, OSiMe$_2$Bu$^t$, S, S, OMe

**Scheme 6.12**

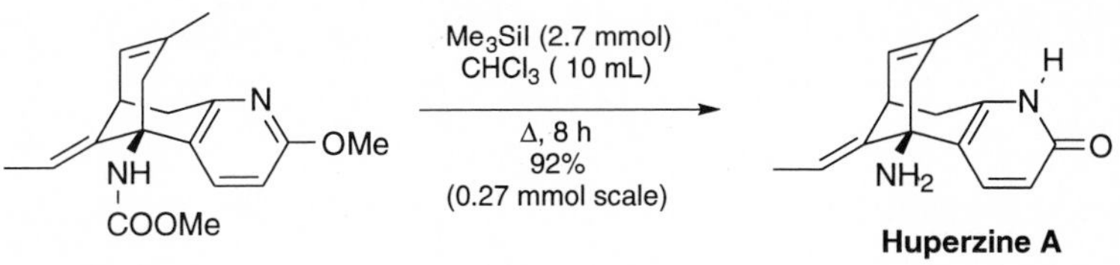

N, OMe, NH, COOMe → $Me_3SiI$ (2.7 mmol), $CHCl_3$ ( 10 mL), Δ, 8 h, 92%, (0.27 mmol scale) → H, N, O, $NH_2$, **Huperzine A**

**Scheme 6.13**

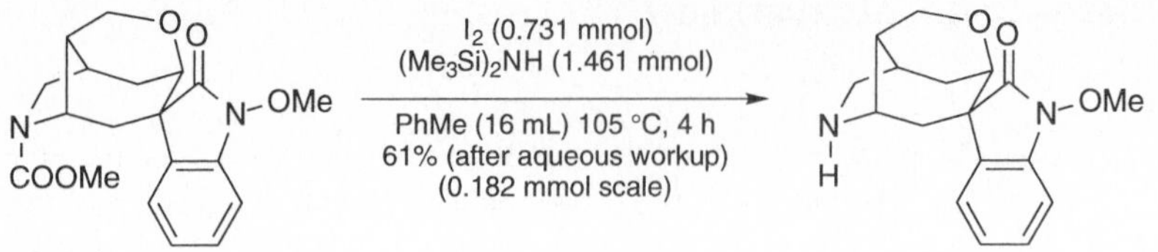

$I_2$ (0.731 mmol)
$(Me_3Si)_2NH$ (1.461 mmol)
PhMe (16 mL) 105 °C, 4 h
61% (after aqueous workup)
(0.182 mmol scale)

**Scheme 6.14**

The last step of a synthesis of the antitumour agent Acivicin used $BCl_3$ to accomplish the simultaneous cleavage of an N–COOMe function and an aminal [Scheme 6.15][46].

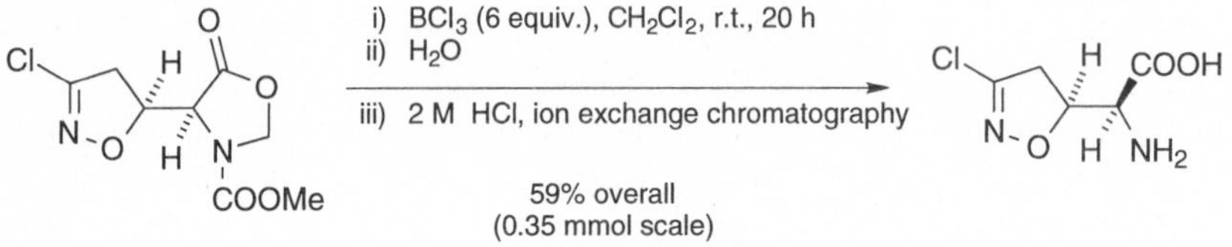

i) $BCl_3$ (6 equiv.), $CH_2Cl_2$, r.t., 20 h
ii) $H_2O$
iii) 2 M HCl, ion exchange chromatography
59% overall
(0.35 mmol scale)

**Scheme 6.15**

**(ii) Formation**

Methyl and ethyl carbamates are typically formed by the reaction of the amine with the corresponding chloroformates in the presence of a base such as $K_2CO_3$ or $NEt_3$.

**(iii) NMR Data**

$^1$H NMR (N–COOMe): $\delta = 3.7$ (3H, s) and N–COOEt, $\delta = 4.4$ (2H, q, $J = 7$ Hz), 1.4 (3H, t, $J = 7$ Hz)
$^{13}$C NMR (N-COOMe): $\delta = 159$ (C=O), 56 (Me)

## 6.2.4 *tert*-Butoxycarbonyl (Boc)

The *tert*-butoxycarbonyl group (abbreviated Boc or *t*-Boc) remains one of the most frequently used amino protecting groups in organic synthesis. Being inert towards catalytic hydrogenolysis and extremely resistant towards basic and nucleophilic reagents makes it an ideal orthogonal partner to benzyl esters and carbamates used in peptide synthesis[47-49]. Some measure of the resistance of hindered carbamates to nucleophilic attack, even by powerful nucleophiles such as organolithium reagents, can be gleaned from the transformations shown in Scheme 6.16. Treatment of dodecyl 2,2,4,4-tetramethyl-1,3-oxazolidine-3-carboxylate (**16.1**) with *s*-BuLi in the presence of the cheap homochiral base (–)-Sparteine effected asymmetric metallation to give an intermediate organolithium reagent which was alkylated *in situ* to give **16.2** in excellent enantiomeric excess[50]. Similarly, Boc-protected pyrolidine **16.3** underwent asymmetric metallation under comparable conditions[51,52].

**(i) Cleavage**

Boc cleavages are conveniently carried out with $CF_3COOH$ either neat or in combination with $CH_2Cl_2$. The reaction is generally complete within 5–10 min at r.t. but complex substrates bearing a large number of Boc groups may require longer[53]. Selective deprotection in the presence of *tert*-butyl ethers and *tert*-butyl esters is generally not possible. The reagent is relatively inert towards benzyloxycarbonyl groups, benzyl esters, and benzyl ethers but a slight loss of benzyloxycarbonyl groups is normally observed[54]. Dilution with water improves selectivity. Since the *tert*-butyl

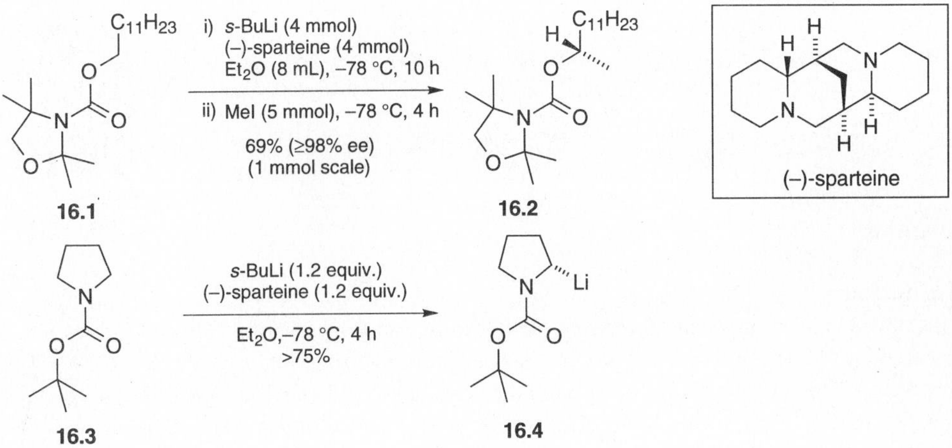

i) *s*-BuLi (4 mmol)
(–)-sparteine (4 mmol)
$Et_2O$ (8 mL), –78 °C, 10 h
ii) MeI (5 mmol), –78 °C, 4 h
69% (≥98% ee)
(1 mmol scale)

**16.1** **16.2**

(–)-sparteine

*s*-BuLi (1.2 equiv.)
(–)-sparteine (1.2 equiv.)
$Et_2O$, –78 °C, 4 h
>75%

**16.3** **16.4**

**Scheme 6.16**

carbocation liberated may attack sensitive groups such as the indole nucleus of tryptophan or the phenol ring of tyrosine, scavengers are usually added such as anisole, thioanisole, thiophenol[55], ethanedithiol, or 1,3-dimethoxybenzene [Scheme 6.17][56]. Since $CF_3COOH$ is volatile, corrosive, and expensive, a cheaper alternative acid might be considered such as 3 M HCl in EtOAc[57] or 10% $H_2SO_4$ in dioxane[58] for large scale deprotections. Anhydrous HF in the presence of anisole has also been recommended for Boc deprotection in peptides[59].

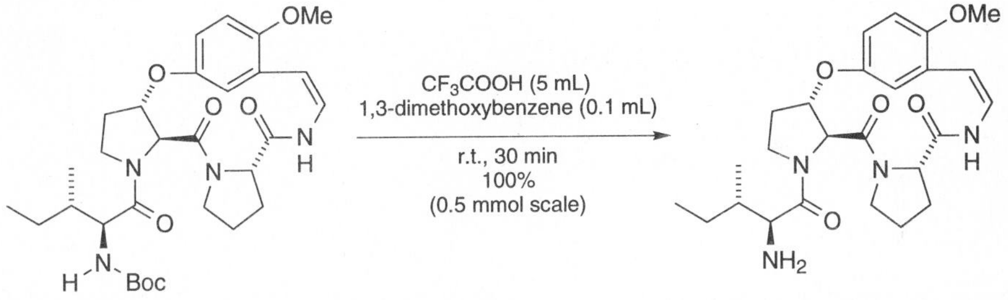

$CF_3COOH$ (5 mL)
1,3-dimethoxybenzene (0.1 mL)
r.t., 30 min
100%
(0.5 mmol scale)

**Scheme 6.17**

A wide range of Lewis acids will induce fragmentation of the Boc group. In a detailed investigation of the selective removal of Boc groups in the presence of benzyloxycarbonyl groups, Schnabel and co-workers[54] found that 0.08 M $BF_3•OEt_2$ in HOAc was especially advantageous for the removal of Boc in peptide derivatives with acid-labile thiol protecting groups such as THP or trityl. Likewise, $BF_3•OEt_2$ can also be used[9] in $CH_2Cl_2$ at 0 °C; however, the gentler reagent $Me_2BBr$ allows selective cleavage of a MOM ether in the presence of a Boc group[60]. Benzyloxycarbonyl groups, which cleave to some extent in neat $CF_3COOH$, are unaffected under these conditions. More recently TMSOTf has been used but this reagent will also prey upon a wide range of other protecting groups[61-63]. Much greater selectivity is observed with *tert*-butyldimethylsilyl triflate (TBSOTf), which accomplishes the selective removal of a Boc group in the presence of a *tert*-butyl ester [64,65]. The cleavage is a two-stage process which generates a labile *O*-silyl carbamate, which must be cleaved in a separate step. Scheme 6.18 shows how the method can be adapted for the interconversion of urethane-type amino protecting groups.

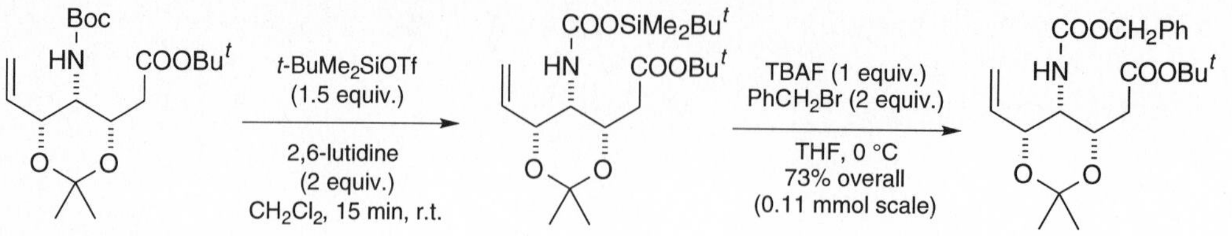

Scheme 6.18

**(ii) Formation**

The instability of *tert*-butyl chloroformate precludes its use for preparing Boc derivatives and so a large number of alternative reagents and methods have been developed of which $Boc_2O$[66] (*tert*-butyl pyrocarbonate, Boc anhydride or di-*tert*-butyl dicarbonate) and 2-(*tert*-butoxycarbonyloxyimino)-2-phenylacetonitrile[67] ("BOC-ON") are favorites. Both reagents are illustrated in Scheme 6.19[68,69] for the large scale protection of phenylalanine. In the presence of DMAP, $Boc_2O$ can be used to protect the nitrogen of amides[70-72] and indoles[73,74].

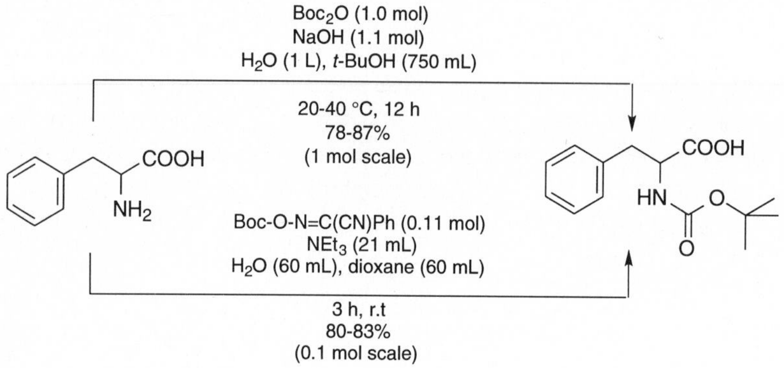

Scheme 6.19

For the simultaneous protection of the thiol and amino functions of cysteine, Kemp and Carey[75] first prepared a thiazolidine ring which was converted to its Boc derivative using $Boc_2O$ in the presence of *i*-$Pr_2NEt$ [Scheme 6.20]. The conditions were critical because thiazolidines substituted at the 2-position are very susceptible to alkaline hydrolysis.

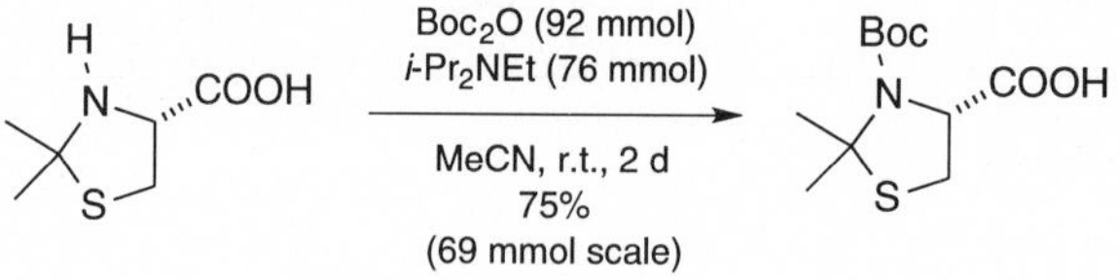

Scheme 6.20

We will now take a slight diversion from the straight and narrow path of Boc protection to look at a recent asymmetric synthesis of $\alpha$-amino acids [Scheme 6.21][76,77]. The key reaction in the 4-step sequence involves the $TiCl_4$ mediated addition of a ketene-*O*-silyl acetal bearing an ephedrine chiral auxiliary to di-*tert*-butyl azodicarboxylate, a commercially available electrophilic aminating agent.

After removal of the Boc groups from the *N,N′*-di-Boc-*α*-hydrazino ester product with $CF_3COOH$ and hydrolysis of the chiral auxiliary, an *α*-hydrazino acid was obtained which underwent hydrogenolysis to the corresponding amino acid.

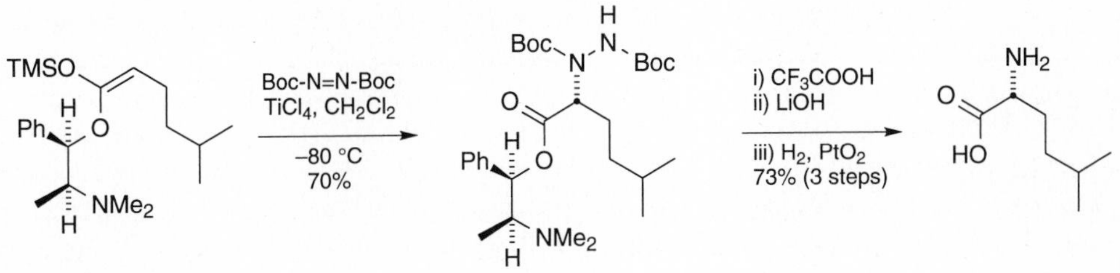

**Scheme 6.21**

**(iii) NMR Data**

$^1H$ NMR: $\delta = 1.4$ (9H, s)

$^{13}C$ NMR: $\delta = 156$ (C=O), 80 (1C), 28 (3C)

### 6.2.5 Benzyloxycarbonyl (Z or Cbz)

The invention of the benzyloxycarbonyl group (abbreviated Z or Cbz) by Bergman and Zervas in 1932[78] was one of the milestones in the development of modern synthetic chemistry. This one group, perhaps more than any other, was the harbinger of the modern era of peptide synthesis and the versatility of the Cbz group ensures it a valued place amongst its many rivals.

**(i) Cleavage**

The Cbz group resides comfortably in three different orthogonal sets since it can be cleaved by dissolving metal reduction, catalytic hydrogenation, and acidolysis. Fortunately, the various conditions required for its removal are mild enough to tolerate a wide range of complementary protecting groups. We begin with dissolving metal reduction which is the least selective of the three modes of cleavage. In the last step of their synthesis of Rhizobitoxine, Keith and co-workers[79] achieved the simultaneous

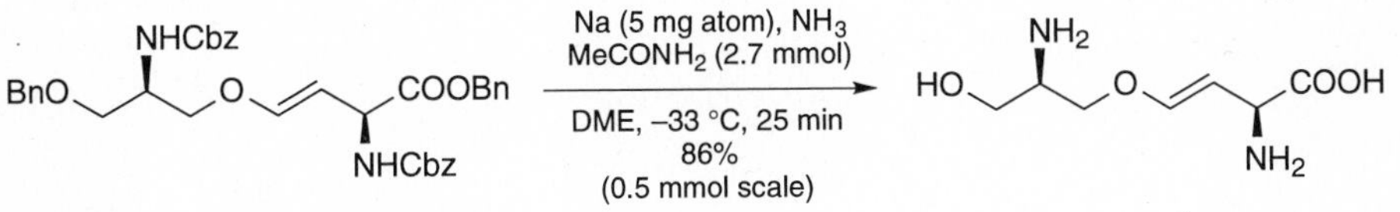

**Scheme 6.22**

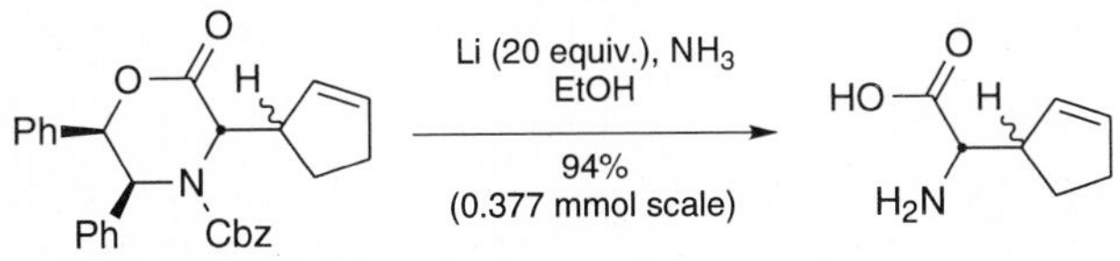

**Scheme 6.23**

removal of a benzyl ether, a benzyl ester, and Cbz groups with Na in liquid ammonia buffered with acetamide [Scheme 6.22]. A similar triple reductive cleavage [Scheme 6.23][80] deprived the 1,4-oxazinan-2-one **23.1** of its homochiral benzylic scaffold to disgorge the desired amino acid fragment **23.2**.

Acidolysis of Cbz groups can be achieved with either protic acids or Lewis acids, and the latitude is wide enough to allow the selective cleavage of Boc groups and *tert*-butyl esters. For example a Cbz group is unscathed in 4 M HCl in dioxane at r.t. for 6 h — conditions which efficiently cleave a *tert*-butyl ester[81] and it survives the conditions used to form *tert*-butyl ethers ($H_2SO_4$, isobutene in $CH_2Cl_2$, r.t., 4 days[82]). An exhaustive list of possible reagents would be too long and so our list will, perforce, be selective. Of the protic acids gaseous HBr in acetic acid is a traditional reagent which is still in use [Scheme 6.24][83,84]. Others include $CF_3SO_3H$[85], $CH_3SO_3H$[86], and 70% HF in pyridine[87,88]. In section 6.2.4 we noted that $CF_3COOH$ slowly cleaves Cbz groups allowing the selective removal of Boc groups. However, Kiso and co-workers[89] noted a marked acceleration in the presence of thioanisole and the method is illustrated in Scheme 4.50 by the simultaneous cleavage of a benzhydryl ester and a Cbz group in the last step of a synthesis of Acivicin[90].

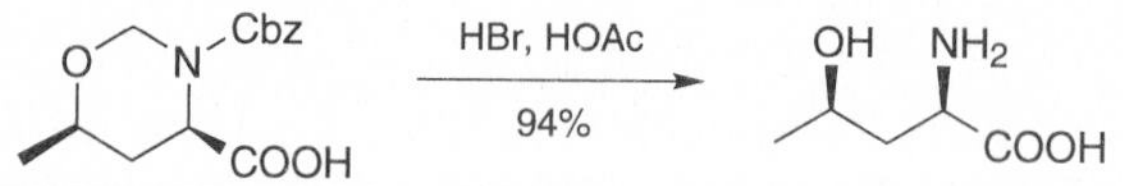
Cbz
O
N
COOH
HBr, HOAc
94%
OH NH2
COOH

**Scheme 6.24**

The accelerating effect of sulfur nucleophiles on the rate of Cbz protonolysis is also observed with Lewis acids. Thus, OBO groups can be formed in the presence of a Cbz group using $BF_3{\bullet}OEt_2$ as the catalyst [Scheme 6.25][91]; similarly, Boc groups can be cleaved in the presence of Cbz groups with the same catalyst in HOAc[54], but in $CH_2Cl_2$, $BF_3{\bullet}OEt_2$ cleaves Cbz groups efficiently if $Me_2S$ is added[92]. In another example, TMSBr in $CH_2Cl_2$ at r.t. deprotects a benzyl ester in the presence of a Cbz group[93], whereas TMSBr in $CF_3COOH$, containing thioanisole accomplished the comprehensive

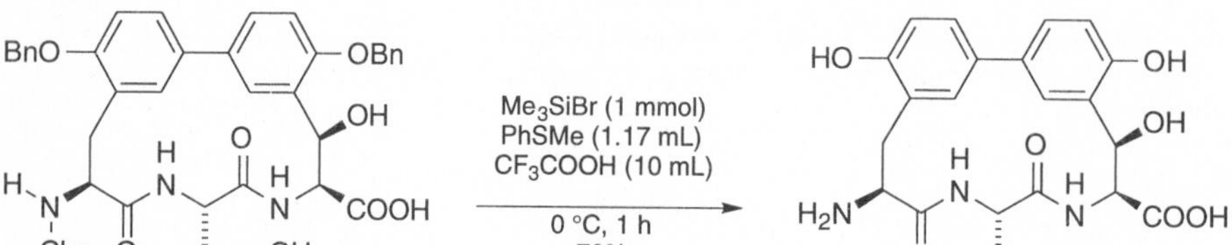
BF3•OEt2 (1.13 mmol)
CH2Cl2 (4.5 mL)
–4 °C, 2 h
84%
(4.53 mmol scale)
NHCbz
N
H
O
NHCbz

**Scheme 6.25**

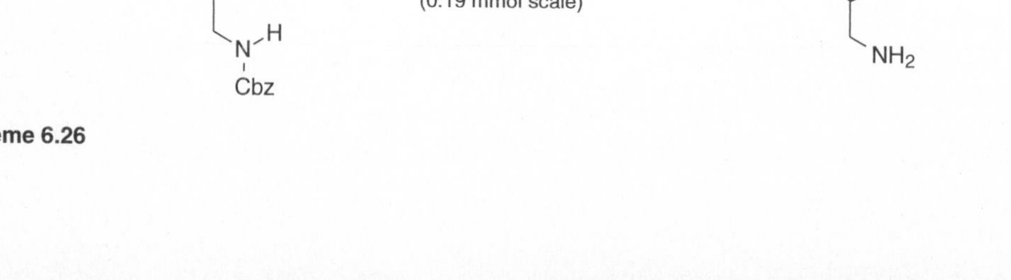
BnO
OBn
OH
H
N
Cbz
COOH
OH
N
H
Cbz
Me3SiBr (1 mmol)
PhSMe (1.17 mL)
CF3COOH (10 mL)
0 °C, 1 h
70%
(0.19 mmol scale)
HO
OH
OH
H2N
COOH
OH
NH2

**Scheme 6.26**

deprotection of two benzyl ethers and two Cbz groups in a recent synthesis of Biphenomycin A [Scheme 6.26][94]. TMSI has also been used to cleave Cbz groups[95,96].

In the concluding steps of a synthesis of antibiotic 593A [Scheme 6.27], Fukuyama and co-workers[97] used one stone to kill four birds: treatment of the intermediate **27.1** with $BCl_3$ in $CH_2Cl_2$ at r.t. not only removed two Cbz groups, but also effected elimination of two ethoxy groups. The resultant imine **27.2** was then reduced to the saturated piperidine rings of the target **27.3**.

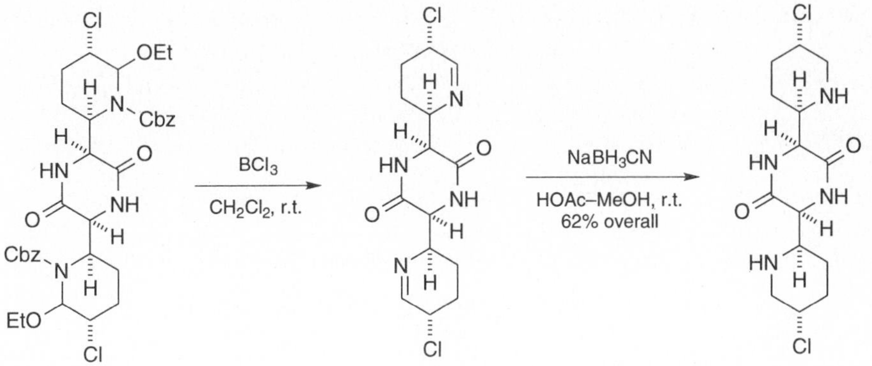

**Scheme 6.27**

Catalytic hydrogenation at atmospheric pressure over Pd–C offers a clean and effcient method for cleaving Cbz groups. The method was recently used to redeem an amino function as a prelude to macrocyclisation with a reactive pentafluorophenol ester *en route* to the naturally occurring angiotensin converting enzyme (ACE) inhibitor K-13 [Scheme 6.28][98]. A similar transformation has been applied to the synthesis of the cyclopeptide alkaloid Nummularine F[99]. In this case Pd black gave better yields than Pd–C owing to adsorption of the product onto the charcoal.

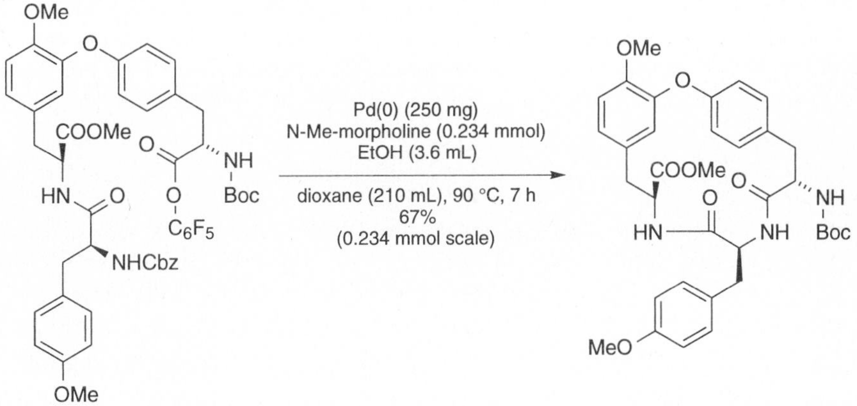

**Scheme 6.28**

Transfer catalytic hydrogenation using 10% Pd-C and cyclohexadiene[100] [Scheme 6.29][101] or ammonium formate[102] as the hydrogen source is generally faster and is now the method of choice. Various solvents can be used but there may be substantial differences in rate. Reduction is fastest in glacial acetic acid which has the advantage that thioethers (but not *S*-benzyl groups) may be present without poisoning the catalyst. The reaction is slower in MeOH, EtOH or DMF and very slow in *i*-PrOH or DMSO. Benzyl ethers and benzyl esters compete.

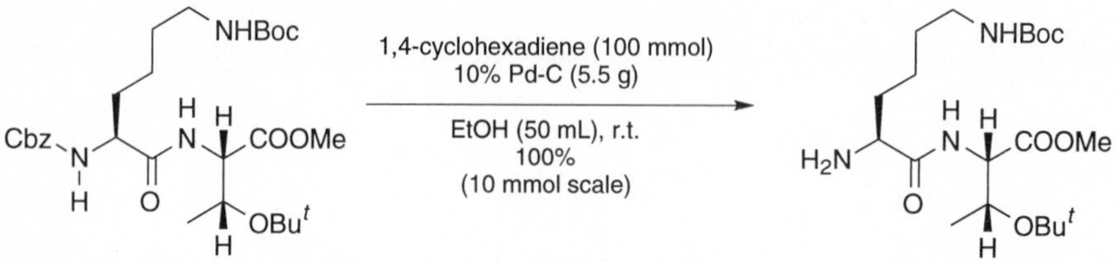

**Scheme 6.29**

In a recent synthesis of the central 1-azabicyclo[3.1.0]hexane fragment **30.3** of the antitumour antibiotic Azinomycin [Scheme 6.30], Coleman and Carpenter[103] effected the liberation of an aziridine ring **30.2** from its Cbz derivative **30.1** using Pd(II) and $Et_3SiH$[64].

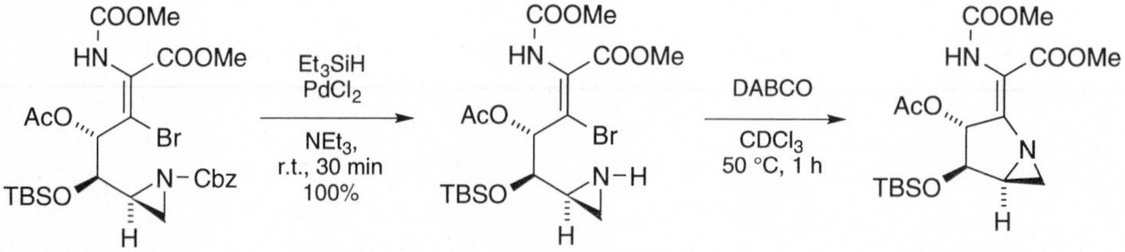

**Scheme 6.30**

The relative reactivity of the Cbz group lends itself to fine tuning by the simple expedient of introducing substituents onto the aromatic ring. Thus 4-nitrobenzyloxycarbonyl groups[104] are stable towards HBr in HOAc but are more readily cleaved by hydrogenolysis. By the same token, enhanced sensitivity to acid is procured by introducing electron-donating substituents onto the ring: 4-methoxybenzyloxycarbonyl groups[105] are cleaved by $CF_3COOH$ at 0 °C. For an extensive compilation of the numerous variants of the Cbz group, the reader should consult the comprehensive review by Wünsch[106].

### (ii) Formation

Benzyl chloroformate, a cheap and commercially available reagent, reacts with an amine in the presence of $NEt_3$, aqueous $NaHCO_3$ or NaOH [Scheme 6.31] to give the Cbz derivative in good yield. The reagent deteriorates on storage and therefore should be freshy distilled *at high vacuum* immediately before use. Variations on the theme of BnOCO–X include reagents where X = *O*-succinimidyl[107], benzotriazolyl[108], imidazolyl[109], and cyano[110]. A recent innovation which offers a

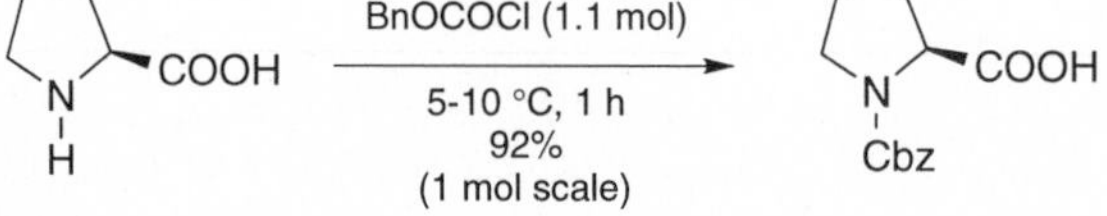

**Scheme 6.31**

versatile method for preparing a wide range of carbamate-type protecting groups including Boc and Fmoc utilises the crystalline reagent 5-norbornene-2,3-dicarboximido chloroformate (**32.1**) (mp 98–100 °C) [Scheme 6.32][111].

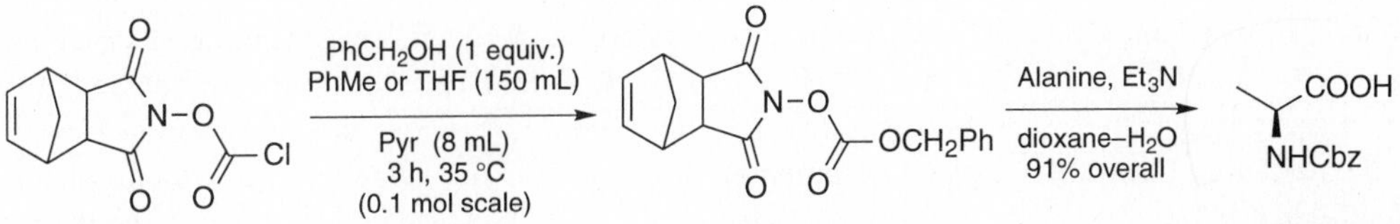

PhCH$_2$OH (1 equiv.), PhMe or THF (150 mL); Pyr (8 mL), 3 h, 35 °C (0.1 mol scale)

Alanine, Et$_3$N, dioxane–H$_2$O, 91% overall

**32.1** **32.2** **32.3**

**Scheme 6.32**

**(iii) NMR Data**

$^1$H NMR: $\delta$ = 7.2 (5H, s), 5.1 (2H, s)

$^{13}$C NMR: $\delta$ = 156 (C=O), 136 (**C**$_{Ar}$-CH$_2$), 125-130 (5C$_{Ar}$), 66 (Ar**C**H$_2$)

## 6.2.6 Allyloxycarbonyl (Aloc)

In section 4.2.6 we introduced allyl esters as protecting groups for carboxylic acids which benefit from easy synthesis and easy removal under essentially neutral conditions by Pd(0)-catalysed allyl transfer. Kunz and Unverzagt[112] showed that these same benefits could be bestowed on the allyloxycarbonyl (Aloc) protection of amines thereby creating new opportunities for the synthesis of hypersensitive target structures such as glycopeptides. Glycopeptides and glycolipids, crucial components of the oligosaccharide coat that adorns the lipid and protein surface components of cell membranes, provide the basis for recognition and discrimination by the immune system. Glycopeptides and glycolipids provide a thin sheath to ward off molecular predators; they determine friend or foe. An understanding of the molecular basis for immune discrimination will eventually unlock the mystery of how cancer cells and viruses, for example, are able to elude destruction and thrive[113]. The chemical dimension of the problem demands the synthesis of structurally homogeneous glycopeptides — a major challenge which has only become feasible in the last decade.

**(i) Cleavage**

We will illustrate the remarkable gentleness of Aloc deprotection from a late step [Scheme 6.33] in the Kunz–Unverzagt synthesis of a fucosyl-chitobiose glycopeptide which constitutes a partial sequence

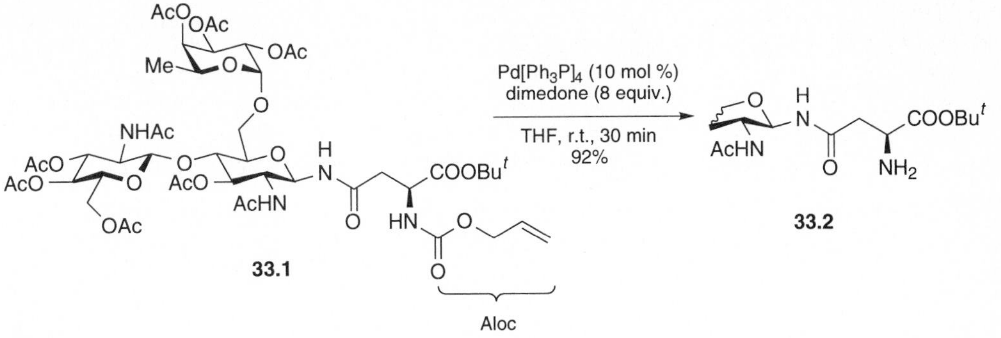

Pd[Ph$_3$P]$_4$ (10 mol %), dimedone (8 equiv.); THF, r.t., 30 min, 92%

**33.1** Aloc **33.2**

**Scheme 6.33**

of a viral envelope protein[114]. By simply treating the complex trisaccharide-asparagine conjugate **33.1** in THF at r.t. with 8 equivalents of dimedone (5,5-dimethyl-1,3-cyclohexanedione) in the presence of 10 mol % of a Pd(0) catalyst, the amino group in the asparagine moiety was liberated to give **33.2** in 92% yield without detriment to the labile *O*-glycosidic link.

The CH-acid dimedone ($pK_a$ = 5.2) was chosen as the allyl transfer reagent in order to facilitate separation from the liberated amine. When used in 7- to 8-fold excess, it also protonates the free amino group ($pK_a$ = 8) to such an extent that the latter cannot function as an allyl acceptor. However, if the reaction times are long, an enamine may form with the dimedone in which case 1,3-dimethylbarbituric acid ($pK_a$ = 4.7) can be used instead[115,116]. It is also noteworthy that the thioether function of methionine is not detrimental to the activity of the Pd catalyst.

In the context of peptide synthesis, some care has to be taken in the choice of orthogonal carboxyl and amino protection to complement the advantages of the Aloc group. Obviously hydrogenolysis and Lewis acidolysis (with, e.g., TMSBr or $BBr_3$) are precluded thereby largely circumscribing benzyl esters and Cbz groups. However, Aloc groups are stable to $CF_3COOH$ under conditions which remove *tert*-butyl esters (r.t., 30 min) or *tert*-butyl ethers (r.t., 16 h). For even greater acid stability, *p*-nitrocinnamyloxycarbonyl (Noc) groups have been recommended[115] as a useful alternative to the unsubstituted Aloc system. The Noc group offers some other advantages as well: it is more highly crystalline and it is UV active, but unlike the Aloc group, it is stable towards Rh(I)-catalysed isomerisation and hydrolysis by which allyl esters are cleaved[117]. Furthermore, Noc groups undergo the Pd-catalysed allyl transfer reaction markedly slower (ca. 8 h) than Aloc groups offering the prospect of selective cleavage of Aloc in the presence of Noc.

A double Aloc deprotection was recently applied to the synthesis of some carbapenem derivatives as shown in Scheme 6.34[118]. The procedure represents a new method for nucleophilic amination in sensitive substrates based on the Mitsunobu reaction.

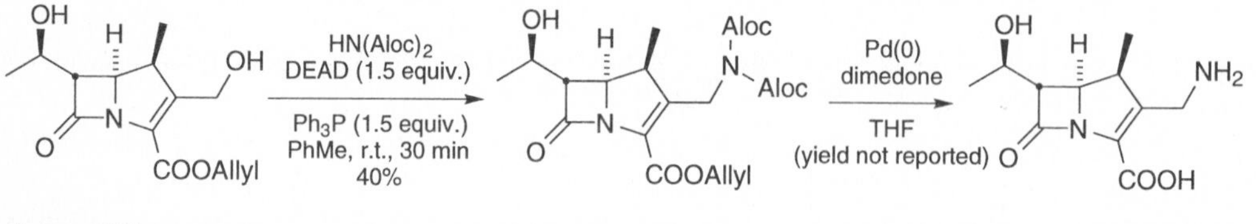

OH H OH N COOAllyl O — $HN(Aloc)_2$, DEAD (1.5 equiv.) / $Ph_3P$ (1.5 equiv.), PhMe, r.t., 30 min, 40% → OH H N(Aloc)Aloc N COOAllyl O — Pd(0), dimedone / THF (yield not reported) → OH H $NH_2$ N COOH O

**Scheme 6.34**

Tributyltin hydride has been described as the allyl-trapping reagent in the deprotection of Aloc groups[119] though its use may be limited by competing enone reduction in certain substrates. A convenient Pd(0)-catalysed transprotection of Aloc to Boc group was recently described by a Dutch

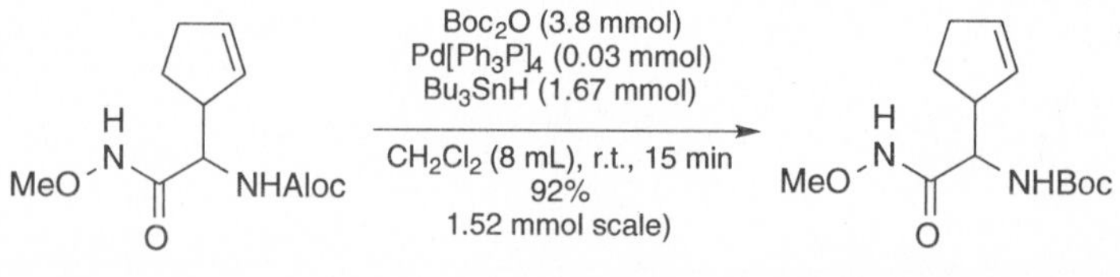

MeO-NH-C(O)-CH(cyclopentenyl)-NHAloc — $Boc_2O$ (3.8 mmol), $Pd[Ph_3P]_4$ (0.03 mmol), $Bu_3SnH$ (1.67 mmol) / $CH_2Cl_2$ (8 mL), r.t., 15 min, 92%, 1.52 mmol scale) → MeO-NH-C(O)-CH(cyclopentenyl)-NHBoc

**Scheme 6.35**

group [Scheme 6.35][120] which uses the tin hydride method. The reaction resembles a similar transprotection strategy based on silanes as the trapping agent in which hydrolytically labile silyl carbamates are formed[64].

Our final example [Scheme 6.36] comes from a synthesis of the alkaloids Crinine and Buphanasine[121]. Aloc deprotection was accomplished using a slight modification of the original conditions developed by Jeffrey and McCombie[122] for the deprotection of allyl esters (see Chapter 4) in which the *potassium salt* of 2-ethylhexanoic acid was used as the allyl trapping agent. The nascent secondary amine cyclised spontaneously to give the desired bicyclic product in good yield. However, these conditions are not generally applicable to Aloc groups because the liberated amino group is wont to react itself with the π-allyl complex intermediate leading to *N*-allylation. Success in the case of the transformation depicted in Scheme 6.37 may simply be due to the desired cyclisation competing with the troublesome side reaction.

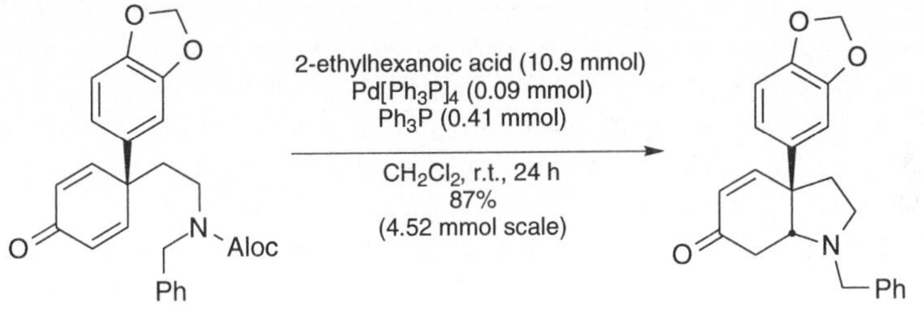

2-ethylhexanoic acid (10.9 mmol)
$Pd[Ph_3P]_4$ (0.09 mmol)
$Ph_3P$ (0.41 mmol)
$CH_2Cl_2$, r.t., 24 h
87%
(4.52 mmol scale)

**Scheme 6.36**

Before we leave Aloc protection, we should mention that *N*-allyl groups can be removed by Rh(I)-catalysed isomerisation to an enamine followed by hydrolysis as illustrated in Scheme 6.37[123,124]. The reaction is analogous to the deprotection of allyl ethers. Aloc groups are easier to introduce and therefore generally preferable.

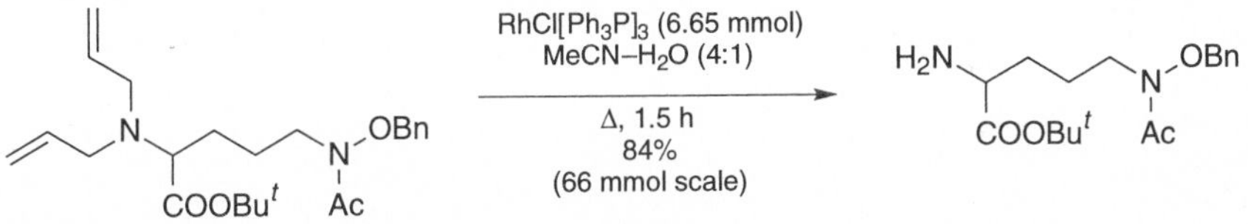

$RhCl[Ph_3P]_3$ (6.65 mmol)
$MeCN–H_2O$ (4:1)
Δ, 1.5 h
84%
(66 mmol scale)

**Scheme 6.37**

**(ii) Formation**

Aloc groups are introduced by *N*-acylation with reagents of the type allyl-O-CO-X in the presence of base as described above for the introduction of Cbz groups.

**(iii) NMR Data**

$^1$H NMR: $\delta = 5.95$ (1H, m, **CH**=$CH_2$), 5.35 (1H, dd, $J = 16$, 2 Hz, CH=**CH**$_A$H$_B$), 5.25 (1H, dd, $J = 10$, 2 Hz, CH=CH$_A$**H**$_B$), 4.55 (2H, dt, $J = 6$, 1 Hz, O-$CH_2$)

$^{13}$C NMR: $\delta = 171$ (C=O), 132 (**CH**=$CH_2$), 118 (CH=**CH**$_2$), 65 (O-$CH_2$)

### 6.2.7 9-Fluorenylmethoxycarbonyl (Fmoc)

The 9-fluorenylmethoxycarbonyl (Fmoc) group is another distinguished contribution from the Carpino laboratory[125,126] to the solution phase synthesis of peptides and latterly it has been adapted to solid phase peptide synthesis too[127]. The Fmoc group is exceptionally stable towards acid; thus, carboxylic acids can be converted to acid chlorides with thionyl chloride[128] or *tert*-butyl esters using $H_2SO_4$ and isobutene[129]. Furthermore, Fmoc groups are unscathed by HBr in HOAc or $CF_3COOH$ thereby enabling the selective deprotection of Cbz and Boc groups. On the debit side is the low solubility of many Fmoc-protected amino acids in common organic solvents and the need for chromatographic separation of the non-volatile by-products from the deprotection step. The use of the Fmoc group in peptide synthesis has been extensively reviewed*.

**(i) Cleavage**

The Fmoc group is a lineal descendent of the $\beta$-nitroethyl acetates whose base-catalysed elimination by an E1cb mechanism (see section 1.2.6), first propounded by Chattaway in 1936[130], laid the foundation for the numerous progeny which have been tailored to amino and carboxyl protection. Fmoc groups undergo rapid non-hydrolytic cleavage on treatment with simple bases (ammonia, piperidine, morpholine), usually in polar solvents such as DMF or MeCN, to liberate the *N*-terminus of a peptide *in the free base form.* The conditions are mild enough to preclude $\beta$-elimination in the sensitive *O*-glycosylserine derivatives [Scheme 6.38][131]. Other bases which have been used include DBU[132] and TBAF[133].

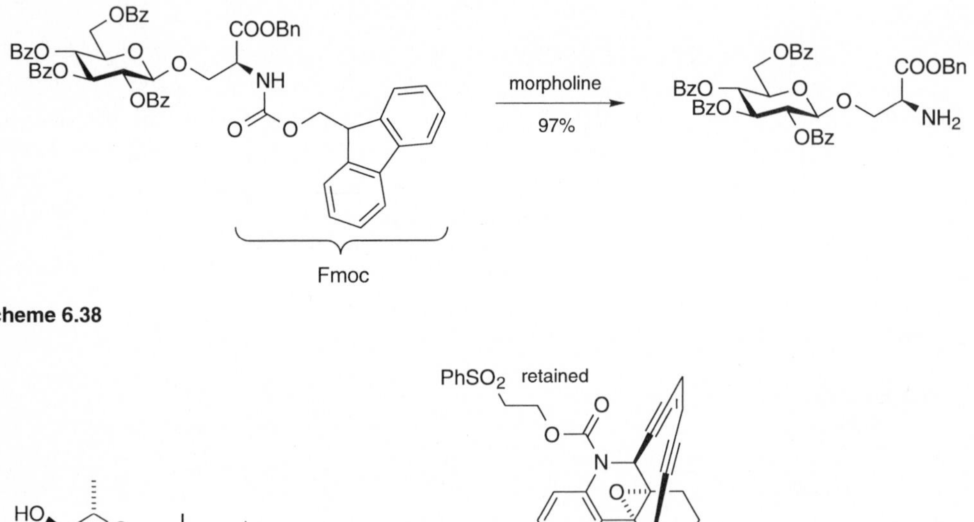

morpholine, 97%

**Scheme 6.38**

PhSO2 retained

Remove with $Et_2NH$–THF (1:2) at 0 °C in 68% yield

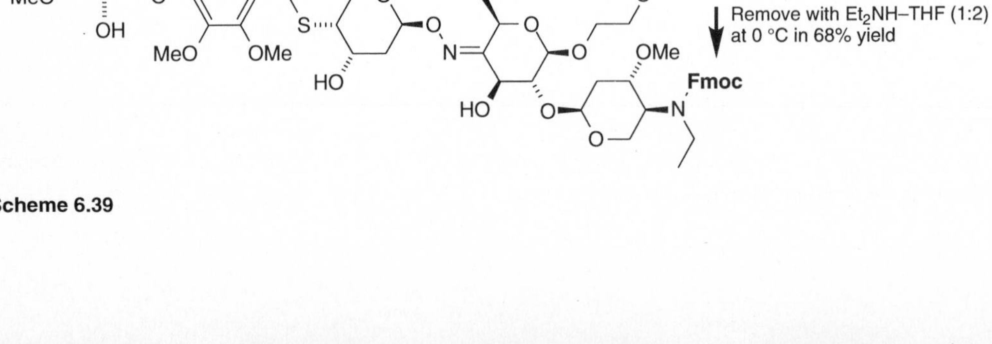

**Scheme 6.39**

In a recent synthesis of a Calicheamicin–Dynemicin hybrid, Nicolaou and co-workers[134] removed an Fmoc group in the presence of the base-sensitive 2-(phenylsulfonyl)ethoxycarbonyl group (*vide infra*) and a thioester using $Et_2NH$ as the base [Scheme 6.39].

The Fmoc group undergoes hydrogenolysis but generally at lower rate than *O*-benzyl based systems and therefore selectivity can be achieved. It is not obvious why the Fmoc system should be labile to hydrogenolysis since it is homobenzylic rather than benzylic. Nor is the reaction unique to Fmoc: 2-phenylethoxycarbonyl groups also undergo hydrogenolysis[135]. The reaction takes place both under the traditional conditions using Pd–C and $H_2$[136] or, more usefully, under transfer catalytic conditions. For example, the concluding step of a synthesis of the endogenous opiate Enkephalin, which is impressive for its speed (3–4 h) and scale (10 mmol), effected simultaneous removal of the *N*-terminal Fmoc group and the *C*-terminal benzyl ester under transfer catalytic conditions [Scheme 6.40][128].

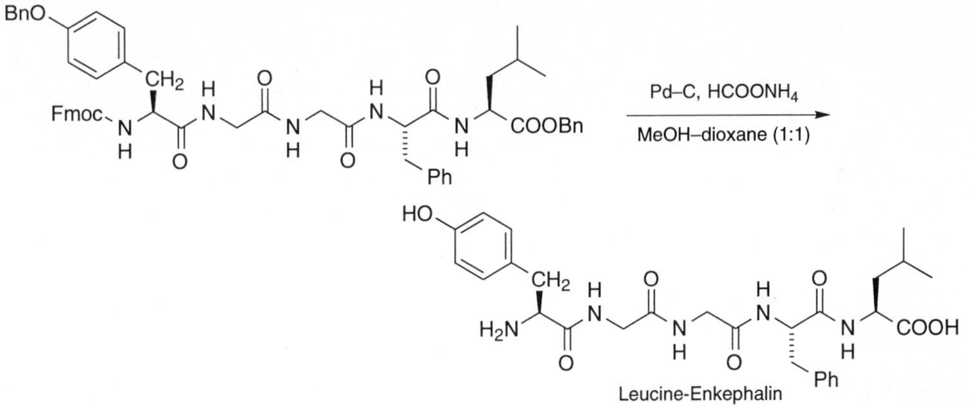

BnO; CH2; Fmoc; N; H; O; COOBn; Ph

Pd–C, HCOONH4

MeOH–dioxane (1:1)

HO; CH2; H2N; COOH; Ph

Leucine-Enkephalin

**Scheme 6.40**

On the other hand, Paulsen and Schultz[137] were able to cleave a benzyl ester selectively from an *O*-glycosylthreonine derivative without harming an Fmoc group using Pd in a mixture of HCOOH and MeOH for 10-12 min [Scheme 6.41]. Longer reaction times led to competing cleavage of the Fmoc group.

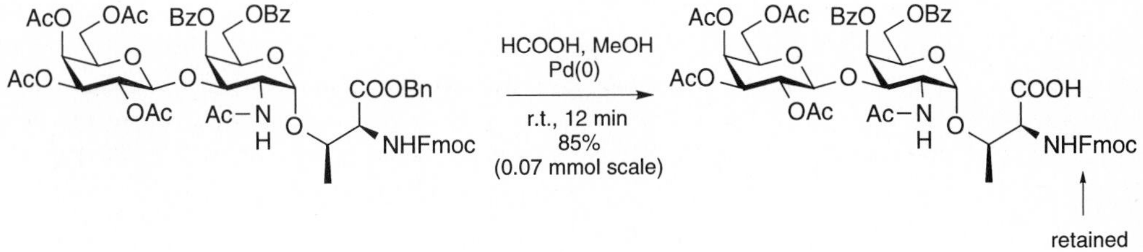

AcO; OAc; BzO; OBz; AcO; O; OAc; Ac–N; H; COOBn; NHFmoc

HCOOH, MeOH
Pd(0)

r.t., 12 min
85%
(0.07 mmol scale)

AcO; OAc; BzO; OBz; AcO; O; OAc; Ac–N; H; COOH; NHFmoc

retained

**Scheme 6.41**

Some measure of the potential selectivity (or caprice) that may be encountered in Fmoc hydrogenolyses comes from a synthesis of the Dolastatins by an Upjohn group[129]. Treatment of the triply protected glutamic acid derivative **42.1** [Scheme 6.42] with ammonium formate in the presence of 10% Pd–C at 5 °C resulted in selective scission of the benzyl ester; however, at room temperature, the Fmoc group was also cleaved.

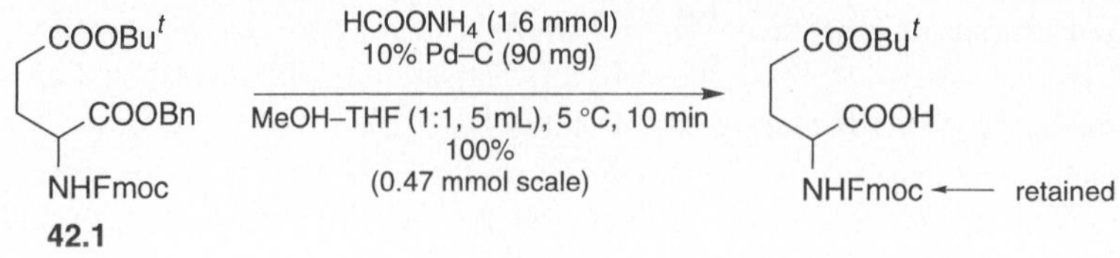

HCOONH$_4$ (1.6 mmol)
10% Pd–C (90 mg)
MeOH–THF (1:1, 5 mL), 5 °C, 10 min
100%
(0.47 mmol scale)

COOBu$^t$ COOBn NHFmoc **42.1** → COOBu$^t$ COOH NHFmoc ← retained

**Scheme 6.42**

There are several other carbamate-type amino protecting groups belonging to the base-catalysed elimination orthogonal set, but these will only be mentioned in passing since they are either pertinent to esoteric needs, or they have been largely superceded by the Fmoc group. We have already encountered the 2-(phenylsulfonyl)ethoxycarbonyl group in Nicolaou's Calicheamicin–Dynemicin hybrid synthesis [Scheme 6.39] and noted that it is stable towards simple amines which remove an Fmoc group. This group and its *p*-toluenesulfonyl analogue (the so-called Tsoc group) are cleaved by 1 M NaOH in under an hour[138]. Musso and co-workers[139] used it in their synthesis of the Muscaflavines, dihydroazepine amino acids from mushrooms [Scheme 6.43].

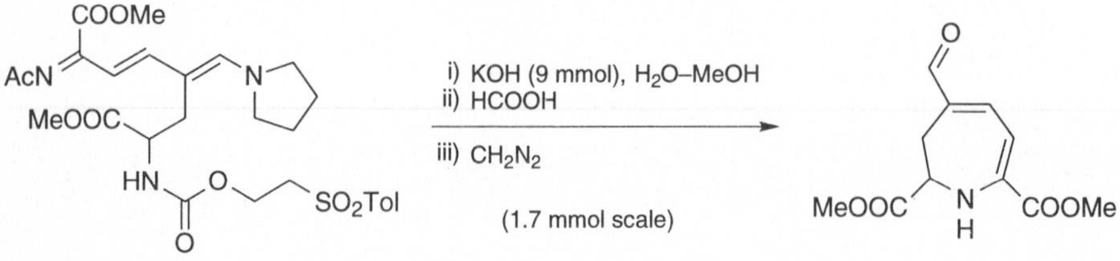

i) KOH (9 mmol), $H_2O$–MeOH
ii) HCOOH
iii) $CH_2N_2$
(1.7 mmol scale)

**Scheme 6.43**

In the interests of enhancing the base lability of the Fmoc group, the aromatic rings have been substituted with electron withdrawing substituents resulting in lability to pyridine[140,141]. Enhanced base lability has also been secured from systems of the type X-$CH_2CH_2$-$OCONR_2$ where X = $Ph_3P$ (cleaved at pH 8.4)[142] and X = 4-pyridyl (cleaved at pH 8.3 after *N*-methylation)[143].

### (ii) Formation

Fmoc groups are usually introduced under Schotten–Baumen conditions [Scheme 6.44] using the commercially available chloroformate (mp 61.5–63 °C), which is readily prepared in 86% yield by the reaction of 9-fluorenylmethanol with phosgene[144] (**HAZARD**). Under these conditions Fmoc dipeptides may be formed *via* mixed anhydrides generated from reaction between the Fmoc amino acid already formed and still unreacted chloroformate, but the problem can be avoided by using pertrimethylsilyl amino acids and base in aprotic solvents[145], or by use of *i*-$Pr_2NEt$ as the base[146]. Alternatively, several less reactive reagents of the type Fmoc-OR have been recommended where R = succinimidyl[107,147,148], benzotriazolyl[107], or pentafluorophenyl[149].

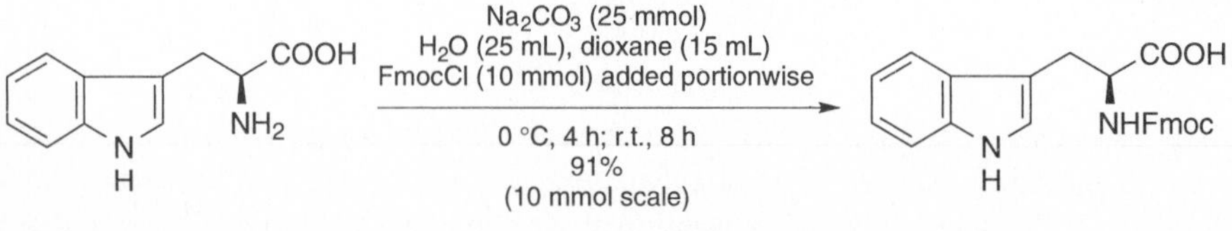

$Na_2CO_3$ (25 mmol)
$H_2O$ (25 mL), dioxane (15 mL)
FmocCl (10 mmol) added portionwise
0 °C, 4 h; r.t., 8 h
91%
(10 mmol scale)

**Scheme 6.44**

**(iii) NMR Data**

$^1$H NMR (DMSO): $\delta$ = 7.75-7.65 (2H, m), 7.6-7.5 (2H, m), 7.3-7.1 (m, 4H), 4.35 (2H, distorted d or m), 4.2 (1H, distorted t or m). The appearance of the Fmoc signals varies significantly in amino acid derivatives.

$^{13}$C NMR (DMSO): $\delta$ = 156 (C=O), 143.9 (2C), 140.8 (2C), 127.7 (2C), 127.2 (2C), 125.3 (2C), 120.2 (2C), 65.7 (1C, $CH_2O$), 46.8 ($CH$-$CH_2O$)

## 6.2.8 2-(Trimethylsilyl)ethoxycarbonyl (Teoc)

The 2-(trimethylsilyl)ethoxycarbonyl (Teoc) group[150] is rapidly growing in popularity as an amino protecting group whose properties and cleavage are reminiscent of those already described for the coeval 2-(trimethylsilyl)ethyl ester group (see section 4.2.4.2).

**(i) Cleavage**

The Teoc group is readily cleaved by TBAF in THF at r.t. in less than an hour[151], or alternatively a more economical method uses tetra-*n*-butylammonium chloride (3 equivalents) and KF•$2H_2O$ (4 equivalents) in MeCN at 45 °C[152]. Obviously, silyl ethers are incompatible as shown in a late step of a recent synthesis of Hitachimycin [Scheme 6.45][153], but the conditions are otherwise mild enough for the construction of very sensitive molecules such as Indolizomycin[154], where other carbamate-type protecting groups proved less satisfactory [Scheme 6.46]. The reaction has also been applied to the *N*-protection of hydroxylamines [152,155].

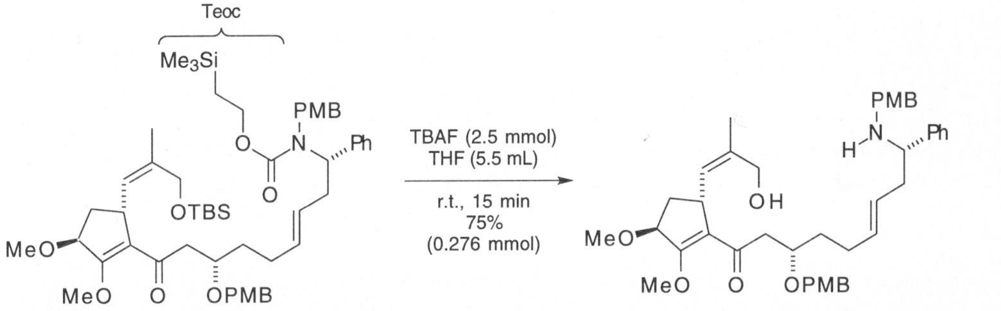

Teoc
$Me_3Si$
PMB
N
Ph
O
O
OTBS
MeO
MeO
O
OPMB
TBAF (2.5 mmol)
THF (5.5 mL)
r.t., 15 min
75%
(0.276 mmol)
PMB
H
N
Ph
OH
MeO
MeO
O
OPMB

**Scheme 6.45**

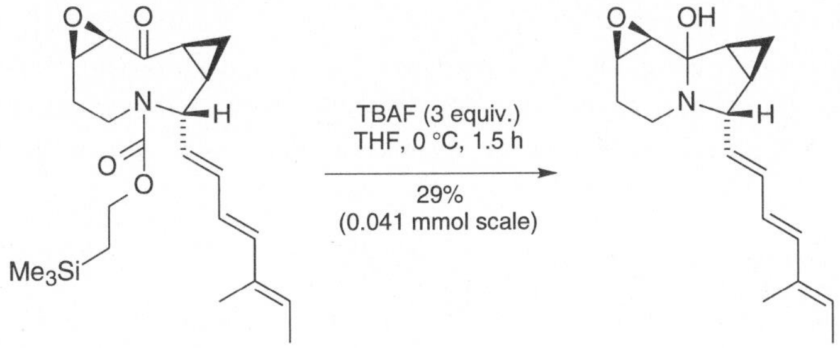

O
O
N
H
O
O
$Me_3Si$
TBAF (3 equiv.)
THF, 0 °C, 1.5 h
29%
(0.041 mmol scale)
O
OH
N
H

**Scheme 6.46**

The Teoc group can also be cleaved under acidic conditions. For example, neat $CF_3COOH$ was used to liberate an amine during a synthesis of Inandenin-12-one [Scheme 6.47][156]. These results suggest that the acid sensitivity of Teoc groups militate against their use in peptide synthesis schemes requiring Boc to be removed before Teoc removal. However, Boc groups can be removed selectively in the presence of Teoc groups by stirring the substrate with one equivalent of PTSA in $Et_2O$–EtOH (6:1) at 60–65 °C for 20 min[157]. There is, as yet, little reported data concerning the lability of Teoc groups to Lewis acidolysis, but the fact that $ZnCl_2$ is effective[158], suggests that the reaction should be easy.

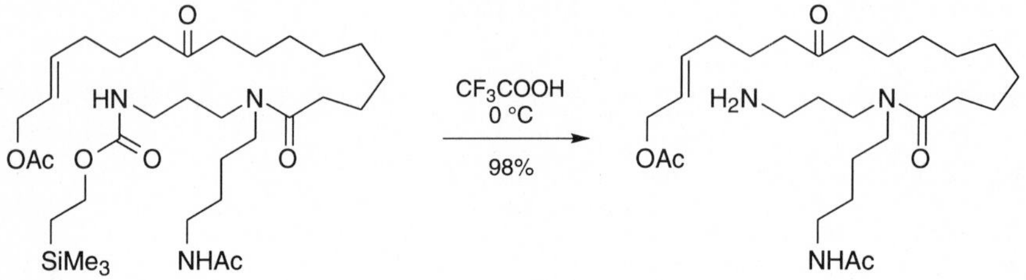

CF3COOH
0 °C
98%
HN
N
O
OAc
SiMe3
NHAc
H2N

**Scheme 6.47**

Another amino protecting group from the same orthogonal set as the Teoc group is the SEM ether. SEM ethers are particularly effective for the protection of pyrroles[159-161] and indoles[162] from which they are readily removed by TBAF as shown in Scheme 6.48.

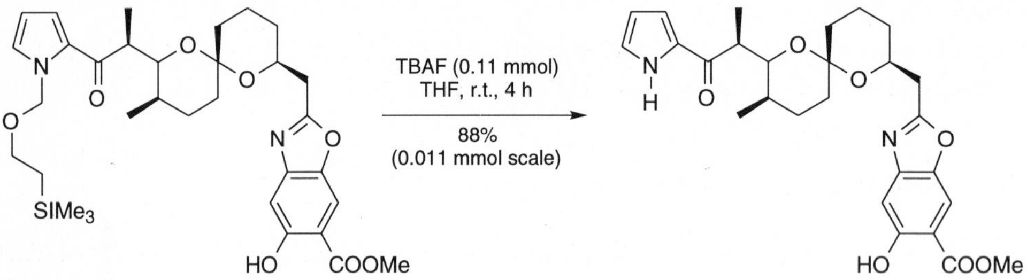

TBAF (0.11 mmol)
THF, r.t., 4 h
88%
(0.011 mmol scale)
SIMe3
HO
COOMe

**Scheme 6.48**

**(ii) Formation**

2-(Trimethylsilyl)ethyl chloroformate is unstable to storage and so should be prepared fresh before use by the reaction of 2-(trimethylsilyl)ethanol with phosgene (**HAZARD**)[163]. The chloroformate reacts with amines in the presence of $K_2CO_3$ [Scheme 6.49][153] or amides in the presence of DMAP [Scheme 6.50][164].

For the sake of increased stability and improved storage, somewhat less reactive alternatives have been prepared from the labile chloroformate. For example, Teoc-*O*-succinimidyl has been used to prepare Teoc-protected amino acids in water or dioxane using $NaHCO_3$ or $NEt_3$ as the base[107,165]

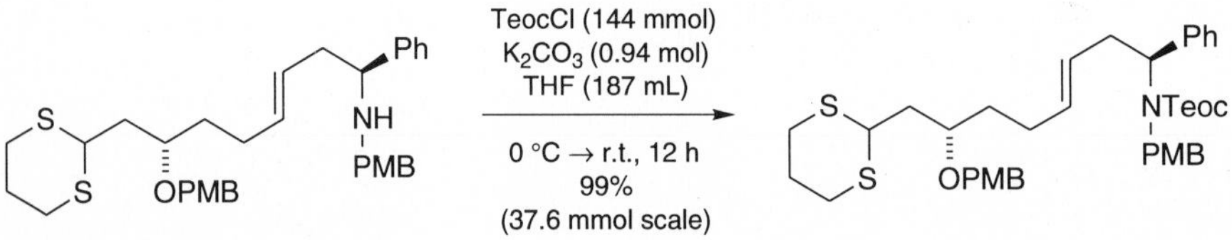

TeocCl (144 mmol)
K2CO3 (0.94 mol)
THF (187 mL)
0 °C → r.t., 12 h
99%
(37.6 mmol scale)
Ph
NH
PMB
OPMB
NTeoc

**Scheme 6.49**

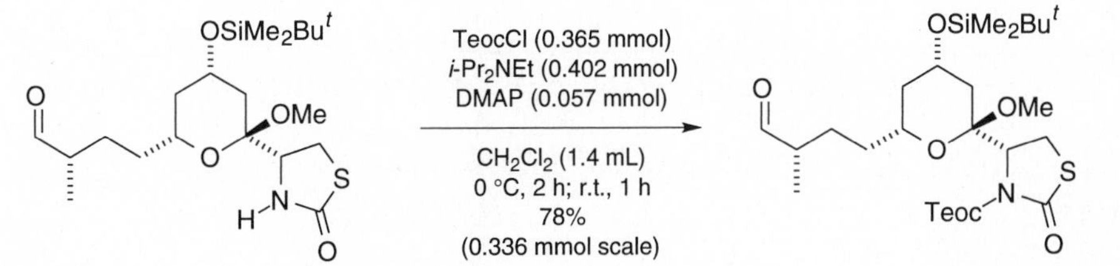

Scheme 6.50

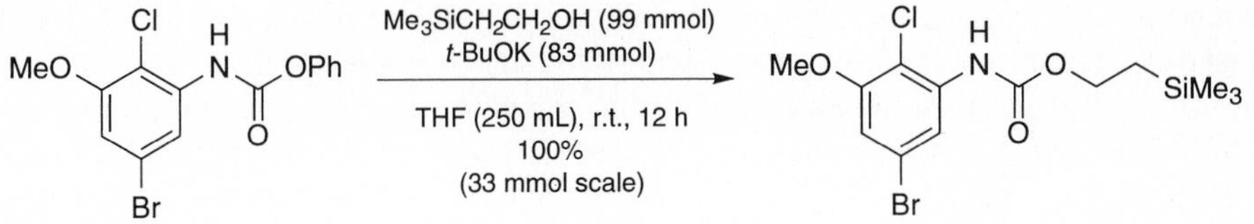

Scheme 6.51

4-Nitrophenyl 2-(trimethylsilyl)ethyl carbonate is a commercially available reagent (Fluka, mp 34–36 °C) which reacts with amino acids in a mixture of aqueous NaOH and *t*-BuOH or dioxane[157,158]. Teoc derivatives can also be prepared by an exchange reaction using an aryl carbamate derivative and 2-(trimethylsilyl)ethanol as shown in Scheme 6.51[166].

**(iii) NMR Data**

$^1$H NMR: $\delta$ = 4.4–4.7 (2H, AA′XX′O-**CH$_2$**CH$_2$-Si), 1.1–1.3 (2H, AA′XX′O-CH$_2$**CH$_2$**-Si), –0.1 (9H, s)
$^{13}$C NMR: $\delta$ = 156 (C=O), 64 (CH$_2$O), 18 (TMSCH$_2$), –3 (TMS)

## 6.2.9 2,2,2-Trichloroethoxycarbonyl (Troc)

The 2,2,2-trichloroethoxycarbonyl (Troc) group belongs to the orthogonal set cleaved by reductive *β*-elimination[167]. The Troc group can be removed in the presence of trifluoroacetamides, Cbz, Boc, Aloc, and Fmoc groups. It easily survives the conditions for removing Boc and Fmoc groups but it is unstable towards hydrogenolysis as used for removal of Cbz groups[168].

**(i) Cleavage**

Troc groups are usually cleaved by zinc in HOAc at r.t. [Scheme 6.52][9] or in EtOH at reflux [Scheme 6.53][169]. The rate of cleavage is pH dependent; thus, cleavage with Zn in aqueous THF is complete in

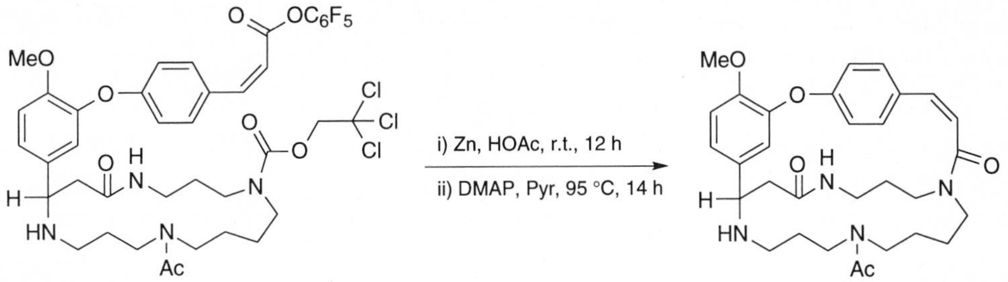

Scheme 6.52

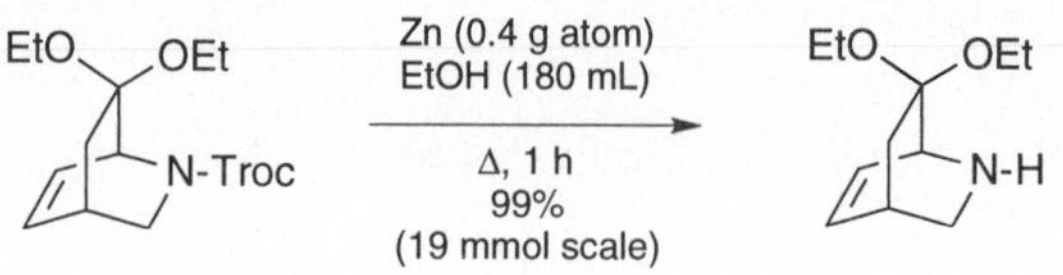
EtO
OEt
N-Troc
Zn (0.4 g atom)
EtOH (180 mL)
Δ, 1 h
99%
(19 mmol scale)
EtO
OEt
N-H

**Scheme 6.53**

30 min at pH 4.2 but requires 18 h at pH 5.5–7.2. Cadmium in 50% AcOH in DMF was found to be superior to zinc for the cleavage of a Troc group from a pentapeptide[168].

A number of more esoteric methods have been devised for the reductive cleavage of Troc groups which the synthetic community have been reluctant to adopt. These include cobalt(I) phthalocyanine[170] and sodium 2-thiophenetellurolate[171], which is readily obtained in a catalytic cycle by the $NaBH_4$ reduction of commercially available bis(2-thienyl)ditelluride. Electrolysis is a method which might be considered in special cases. For example, cleavage of the Troc groups from **54.1** [Scheme 6.54][172] with Zn would also lead to cleavage of the N–N bond whereas electrolysis produced the hydrazine **54.2** in greater than 72% yield.

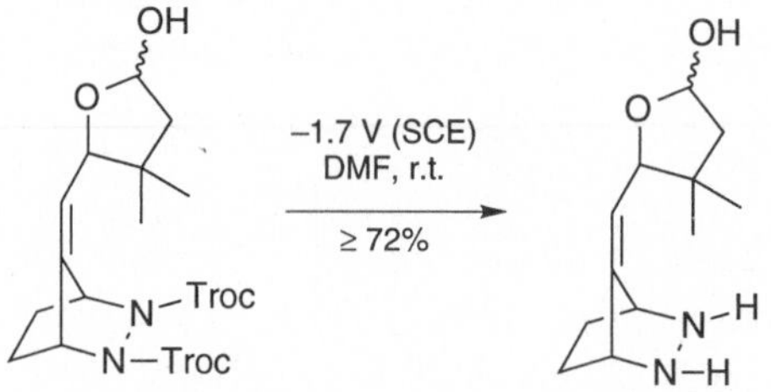
OH
O
N–Troc
N–Troc
–1.7 V (SCE)
DMF, r.t.
≥ 72%
OH
O
N–H
N–H

**54.1** **54.2**

**Scheme 6.54**

**(ii) Formation**

Troc groups are formed by reaction of compounds of the type $Cl_3CCH_2$-OCO-X (X = Cl[173] or *O*-succinimidyl[107,148]) with the amine in the presence NaOH, pyridine, or $Na_2CO_3$. A reaction of strategic importance in the elaboration of indole alkaloids involes *N*-acylation of the *β*-carboline moiety with TrocCl followed by assisted cleavage from the indole *N*[174]. The reaction figured prominently in a recent synthesis or Strychnine [Scheme 6.55][175] and analogous processes using other chloroformates were elegantly applied to the synthesis of the antitumour bisindole Vinblastine by Magnus and co-workers[176].

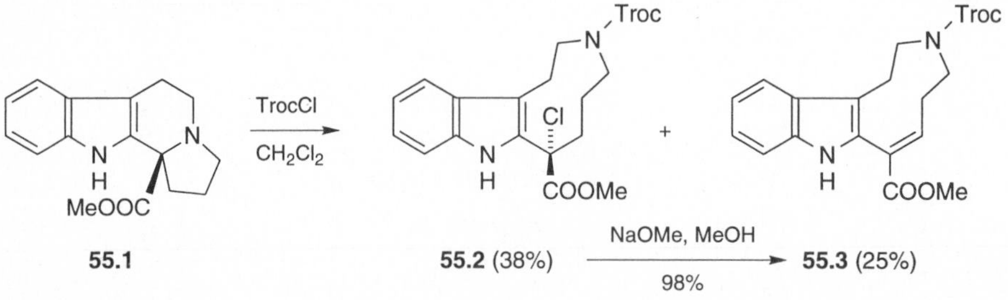
N
H
N
MeOOC
TrocCl
CH2Cl2
Troc
N
Cl
N
H
COOMe
+
Troc
N
N
H
COOMe
NaOMe, MeOH
98%

**55.1** **55.2** (38%) **55.3** (25%)

**Scheme 6.55**

**(iii) NMR Data**
$^{1}H$ NMR: $\delta = 4.7$ (s or AB system)
$^{13}C$ NMR: $\delta = 153$ (C=O), 95 ($Cl_3C$), 75 ($CH_2$)

# 6.3 *N*-Sulfonyl Derivatives

Sulfonamides are amongst the most stable of the nitrogen protecting groups. They have the added bonus of being highly crystalline and less susceptible to nucleophilic attack than the more common carbamate type protecting groups. The ease of cleavage depends on the structure of the amine. Sulfonamide derivatives of weakly basic amines such as indoles, pyrroles, and imidazoles can be cleaved by simple basic hydrolysis, whereas primary or secondary amine sulfonamides require strongly reducing conditions.

## 6.3.1 *N*-Sulfonyl Derivatives of Indoles, Pyrroles, and Imidazoles

Sulfonamide derivatives are frequently used to circumvent two serious problems associated with nitrogen heterocycles. First, the powerful electron-withdrawing effect of the arylsulfonyl group reduces the high susceptibility of the pyrrole and indole nucleus to electrophilic attack and oxidation. Secondly, replacement of the acidic N–H group in indoles, pyrroles, carbazoles, and imidazoles by an arylsulfonyl group allows the lithiation of the heterocyclic nucleus.

**(i) Cleavage**
Ellipticine derivatives are used to treat advanced breast cancer, myeloblastic leukemia, and some solid tumours. In an extensive programme aimed at the synthesis of various Ellipticine analogues, Gribble and co-workers[177,178] used the *N*-phenylsulfonyl protecting group to accomplish directed lithiation of the 2-position of the indole nucleus [Scheme 6.56]. The phenylsulfonyl group was later removed by simple treatment with $K_2CO_3$ in refluxing aqueous MeOH. No doubt, the acyl group in the 2-position of intermediate **56.4** conferred upon it an added measure of lability, as indicated by the harsher basic conditions required to remove the phenylsulfonyl group in the transformation depicted in Scheme 6.57[179].

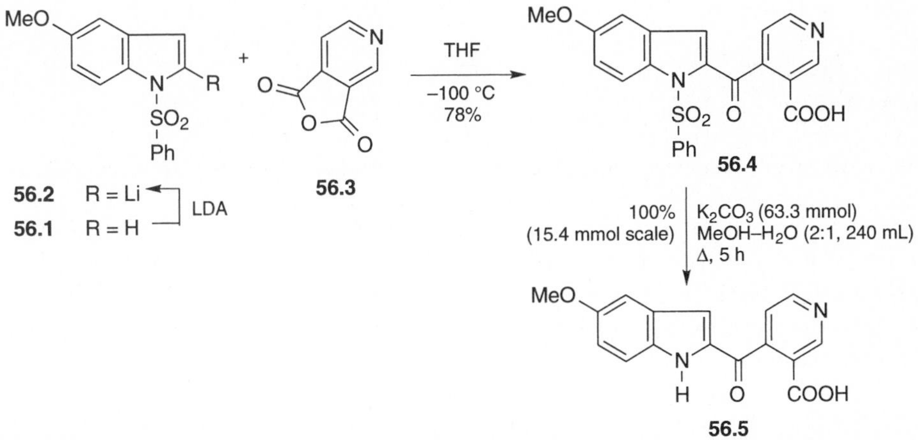

**Scheme 6.56**

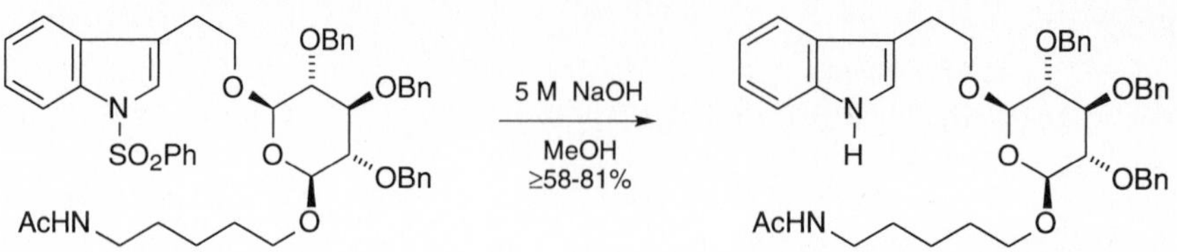

Scheme 6.57

*N*-Sulfonyl groups can be removed from imidazoles by acid-catalysed hydrolysis. Thus, in a recent synthesis of the $\alpha_2$-adrenergic agonist Medetomidine [Scheme 6.58][180], two adjacent acidic hydrogens on an imidazole nucleus had to be blocked. An *N,N*-dimethylsulfamoyl group removed an acidic N–H and activated the imidazole nucleus towards directed lithiation whilst a *tert*-butyldimethylsilyl group was used to block the 2-position of the imidazole ring so that directed lithiation-acylation could take place at the less acidic C-5 position. Both the silicon and sulfamoyl protecting groups were then excised in excellent yield by brief heating in 1.5 M HCl.

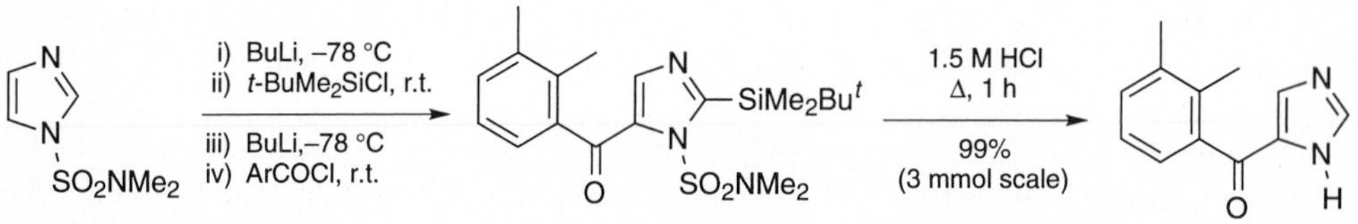

Scheme 6.58

The *N*-phenylsulfonyl group in carbazole **59.1** was removed under reductive conditions using buffered Na–Hg [Scheme 6.59][178] and similar conditions have been used to cleave arylsulfonyl groups from indoles[181].

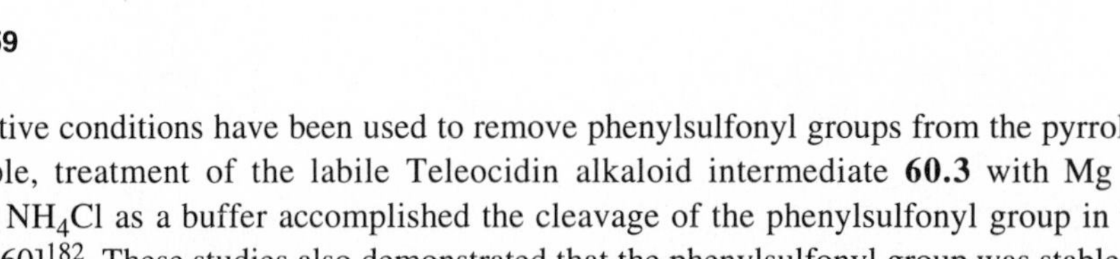

Scheme 6.59

Mild reductive conditions have been used to remove phenylsulfonyl groups from the pyrrole nucleus. For example, treatment of the labile Teleocidin alkaloid intermediate **60.3** with Mg in MeOH containing $NH_4Cl$ as a buffer accomplished the cleavage of the phenylsulfonyl group in 78% yield [Scheme 6.60][182]. These studies also demonstrated that the phenylsulfonyl group was stable enough to (a) withstand the halogen–metal exchange of bromopyrrole **60.1** using BuLi and (b) deactivate the pyrrole nucleus sufficiently to allow Friedel–Crafts acylation with acid chlorides in the presence of $BF_3$•$OEt_2$[183].

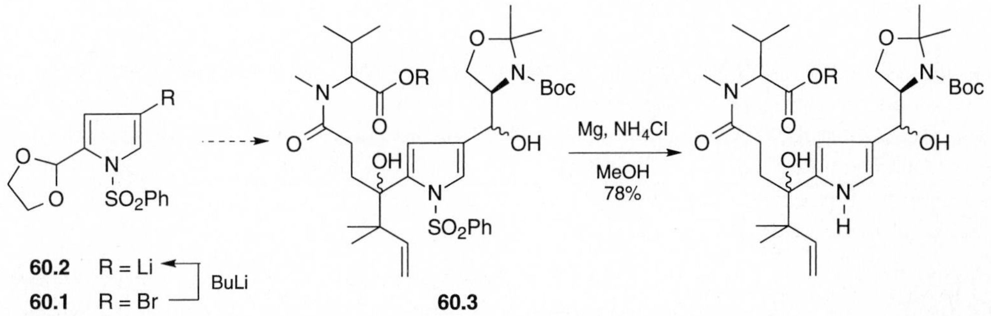
R
O
O
N
SO2Ph
OR
O
N
Boc
N
O
O
OH
OH
N
SO2Ph
Mg, NH4Cl
MeOH
78%
N
H
60.2 R = Li
60.1 R = Br
BuLi
60.3

**Scheme 6.60**

### (ii) Formation

Reaction of the parent heterocycle with arenesulfonyl chorides in the presence of a suitable base constitutes the only widely applied method for the *N*-arylsulfonylation of indoles, pyrroles, imidazoles, etc. We will cite three procedures using indoles to illustrate the variety of conditions. Typical bases include BuLi [Scheme 6.61][177] or a metal hydride [Scheme 6.62][184] in THF, or phase transfer catalysis using NaOH in aqueous $CH_2Cl_2$ [Scheme 6.63][177].

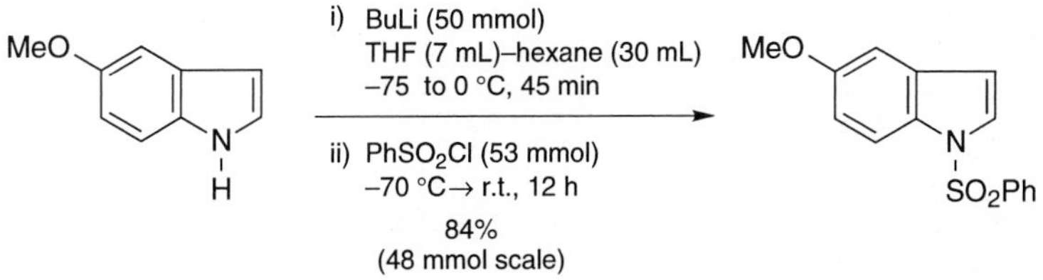
MeO
N
H
i) BuLi (50 mmol)
THF (7 mL)–hexane (30 mL)
–75 to 0 °C, 45 min
ii) PhSO2Cl (53 mmol)
–70 °C→ r.t., 12 h
84%
(48 mmol scale)
MeO
N
SO2Ph

**Scheme 6.61**

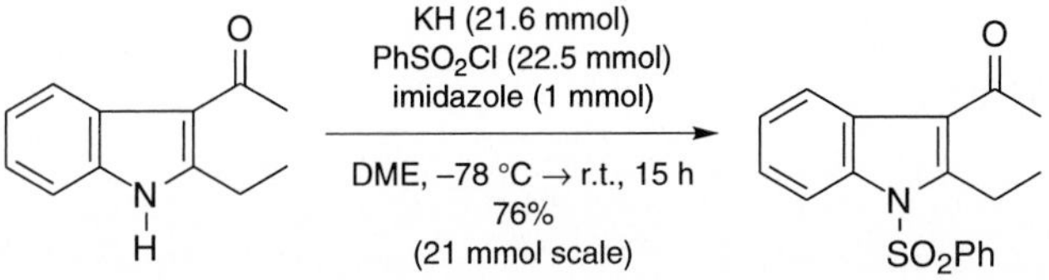
O
N
H
KH (21.6 mmol)
PhSO2Cl (22.5 mmol)
imidazole (1 mmol)
DME, –78 °C → r.t., 15 h
76%
(21 mmol scale)
O
N
SO2Ph

**Scheme 6.62**

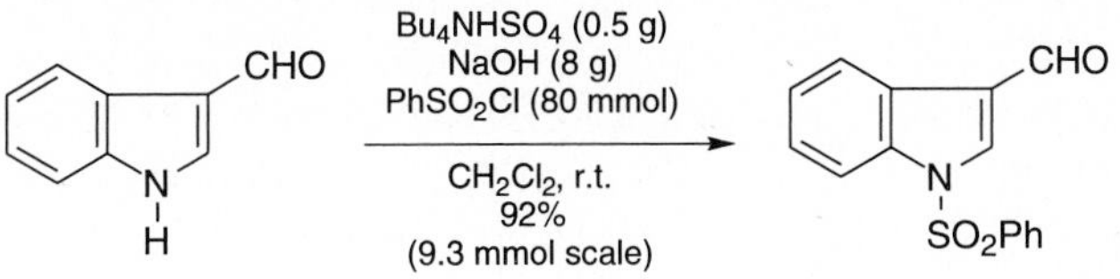
CHO
N
H
Bu4NHSO4 (0.5 g)
NaOH (8 g)
PhSO2Cl (80 mmol)
CH2Cl2, r.t.
92%
(9.3 mmol scale)
CHO
N
SO2Ph

**Scheme 6.63**

## 6.3.2 *N*-Sulfonyl Derivatives of Primary and Secondary Amines

Arylsulfonyl groups are highly effective protecting groups for a wide range of amine derivatives including the amino functions of $\alpha$-amino acids[185]. They were first used in 1915 by Fischer[186], are stable to most reaction conditions, provide a strong chromophore, and have been especially useful when a carboxyl group is destined to undergo reaction with an organometallic reagent. Unfortunately, these protecting groups are troublesome to remove and since hydrolysis requires drastic conditions, reductive methods are usually employed[187].

### (i) Cleavage

Reductive cleavage with Na or Li in liquid ammonia with or without an added proton source is an early and effective method for cleaving arenesulfonamides[188,189] though the yields seldom exceed 80%. A synthesis of an immunomodulatory peptide [Scheme 6.64][190] illustrates the method and is noteworthy for its use of an *N*-tosyl group to facilitate the nucleophilic cleavage of an aziridine. Na–Hg in a protic solvent[191,192] may also be used but the method is obviously less convenient for large scale work.

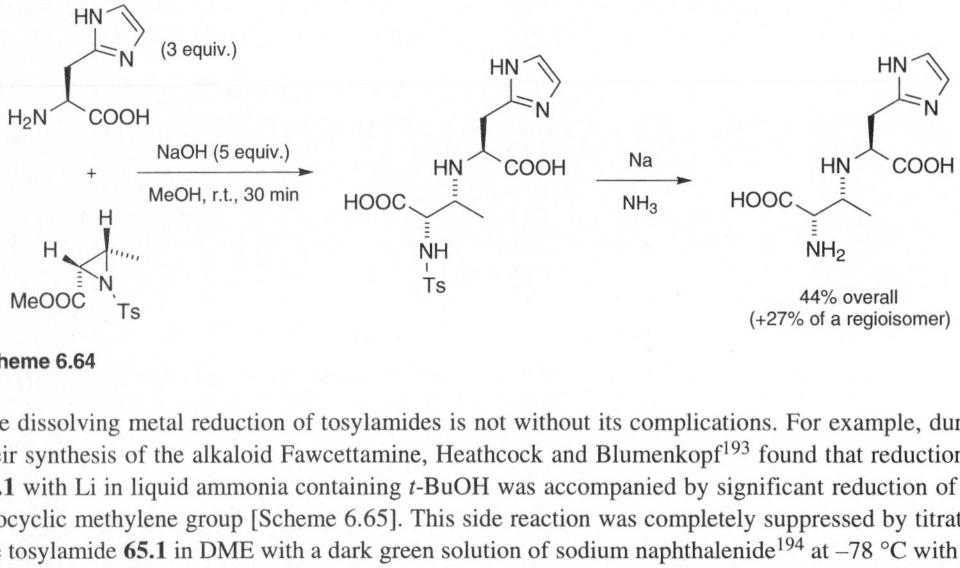

HN
N
(3 equiv.)
$H_2N$
COOH
+
NaOH (5 equiv.)
MeOH, r.t., 30 min
H
H
N
MeOOC
Ts
HN
N
HN
COOH
HOOC
NH
Ts
Na
$NH_3$
HN
N
HN
COOH
HOOC
$NH_2$
44% overall
(+27% of a regioisomer)

**Scheme 6.64**

The dissolving metal reduction of tosylamides is not without its complications. For example, during their synthesis of the alkaloid Fawcettamine, Heathcock and Blumenkopf[193] found that reduction of **65.1** with Li in liquid ammonia containing *t*-BuOH was accompanied by significant reduction of the exocyclic methylene group [Scheme 6.65]. This side reaction was completely suppressed by titrating the tosylamide **65.1** in DME with a dark green solution of sodium naphthalenide[194] at –78 °C with the end point occurring after the addition of approximately 1 equivalent of reducing agent. The method is practical: Scheme 6.66 shows an efficient, large scale reductive cleavage of a tosylamide in the presence of *p*-methoxybenzyl and dithiane groups[153]. Similarly, sodium anthracenide in DME[195] was recently used to cleave a *p*-methoxybenzenesulfonamide during a synthesis of Strychnine[175].

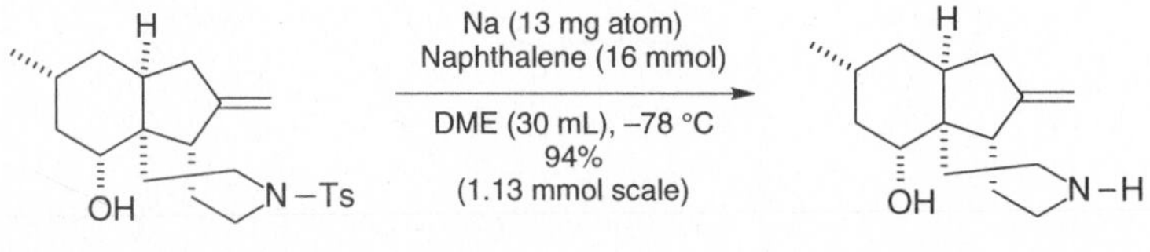

H
OH
N–Ts
Na (13 mg atom)
Naphthalene (16 mmol)
DME (30 mL), –78 °C
94%
(1.13 mmol scale)
H
OH
N–H

**65.1** **65.2**

**Scheme 6.65**

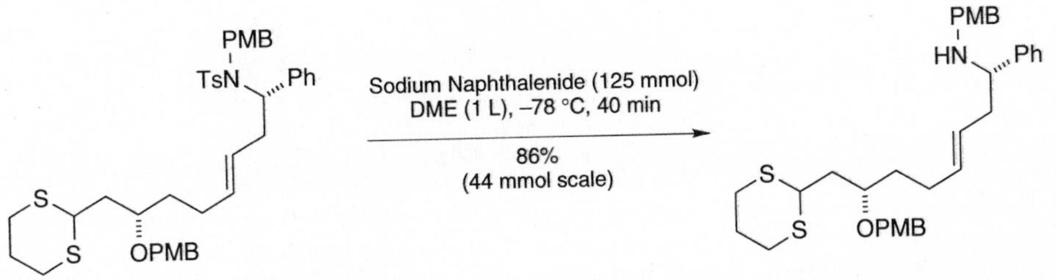
PMB
TsN
Ph
S
S
OPMB
Sodium Naphthalenide (125 mmol)
DME (1 L), –78 °C, 40 min
86%
(44 mmol scale)
PMB
HN
Ph
S
S
OPMB

**Scheme 6.66**

Pennies from heaven are rare, but a recent synthesis of the morphine skeleton benefited from an unusual and fortuitous cyclisation reaction, which occurred during the reductive deprotection of a tosylamide [Scheme 6.67] accomplishing thereby the construction of the fifth and final ring[196].

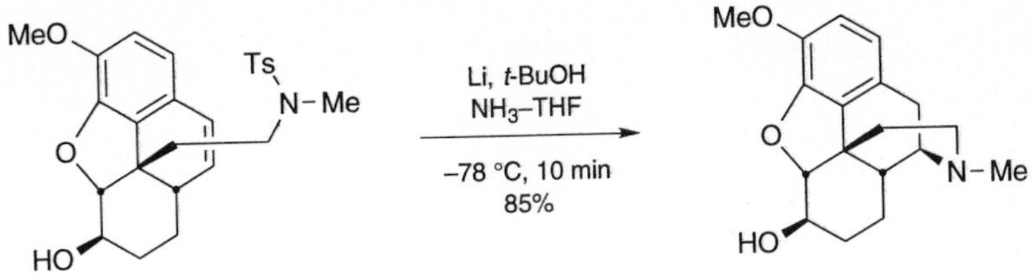
MeO
Ts
N–Me
O
HO
Li, t-BuOH
NH3–THF
–78 °C, 10 min
85%
MeO
O
N–Me
HO

**Scheme 6.67**

In certain cases metal hydrides can be used to cleave sulfonamides but the method is rarely used. Our single example comes from a synthesis of analogues of CC-1065 — an antitumour antibiotic which acts by sequence selective DNA minor groove alkylation [Scheme 6.68][197]. By heating the secondary tosylamide derivative **68.1** with $NaAlH_2(OCH_2CH_2OMe)_2$ in a mixture of DME and toluene at 100 °C for 4 h, the tosyl group was cleaved[198] to give the desired secondary amine **68.2** in 71% yield.

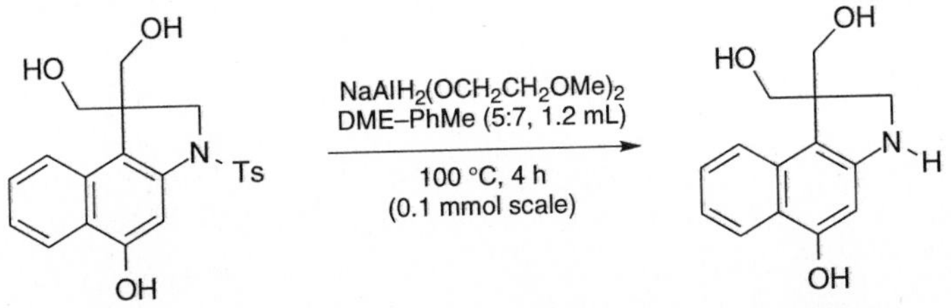
OH
HO
N
Ts
OH
NaAlH2(OCH2CH2OMe)2
DME–PhMe (5:7, 1.2 mL)
100 °C, 4 h
(0.1 mmol scale)
OH
HO
N
H
OH

**68.1** **68.2**

**Scheme 6.68**

Rapoport and co-workers[199] found that HBr in acetic acid (a reductive process)[200,201] gave better yields than $Na–NH_3$ for the removal of *N*-arylsulfonyl groups from $\alpha$-amino acids. A study of substituent effects on the ease of cleavage revealed that alkyl groups on the aryl ring of the sulfonamides lowered the reduction potential; consequently tosyl groups cleaved more readily than phenylsulfonyl, and 2,4-dimethylphenylsulfonyl was better yet. The fine tuning of reduction potential was useful for optimising the cleavage of arenesulfonamides by electrolysis[202] in MeCN using a Hg pool cathode and Pt anode, with phenol as the proton source and 0.1 M $Et_4NBr$ as anolyte. The method employing HBr is illustrated by a transformation in a synthesis of the alkaloid Pilocarpine [Scheme 6.69][203].

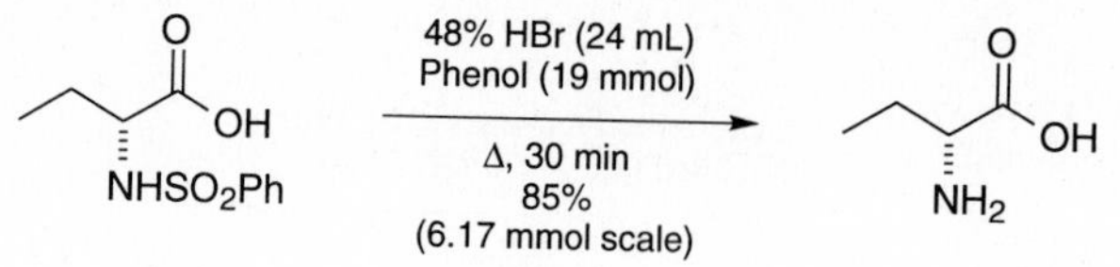

NHSO2Ph
OH
48% HBr (24 mL)
Phenol (19 mmol)
Δ, 30 min
85%
(6.17 mmol scale)
NH2
OH


**Scheme 6.69**

Two groups employed the *p*-methylbenzylsulfonyl group[204] for the synthesis of the antibiotic Actinobolin[205,206]. In the synthesis by Weinreb and co-workers [Scheme 6.70][205] deprotection of a single amino function took place on treatment of a *p*-methylbenzylsulfonamide with neat anhydrous HF in the presence of anisole. A more convenient procedure employs 70% HF in pyridine containing 2 equivalents of anisole[207] to accomplish the deprotection of a mesitylenesulfonamide [Scheme 6.71][208].

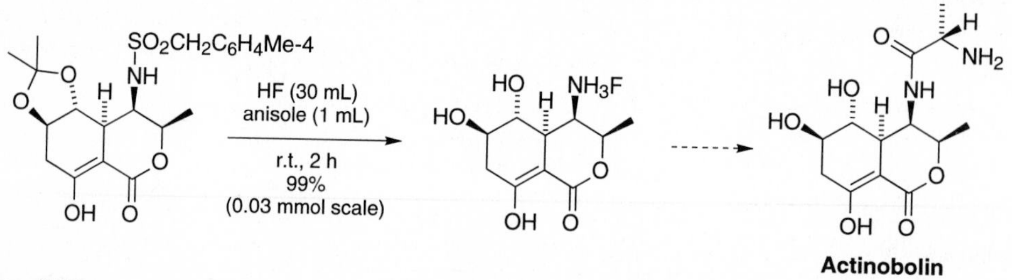

SO2CH2C6H4Me-4
NH
HF (30 mL)
anisole (1 mL)
r.t., 2 h
99%
(0.03 mmol scale)
NH3F
Actinobolin


**Scheme 6.70**

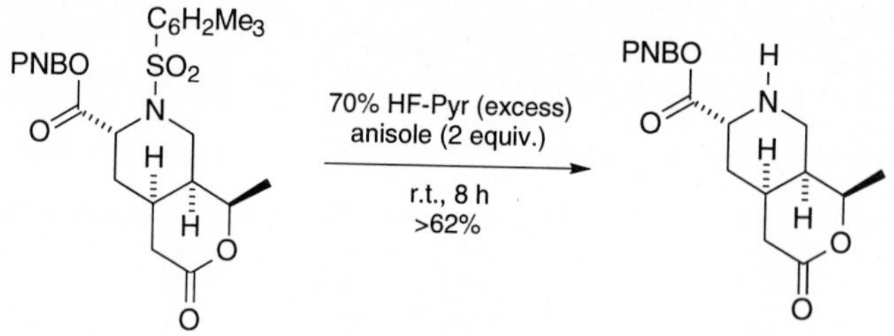

C6H2Me3
SO2
PNBO
70% HF-Pyr (excess)
anisole (2 equiv.)
r.t., 8 h
>62%


**Scheme 6.71**

### (ii) Formation

*N*-Sulfonylation of primary and secondary amines is usually accomplished with the appropriate sulfonyl chloride in the presence of a suitable base. Both anhydrous [Scheme 6.72][188] and aqueous conditions [Scheme 6.73][199] have been employed.

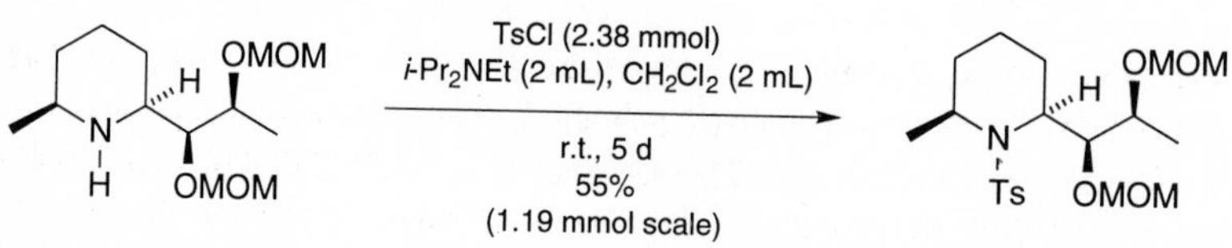

OMOM
TsCl (2.38 mmol)
i-Pr2NEt (2 mL), CH2Cl2 (2 mL)
r.t., 5 d
55%
(1.19 mmol scale)
Ts


**Scheme 6.72**

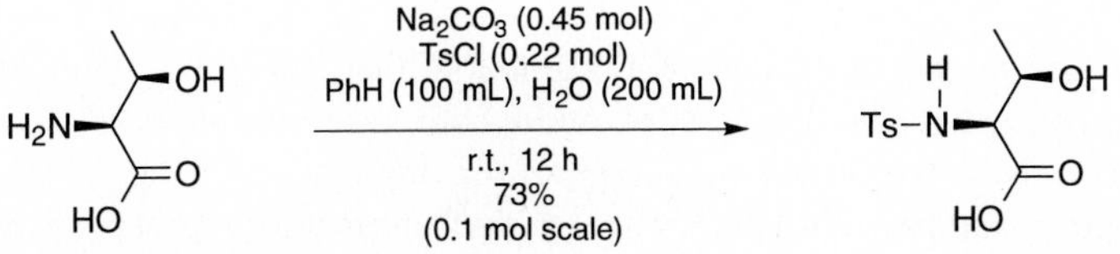
Na2CO3 (0.45 mol)
TsCl (0.22 mol)
PhH (100 mL), H2O (200 mL)
r.t., 12 h
73%
(0.1 mol scale)
H2N
OH
O
HO
Ts–N
H

**Scheme 6.73**

An important consideration in the use of tosyl groups to protect primary amines is the fact that the powerful electron withdrawing effect of the tosyl group greatly increases the acidity of the remaining hydrogen and under basic conditions the resultant sulfonamide may serve as a nucleophile in a competing cyclisation[209] or fragmentation processes[210,211]. However, Schemes 6.74[193] and 6.75[212] illustrate the deliberate use of sulfonyl activation to accomplish cyclisation. The latter reaction is notable for the ease of cyclisation under Hassner *esterification* conditions. Alkylation of tosylamides can be accomplished under Mitsunobu conditions [213,214], or via reaction of tosylates with the cesium derivatives prepared from reaction of sulfonamides with $Cs_2CO_3$. The latter reaction provides an effective synthesis of macrocyclic polyamines[215,216].

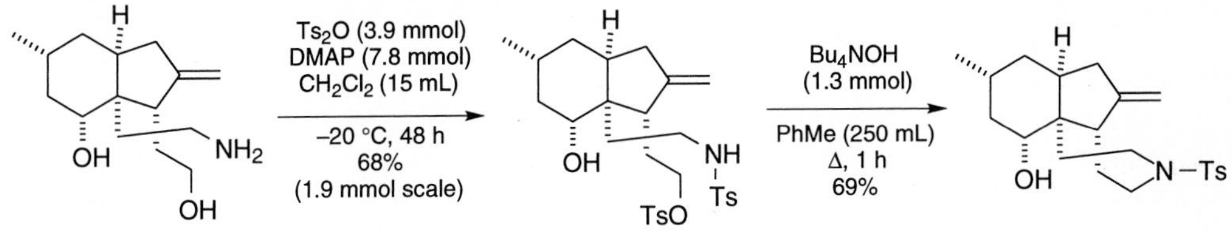
Ts2O (3.9 mmol)
DMAP (7.8 mmol)
CH2Cl2 (15 mL)
–20 °C, 48 h
68%
(1.9 mmol scale)
Bu4NOH
(1.3 mmol)
PhMe (250 mL)
Δ, 1 h
69%
H
OH
NH2
OH
NH
Ts
TsO
N–Ts

**Scheme 6.74**

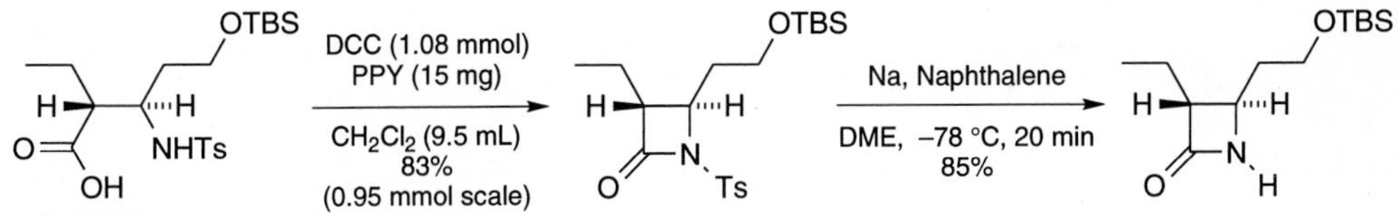
DCC (1.08 mmol)
PPY (15 mg)
CH2Cl2 (9.5 mL)
83%
(0.95 mmol scale)
Na, Naphthalene
DME, –78 °C, 20 min
85%
OTBS
H
O
OH
NHTs
N
Ts

**Scheme 6.75**

**(iii) NMR Data**

$^1$H NMR (tosylamide): $\delta = 7.7$ and 7.3 (2H each, d, $J = 8.1$ Hz), 1.7 (3H, s)

## 6.3.3 β-(Trimethylsilyl)ethanesulfonyl (SES) Derivatives

Our last sulfonyl protecting group deserves to be better known. The β-(trimethylsilyl)ethanesulfonyl group (abbreviated SES)[217] is every bit as stable as the arylsulfonyl groups discussed above: it is impervious to the ravages of trifluoroacetic acid, hot 6 M HCl, $BF_3{\bullet}OEt_2$, or 40% HF in EtOH. Nevertheless, under very specific *non-reductive* conditions, it can be cleaved. The link with β-(trimethylsilyl)ethyl esters, Teoc, and SEM ethers hardly needs comment.

**(i) Cleavage**

We will use a single example to illustrate the potential of the SES group. Weinreb and co-workers planned to adapt their successful synthesis of Actinobolin (see Scheme 6.70 above) to the related antibiotic Bactobolin. Despite the seemingly insignificant difference between the two structures – just one extra methyl group and two chlorines — the *p*-methylbenzylsulfonyl group, intended as protection for the lone amino group in both structures, could not be removed in the closing stages of the Bactobolin synthesis. However, after gentle warming with TBAF, the corresponding *β*-(trimethylsilyl)ethanesulfonamide was succesfully cleaved [Scheme 6.76][205]. CsF in DMF at 95 °C can also be used.

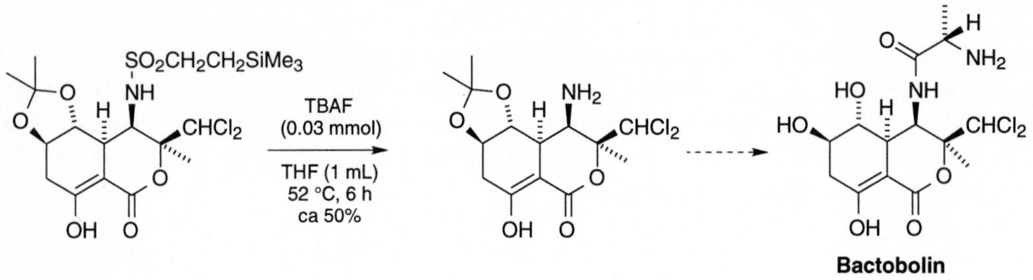

$SO_2CH_2CH_2SiMe_3$ NH H $CHCl_2$ OH O — TBAF (0.03 mmol), THF (1 mL), 52 °C, 6 h, ca 50% → $NH_2$ H $CHCl_2$ OH O - - - → HO HO NH H $NH_2$ H $CHCl_2$ OH O **Bactobolin**

**Scheme 6.76**

**(ii) Formation**

SES amides are usually prepared by reaction of the amine with *β*-(trimethylsilyl)ethanesulfonyl chloride in the presence of a suitable base. The sulfonyl chloride is a stable liquid (bp 60 °C / 0.1 mm Hg) prepared in two steps from vinyltrimethylsilane[217].

## 6.4 *N*-Sulfenyl Derivatives

Sulfenamides* are much more labile than sulfonamides and their use is usually reserved for substrates which require deprotection under exceptionally mild conditions. They are sensitive to acid as well as attack by a host of nucleophiles. Several sulfenyl groups have been recommended for the protection of the amino function including tritylsulfenyl[218-220], *o*-nitrophenylsulfenyl[218,221,222], 2,4-dinitrophenylsulfenyl[223], pentachlorophenylsulfenyl[223], and the very acid sensitive 2-nitro-4-methoxyphenylsulfenyl group[224]. However, only the *o*-nitrophenylsulfenyl (abbreviated Nps) group has found general favor, e.g. in peptide chemistry[216-220] or for the protection of the imide and lactam functions of thymine, uracil, and guanine nucleosides[225].

**(i) Cleavage**

The N–S bond of the *o*-nitrophenylsulfenyl group is readily cleaved by a wide range of nucleophiles including sulfites, thiosulfate, dithionite, hydrogen sulfide, HCN, HI, $HN_3$, thiourea, thioacetamide, thioglycolic acid, mercaptoethanol, thiophenols, dithioerythritol, and 1-hydroxybenzotriazole[226]. Scheme 6.77 illustrates the cleavage of an *o*-nitrophenylsulfenamide in a synthesis of the *β*-lactone Obafluorin using thiocresol in the presence of PTSA[227]. Isolation of the mixed disulfide supports the previously proposed deprotection mechanism[228] involving nucleophilic attack by the aromatic thiol group on the sulfenyl sulfur atom. However, the overall process may be considerably more complicated because reagent concentration appears to be a critical factor for the success of the reaction.

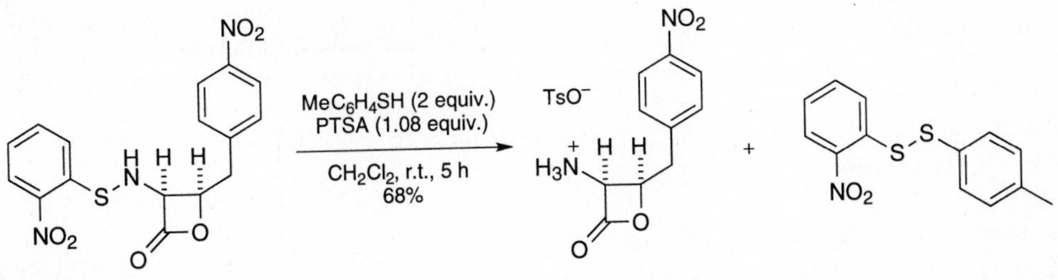

$MeC_6H_4SH$ (2 equiv.)
PTSA (1.08 equiv.)
$CH_2Cl_2$, r.t., 5 h
68%

**Scheme 6.77**

2,2,2-Trifluoro-1,1-diphenylethanesulfenamides [$CF_3C(Ph_2)SNR_2$] have recently been recommended[229] as alternative *N*-protecting groups which are stable to air, $LiAlH_4$, $NaBH_4$, aq. HCl–THF at r.t., 1 M NaOH, $Bu_3SnH$, and $Ac_2O$. They are more stable to acid than tritylsulfenamides owing to the electron withdrawing effect of the $CF_3$ group. However, they are cleaved with Na in liquid ammonia or anhydrous HCl in ether, as shown by a deprotection step of a key intermediate which formerly served[230] in a synthesis of aminoglycoside antibiotics [Scheme 6.78].

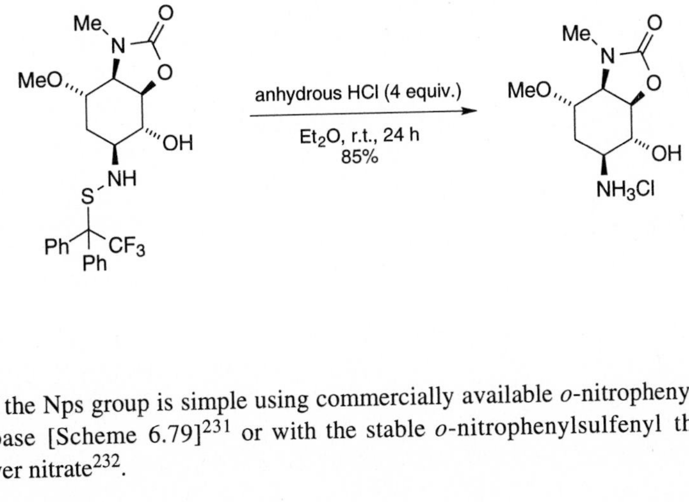

anhydrous HCl (4 equiv.)
$Et_2O$, r.t., 24 h
85%

**Scheme 6.78**

**(ii) Formation**

Introduction of the Nps group is simple using commercially available *o*-nitrophenylsulfenyl chloride and aqueous base [Scheme 6.79][231] or with the stable *o*-nitrophenylsulfenyl thiocyanate in the presence of silver nitrate[232].

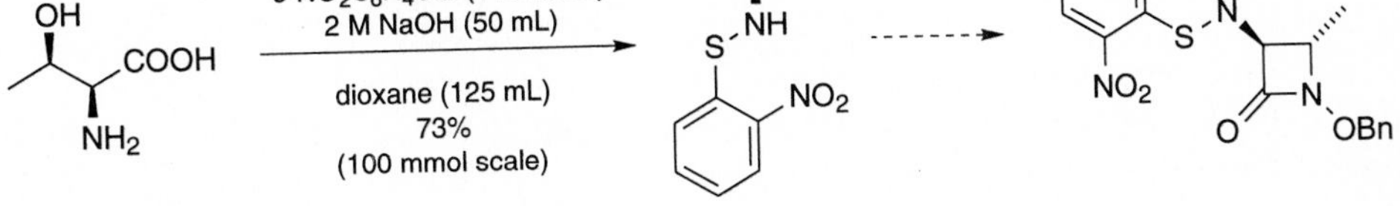

*o*-$NO_2C_6H_4SCl$ (110 mmol)
2 M NaOH (50 mL)
dioxane (125 mL)
73%
(100 mmol scale)

**Scheme 6.79**

## 6.5 *N*-Alkyl Derivatives

All of the foregoing *N*-protecting groups (except sulfenamides) render the nitrogen atom essentially non-basic. In cases where a synthetic endeavour simply requires the suppression of *N*-acylation or proton abstraction from a primary or secondary amine, it may be useful to protect the nitrogen as a tertiary amine.

### 6.5.1 *N,O*-Acetals

*N,O*-Acetals* are extremely sensitive to hydrolysis and therefore seldom used to protect simple amines but MOM[233], BOM[234], and SEM[235,236] groups have been occasionally employed to protect the nitrogen of indoles, imidazoles, and pyrroles [Scheme 6.47]. However, *N,O*-acetals do find use for the simultaneous protection of amino and hydroxyl groups in 2-aminoethanol and 3-aminopropanol derivatives.

**(i) Cleavage**

During their synthesis of Biphenomycin A, Schmidt and co-workers[94] hydrolysed an *N,O*-acetal derivative [Scheme 6.80] by gentle heating in aqueous acetic acid and these conditions are typical of the ease of hydrolysis even when the basicity of the nitrogen atom is diminished by incorporation into a carbamate residue. Proximate electron withdrawing groups increase the stability of oxazolidine carbamates relative to similar systems not so endowed, enabling selective cleavage [Scheme 6.81][237].

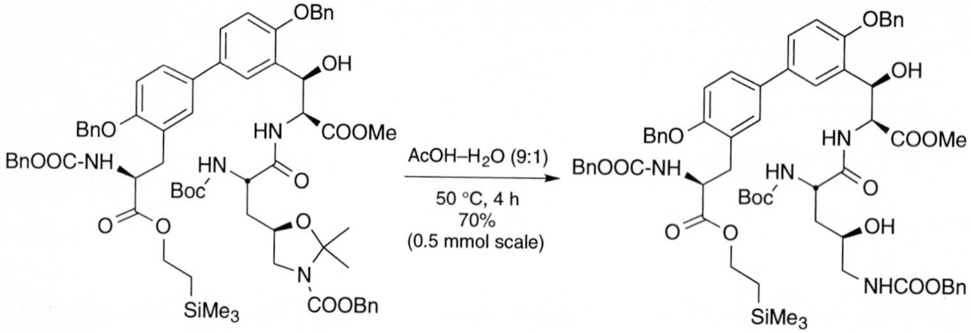

Scheme 6.80

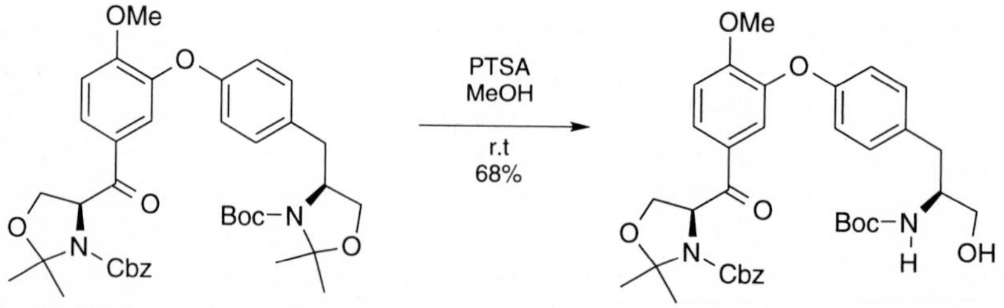

Scheme 6.81

**(ii) Formation**

3-Amino-1-propanols react readily with aldehydes and ketones in the absence of an acid catalyst to give tetrahydro-1,3-oxazines in good yield. The corresponding 5-membered oxazolidines are more difficult to form[238]. The ease of formation and hydrolysis of an *N,O*-acetal was a prominent feature in the resolution of tetrahydropyran-3-carboxaldehyde [Scheme 6.82][239]. Starting with racemic aldehyde **82.1**, a diastereomeric mixture of *N,O*-acetals **82.2** was formed in quantitative yield on condensation with (+)-Ephedrine. The camphorsulfonic acid salt of one of the diastereoisomers crystallised from solution and subsequent hydrolysis to the desired enantiomer (3*R*)-**82.1** occurred on treatment with water.

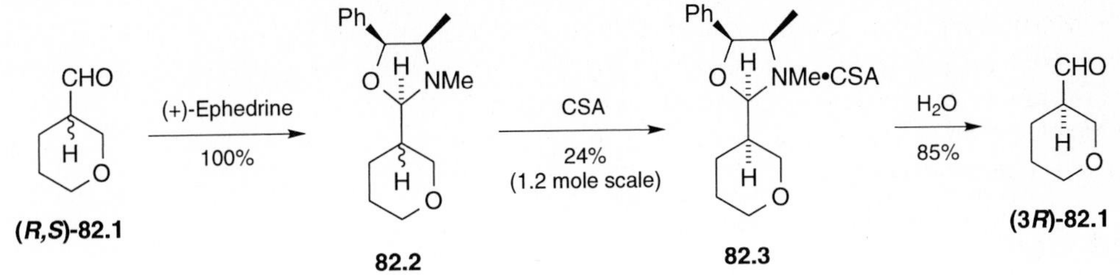

**Scheme 6.82**

Simultaneous protection of a *β*-lactam NH and a side-chain hydroxyl is a common tactic in the *β*-lactam literature[240-245] and Scheme 6.83[246] illustrates how the requisite tetrahydro-1,3-oxazine ring can be created by an acetal exchange process using $BF_3$•$OEt_2$ as the catalyst. In the case of acyl-protected oxazolidines, PTSA-catalysed acetal exchange has been employed[237,247,248].

**Scheme 6.83**

Like their ubiquitous *O,O*-acetal relatives, *N,O*-acetals can exceed the narrow bounds of passive protection and participate in synthetic operations of far greater significance. We will illustrate the point by the use of bicyclic *N,O*-acetals **84.2** and **84.3** prepared from (*R*)-phenylglycinol (**84.1**) for the asymmetric synthesis of 2-substituted piperidines[249] and pyrrolidines[250] [Scheme 6.84]. Both routes have in common the lability of a C–O bond to nucleophilic cleavage as well as the lamentable destruction of stereogenicity in the hydrogenolysis needed to relinquish the desired secondary amine products **84.4** and **84.5**.

Another bicyclic system, the homochiral *N,O*-acetal **85.1**, provides a useful platform for the enantioselective *α*-alkylation of proline [Scheme 6.85][251-254] without the use of a chiral auxiliary, and is another example of "self-reproduction of chirality" (cf. Scheme 5.27). When **85.1** was treated with LDA at –78 °C, a non-racemic enolate resulted which underwent diastereoselective reaction with a variety of electrophiles cis to the *tert*-butyl group (*re*-facial). Subsequent cleavage of the acetal moiety proved unexpectedly refractory: reflux in 48% HBr for 12 h was required to cleave the acetal in **85.2**, whereas the parent acetal **85.1** hydrolysed on brief contact with atmospheric moisture.

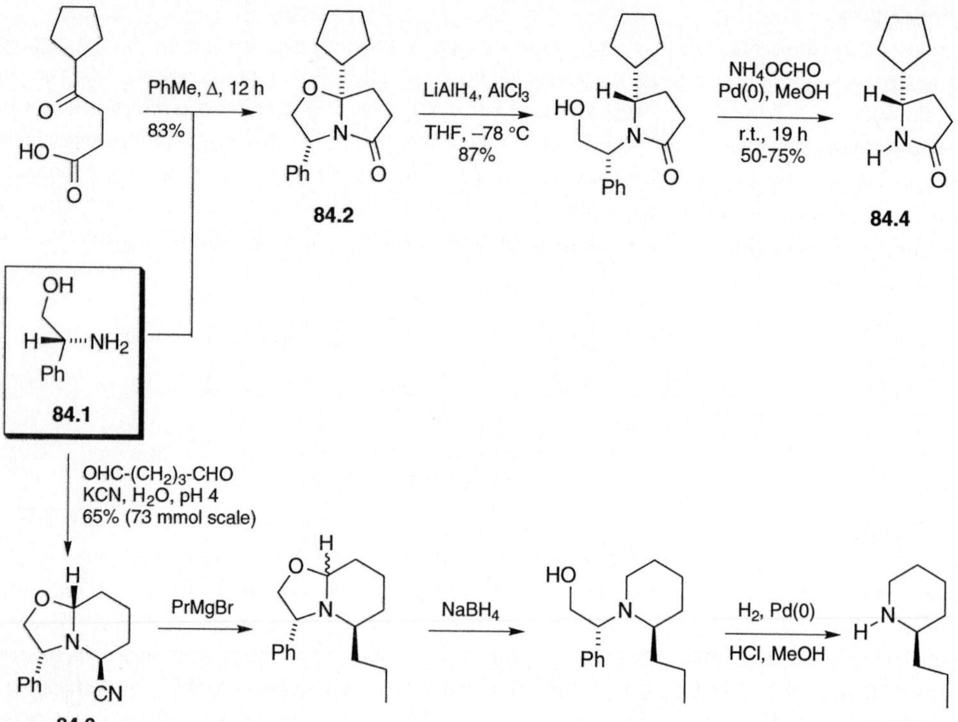

Scheme 6.84

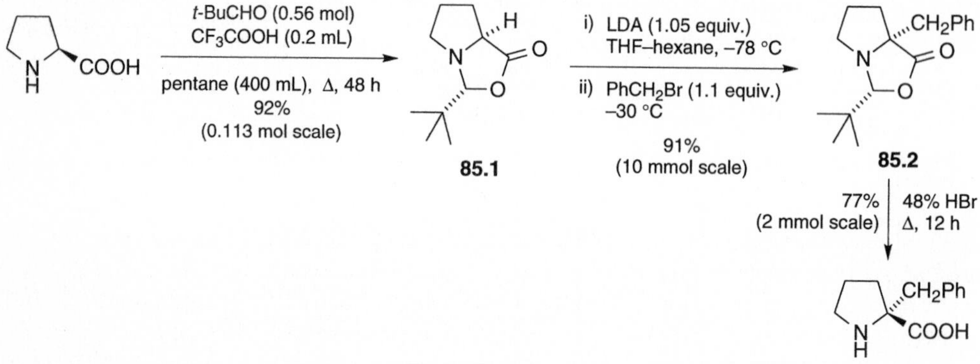

Scheme 6.85

## 6.5.2 Benzyl and Benzhydryl

Benzyl groups are the second fiddles of the amine protection repertoire and they are especally useful when a substrate is to be subjected to powerful organometallic reagents or metal hydrides which might attack a carbamate. Benzylamines are less susceptible to catalytic hydrogenolysis than benzyl ethers, benzyl esters, or Cbz groups and they are not generally cleaved by Lewis acids under preparatively useful conditions. We have already shown that a benzyl ether can be cleaved in the presence of a benzylamine using $BF_3$ in the presence of EtSH [Scheme 2.66].

**(i) Cleavage**

Catalytic hydrogenolysis of benzylamines generally requires higher catalyst loading than the benzyl ethers, and even then higher pressure and/or temperature is often required. For example in the last step of a synthesis of Bao Gong Teng A, an alkaloid from a Chinese herb, which is effective in treating glaucoma, a tertiary *N*-benzyl group required 3 atmospheres of hydrogen at 60 °C to achieve hydrogenolysis [Scheme 6.86][255].

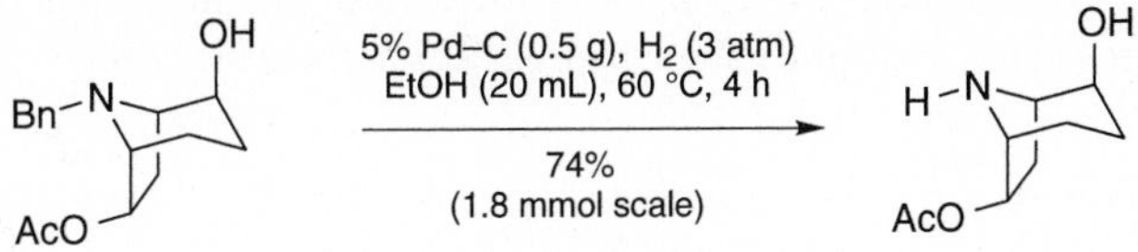

OH
Bn–N
AcO
5% Pd–C (0.5 g), H2 (3 atm)
EtOH (20 mL), 60 °C, 4 h
74%
(1.8 mmol scale)
OH
H–N
AcO

**Scheme 6.86**

The ease of cleavage depends on steric hindrance about the nitrogen atom and substituents on the aromatic ring. A case in point is the hydrogenolysis of homochiral benzylic amine **87.3** which was secured in one highly diastereoselective step[256] by conjugate addition of the homochiral lithium amide derivative **87.1** to the cinnamate ester **87.2**. We found that the less hindered *N*-benzyl was easily removed first when **87.3** was treated with 5% $Pd(OH)_2$–C to give **87.4** in ca. 80% yield. However, removal of the remaining phenylethyl group was much slower and was only achieved efficiently by using 20% $Pd(OH)_2$–C under 3 atmospheres of hydrogen to give the *β*-tyrosine derivative **87.5** in 97% yield. The selectivity of the second reduction is in accord with Bringmann's observation[257] that benzylamines with alkoxy substituents on the aromatic ring are remarkably inert to hydrogenolysis. The effect cannot be simply electronic in nature since amino groups on the aromatic ring *promote* hydrogenolysis in certain cases.

Hydrogenolysis of benzylamines can be facilitated by acid as illustrated in Scheme 6.88[258].

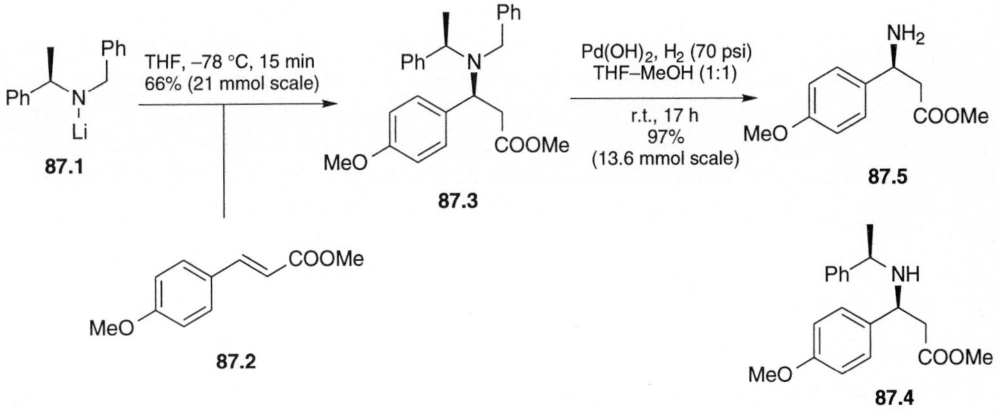

Ph
Ph
N
Li
87.1
THF, –78 °C, 15 min
66% (21 mmol scale)
COOMe
MeO
87.2
Ph
Ph
N
COOMe
MeO
87.3
Pd(OH)2, H2 (70 psi)
THF–MeOH (1:1)
r.t., 17 h
97%
(13.6 mmol scale)
NH2
COOMe
MeO
87.5
Ph
NH
COOMe
MeO
87.4

**Scheme 6.87**

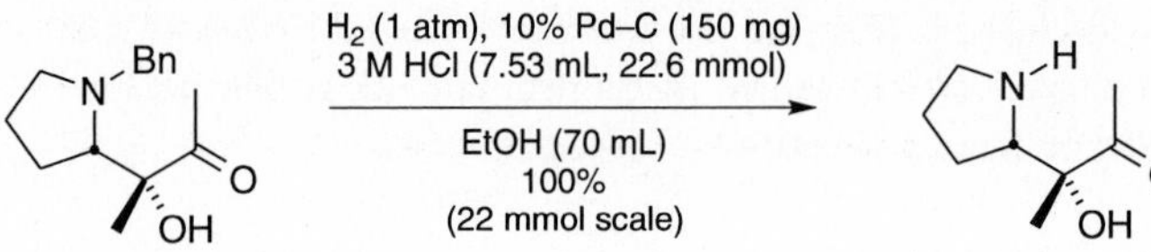

Bn
N
O
OH
$H_2$ (1 atm), 10% Pd–C (150 mg)
3 M HCl (7.53 mL, 22.6 mmol)
EtOH (70 mL)
100%
(22 mmol scale)
H
N
O
OH


**Scheme 6.88**

Transfer catalytic hydrogenation (see Chapter 2) is especially successful for the hydrogenolysis of benzylamines. For example, Jacobi and co-workers[259] required the hydrogenolysis of a benzylamine in the presence of a dithiane moiety for their elegant and concise synthesis of the potent paralytic agent Saxitoxin. Success was eventually achieved using Pd black and formic acid as the hydrogen source [Scheme 6.89]. Benzhydrylamines, which are usually cleaved more easily than benzylamines, have also been removed using transfer catalytic hydrogenation [Scheme 6.90][260].

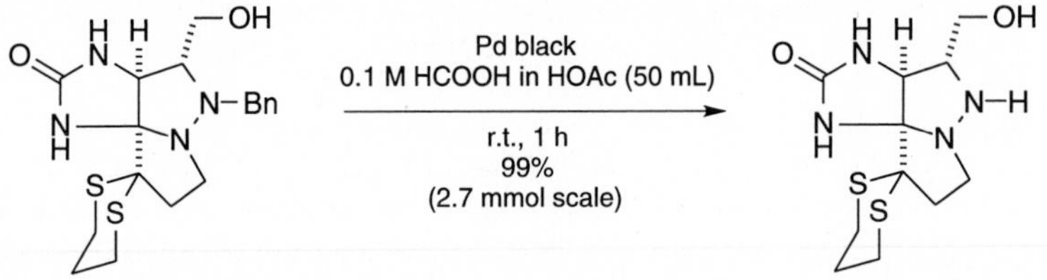

Pd black
0.1 M HCOOH in HOAc (50 mL)
r.t., 1 h
99%
(2.7 mmol scale)
N–Bn
N–H
OH


**Scheme 6.89**

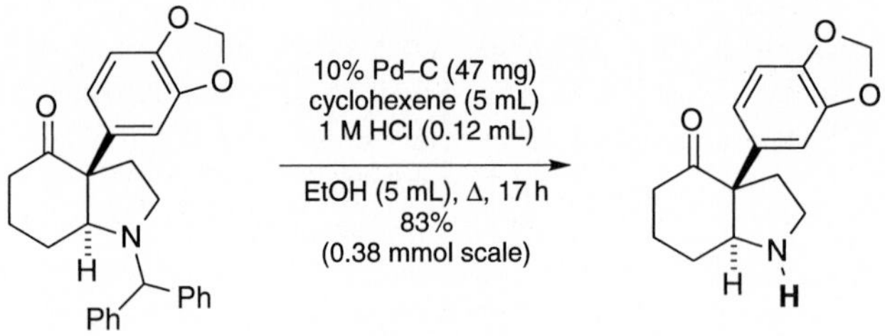

10% Pd–C (47 mg)
cyclohexene (5 mL)
1 M HCl (0.12 mL)
EtOH (5 mL), Δ, 17 h
83%
(0.38 mmol scale)
Ph
Ph


**Scheme 6.90**

*N*-Benzyl groups inert to hydrogenolysis can be cleaved by a two step procedure which is a variant of the classical von Braun degradation. The benzylamine is simply treated with 2,2,2-trichloroethoxycarbonyl chloride[261] or *β*-(trimethylsilyl)ethoxycarbonyl chloride[262] to give the corresponding Troc or Teoc derivatives and benzyl chloride as a by-product. The carbamates are then cleaved in the usual way as discussed in sections 6.2.8 and 6.2.9. The procedure is illustrated in Scheme 6.91[261]. Benzylamines are cleaved about 20 times faster than methylamines.

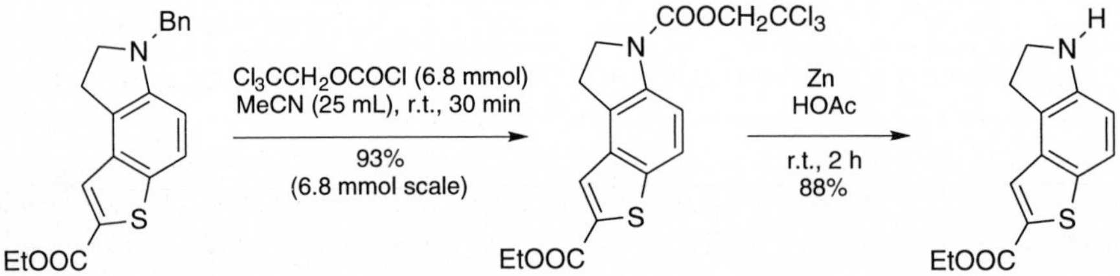

Bn
N
S
EtOOC
$Cl_3CCH_2OCOCl$ (6.8 mmol)
MeCN (25 mL), r.t., 30 min
93%
(6.8 mmol scale)
$COOCH_2CCl_3$
N
S
EtOOC
Zn
HOAc
r.t., 2 h
88%
H
N
S
EtOOC


**Scheme 6.91**

The N–H bond of amides has a $pK_a$ of 15–18 and therefore has an acidity comparable to alcohols. Protection of the N–H bond by *N*-benzylation is a common play. Schemes 6.92 and 6.93 exemplify catalytic hydrogenolyses of *N*-benzylamides which were steps in recent syntheses of Huperzine A[42] (anticholinesterase inhibitor) and AI-77-B[263] (gastroprotective agent) respectively. However, a number of authors have commented that *N*-benzylamide groups can be difficult to hydrogenolyse and therefore Schemes 6.92 and 6.93 probably represent particularly favorable cases.

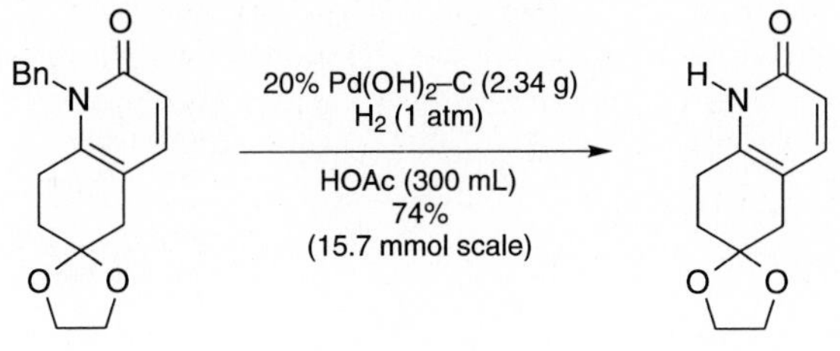

20% $Pd(OH)_2$–C (2.34 g)
$H_2$ (1 atm)
HOAc (300 mL)
74%
(15.7 mmol scale)

**Scheme 6.92**

$H_2N\text{-}NH_2 \cdot H_2O$
5% Pd–C
MeOH, r.t.
95%

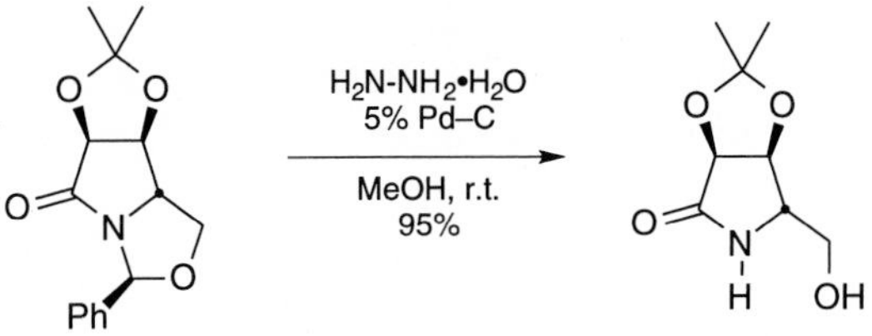

**Scheme 6.93**

An alternative common method uses Na in liquid ammonia to cleave *N*-benzylamides under conditions mild enough to tolerate *β*-lactam rings[264-266], as shown in Scheme 6.94[267].

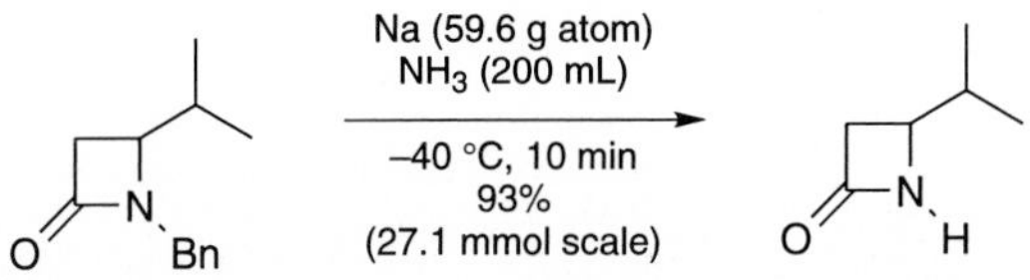

Na (59.6 g atom)
$NH_3$ (200 mL)
–40 °C, 10 min
93%
(27.1 mmol scale)

**Scheme 6.94**

In a practical synthesis of D-*erythro*-sphingosine (**95.4**) [Scheme 6.95], Vasella and co-workers[268] used the Sharpless asymmetric epoxidation reaction to create two stereogenic centres in the epoxy alcohol **95.1**. The sodio derivative of the alcohol was then treated with benzyl isocyanate to generate an adduct, which underwent ring opening with formation of the *N*-benzyl oxazolidinone **95.2**. Dissolving metal reduction under carefully controlled conditions using Li in $EtNH_2$ in the presence of a proton source simultaneously reduced the alkyne and removed the *N*-Bn group to give **95.3**, from which D-*erythro*-sphingosine was obtained by simple hydrolysis.

Taken together, the caprice of hydrogenolysis and the vulnerability of many functional groups under dissolving metal conditions have stimulated the search for substituted benzylic groups which might be cleaved under gentle conditions. *p*-Methoxybenzyl (PMB) groups, whose oxidative cleavage provided an important and reliable deprotection method for alcohols (see section 2.4.3) have also been adapted

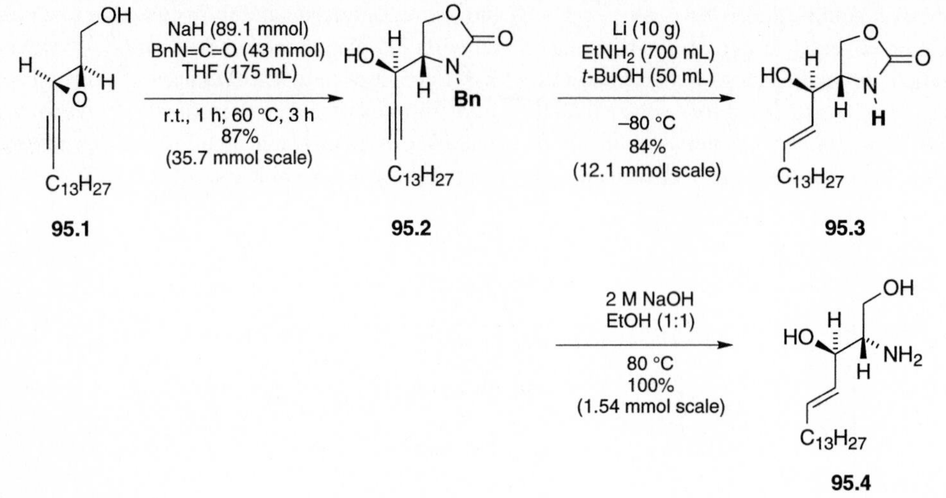

Scheme 6.95

to amide protection. The *O*-PMB group can be selectively removed in the presence of an *N*-PMB group using DDQ but the *N*-PMB group can be severed using ceric ammonium nitrate (CAN) [Scheme 6.96].

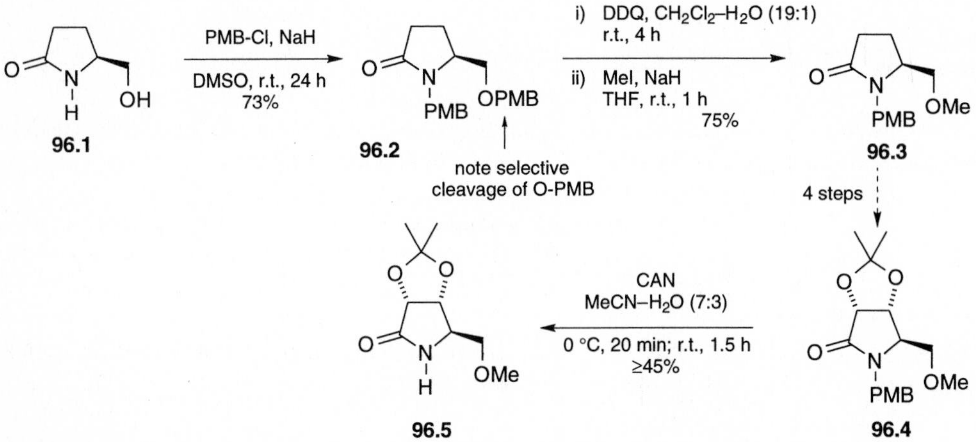

Scheme 6.96

There are a number of variations on the oxidative cleavage of alkoxy-activated arenes which have been particularly useful in $\beta$-lactam chemistry. CAN has been used to deprotect di-(*p*-methoxyphenyl)methyl [Scheme 6.97][269] and *p*-methoxyphenyl groups[270-274] [Scheme 6.98][275] whereas potassium peroxydisulfate has been extensively employed for the cleavage of 2,4-dimethoxybenzyl groups[276,277] [Scheme 6.99][278].

There are two final methods for cleaving electron rich benzyl groups which deserve illustration. The first method simply involves treating a PMB-protected amide with neat trifluoroacetic acid as shown in the last step of Smith's synthesis of Hitachimycin [Scheme 6.100][153]. A similar cleavage of a 2,4-

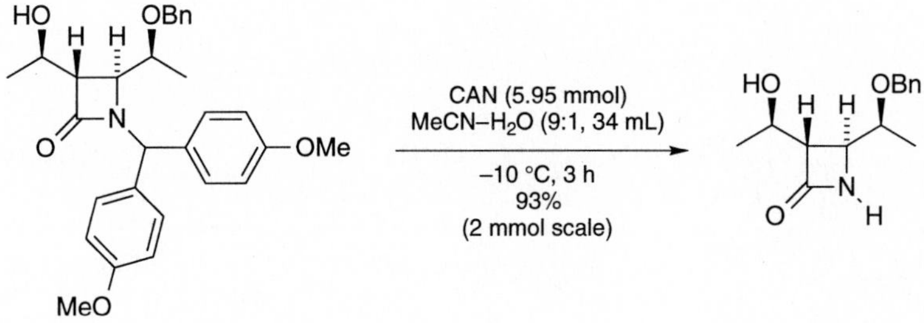

CAN (5.95 mmol)
MeCN–$H_2O$ (9:1, 34 mL)
−10 °C, 3 h
93%
(2 mmol scale)

**Scheme 6.97**

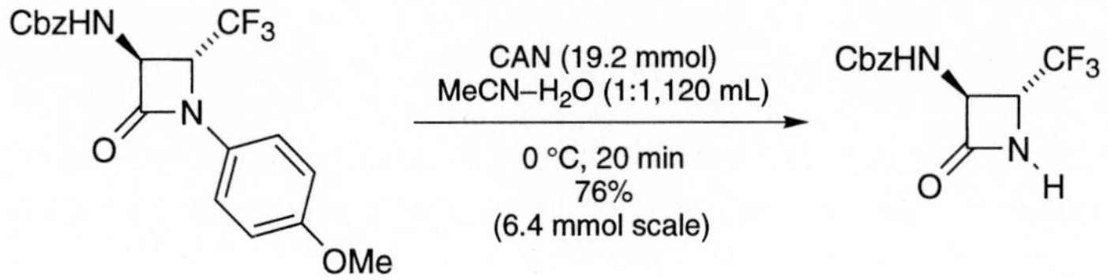

CAN (19.2 mmol)
MeCN–$H_2O$ (1:1,120 mL)
0 °C, 20 min
76%
(6.4 mmol scale)

**Scheme 6.98**

$K_2S_2O_6$ (40 mmol)
$KH_2PO_4$ (80 mmol)
MeCN–$H_2O$ (1:1,160 mL)
85–90 °C, 6 h
65%
(20 mmol scale)

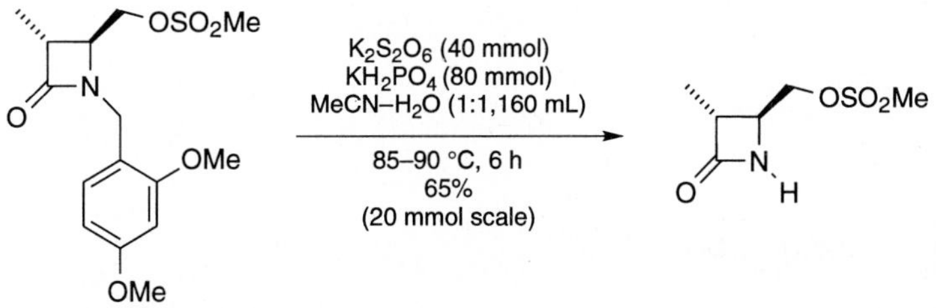

**Scheme 6.99**

dimethoxybenzyl group terminated DeShong's synthesis of Tirandamycin B[279]. The second method, taken from a synthesis of Latrunculin[164] [Scheme 6.101], is a carbanion-mediated oxidative deprotection of non-enolisable benzylated amides[280-282]. *N*-Benzyl or PMB amides react with *t*-BuLi to generate dipole stabilised benzylic carbanions which can be oxidised with either molecular oxygen or MoOPh ($MoO_5$•HMPA•pyridine). The resulting hemiaminals collapse releasing benzaldehyde and the free amine. The Latrunculin synthesis is also noteworthy for the cleavage of an *N*-PMB group from a thiazolidinone using CAN.

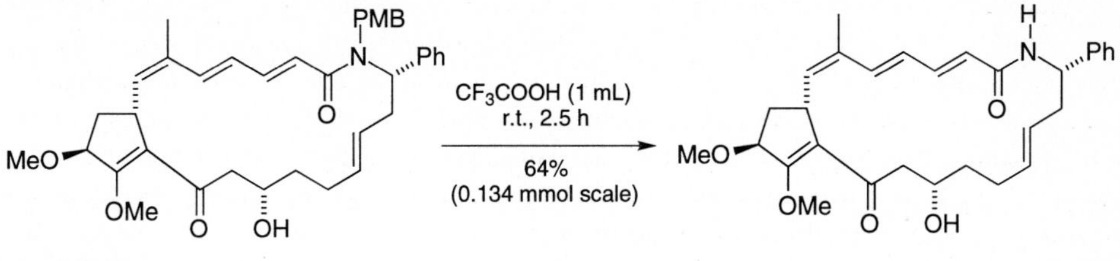

$CF_3COOH$ (1 mL)
r.t., 2.5 h
64%
(0.134 mmol scale)

**Scheme 6.100**

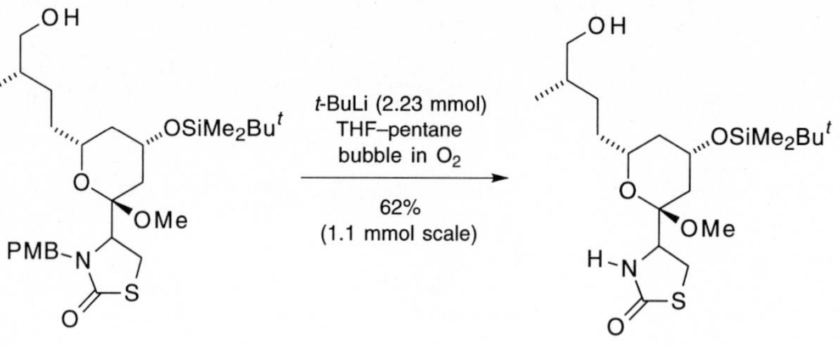
OH
OSiMe2But
O
OMe
PMB
N
S
O
t-BuLi (2.23 mmol)
THF–pentane
bubble in O2
62%
(1.1 mmol scale)
OH
OSiMe2But
O
OMe
H
N
S
O

Scheme 6.101

**(ii) Formation**

Alkylation of amines and amides with benzylic halides is an early and useful method for the protection of amines and amides. Primary amines can alkylate twice to give the *N,N*-dibenzyl derivative but severe steric congestion prevents quaternisation [Scheme 6.102][188].

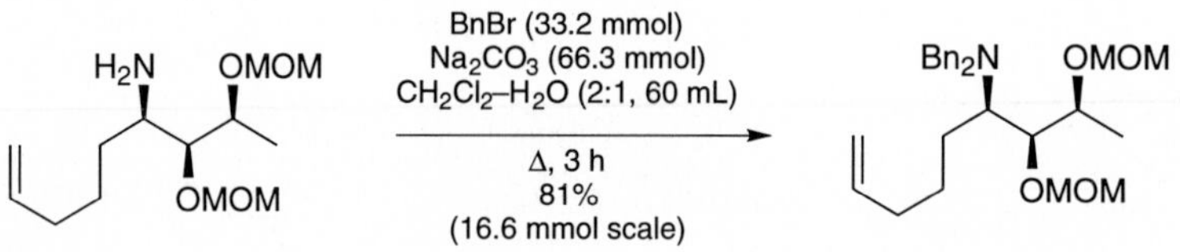
H2N
OMOM
OMOM
BnBr (33.2 mmol)
Na2CO3 (66.3 mmol)
CH2Cl2–H2O (2:1, 60 mL)
Δ, 3 h
81%
(16.6 mmol scale)
Bn2N
OMOM
OMOM

Scheme 6.102

Although amide anions are ambident nucleophiles, they tend to alkylate preferentially at nitrogen. Sodium hydride is the usual base and dipolar aprotic solvents (DMSO or DMF) are used to secure a favorable rate as shown in Scheme 6.103[164].

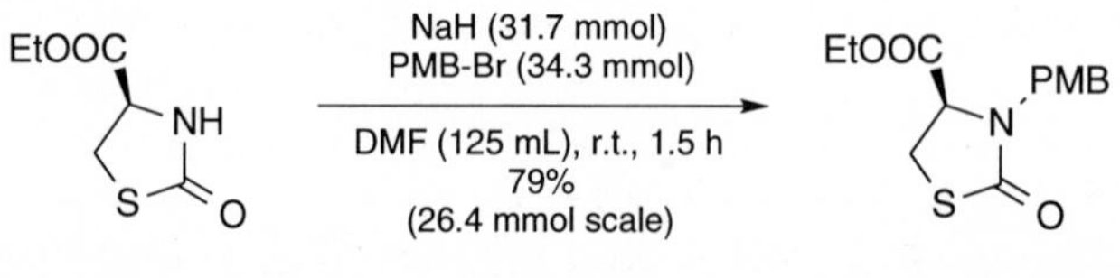
EtOOC
NH
S
O
NaH (31.7 mmol)
PMB-Br (34.3 mmol)
DMF (125 mL), r.t., 1.5 h
79%
(26.4 mmol scale)
EtOOC
PMB
N
S
O

Scheme 6.103

Reductive alkylation of amines* is especially convenient for the large scale monoalkylation of amines, and the method has been widely applied to the introduction of Bn and PMB protecting groups. There are two steps: first the primary amine and benzaldehyde or anisaldehyde are heated in benzene or toluene with azeotropic removal of water to give an imine, which is then subsequently reduced either catalytically or, more often, with a metal hydride. Either $NaBH_4$ or $NaBH_3CN$ can be used. Although $NaBH_3CN$ is more expensive, it has the advantage that all common functionalities including aldehydes and ketones are impervious to attack at the slightly acidic pH 5 required for reduction[283]. Consequently in some cases both imine formation and reduction can be conducted *in situ.* Scheme 6.104[153] illustrates the formation of a PMB protected amine by the reductive alkylation method.

During a synthesis of the alkaloid (–)-Ancistrocladine, Bringmann and co-workers[284] devised an exceptionally simple method for the asymmetric synthesis of primary amines, which very effectively combines reductive alkylation with hydrogenolysis to accomplish chirality transfer. The three step

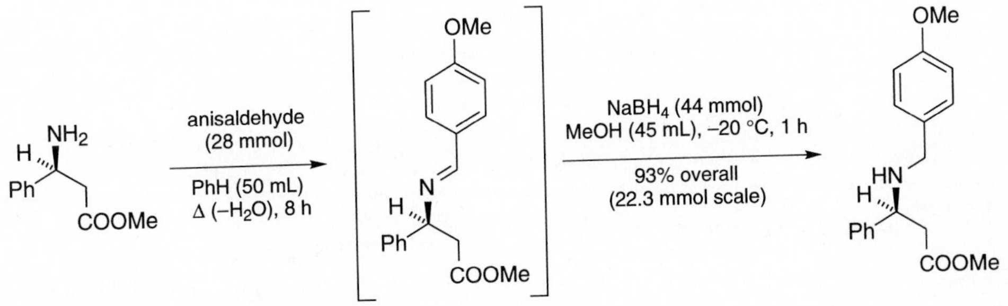

Scheme 6.104

procedure [Scheme 6.105] begins with imine formation between ketone **105.1** and (*S*)-phenethylamine, which is cheap and commercially available in bulk quantities. The imine intermediate is catalytically reduced over W2 Raney Ni at elevated pressure to give secondary amine **105.2** in 92% d.e. Finally hydrogenolysis selectively destroys the C–N bond derived from (S)-phenethylamine[257] to give amine **105.3** in 84% overall yield.

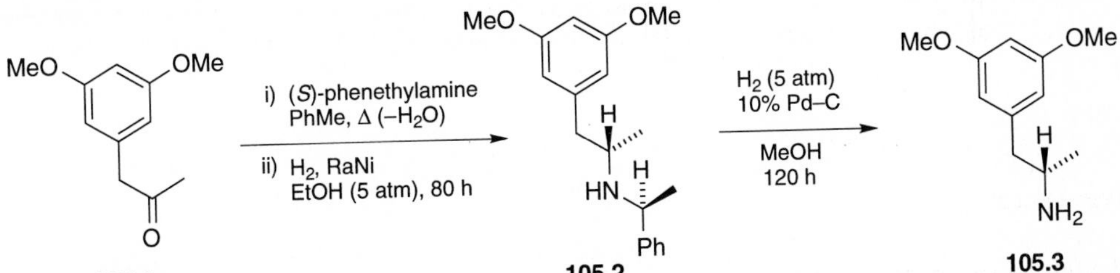

Scheme 6.105

**(iii) NMR Data**
$^1$H NMR: $\delta = 3.7$–$3.8$ (2H, s, $ArCH_2N$)
$^{13}$C NMR:
*t*-Bu-NHCH$_2$Ph $\delta = 47.2$ ($Ar\mathbf{C}H_2$), 50.6 ($\mathbf{C}Me_3$), 29.15 ($C\mathbf{Me}_3$)
*i*-Pr-NHCH$_2$Ph $\delta = 48.15$ ($Ar\mathbf{C}H_2$), 51.7 ($\mathbf{C}HMe_2$), 23.0 ($CH\mathbf{Me}_2$)
*n*-Bu-NHCH$_2$Ph $\delta = 54.1$ ($ArCH_2$), 49.2 ($NCH_2$), 32.3 ($NCH_2CH_2$), 20.5 ($NCH_2CH_2CH_2$), 14.0 (Me)
Me$_2$NCH$_2$Ph $\delta = 64.4$ ($ArCH_2$), 45.3 (Me)

## 6.5.3 Trityl and 9-Phenylfluorenyl

The trityl group has already made a brief appearance as a protector for hydroxyl functions (section 2.4.4) and it has also been used for primary amines and amides. Sheehan employed the trityl group in his synthesis of Penicillin G, and it is occasionally used to protect amino acids, though its steric bulk and high acid lability is detrimental to peptide coupling.

**(i) Cleavage**
The trityl group has recently been recommended[285] for the protection of side chain carboxamide functions in asparagine and glutamine from which it can be readily removed by $CF_3COOH$, though it survives strong aqueous mineral acids. By way of contrast, trityl derivatives of amines are cleaved in

80% acetic acid. Like the trityl ethers, trityl amines are hydrogenolysed very slowly. Both the formation and cleavage of trityl amines is illustrated in Scheme 6.106[286].

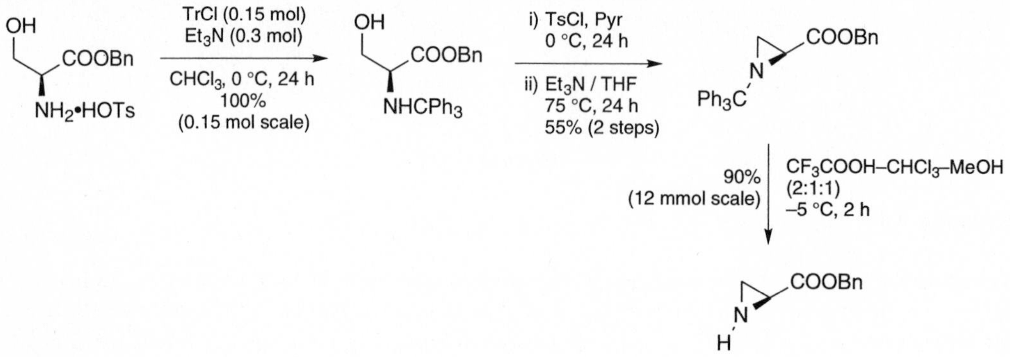

OH COOBn $NH_2$•HOTs — TrCl (0.15 mol), $Et_3N$ (0.3 mol), $CHCl_3$, 0 °C, 24 h, 100% (0.15 mol scale) → OH COOBn $NHCPh_3$ — i) TsCl, Pyr 0 °C, 24 h; ii) $Et_3N$ / THF 75 °C, 24 h; 55% (2 steps) → COOBn N $Ph_3C$ — $CF_3COOH–CHCl_3–MeOH$ (2:1:1) −5 °C, 2 h; 90% (12 mmol scale) → COOBn N H

**Scheme 6.106**

The enhanced acid-lability of monomethoxytrityl groups was an asset exploited in a recent synthesis of the sensitive Azinomycin derivative shown in Scheme 6.107[287].

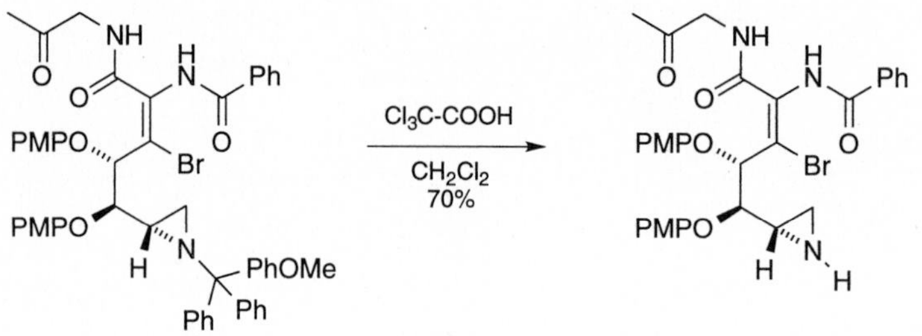

$Cl_3C$-COOH, $CH_2Cl_2$, 70%

**Scheme 6.107**

The 9-phenylfluoren-9-yl group (abbreviated PhFl) has been exploited extensively by Rapoport and co-workers for the protection of primary and secondary amines. Its hydrophobicity, steric bulk, and ease of introduction are similar to the trityl group, but the PhFl group is about 6000 times more stable to acid than the trityl group owing to the antiaromatic character of the fluorenyl carbocation[288]. *N*-(9-Phenylfluoren-9-yl)-$\alpha$-amino aldehydes and ketones derived from the corresponding $\alpha$-amino acids maintain configurational stability during silica gel chromatography, Wittig olefination, or reaction with organometallic reagents. The advantages of steric bulk are evident in the selective alkylation of aspartate shown in Scheme 6.108[289], which is taken from a synthesis of Vincamine, a cerebral vasodilator. The 9-phenylfluorenyl group was used to insulate the $\alpha$-position of aspartate derivative **108.1** by obstructing proton abstraction leading to exclusive formation and alkylation of the $\beta$-enolate to give **108.2**. Neither racemisation nor *N*-alkylation competed. However, conversion of the chloride in **108.2** to the iodide led to intramolecular *N*-alkylation to give **108.3**. A second $\beta$-alkylation, this time with EtI, followed by hydrogenolysis of the PhFl group gave the requisite pipecolate derivative **108.5**. These transformations are but one facet of the versatility of PhFl protection in the elaboration of amino acid derivatives, and further applications to the synthesis of other unusual amino acids[290-293], $\alpha$-amino aldehydes[290,294-296], indole alkaloids[289,297,298], and $\alpha$-alkyl branched carboxylic acids[296] have been amply recorded by the Rapoport group.

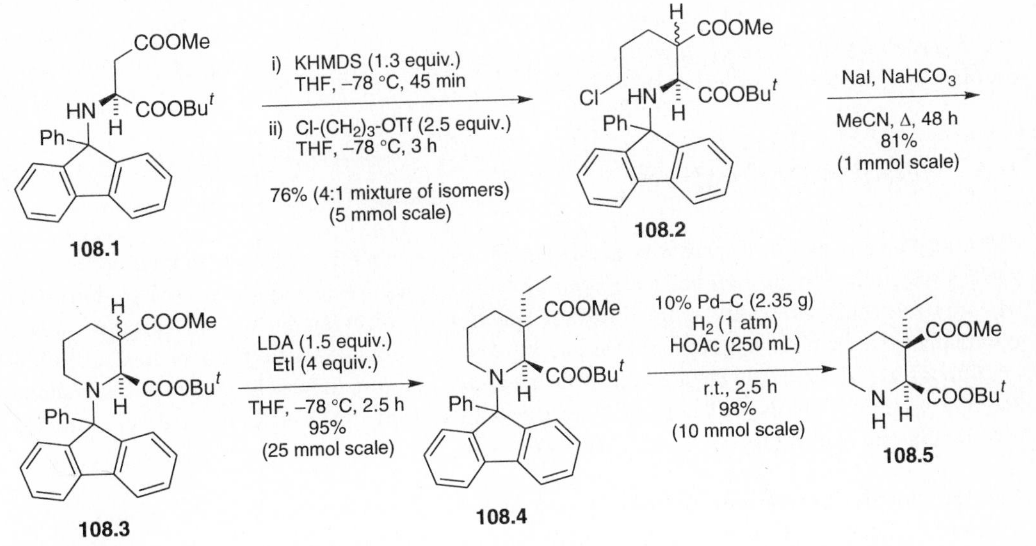
COOMe
HN
COOBu$^t$
Ph
H
108.1
i) KHMDS (1.3 equiv.)
THF, –78 °C, 45 min
ii) Cl-(CH$_2$)$_3$-OTf (2.5 equiv.)
THF, –78 °C, 3 h
76% (4:1 mixture of isomers)
(5 mmol scale)
H
COOMe
Cl
HN
COOBu$^t$
Ph
H
108.2
NaI, NaHCO$_3$
MeCN, Δ, 48 h
81%
(1 mmol scale)
H
COOMe
N
COOBu$^t$
Ph
H
108.3
LDA (1.5 equiv.)
EtI (4 equiv.)
THF, –78 °C, 2.5 h
95%
(25 mmol scale)
COOMe
N
COOBu$^t$
Ph
H
108.4
10% Pd–C (2.35 g)
H$_2$ (1 atm)
HOAc (250 mL)
r.t., 2.5 h
98%
(10 mmol scale)
COOMe
N
COOBu$^t$
H H
108.5

Scheme 6.108

The PhFl group can also be cleaved by acidolysis with $CF_3COOH$[213,291,294,297,298] or dissolving metal reduction [Scheme 6.109][290].

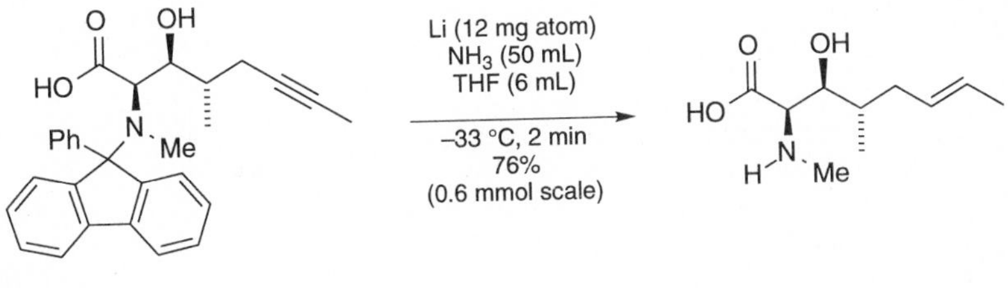
O
OH
HO
N
Ph
Me
Li (12 mg atom)
NH$_3$ (50 mL)
THF (6 mL)
–33 °C, 2 min
76%
(0.6 mmol scale)
O
OH
HO
H
N
Me

Scheme 6.109

### (ii) Formation

PhFl groups are introduced by reaction of the free amine with 9-bromo-9-phenylfluorene in the presence of $K_3PO_4$ and $Pb(NO_3)_2$ which is present to scavenge bromide ions. Free hydroxyl and carboxyl groups cannot be present and reagents, solvents, and glassware should be scrupulously dried. Scheme 6.110[289] illustrates the method applied to the aspartic acid derivative, which served as starting

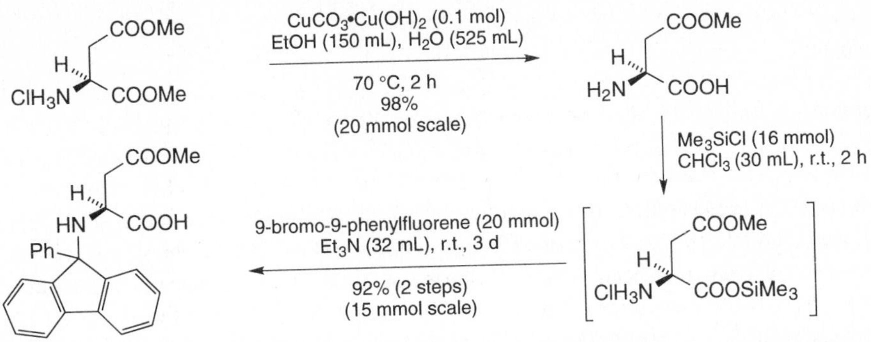
COOMe
H
ClH$_3$N
COOMe
CuCO$_3$•Cu(OH)$_2$ (0.1 mol)
EtOH (150 mL), H$_2$O (525 mL)
70 °C, 2 h
98%
(20 mmol scale)
COOMe
H
H$_2$N
COOH
Me$_3$SiCl (16 mmol)
CHCl$_3$ (30 mL), r.t., 2 h
COOMe
H
ClH$_3$N
COOSiMe$_3$
9-bromo-9-phenylfluorene (20 mmol)
Et$_3$N (32 mL), r.t., 3 d
92% (2 steps)
(15 mmol scale)
COOMe
H
HN
COOH
Ph

Scheme 6.110

material for the sequence described in Scheme 6.108. The method illustrates the selective hydrolysis of the $\alpha$-carboxyl function of dimethyl aspartate using Cu(II) and the temporary protection of the resultant carboxylic acid as the labile TMS ester.

## 6.6 *N*-Silyl Derivatives

The high acid and moisture sensitivity of silylamines has been a major obstacle to their use in amino group protection. Primary amines have been protected as their *tert*-butyldiphenylsilyl derivatives whose steric bulk militates against further reaction to form an *N,N*-disilylamine[299]. *tert*-Butyldiphenylsilylamines have remarkable stability towards strongly basic conditions (e.g., 20% KOH in refluxing MeOH) but they cannot even withstand brief exposure to 80% HOAc at r.t. If silylamines are used at all in synthesis, they are usually prepared under anhydrous conditions, and used for one or two steps before their flimsy services are dispensed with.

The C-alkylation of glycine provides an expeditious method for synthesising $\alpha$-amino acids. Obviously, glycine itself is unsuitable as a substrate owing to the acidity of the amino protons. The protection of glycine so that enolate chemistry can be performed is the first of two situations in which *N*-silyl protection has proved generally useful. A synthesis of the desferri form of $\delta_1$ Albomycin [Scheme 6.111][300] illustrates the principal and bears witness to the high hydrolytic lability of the N–Si bond. Further examples of protecting group innovation in the *C*-alkylation of glycine are given below in sections 6.7.2 and 6.7.3.

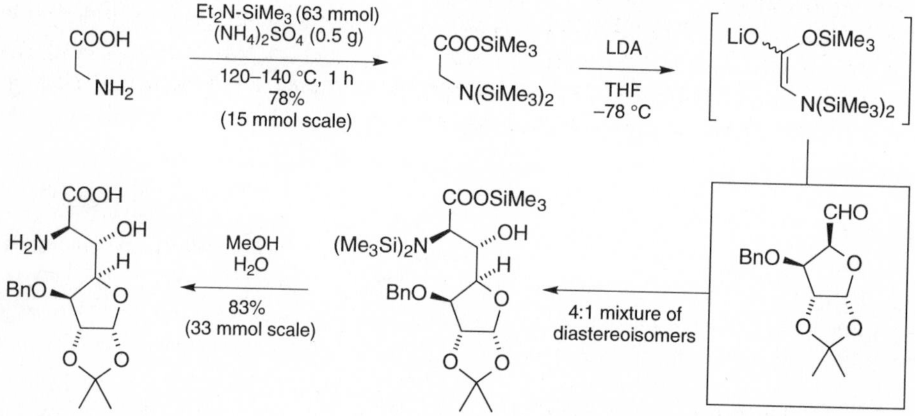

**Scheme 6.111**

For purposes of synthetic efficiency, amino groups are better protected as their "STABASE" (2,2,5,5-tetramethyl-1-aza-2,5-disilacyclopentane) derivatives[301]. STABASE derivatives are stable to strong bases such as LDA or alkyl-lithium reagents but they are susceptible to ready cleavage by oxygen nucleophiles (e.g. water) so they, too, are only really useful for temporary protection of amino groups. Such is their lability that even flash chromatography is inadvisable. A premier application of STABASE derivatives has been in the synthesis of 3-amino-$\beta$-lactams via ester enolate-imine condensations* wherein glycine is once again the target for protection [Scheme 6.112][302]. The stereochemistry of the reaction strongly depends on the metal cation present in the reaction medium.

When the lithium enolate **112.2** of STABASE-protected glycine **112.1** condenses with the *N*-trimethylsilylimine of thiophene-2-carboxaldehyde **112.3**, the *cis*-$\beta$-lactam **112.4** is the major product (cis:trans = 96:6) whereas addition of zinc ions to enolate **112.2** prior to addition of the imine gave the *trans*-$\beta$-lactam **112.5** predominantly.

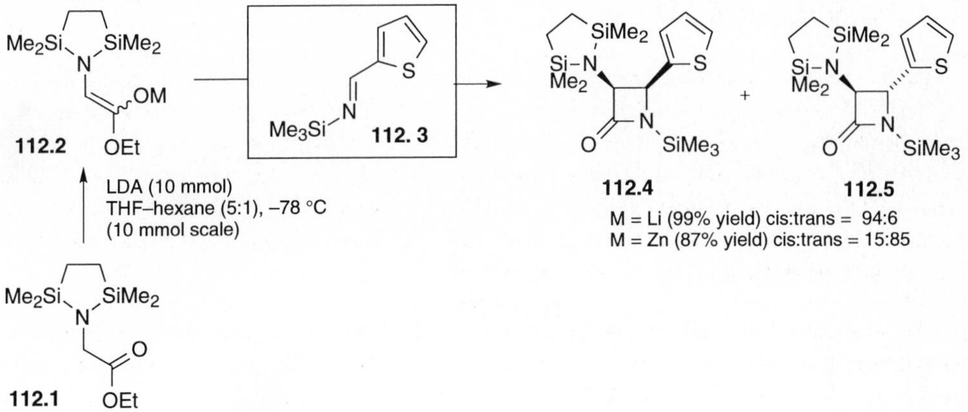

**Scheme 6.112**

For the sake of greater stability towards acid and chromatography, the benzostabase protecting group was developed by Davis and co-workers[303]. Benzostabase derivatives are prepared by the Pd(0)-catalysed dehydrogenative silylation of a primary amine using 1,2-bis(dimethylsilyl)benzene[304] [Scheme 6.113], whereas STABASE derivatives are prepared by heating the amine with 1,2-bis[(dimethylamino)dimethylsilyl]ethane[305] in the presence of a catalytic amount of $ZnI_2$ [Scheme 6.114][306]. Benzostabase derivatives of anilines are appreciably more stable than the parent STABASE analogues but aliphatic amines show negligible improvement.

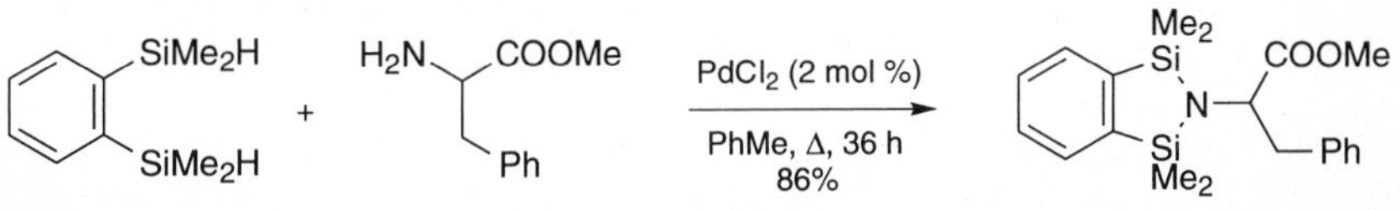

**Scheme 6.113**

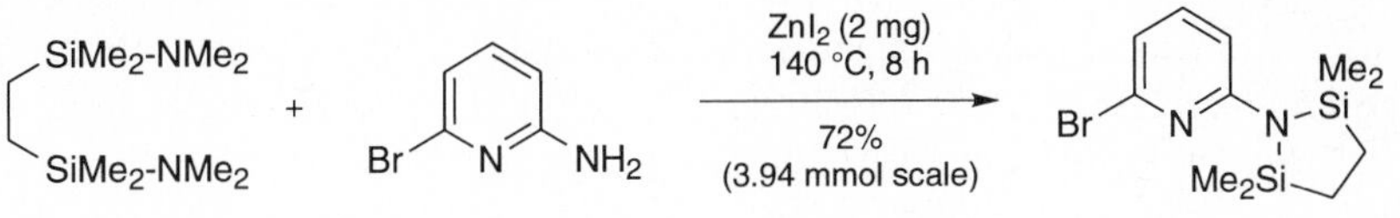

**Scheme 6.114**

The Merck synthesis of Thienamycin[307] provides a particularly cogent example of the value of *N*-silyl groups for the synthesis of $\beta$-lactams. In order to create a $\beta$-lactam ring [Scheme 6.115], the *N*-trimethylsilyl derivative **115.2** of dibenzyl aspartate **115.1** was treated with *t*-BuMgCl to give **115.3**. Subsequent desilylation took place on brief treatment with mild acid to give **115.4**. In order to prime the structure for further elaboration, the $\beta$-lactam N–H was protected as a *tert*-butyldimethylsilyl derivative as shown.

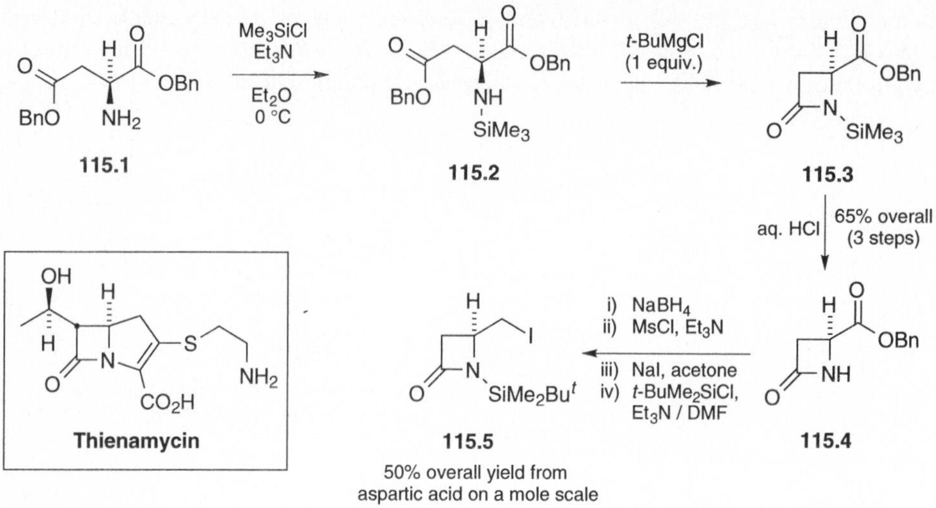

Scheme 6.115

The *tert*-butyldimethylsilyl group in **115.5** was stable enough to survive a number of further transformations deploying reagents such as LDA, $KBH(s\text{-}Bu)_3$, $HgCl_2$, and $H_2O_2$ leading up to intermediate **116.1** [Scheme 6.116] from which the *tert*-butyldimethylsilyl group was removed using methanolic HCl as a prelude to closure of the 5-membered ring. Other reagents which have been employed to remove silyl groups from a $\beta$-lactam nitrogen are TBAF[308], HF–pyridine in MeCN[309], 5% HF in MeCN[310], and 0.04 M KF in MeOH[311].

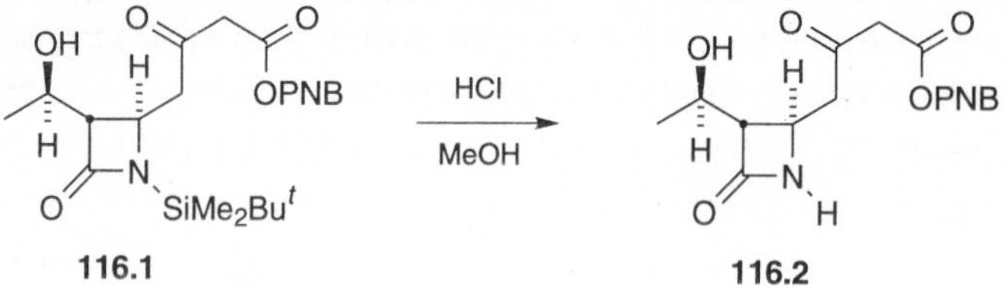

Scheme 6.116

Trialkylsilyl groups are gaining in popularity for the temporary protection of the common nitrogen heterocycles such as pyrroles and indoles[312], whose lower basicity compared with aliphatic alkylamines leads to greater hydrolytic and chromatographic stability. For example the TIPS group served as a useful block for the pyrrole N–H during a synthesis of 3-aryl pyrroles from pyrrole itself [Scheme 6.117][313]. The steric bulk of the TIPS group helped direct electrophilic iodination of pyrrole away from the normal 2-position to the 3-position. Similarly, protection of the indole N–H with a TBS group was required in order to accomplish the *N*-methylation of *N*-Boc-tryptophan [Scheme 6.118][314].

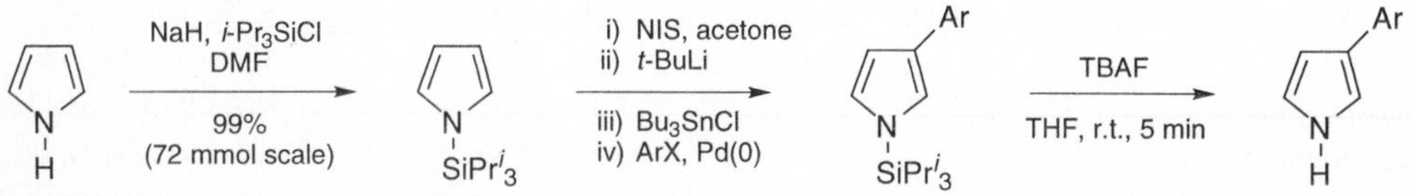

Scheme 6.117

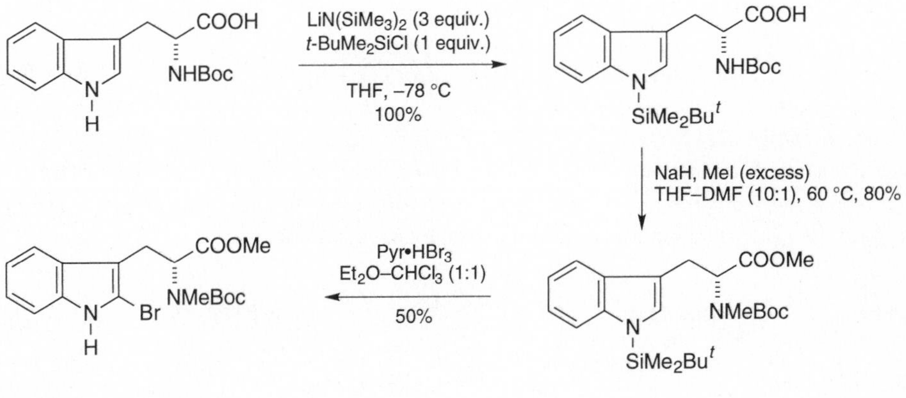

COOH
NHBoc
N
H
LiN(SiMe3)2 (3 equiv.)
t-BuMe2SiCl (1 equiv.)
THF, –78 °C
100%
COOH
NHBoc
N
SiMe2Bu^t
NaH, MeI (excess)
THF–DMF (10:1), 60 °C, 80%
COOMe
NMeBoc
N
SiMe2Bu^t
Pyr•HBr3
Et2O–CHCl3 (1:1)
50%
COOMe
NMeBoc
N
H
Br

**Scheme 6.118**

# 6.7 Imine Derivatives

The double bond of the imine function allows for the simultaneous protection for both N–H bonds of a primary amine. Imines are generally stable towards strongly basic conditions but they are labile to aqueous acid.

## 6.7.1 *N*-Silyl Imines

Perhaps the most important chemical role that *N*-silyl imines have performed to date has been as substrates for the synthesis of *β*-lactams via ester enolate condensation[315]. The high hydrolytic lability of the *N*-trimethylsilyl adducts ensures rapid and efficient deprotection on aqueous workup as illustrated in Scheme 6.119[316]. *N*-Silylimines are prepared by the reaction of lithium hexamethyldisilazide with an aldehyde[317].

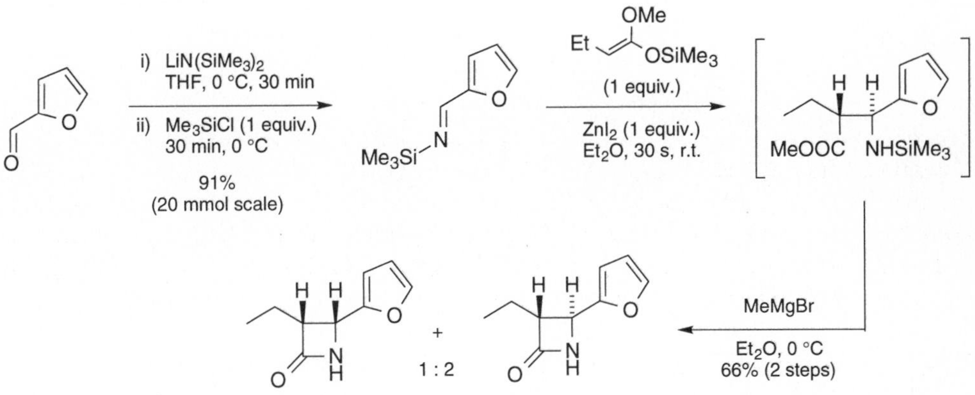

O
O
i) LiN(SiMe3)2
THF, 0 °C, 30 min
ii) Me3SiCl (1 equiv.)
30 min, 0 °C
91%
(20 mmol scale)
Me3Si
N
O
OMe
Et
OSiMe3
(1 equiv.)
ZnI2 (1 equiv.)
Et2O, 30 s, r.t.
H H
O
MeOOC NHSiMe3
MeMgBr
Et2O, 0 °C
66% (2 steps)
H H
O
N
H
O
+
1 : 2

**Scheme 6.119**

## 6.7.2 *N*-Bis(methylthio)methyleneamines

*N*-Bis(methylthio)methyleneamines were first introduced by Hoppe and Beckmann[318] for the synthesis of $\alpha$-amino acids from glycine and in recent years asymmetric variants have come into prominence as shown in Scheme 6.120[208,319]. The *N*-Bis(methylthio)methyleneamines are prepared by reaction of a primary amine with carbon disulfide in the presence of $Et_3N$ and MeI to form a methyl dithiocarbamate derivative, which is then *S*-alkylated with a second equivalent of MeI in the presence of $K_2CO_3$ as the base. Deprotection is accomplished by aqueous acid.

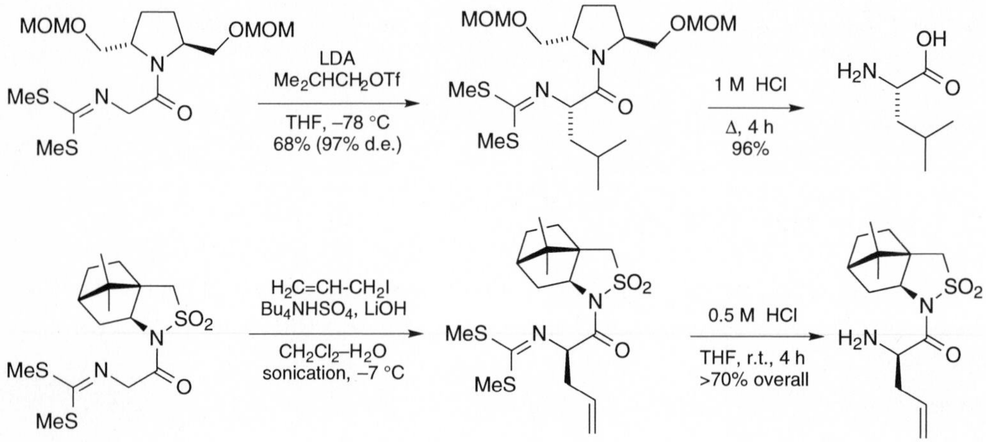

MOMO OMOM N MeS N O MeS — LDA, $Me_2CHCH_2OTf$ / THF, –78 °C, 68% (97% d.e.) → MOMO OMOM N MeS N O MeS — 1 M HCl / Δ, 4 h, 96% → OH $H_2N$ O

SO2 N MeS N O MeS — $H_2C{=}CH{-}CH_2I$, $Bu_4NHSO_4$, LiOH / $CH_2Cl_2$–$H_2O$, sonication, –7 °C → SO2 N MeS N O MeS — 0.5 M HCl / THF, r.t., 4 h, >70% overall → SO2 N $H_2N$ O

**Scheme 6.120**

## 6.7.3 *N*-Diphenylmethyleneamines

### (i) Cleavage

*N*-Diphenylmethyleneamines are valuable for the protection of amino groups in the synthesis and elaboration of $\alpha$-amino acids. For example, O'Donnell and Bennett recently showed that the prochiral

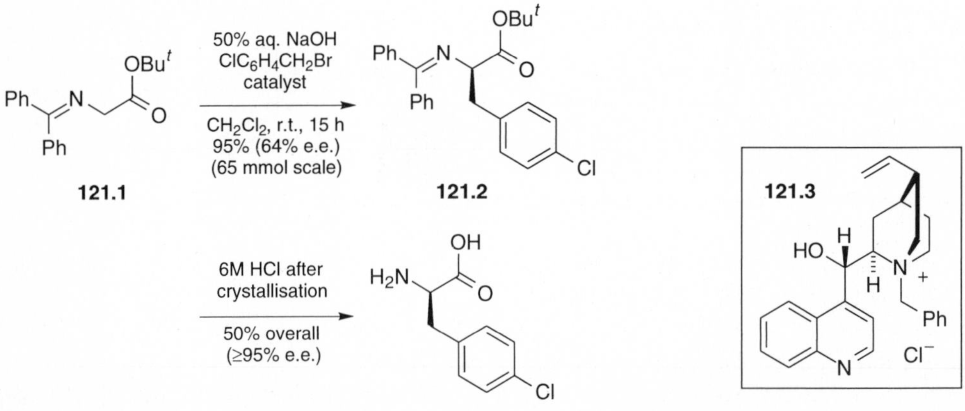

Ph N Ph OBu$^t$ O **121.1** — 50% aq. NaOH, $ClC_6H_4CH_2Br$, catalyst / $CH_2Cl_2$, r.t., 15 h, 95% (64% e.e.) (65 mmol scale) → Ph N Ph OBu$^t$ O Cl **121.2**

— 6M HCl after crystallisation / 50% overall (≥95% e.e.) → $H_2N$ OH O Cl

**121.3** HO H H N + Ph N $Cl^-$

**Scheme 6.121**

*N*-diphenylmethylene derivative of glycine (**121.1**) underwent asymmetric alkylation under phase transfer catalysis using a quaternised Cinchona alkaloid (**1213**) as the catalyst [Scheme 6.121][320]. Although the e.e. of the alkylation was modest (64%), the racemate easily crystallised allowing isolation of the desired (*R*)-enantiomer **121.2** of *p*-chlorophenylalanine in greater than 90% e.e. Deprotection was achieved by simple hydrolysis with aqueous HCl.

Sequential catalytic hydrogenation of the imine bond followed by hydrogenolysis of the resultant *N*-benzhydrylamine is an alternative method of deprotection which was used recently in a solid phase synthesis of *O*-glycosyl cyclic Enkephalin analogues [Scheme 6.122][321]. Note the comprehensive deprotection of 4 benzyl ethers, a benzhydryl ester, and an *N*-diphenylmethyleneamine in a single operation. *N*-Diphenylmethyleneamine derivatives of serine have also been applied to the synthesis of Sphingosines[322].

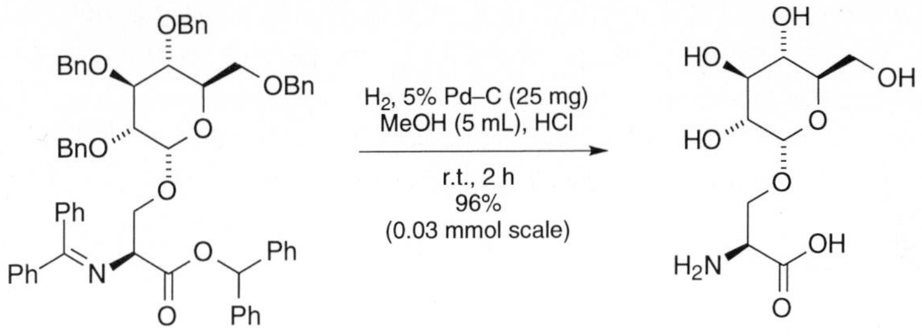

$H_2$, 5% Pd–C (25 mg)
MeOH (5 mL), HCl
r.t., 2 h
96%
(0.03 mmol scale)

Scheme 6.122

**(ii) Formation**

*N*-Diphenylmethyleneamines are easily prepared by gently heating the primary amine in $CH_2Cl_2$ with one equivalent of benzophenone imine [Scheme 6.123][321,323]. The equilibrium in this exchange process is driven to the right by the formation of gaseous ammonia.

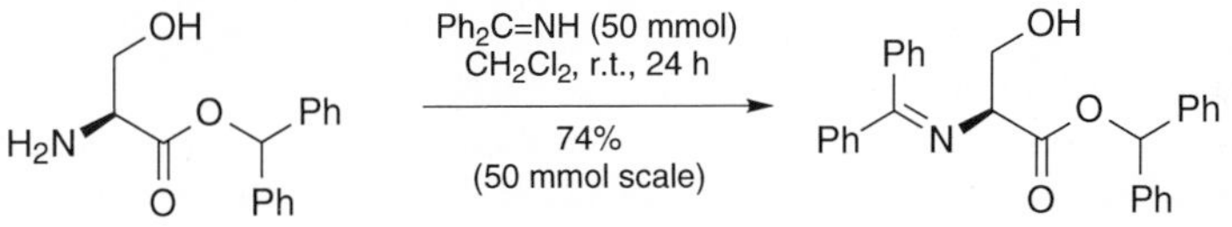

$Ph_2C{=}NH$ (50 mmol)
$CH_2Cl_2$, r.t., 24 h
74%
(50 mmol scale)

Scheme 6.123

# 6.8 Reviews

## 6.8.1 General Reviews on Amino Protecting Groups

1 Protection of the Amino Group. Wolman, Y. In *The Chemistry of the Amino Group*, Patai, S., Ed.; Wiley: New York, 1968; Chapter 11.
2 Protection of N–H Bonds and $NR_3$. Barton, J. W. In *Protective Groups in Organic Chemistry*; McOmie, J. F. W., Ed.; Plenum: London, 1973; Chapter 2.
3 Protection for the Amino Group. Greene, T.W.; Wuts, P. G. M. *Protective Groups in Organic Synthesis,* 2nd ed.; Wiley: New York, 1991; Chapter 7.

4 Protecting Groups. Kunz, H.; Waldmann, H. In *Comprehensive Organic Synthesis*; Trost, B. M.; Fleming, I., Eds.; Pergamon Press: Oxford, 1991; Chapter 6.3.1.

### 6.8.2 Reviews Concerning *N*-Protection in the Synthesis of $\alpha$-Amino Acids

Section 1.6.2 contains a list of references on the synthesis of peptides, glycopeptides, and oligonucleotides which is relevant to the problem of *N*-protection. In addition, the Royal Society of Chemistry publishes annual reviews in its *Specialist Periodical Reports* series entitled *Amino Acids and Peptides* covering the literature since 1968.

1 Blockierung und Schutz der $\alpha$-Amino Funktion. Wünsch, E. In *Synthese von Peptiden I, Houben–Weyl,* 4th ed., Vol. 15/1; Wünsch, E., Ed.; Thieme: Stuttgart, 1974; pp 46-308.
2 Amine Protecting Groups. Geiger, R.; König, W. In *The Peptides;* Vol. 3; Gross, E.; Meienhofer, J., Eds.; Academic Press: New York, 1981; Chapter 1.
3 The Fluorenylmethoxycarbonyl Amino Protecting Group. Atherton, E.; Sheppard, R. C. In *The Peptides*; Vol. 9; Udenfreund, S.; Meienhofer, J., Eds.; Academic Press: London, 1987; Chapter 1.
4 New Amino Protecting Groups in Organic Synthesis. Carpino, L. A. *Acc. Chem. Res.* **1973**, *6*, 191.
5 The 9-Fluorenylmethoxycarbonyl Family of Base-Sensitive Amino-Protecting Groups. Carpino, L. A. *Acc. Chem. Res.* **1987**, *20*, 401.
6 *Synthesis of Optically Active $\alpha$-Amino Acids.* Williams, R. M.; Pergamon: Oxford, 1989.

### 6.8.3 Reviews Concerning *N*-Protection in the Synthesis of $\alpha$-Amino Aldehydes and Ketones

1 Optically Active N-Protected $\alpha$-Amino Aldehydes in Organic Synthesis. Jurczak, J; Golebiowski, A. *Chem. Rev.* **1989**, *89*, 51.
2 Synthesis of $\alpha$-Amino Aldehydes and $\alpha$-Amino Ketones. Fisher, L. E.; Muchowski, J. M. *Org. Prep. Proc. Int.* **1990**, *22*, 399.

### 6.8.4 Reviews Concerning *N*-Protection in the Synthesis of $\beta$-Lactams

1 Naturally Occurring $\beta$-Lactams. Southgate, R.; Elson, S. In *Progress in the Chemistry of Organic Natural Products* **1985**, *47*, 1.
2 Penicillins. McGregor, D. N. *Comprehensive Heterocyclic Chemistry* **1987**, *7*, 299.
3 Cephalosporins. Holden, K. G. *Comprehensive Heterocyclic Chemistry* **1987**, *7*, 285.
4 Syntheses of 3-Amino-2-azetidinones: A Literature Survey. van der Steen, F. H.; van Koten, G. *Tetrahedron* **1991**, *47*, 7503.
5 The Ester Enolate-Imine Condensation Route to $\beta$-Lactams. Hart, D. J.; Ha, D.-C. *Chem. Rev.* **1989**, *89*, 1447.
6 $\alpha$-, $\beta$-Lactame und Derivate. Backes, J. *Houben Weyl*, 4th ed., Vol E16b; Klamann, D., Ed.; Thieme: Stuttgart, 1991.

### 6.8.5 Reviews Concerning *N*-Functional Groups Pertinent to Their Role as Protecting Groups


1 Esters of Carbamic Acids. Adams, P.; Baron, F. A. *Chem. Rev.* **1965**, *65*, 567.

2 The Gabriel Synthesis of Primary Amines. Gibson, M. S.; Bradshaw, R. W. *Angew. Chem. Int. Ed. Engl.* **1968**, *7*, 919.

3 The Chemistry of Sulfenamides. Crane, L.; Raban, M. *Chem. Rev.* **1989**, *89*, 689.

4 Reduction of C=N to CH–NH by Metal Hydrides. Hutchins, R. O.; Hutchins, M. K. *Comp. Org. Synth.* **1991**, *8*, 25.

5 *O,N*-Acetale. Rasshofer, W. *Houben–Weyl*, 4th ed., Vol 14a/2; Hagemann, H. Klamann, D., Eds.; Thieme: Stuttgart, 1991.


## References


1 Pinto, B. M.; Reimer, K. B.; Morissette, D. G.; Bundle, D. R. *J. Org. Chem.* **1989**, *54*, 2650.

2 Kukolja, S.; Lammert, S. R. *J. Am. Chem. Soc.* **1975**, *97*, 5582.

3 Freidinger, R. M.; Hinkle, J. S.; Perlow, D. S.; Arison, B. H. *J. Org. Chem.* **1983**, *48*, 77.

4 Kidd, D. A. A.; King, F. E. *Nature (London)* **1948**, *162*, 776.

5 Lemieux, R. U.; Takeda, T.; Young, B. Y. *ACS Symp. Ser.* **1976**, *39*, 90.

6 Nicolaou, K. C.; Bockovich, N. J.; Carcanague, D. R.; Hummel, C. W.; Even, L. F. *J. Am. Chem. Soc.* **1992**, *114*, 8701.

7 Gabriel, S. *Ber. Dtsch. Chem. Ges.* **1887**, *20*, 2224.

8 Ing, H. R.; Manske, R. F. H. *J. Chem. Soc.* **1926**, 2348.

9 Wasserman, H. H.; Robinson, R. P.; Carter, C. G. *J. Am. Chem. Soc.* **1983**, *105*, 1697.

10 Smith, A. L.; Hwang, C.-K.; Pitsinos, E.; Scarlato, G. R.; Nicolaou, K. C. *J. Am. Chem. Soc.* **1992**, *114*, 3134.

11 Kamiya, T.; Hashimoto, M.; Nakaguchi, O.; Oku, T. *Tetrahedron* **1979**, *35*, 323.

12 Dasgupta, F.; Garegg, P. J. *J. Carbohydr. Res.* **1988**, *7*, 701.

13 Sheehan, J. C.; Guziec, F. S. *J. Org. Chem.* **1973**, *38*, 3034.

14 Hoogwater, D. A.; Reinhoudt, D. N.; Lie, T. S.; Gunneweg, J. J.; Beyerman, H. C. *Recl. Trav. Chim. Pays-Bas* **1973**, *92*, 819.

15 Kelly, R. C.; Schletter, I.; Stein, S. J.; Wierenga, W. *J. Am. Chem. Soc.* **1979**, *101*, 1054.

16 Nefkens, G. H. L.; Tesser, G. I.; Nivard, R. J. F. *Recl. Trav. Chim. Pays-Bas* **1960**, *79*, 688.

17 Wade, P. A.; Singh, S. M.; Pillay, M. K. *Tetrahedron* **1984**, *40*, 601.

18 Mitsunobu, O.; Wada, M.; Sano, T. *J. Am. Chem. Soc.* **1972**, *94*, 679.

19 Mulzer, J.; Angermann, A.; Schubert, B.; Seilz, C. *J. Org. Chem.* **1986**, *51*, 5294.

20 Sachdev, H. S.; Starkovsky, N. A. *Tetrahedron Lett.* **1969**, 733.

21 Tsushima, T.; Kawada, K.; Ishihara, S.; Uchida, N.; Shiratori, O.; Higaki, J.; Hirata, M. *Tetrahedron* **1988**, *44*, 5375.

22 Waldmann, H. *Liebigs Ann. Chem.* **1988**, 1175.

23 Fischer, E.; Otto, E. *Ber. Dtsch. Chem. Ges.* **1903**, *36*, 2106.

24 Schmidt, U.; Wild, J. *Liebigs Ann. Chem.* **1985**, 1882.

25 Steglich, W.; Batz, H.-G. *Angew. Chem. Int. Ed. Engl.* **1971**, *10*, 75.

26 Allmendinger, T.; Rihs, G.; Wetter, H. *Helv. Chim. Acta* **1988**, *71*, 395.

27 Baldwin, J. E.; Otsuka, M.; Wallace, P. M. *Tetrahedron* **1986**, *42*, 3097.

28 Weygand, F.; Czendes, E. *Angew. Chem.* **1952**, *64*, 136.

29 Weygand, F.; Geiger, R. *Chem. Ber.* **1956**, *89*, 647.

30 Boger, D. L.; Yohannes, D. *J. Org. Chem.* **1989**, *54*, 2498.

31 Bergeron, R. J.; McManis, J. S. *J. Org. Chem.* **1988**, *53*, 3108.

32 Baker, R.; Castro, J. L. *J. Chem. Soc., Perkin Trans. 1* **1990**, 47.

33 Fischer, E.; Bergell, P. *Ber. Dtsch. Chem. Ges.* **1903**, *36*, 2592.

34 Falck, J. R.; Manna, S.; Mioskowski, C. *J. Am. Chem. Soc.* **1983**, *105*, 631.

35 Lenz, G. R. *J. Org. Chem.* **1988**, *53*, 4447.

36 Wenkert, E.; Hudlicky, T.; Showalter, H. D. H. *J. Am. Chem. Soc.* **1978**, *100*, 4893.

37 Wovkulich, P. M.; Uskokovic, M. R. *Tetrahedron* **1985**, *41*, 3455.

38 Kozikowski, A. P.; Greco, M. N. *J. Org. Chem.* **1984**, *49*, 2310.
39 Corey, E. J.; Weigel, L. O.; Floyd, D.; Bock, M. G. *J. Am. Chem. Soc.* **1978**, *100*, 2916.
40 Lott, R. S.; Chauhan, V. S.; Stammer, C. H. *J. Chem. Soc., Chem. Commun.* **1979**, 495.
41 Raucher, S.; Bray, B. L.; Lawrence, R. F. *J. Am. Chem. Soc.* **1987**, *109*, 442.
42 Kozikowski, A. P.; Xia, Y.; Reddy, E. R.; Tückmantel, W.; Hanin, I.; Tang, X. C. *J. Org. Chem.* **1991**, *56*, 4636.
43 Cooper, J.; Knight, D. W.; Gallagher, P. T. *J. Chem. Soc., Perkin Trans. 1* **1991**, 705.
44 Wender, P. A.; Schaus, J. M.; White, A. W. *J. Am. Chem. Soc.* **1980**, *102*, 6157.
45 Kende, A. S.; Luzzio, M. J.; Mendoza, J. S. *J. Org. Chem.* **1990**, *55*, 918.
46 Mzengeza, S.; Whitney, R. A. *J. Org. Chem.* **1988**, *53*, 4074.
47 McKay, F. C.; Albertson, N. F. *J. Am. Chem. Soc.* **1957**, *79*, 4686.
48 Anderson, G. W.; McGregor, A. C. *J. Am. Chem. Soc.* **1957**, *79*, 6180.
49 Carpino, L. A. *J. Am. Chem. Soc.* **1957**, *79*, 98.
50 Hintze, F.; Hoppe, D. *Synthesis* **1992**, 1216.
51 Kerrick, S. T.; Beak, P. *J. Am. Chem. Soc.* **1991**, *113*, 9708.
52 Gallagher, D. J.; Kerrick, S. T.; Beak, P. *J. Am. Chem. Soc.* **1992**, *114*, 5872.
53 Sakai, N.; Ohfune, Y. *J. Am. Chem. Soc.* **1992**, *114*, 998.
54 Schnabel, E.; Klostermeyer, H.; Berndt, H. *Liebigs Ann. Chem.* **1971**, *749*, 90.
55 Wünsch, E.; Jaeger, E.; Kisfaludy, L.; Löw, M. *Angew. Chem. Int. Ed. Engl.* **1977**, *16*, 317.
56 Schmidt, U.; Lieberknecht, A.; Bökens, H.; Griesser, H. *J. Org. Chem.* **1983**, *48*, 2680.
57 Stahl, G. L.; Walter, R.; Smith, C. W. *J. Org. Chem.* **1978**, *43*, 2285.
58 Houghton, R. A.; Beckman, A.; Ostresh, J. M. *Int. J. Pept. Protein Res.* **1986**, *27*, 653.
59 Yamashiro, D.; Blake, J.; Li, C. H. *J. Am. Chem. Soc.* **1972**, *94*, 2855.
60 Li, W.-R.; Ewing, W. R.; Harris, B. D.; Joullié, M. M. *J. Am. Chem. Soc.* **1990**, *112*, 7659.
61 Schmidt, U.; Utz, R.; Lieberknecht, A.; Griesser, H.; Potzolli, B.; Bahr, J.; Wagner, K.; Fischer, P. *Synthesis* **1987**, 236.
62 Vorbrüggen, H.; Krolikiewicz, K. *Angew. Chem. Int. Ed. Engl.* **1975**, *14*, 818.
63 Hamada, Y.; Shioiri, T. *J. Org. Chem.* **1986**, *51*, 5489.
64 Sakaitani, M.; Ohfune, Y. *J. Org. Chem.* **1990**, *55*, 870.
65 Hirai, Y.; Terada, T.; Okaji, Y.; Yamazaki, T.; Momose, T. *Tetrahedron Lett.* **1990**, *31*, 4755.
66 Pope, B. M.; Yamamoto, Y.; Tarbell, D. S. *Org. Synth. Coll. Vol. VI* **1988**, 418.
67 Itoh, M.; Hagiwara, D.; Kamiya, T. *Bull. Chem. Soc. Jpn.* **1977**, *50*, 718.
68 Keller, O.; Keller, W. E.; van Look, G.; Wersin, G. *Org. Synth. Coll. Vol. VII* **1990**, 70.
69 Paleveda, W. J.; Holly, F. W.; Veber, D. F. *Org. Synth. Coll. Vol. VII* **1990**, 75.
70 Ohfune, Y.; Tomita, M. *J. Am. Chem. Soc.* **1982**, *104*, 3511.
71 Flynn, D. L.; Zelle, R. E.; Grieco, P. A. *J. Org. Chem.* **1983**, *48*, 2424.
72 Grehn, L.; Gunnarsson, K.; Ragnarsson, V. *Acta Chem. Scand.* **1986**, *B40*, 745.
73 Hanessian, S.; Faucher, A.-M. *J. Org. Chem.* **1991**, *56*, 2947.
74 Bourne, G. T.; Horwell, D. C.; Pritchard, M. C. *Tetrahedron* **1991**, *47*, 4763.
75 Kemp, D. S.; Carey, R. I. *J. Org. Chem.* **1989**, *54*, 3640.
76 Gennari, C.; Colombo, L.; Bertolini, G. *J. Am. Chem. Soc.* **1986**, *108*, 6394.
77 Evans, D. A.; Britton, T. C.; Dorow, R. L.; Dellaria, J. F. *Tetrahedron* **1988**, *44*, 5525.
78 Bergman, M.; Zervas, L. *Ber. Dtsch. Chem. Ges.* **1932**, *65*, 1192.
79 Keith, D. D.; Tortora, J. A.; Ineichen, K.; Leimgruber, W. *Tetrahedron* **1975**, *31*, 2633.
80 Williams, R. M.; Sinclair, P. J.; Zhai, D.; Chen, D. *J. Am. Chem. Soc.* **1988**, *110*, 1547.
81 Valerio, R. M.; Alewood, P. F.; Johns, R. B. *Synthesis* **1988**, 786.
82 Wünsch, E.; Jentsch, J. *Chem. Ber.* **1964**, *97*, 2490.
83 Harding, K. E.; Marman, T. H.; Nam, D.-H. *Tetrahedron Lett.* **1988**, *29*, 1627.
84 Bodanszky, M.; du Vigneaud, V. *J. Am. Chem. Soc.* **1959**, *81*, 5688.
85 Yajima, H.; Fujii, N.; Ogawa, H.; Kawatani, H. *J. Chem. Soc., Chem. Commun.* **1974**, 107.
86 Yajima, H.; Ogawa, H.; Sakurai, H. *J. Chem. Soc., Chem. Commun.* **1977**, 909.
87 Matsuura, S.; Niu, C.-H.; Cohen, J. S. *J. Chem. Soc., Chem. Commun.* **1976**, 451.
88 Sakakibara, S.; Shimonishi, Y. *Bull. Chem. Soc. Jpn.* **1965**, *38*, 1412.
89 Kiso, Y.; Ukawa, K.; Akita, T. *J. Chem. Soc., Chem. Commun.* **1980**, 101.
90 Silverman, R. B.; Holladay, M. W. *J. Am. Chem. Soc.* **1981**, *103*, 7357.
91 Winkler, J. D.; Scott, R. D.; Williard, P. G. *J. Am. Chem. Soc.* **1990**, *112*, 8971.
92 Sánchez, I. H.; López, F. J.; Soria, J. J.; Larraza, M. I.; Flores, H. J. *J. Am. Chem. Soc.* **1983**, *105*, 7640.
93 Walker, D. M.; McDonald, J. F.; Logusch, E. W. *J. Chem. Soc., Chem. Commun.* **1987**, 1710.
94 Schmidt, U.; Leitenberger, V.; Griesser, H.; Schmidt, J.; Meyer, R. *Synthesis* **1992**, 1248.
95 Ihara, M.; Taniguchi, N.; Noguchi, K.; Fukumoto, K.; Kametani, T. *J. Chem. Soc., Perkin Trans. 1* **1988**, 1277.

96 Bolós, J.; Perez-Beroy, A.; Gubert, S.; Anglada, L.; Sacristán, A.; Ortiz, J. A. *Tetrahedron* **1992**, *48*, 9567.
97 Fukuyama, T.; Frank, R. K.; Jewell, C. F. *J. Am. Chem. Soc.* **1980**, *102*, 2122.
98 Evans, D. A.; Ellman, J. A. *J. Am. Chem. Soc.* **1989**, *111*, 1063.
99 Heffner, R. J.; Jiang, J.; Joullié, M. M. *J. Am. Chem. Soc.* **1992**, *114*, 10181.
100 Felix, A. M.; Heimer, E. P.; Lambros, T. J.; tzougraki, C.; Meienhofer, J. *J. Org. Chem.* **1978**, *43*, 4194.
101 Bodanszky, M.; Bodanszky, A. *The Practice of Peptide Synthesis*; Springer-Verlag: Berlin, 1984, pp 158.
102 Anwer, M. K.; Spatola, A. F. *Synthesis* **1980**, 929.
103 Coleman, R. S.; Carpenter, A. J. *J. Org. Chem.* **1992**, *57*, 5813.
104 Shields, J. E.; Carpenter, F. H. *J. Am. Chem. Soc.* **1961**, *83*, 3066.
105 Weygand, F.; Hunger, K. *Chem. Ber.* **1962**, *95*, 1.
106 Wünsch, E. *Methoden Org. Chem. (Houben-Weyl)* **1974**, *15/1*, 46.
107 Paquet, A. *Can. J. Chem.* **1982**, *60*, 976.
108 Wünsch, E.; Graf, W.; Keller, O.; Keller, W.; Wersin, G. *Synthesis* **1986**, 958.
109 Sharma, S. K.; Miller, M. J.; Payne, S. M. *J. Med. Chem.* **1989**, *32*, 357.
110 Murahashi, S.; Naota, T.; Nakajima, N. *Chem. Lett.* **1987**, 879.
111 Henklein, P.; Heyne, H.-U.; Halatsch, W.-R.; Niedrich, H. *Synthesis* **1987**, 166.
112 Kunz, H.; Unverzagt, C. *Angew. Chem. Int. Ed. Engl.* **1984**, *23*, 436.
113 Lemieux, R. U. *Chem. Soc. Rev.* **1989**, *18*, 347.
114 Kunz, H.; Unverzagt, C. *Angew. Chem. Int. Ed. Engl.* **1988**, *27*, 1697.
115 Kunz, H.; März, J. *Angew. Chem. Int. Ed. Engl.* **1988**, *27*, 1375.
116 Paulsen, H.; Merz, G.; Brockhausen, I. *Liebigs Ann. Chem.* **1990**, 719.
117 Waldmann, H.; Kunz, H. *Liebigs Ann. Chem.* **1983**, 1712.
118 Arnould, J. C.; Landlier, F.; Pasquet, M. J. *Tetrahedron Lett.* **1992**, *33*, 7133.
119 Dangles, O.; Guibé, F.; Balavoine, G.; Lavielle, S.; Marquet, A. *J. Org. Chem.* **1987**, *52*, 4984.
120 Roos, E. C.; Mooiweer, H. H.; Hiemstra, H.; Speckamp, W. N.; Kaptein, B.; Boesten, W. H. J.; Kamphuis, J. *J. Org. Chem.* **1992**, *57*, 6769.
121 Martin, S. F.; Campbell, C. L. *J. Org. Chem.* **1988**, *53*, 3184.
122 Jeffrey, P. D.; McCombie, S. W. *J. Org. Chem.* **1982**, *47*, 587.
123 Benz, G. *Liebigs Ann. Chem.* **1984**, 1424.
124 Laguzza, B. C.; Ganem, B. *Tetrahedron Lett.* **1981**, *22*, 1483.
125 Carpino, L. A.; Han, G. Y. *J. Am. Chem. Soc.* **1970**, *92*, 5748.
126 Carpino, L. A.; Sadat-Aalaee, D.; Beyermann, M. *J. Org. Chem.* **1990**, *55*, 1673.
127 Atherton, E.; Logan, C. J.; Shappard, R. C. *J. Chem. Soc., Perkin Trans. 1* **1981**, 538.
128 Carpino, L. A.; Cohen, B. J.; Stephens, K. E.; Sadat-Aalaee, S. Y.; Tien, J.-H.; Langridge, D. C. *J. Org. Chem.* **1986**, *51*, 3732.
129 Kelly, R. C.; Gebhard, I.; Wicnienski, N. *J. Org. Chem.* **1986**, *51*, 4590.
130 Chattaway, F. D. *J. Chem. Soc.* **1936**, 355.
131 Schultheiss-Reimann, P.; Kunz, H. *Angew. Chem. Int. Ed. Engl.* **1983**, *22*, 62.
132 Schmidt, U.; Mundinger, K.; Mangold, R.; Lieberknecht, A. *J. Chem. Soc., Chem. Commun.* **1990**, 1216.
133 Ueki, M.; Amemiya, M. *Tetrahedron Lett.* **1987**, *28*, 6617.
134 Nicolaou, K. C.; Schreiner, E. P.; Iwabuchi, Y.; Suzuki, T. *Angew. Chem. Int. Ed. Engl.* **1992**, *31*, 340.
135 Carpino, L. A.; Tunga, A. *J. Org. Chem.* **1986**, *51*, 1930.
136 Atherton, E.; Bury, C.; Sheppard, R. C.; Williams, B. J. *Tetrahedron Lett.* **1979**, 3041.
137 Paulsen, H.; Schultz, M. *Liebigs Ann. Chem.* **1986**, 1435.
138 Kader, A. T.; Stirling, C. J. M. *J. Chem. Soc.* **1964**, 258.
139 Barth, H.; Burger, G.; Döpp, H.; Kobayashi, M.; Musso, H. *Liebigs Ann. Chem.* **1981**, 2164.
140 Carpino, L. A. *J. Org. Chem.* **1980**, *45*, 4250.
141 Carpino, L. A.; Gao, H.-S.; Ti, G.-S.; Segev, D. *J. Org. Chem.* **1989**, *54*, 5887.
142 Kunz, H.; Lerchen, H.-G. *Angew. Chem. Int. Ed. Engl.* **1984**, *23*, 808.
143 Kunz, H.; Birnbach, S. *Angew. Chem. Int. Ed. Engl.* **1986**, *25*, 360.
144 Carpino, L. A.; Han, G. Y. *J. Org. Chem.* **1972**, *37*, 3404.
145 Bolin, D. R.; Sytwu, I.-I.; Humiec, F.; Meienhofer, J. *Int. J. Pept. Protein Res.* **1989**, *33*, 353.
146 Chen, F. M. F.; Benoiton, N. L. *Can. J. Chem.* **1987**, *65*, 1224.
147 Hoogerhout, P.; Guis, C. P.; Erkelens, C.; Bloemhoff, W.; Kerling, K. E. T.; van Boom, J. H. *Recl. Trav. Chim. Pays-Bas* **1985**, *104*, 54.
148 Lapatsanis, L.; Milias, G.; Froussios, K.; Kolovos, M. *Synthesis* **1983**, 671.
149 Schön, I.; Kisfaludy, L. *Synthesis* **1986**, 303.
150 Carpino, L. A.; Tsao, J.-H.; Ringsdorf, H.; Fell, E.; Hettrich, G. *J. Chem. Soc., Chem. Commun.* **1978**, 358.
151 Carpino, L. A.; Sau, A. C. *J. Chem. Soc., Chem. Commun.* **1979**, 514.

152 Van Maarseveen, J. H.; Scheeren, H. W.; Kruse, C. G. *Tetrahedron* **1993**, *49*, 2325.
153 Smith, A. B.; Rano, T. A.; Chida, N.; Sulikowski, G. A.; Wood, J. L. *J. Am. Chem. Soc.* **1992**, *114*, 8008.
154 Kim, G.; Chu-Moyer, M. Y.; Danishefsky, S. J.; Schulte, G. K. *J. Am. Chem. Soc.* **1993**, *115*, 30.
155 Halcomb, R. L.; Wittman, M. D.; Olson, S. H.; Danishefsky, S. J.; Golik, J.; Wong, H.; Vyas, D. *J. Am. Chem. Soc.* **1991**, *113*, 5080.
156 Trost, B. M.; Cossy, J. *J. Am. Chem. Soc.* **1982**, *104*, 6881.
157 Rosowsky, A.; Wright, J. E. *J. Org. Chem.* **1983**, *48*, 1539.
158 Wünsch, E.; Moroder, L.; Keller, O. *Hoppe-Seyler's Z. Physiol. Chem.* **1981**, *362*, 1289.
159 Díez-Martin, D.; Kotecha, N. R.; Ley, S. V.; Mantegani, S.; Menéndez, J. C.; Organ, H. M.; White, A. D.; Banks, B. J. *Tetrahedron* **1992**, *48*, 7899.
160 Edwards, M. P.; Ley, S. V.; Lister, S. G.; Palmer, B. D.; Williams, D. J. *J. Org. Chem.* **1984**, *49*, 3503.
161 Edwards, M. P.; Doherty, A. M.; Ley, S. V.; Organ, H. M. *Tetrahedron* **1986**, *42*, 3723.
162 Ley, S. V.; Smith, S. C.; Woodward, P. R. *Tetrahedron* **1992**, *48*, 1145.
163 Kozyukov, V. P.; Skeludyakov, V. D.; Mironov, V. F. *Zh. Obshch. Chim.* **1968**, *38*, 1179.
164 Smith, A. B.; Leahy, J. W.; Noda, I.; Remiszewski, S. W.; Liverton, N. J.; Zibuck, R. *J. Am. Chem. Soc.* **1992**, *114*, 2995.
165 Shute, R. E.; Rich, D. H. *Synthesis* **1987**, 346.
166 Meyers, A. I.; Babiak, K. A.; Campbell, A. L.; Comins, D. L.; Fleming, M. P.; Henning, R.; Heuschmann, M.; Hudspeth, J. P.; Kane, J. M.; Reider, P. J.; Roland, D. M.; Shimizu, K.; Tomioka, K.; Walkup, R. D. *J. Am. Chem. Soc.* **1983**, *105*, 5015.
167 Windholz, T. B.; Johnston, D. B. R. *Tetrahedron Lett.* **1967**, 2555.
168 Hancock, G.; Galpin, I. J.; Morgon, B. A. *Tetrahedron Lett.* **1982**, *23*, 249.
169 Baxter, E. W.; Labaree, D.; Ammon, H. L.; Mariano, P. S. *J. Am. Chem. Soc.* **1990**, *112*, 7682.
170 Eckert, H.; Ugi, I. *Liebigs Ann. Chem.* **1979**, 278.
171 Lakshmikantham, M. V.; Jackson, Y. A.; Jones, R. J.; O'Malley, G. J.; Ravichandran, K.; Cava, M. P. *Tetrahedron Lett.* **1986**, *27*, 4687.
172 Van Hijfte, L.; Little, R. D. *J. Org. Chem.* **1985**, *50*, 3940.
173 Carson, J. F. *Synthesis* **1981**, 268.
174 Takayama, H.; Odaka, H.; Aimi, N.; Sakai, S. *Tetrahedron Lett.* **1990**, *31*, 5483.
175 Magnus, P.; Giles, M.; Bonnert, R.; Johnson, G.; McQuire, L.; Deluca, M.; Merritt, A.; Kim, C. S.; Vicker, N. *J. Am. Chem. Soc.* **1993**, *115*, 8116.
176 Magnus, P.; Mendoza, J. S.; Stamford, A.; Ladlow, M.; Willis, P. *J. Am. Chem. Soc.* **1992**, *114*, 10232.
177 Gribble, G. W.; Saulnier, M. G.; Obaza-Nutaitis, J. A.; Ketcha, D. M. *J. Org. Chem.* **1992**, *57*, 5891.
178 Gribble, G. W.; Keavy, D. J.; Davis, D. A.; Saulnier, M. G.; Pelcman, B.; Barden, T. C.; Sibi, M. P.; Olson, E. R.; BelBruno, J. J. *J. Org. Chem.* **1992**, *57*, 5878.
179 Hirschmann, R.; Nicolaou, K. C.; Pietranico, S.; Salvino, J.; Leahy, E. M.; Sprengeler, P. A.; Furst, G.; Smith, A. B.; Strader, C. D.; Cascieri, M. A.; Candelore, M. R.; Donaldson, C.; Vale, W.; Maechler, L. *J. Am. Chem. Soc.* **1992**, *114*, 9217.
180 Kudzma, L. V.; Turnbull, S. P. *Synthesis* **1991**, 1021.
181 Beerli, R.; Borschberg, H.-J. *Helv. Chim. Acta* **1992**, *74*, 110.
182 Okabe, K.; Natsume, M. *Tetrahedron* **1991**, *47*, 7615.
183 Muratake, H.; Natsume, M. *Tetrahedron* **1991**, *47*, 8535.
184 Sha, C.-K.; Yang, J.-F. *Tetrahedron* **1992**, *48*, 10645.
185 Schönheimer, R. *Hoppe-Seyler's Z. Physiol. Chem* **1926**, *154*, 203.
186 Fischer, E.; Livschitz, W. *Ber. Dtsch. Chem. Ges.* **1915**, *48*, 360.
187 du Vigneaud, V.; Behrens, O. K. *J. Biol. Chem.* **1937**, *117*, 27.
188 Yamazaki, N.; Kibayashi, C. *J. Am. Chem. Soc.* **1989**, *111*, 1396.
189 Schultz, A. G.; McCloskey, P. J.; Court, J. J. *J. Am. Chem. Soc.* **1987**, *109*, 6493.
190 Shigematsu, N.; Setoi, H.; Uchida, I.; Shibata, T.; Terano, H.; Hashimoto, M. *Tetrahedron Lett.* **1988**, *29*, 5147.
191 Birkinshaw, T. N.; Holmes, A. B. *Tetrahedron Lett.* **1987**, *28*, 813.
192 Tanner, D.; Ming, H. H.; Bergdahl, M. *Tetrahedron Lett.* **1988**, *29*, 6493.
193 Heathcock, C. H.; Blumenkopf, T. A.; Smith, K. M. *J. Org. Chem.* **1989**, *54*, 1548.
194 Ji, S.; Gortler, L. B.; Waring, A.; Battisti, A.; Bank, S.; Closson, W. D.; Wriede, P. *J. Am. Chem. Soc.* **1967**, *89*, 5311.
195 Saboda Quaal, K.; Ji, S.; Kim, Y. M.; Closson, W. D.; Zubieta, J. A. *J. Org. Chem.* **1978**, *43*, 1311.
196 Parker, K. A.; Fokas, D. *J. Am. Chem. Soc.* **1992**, *114*, 9688.
197 Boger, D. L.; Palanki, M. S. S. *J. Am. Chem. Soc.* **1992**, *114*, 9318.
198 Gold, E. H.; Babad, E. *J. Org. Chem.* **1972**, *37*, 2208.
199 Roemmele, R. C.; Rapoport, H. *J. Org. Chem.* **1988**, *53*, 2367.

200 Snyder, H. R.; Heckert, R. E. *J. Am. Chem. Soc.* **1952**, *74*, 2006.
201 Weisblat, D. I.; Magerlein, B. J.; Myers, D. R. *J. Am. Chem. Soc.* **1953**, *75*, 3630.
202 Horner, L.; Neumann, H. *Chem. Ber.* **1965**, *98*, 3462.
203 Compagnone, R. S.; Rapoport, H. *J. Org. Chem.* **1986**, *51*, 1713.
204 Fukuda, T.; Kitada, C.; Fujima, M. *J. Chem. Soc., Chem. Commun.* **1978**, 220.
205 Garigipati, R. S.; Tschaen, D. M.; Weinreb, S. M. *J. Am. Chem. Soc.* **1990**, *112*, 3475.
206 Yoshioka, M.; Nakai, N.; Ohno, M. *J. Am. Chem. Soc.* **1984**, *106*, 1133.
207 Yajima, H.; Takeyama, M.; Kanaki, J.; Nishimura, D.; Fujino, M. *Chem. Pharm. Bull.* **1978**, *26*, 3752.
208 Oppolzer, W.; Bienaymé, H.; Genevois-Borella, A. *J. Am. Chem. Soc.* **1991**, *113*, 9660.
209 Rudinger, J. In *The Chemistry of Polypeptides*; P. G. Katsoyannis, Ed.; Plenum Press: New York, 1973; pp 87.
210 Beecham, A. F. *J. Am. Chem. Soc.* **1957**, *79*, 3257.
211 Wiley, R. H.; Davis, R. P. *J. Am. Chem. Soc.* **1954**, *76*, 3496.
212 Tanner, D.; Somfai, P. *Tetrahedron* **1988**, *44*, 619.
213 Jones, R. J.; Rapoport, H. *J. Org. Chem.* **1990**, *55*, 1144.
214 Henry, J. R.; Marcin, L. R.; McIntosh, M. C.; Scola, P. M.; Harris, G. D.; Weinreb, S. M. *Tetrahedron Lett.* **1989**, *30*, 5709.
215 Heyer, D.; Lehn, J.-M. *Tetrahedron Lett.* **1986**, *27*, 5869.
216 Marecek, J. F.; Burrows, C. J. *Tetrahedron Lett.* **1986**, *27*, 5943.
217 Weinreb, S. M.; Demko, D. M.; Lessen, T. A.; Demers, J. P. *Tetrahedron Lett.* **1986**, *27*, 2099.
218 Zervas, L.; Borovas, D.; Gazis, E. *J. Am. Chem. Soc.* **1963**, *85*, 3660.
219 Branchaud, B. P. *J. Org. Chem.* **1983**, *48*, 3538.
220 Burnett, D. A.; Hart, D. J.; Liu, J. *J. Org. Chem.* **1986**, *51*, 1929.
221 Goerdeler, J.; Holst, A. *Angew. Chem.* **1959**, *71*, 775.
222 Romani, S.; Bovermann, G.; Moroder, L.; Wünsch, E. *Synthesis* **1985**, 512.
223 Kessler, W.; Iselin, B. *Helv. Chim. Acta* **1966**, *49*, 1330.
224 Wolman, Y. *Isr. J. Chem.* **1967**, *5*, 231.
225 Sekine, M. *J. Org. Chem.* **1989**, *54*, 2321.
226 Bodanszky, M.; Bednarek, M. A.; Bodanszky, A. *Int. J. Peptide Protein Res.* **1982**, *20*, 387.
227 Lowe, C.; Pu, Y.; Vederas, J. C. *J. Org. Chem.* **1992**, *57*, 10.
228 Juillerat, M.; Bargetzi, H. *Helv. Chim. Acta* **1976**, *59*, 855.
229 Netscher, T.; Weller, T. *Tetrahedron* **1991**, *47*, 8145.
230 Seitz, B.; Kühlmeyer, R.; Weiler, R.; Meier, T.; Ludin, C.; Schwesinger, R.; Knothe, L.; Prinzbach, H. *Chem. Ber.* **1989**, *122*, 1745.
231 Gordon, E. M.; Ondetti, M. A.; Pluscec, J.; Cimarusti, C. M.; Bonner, D. P.; Sykes, R. B. *J. Am. Chem. Soc.* **1982**, *104*, 6053.
232 Savrda, J.; Vehrat, D. H. *J. Chem. Soc. (C)* **1970**, 2180.
233 Sundberg, R. J.; Russell, H. F. *J. Org. Chem.* **1973**, *38*, 3324.
234 Brown, T. D.; Jones, J. H.; Richards, J. D. *J. Chem. Soc., Perkin Trans. 1* **1982**, 1553.
235 Semmelhack, M. F.; Rhee, H. *Tetrahedron Lett.* **1993**, *49*, 1399.
236 Whitten, J. P.; Matthews, D. P.; McCarthy, J. R. *J. Org. Chem.* **1986**, *51*, 1891.
237 Feng, X.; Olsen, R. K. *J. Org. Chem.* **1992**, *59*, 5811.
238 Bergmann, E.; Kaluszyner, A. *Recl. Trav. Chim. Pays-Bays* **1959**, *78*, 315.
239 Urban, F. J.; Moore, B. S. *Tetrahedron Asymmetry* **1992**, *3*, 731.
240 Bouffard, F. A.; Johnston, D. B. R.; Christensen, B. G. *J. Org. Chem.* **1980**, *45*, 1130.
241 Shih, D. H.; Fayter, J. A.; Cama, L. D.; Christensen, B. G.; Hirsgfield, J. *Tetrahedron Lett.* **1985**, *26*, 583.
242 Fleming, I.; Kilburn, J. D. *J. Chem. Soc., Chem. Commun.* **1986**, 1198.
243 Greenlee, M. L.; DiNinno, F. P.; Salzmann, T. N. *Heterocycles* **1989**, *28*, 195.
244 Bateson, J.; Quinn, A. M.; Southgate, R. *J. Chem. Soc., Chem. Commun.* **1986**, 1156.
245 Ponsford, R. J.; Southgate, R. *J. Chem. Soc., Chem. Commun.* **1980**, 1085.
246 Fuentes, L. M.; Shinkai, I.; King, A.; Purick, R.; Reamer, R. A.; Schmitt, S. M.; Cama, L.; Christensen, B. G. *J. Org. Chem.* **1987**, *52*, 2563.
247 Garner, P.; Park, J. M. *J. Org. Chem.* **1987**, *52*, 2361.
248 Thaisrivongs, S.; Pals, D.; Kroll, L.; Turner, S.; Han, F.-S. *J. Med. Chem.* **1987**, *30*, 976.
249 Bonin, M.; Grierson, D. S.; Roger, J.; Husson, H.-P. *Org. Synth.* **1991**, *70*, 54.
250 Meyers, A. I.; Burgess, L. E. *J. Org. Chem.* **1991**, *56*, 2294.
251 Seebach, D.; Boes, M.; Naef, R.; Schweizer, W. B. *J. Am. Chem. Soc.* **1983**, *105*, 5390.
252 Seebach, D.; Vettiger, T.; Müller, H.-M.; Plattner, D. A.; Petter, W. *Justus Liebigs Ann. Chem.* **1990**, 687.
253 Thaisrivongs, S. *J. Med. Chem.* **1985**, *30*, 536.
254 Williams, R. M.; Glinka, T.; Kwast, E. *J. Am. Chem. Soc.* **1988**, *110*, 5927.

255 Jung, M. E.; Longmei, Z.; Tangsheng, P.; Huiyan, Z.; Yan, L.; Jingyu, S. *J. Org. Chem.* **1992**, *57*, 3528.
256 Davies, S. G.; Ichihara, O. *Terahedron Asymmetry* **1991**, *2*, 183.
257 Bringmann, G.; Geisler, J.-P.; Geuder, T.; Künkel, G.; Kinzinger, L. *Liebigs Ann. Chem.* **1990**, 795.
258 Goldstein, S. W.; Overman, L. E.; Rabinowitz, M. H. *J. Org. Chem.* **1992**, *57*, 1179.
259 Jacobi, P. A.; Martinelli, M. J.; Polanc, S. *J. Am. Chem. Soc.* **1984**, *106*, 5594.
260 Overman, L. E.; Mendelson, L. T.; Jacobsen, E. J. *J. Am. Chem. Soc.* **1983**, *105*, 6629.
261 Rawal, V. H.; Jones, R. J.; Cava, M. P. *J. Org. Chem.* **1987**, *52*, 19.
262 Campbell, A. L.; Pilipauskas, D. R.; Khanna, I. K.; Rhodes, R. A. *Tetrahedron Lett.* **1987**, *28*, 2331.
263 Hamada, Y.; Hara, O.; Kawai, A.; Kohno, Y.; Shioiri, T. *Tetrahedron, 1991* **1991**, *47*, 8635.
264 Annis, G. D.; Hebblethwaite, E. M.; Hodgson, S. T.; Hollishead, D. M.; Ley, S. V. *J. Chem. Soc., Perkin Trans. 1* **1983**, 2851.
265 Shibasaki, M.; Ishida, Y.; Iwasaki, G.; Iimori, T. *J. Org. Chem.* **1987**, *52*, 3488.
266 Yamada, T.; Suzuki, H.; Mukaiyama, T. *Chem. Lett.* **1987**, 293.
267 Williams, R. M.; Lee, B. H.; Miller, M. M.; Anderson, O. P. *J. Am. Chem. Soc.* **1989**, *111*, 1073.
268 Julina, R.; Herzig, T.; Bernet, B.; Vasella, A. *Helv. Chim. Acta* **1986**, *69*, 368.
269 Ito, Y.; Kobayashi, Y.; Kawabata, T.; Takase, M.; Terashima, S. *Tetrahedron* **1989**, *45*, 5767.
270 Manhas, M. S.; Hedge, V. R.; Wagle, D. R.; Bose, A. K. *J. Chem. Soc., Perkin Trans. 1* **1985**, 2045.
271 Shiozaki, M.; Masuko, H. *Bull. Chem. Soc. Jpn.* **1987**, *60*, 645.
272 Ihara, M.; Haga, Y.; Yonekura, M.; Ohsawa, T.; Fukumoto, K.; Kametani, T. *J. Am. Chem. Soc.* **1983**, *105*, 7345.
273 Hanessian, S.; Couture, C.; Wyss, H. *Can. J. Chem.* **1985**, *63*, 3613.
274 Greenlee, M. L.; Di Ninno, F. P.; Salzmann, T. *Heterocycles* **1989**, *28*, 195.
275 Guanti, G.; Banfi, L.; Narisano, E.; Scolastico, C.; Bossone, E. *Synthesis* **1985**, 609.
276 Fetter, J.; Lampert, K.; Kajtár-Peredy, M.; Simig, G. *J. Chem. Soc., Perkin Trans. 1* **1988**, 1135.
277 Shiozaki, M.; Ishida, N.; Hiraoka, T.; Maruyama, H. *Tetrahedron* **1984**, *40*, 1795.
278 Bertha, F.; K., L.; Kajtar-Peredy, M. *Acta Chem. Acad. Sci. Hung.* **1985**, *120*, 111.
279 Shimshock, S. J.; Waltermire, R. E.; DeShong, P. *J. Am. Chem. Soc.* **1991**, *113*, 8791.
280 Williams, R. M.; Kwast, E. *Tetrahedron Lett.* **1989**, *30*, 451.
281 Yoshimura, J.; Yamamura, M.; Suzuki, T.; Hashimoto, H. *Chem. Lett.* **1983**, 1011.
282 Williams, R. M.; Armstrong, R. M.; Dung, J.-S. *J. Med. Chem.* **1985**, *28*, 733.
283 Orlemans, E. O. M.; Schreuder, A. H.; Conti, P. G. M.; Verboom, W.; Reinhoudt, D. N. *Tetrahedron* **1987**, *43*, 3817.
284 Bringmann, G.; Jansen, J. R.; Rink, H.-P. *Angew. Chem. Int. Ed. Engl.* **1986**, *25*, 913.
285 Sieber, P.; Riniker, B. *Tetrahedron Lett.* **1991**, *32*, 739.
286 Nakajima, K.; Takai, F.; Tanaka, T.; Okawa, K. *Bull. Chem. Soc. Jpn.* **1978**, *51*, 1577.
287 Armstrong, R. W.; Moran, E. J. *J. Am. Chem. Soc.* **1992**, *114*, 371.
288 Bolton, R.; Chapman, N. B.; Shorter, J. *J. Chem. Soc.* **1964**, 1895.
289 Gmeiner, P.; Feldman, P. L.; Chu-Moyer, M. Y.; Rapoport, H. *J. Org. Chem.* **1990**, *55*, 3068.
290 Lubell, W. D.; Jamison, T. F.; Rapoport, H. *J. Org. Chem.* **1990**, *55*, 3511.
291 Koskinen, A. M. P.; Rapoport, H. *J. Org. Chem.* **1989**, *54*, 1859.
292 Wolf, J.-P.; Rapoport, H. *J. Org. Chem.* **1989**, *54*, 3164.
293 Gerspracher, M.; Rapoport, H. *J. Org. Chem.* **1991**, *56*, 3700.
294 Lubell, W. D.; Rapoport, H. *J. Am. Chem. Soc.* **1987**, *109*, 236.
295 Lubell, W. D.; Rapoport, H. *J. Org. Chem.* **1989**, *54*, 3824.
296 Lubell, W. D.; Rapoport, H. *J. Am. Chem. Soc.* **1988**, *110*, 7447.
297 Christie, B. D.; Rapoport, H. *J. Org. Chem.* **1985**, *50*, 1239.
298 Feldman, P. L.; Rapoport, H. *J. Org. Chem.* **1986**, *51*, 3882.
299 Overman, L. E.; Okazaki, M. E.; Mishra, P. *Tetrahedron Lett.* **1986**, *27*, 4391.
300 Paulsen, H.; Brieden, M.; Benz, G. *Liebigs Ann. Chem.* **1987**, 565.
301 Djuric, S.; Venit, J.; Magnus, P. D. *Tetrahedron Lett.* **1981**, *22*, 1787.
302 van der Steen, F. H.; Kleijn, H.; Spek, A. L.; van Koten, G. *J. Org. Chem.* **1991**, *56*, 5868.
303 Bonar-Law, R. P.; Davis, A. P.; Dorgan, B. J. *Tetrahedron Lett.* **1990**, *31*, 6721.
304 Fink, W. *Helv. Chim. Acta* **1974**, *57*, 1010.
305 Guggenheim, T. L. *Tetrahedron Lett.* **1984**, *25*, 1253.
306 Deshayes, K.; Broene, R. D.; Chao, I.; Knobler, C. B.; Diederich, F. *J. Org. Chem.* **1991**, *56*, 6787.
307 Salzmann, T. N.; Ratcliffe, R. W.; Christensen, B. G.; Bouffard, F. A. *J. Am. Chem. Soc.* **1980**, *102*, 6161.
308 Haruta, J. I.; Nishi, K.; Kikuchi, K.; Marsuda, S.; Tamura, Y.; Kita, Y. *Chem. Pharm. Bull.* **1989**, *37*, 2338.
309 Wasserman, H. H.; Han, W. T. *J. Am. Chem. Soc.* **1985**, *107*, 1445.
310 Dumas, F.; D'Angelo, J. *Tetrahedron Lett.* **1986**, *27*, 3725.
311 Shibasaki, M.; Ishida, Y.; Okabe, N. *Tetrahedron Lett.* **1985**, *26*, 2217.

312 Lipshutz, B. H.; Huff, B.; Hagen, W. *Tetrahedron Lett.* **1988**, *29*, 3411.
313 Alvarez, A.; Guzmán, A.; Ruiz, A.; Velarde, E.; Muchowski, J. M. *J. Org. Chem.* **1992**, *57*, 1653.
314 Grieco, P. A.; Hon, Y. S.; Perez-Medrano, A. *J. Am. Chem. Soc.* **1988**, *110*, 1630.
315 Cainelli, G.; Panunzio, M.; Giacomini, D.; Martelli, G.; Spunta, G. *J. Am. Chem. Soc.* **1988**, *110*, 6879.
316 Colvin, E. W.; McGarry, D.; Nugent, M. J. *Tetrahedron* **1988**, *44*, 4157.
317 Krüger, C.; Rochow, E. G.; Wannagat, U. *Chem. Ber.* **1963**, *96*, 2132.
318 Hoppe, D.; Beckmann, L. *Liebigs Ann. Chem.* **1979**, 2066.
319 Ikegami, S.; Uchiyama, H.; Hayama, T.; Katsuki, T.; Yamaguchi, M. *Tetrahedron* **1988**, *44*, 5333.
320 O'Donnell, M. J.; Bennett, W. D.; Wu, S. *J. Am. Chem. Soc.* **1989**, *111*, 2353.
321 Polt, R.; Szabó, L.; Treiberg, J.; Li, Y.; Hruby, V. J. *J. Am. Chem. Soc.* **1992**, *114*, 10249.
322 Polt, R.; Peterson, M. A.; De Young, L. *J. Org. Chem.* **1992**, *57*, 5469.
323 O'Donnell, M. J.; Polt, R. L. *J. Org. Chem.* **1982**, *47*, 2663.

# Chapter 7 Epilogue

*Necessity is the mother of invention*

*Roger Ascham, Toxophilus, 1545*

# 7.1 Introduction

The Erythromycin macrolides are a family of clinically important broad spectrum antibiotics isolated from *Streptomyces erythreus*. Since the total synthesis of Erythromycin A (**1.1a**) was completed by Woodward and his associates in 1981[1], a number of laboratories have risen to the challenge implicit in its stereochemical complexity. Thus, there are several syntheses of the aglycones Erythronolide A (**1.2a**)[2-8], Erythronolide B (**1.2b**)[6,7,9-12], and of 6-Deoxyerythronolide B (**1.2c**)[13-15]. In addition, a number of approaches to fragments[15-25] and advanced synthetic intermediates have been reported including (9*S*)-Dihydroerythronolide A (**1.3**)[26-28] and various seco acid derivatives of Erythronolide B[29] and (9*S*)-Dihydroerythronolide A[30-34]. The history of Erythromycin synthesis in particular, and macrolide synthesis* in general, reflects the parallel development of acyclic stereocontrol and therefore constitutes a very important chapter in the chronicle of modern organic synthesis. It is also a lesson in the trials and tribulations of protecting group methodology, and for that reason, we will close our survey with a case study taken from the copious Erythromycin literature.

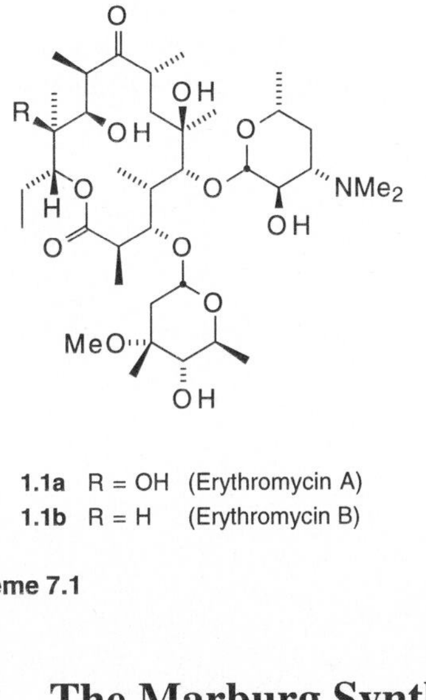

**1.1a** R = OH (Erythromycin A)
**1.1b** R = H (Erythromycin B)

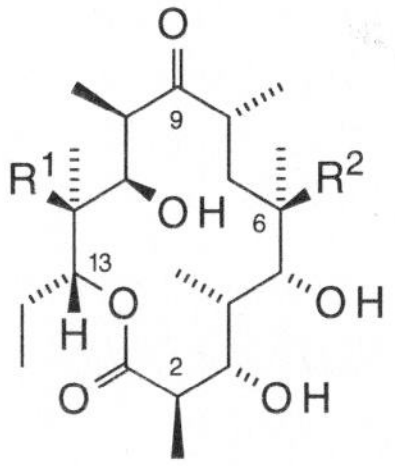

**1.2a** $R^1 = R^2 = OH$ (Erythronolide A)
**1.2b** $R^1 = H, R^2 = OH$ (Erythronolide B)
**1.2c** $R^1 = R^2 = H$ (6-Deoxyerythronolide B)

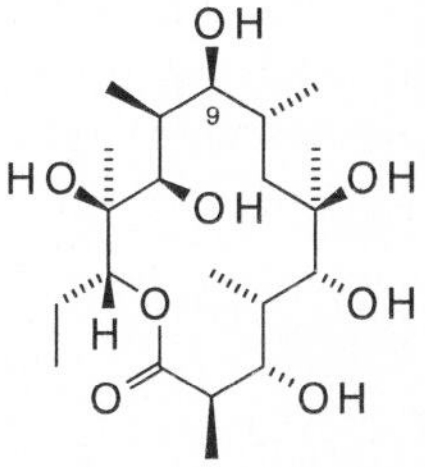

**1.3** [(9*S*)-Dihydroerythronolide A]

**Scheme 7.1**

# 7.2 The Marburg Synthesis of (9*S*)-Dihydroerythronolide A

"Synthesis must always be carried out by plan, and the synthetic frontier can be defined only in terms of the degree to which realistic planning is possible, utilizing all of the intellectual and physical tools available." There is a paradox in R. B. Woodward's dictum[35]: all syntheses require a plan but no synthesis goes according to plan. More often than not, a grand synthetic plan will be compromised by an ugly little fact — an oxidation that cannot be controlled; a reduction that will not initiate; a protecting group that remains steadfastly bound to its charge. A good synthetic plan must anticipate problems and incorporate avenues of escape from the inexorable threat of failure and to reach the target, there must be many arrows in the quiver. The point is aptly illustrated by the synthesis of (9*S*)-

Dihydroerythronolide A (**1.3**) reported early in 1993 by R. W. Hoffmann and his associates[36] at the University of Marburg — a synthesis which very nearly foundered because of the vagaries of a carefully wrought protecting group strategy.

The Marburg synthesis of (9*S*)-Dihydroerythronolide A is linear (23 steps and 16 isolated intermediates) and is notable for its economy of means: all 11 stereogenic centers were constructed using only Sharpless asymmetric epoxidation* (C6, C12, C13) and crotylboration* reactions (C2–C5, C8–C11) as outlined in the synthetic plan depicted in Scheme 7.2. The synthesis is conveniently divided into 4 phases which will be discussed in more detail below; however, for present purposes, we would like to view the synthesis through the prism of protecting group strategy. It is in phase 3, the conversion of **2.4** to **2.5**, that the factors governing the choice of protecting groups become most apparent. In order to accomplish the oxidation of the alkene in **2.4** to the C1 carboxyl function, it was necessary to protect all 5 of the secondary hydroxyl functions in the chain. This was most economically accomplished using only two types of acetal protecting group. The choice of protecting groups was significant: in order to promote macrolactonisation, it was essential that (a) the hydroxyl and carboxyl groups in intermediate **2.5** occupy similar conformational space and (b) the chain have restricted degrees of rotational freedom. *p*-Methoxybenzylidene protection (see section 3.2.1) of the secondary hydroxyls at C3, C5, C9, and C11 accomplished both tasks. The choice of the cyclopentylidene acetal (see section 3.2.3) as a protector for the C12 and C13 hydroxyls was governed by the need for *selective* hydrolysis in the presence of the two *p*-methoxybenzylidene acetals so that the C13 secondary hydroxyl could participate in macrolactonisation. Added urgency for the correct choice of protecting group regime came from the experience of the Woodward team who had previously showed that only one among 17 variously protected seco acid derivatives cyclised to give 9-Dihydroerythronolide in reasonable yield[1].

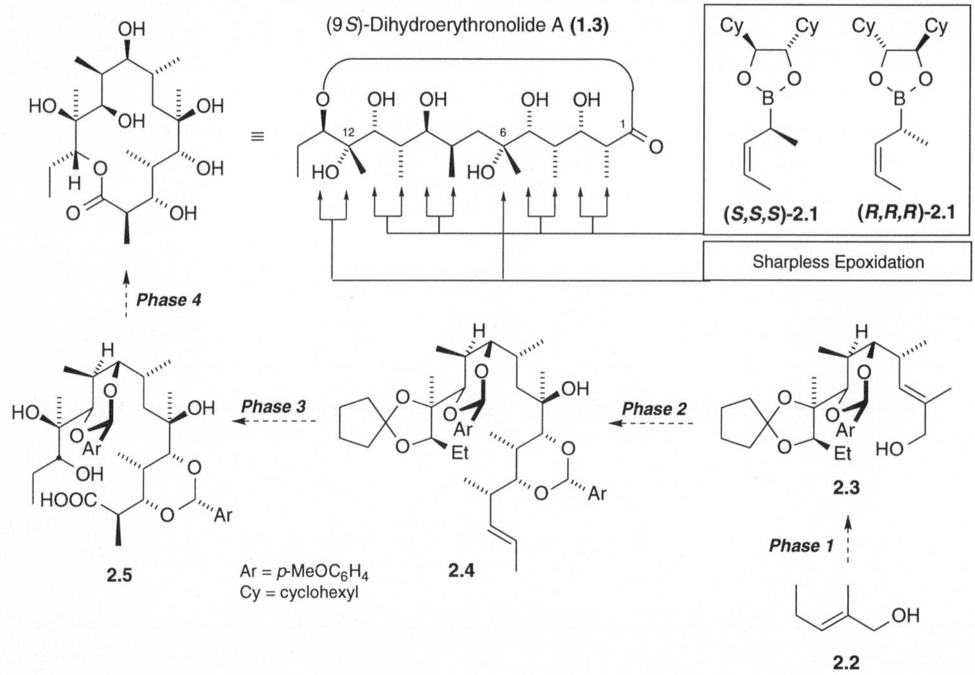

**Scheme 7.2**

*Phase 1.* The synthesis began (Scheme 7.3) with the simple allylic alcohol **2.2**. Sharpless asymmetric epoxidation introduced the first two stereogenic centers ar C12 and C13 which were transformed to the cyclopentylidene acetal **3.1** in two steps[37]. A crotylboration reaction of aldehyde **3.1** with (*S*,*S*,*S*)-1-methyl-2-butenylboronic ester [**(*S*,*S*,*S*)-2.1**] led to alcohol **3.2** in 81% yield (*de*> 96%). The hydroxyl group was protected as the PMB ether whereupon ozonolysis provided aldehyde **3.3**, which underwent a second crotylboration, this time with **(*R*,*R*,*R*)-2.1**, to give the homoallylic alcohol **3.4** with > 95% *de*. The hydroxyl group in **3.4** was then protected by oxidative cyclisation (DDQ) of the PMB ether[38] to the *kinetic p*-methoxybenzylidene acetal **3.6** (*vide infra*). Phase 1 was completed by standard chain extension chemistry to give the allylic alcohol **2.3**.

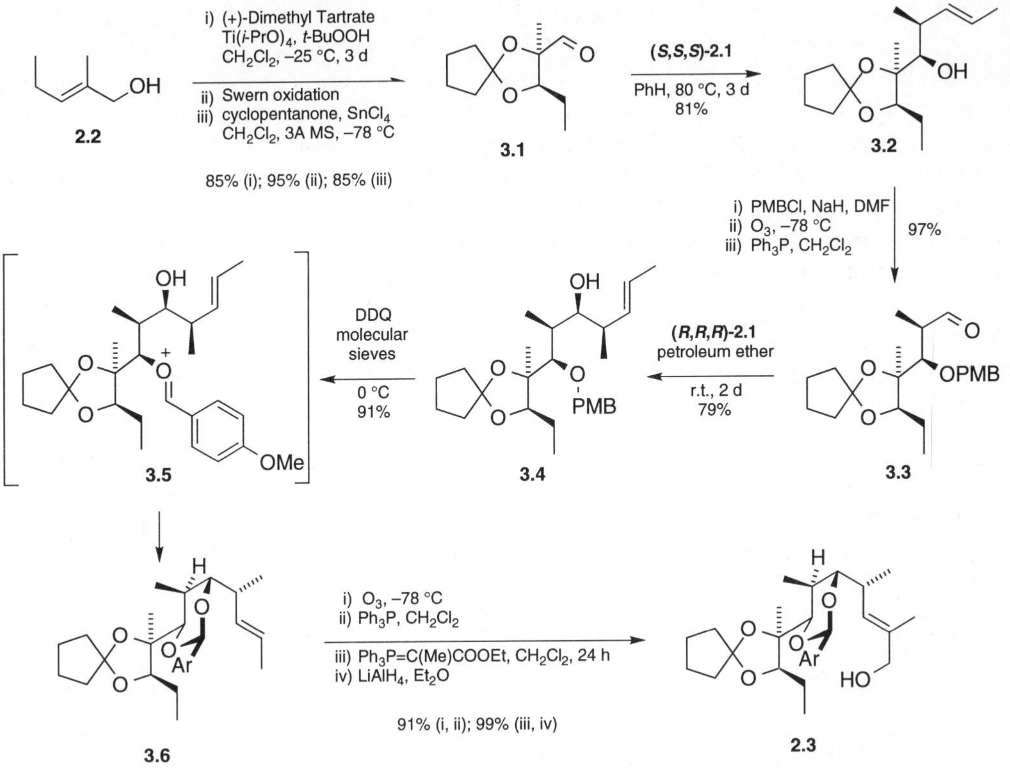

2.2 → 3.1: i) (+)-Dimethyl Tartrate, Ti(*i*-PrO)$_4$, *t*-BuOOH, $CH_2Cl_2$, –25 °C, 3 d; ii) Swern oxidation; iii) cyclopentanone, $SnCl_4$, $CH_2Cl_2$, 3A MS, –78 °C; 85% (i); 95% (ii); 85% (iii)

3.1 → 3.2: (*S,S,S*)-2.1, PhH, 80 °C, 3 d, 81%

3.2 → 3.3: i) PMBCl, NaH, DMF; ii) $O_3$, –78 °C; iii) $Ph_3P$, $CH_2Cl_2$; 97%

3.3 → 3.4: (*R,R,R*)-2.1, petroleum ether, r.t., 2 d, 79%

3.4 → 3.5: DDQ, molecular sieves, 0 °C, 91%

3.6 → 2.3: i) $O_3$, –78 °C; ii) $Ph_3P$, $CH_2Cl_2$; iii) $Ph_3P$=C(Me)COOEt, $CH_2Cl_2$, 24 h; iv) $LiAlH_4$, $Et_2O$; 91% (i, ii); 99% (iii, iv)

**Scheme 7.3**

*Phase 2.* Phase 2 of the synthesis (Scheme 7.4) was operationally a reprise of Phase 1; viz., an allylic alcohol (**2.3**) was subjected to Sharpless asymmetric epoxidation and two crotylboration reactions with reagent **(S,S,S)-2.1**. Further functional group transformations gave intermediate **2.4** harbouring all 11 stereogenic centers of the final product in the correct relative and absolute configuration.

As presented here, the synthesis, which appears so well-planned and unproblematic, was actually close to failure at a number of stages. The first critical point was the oxidation of **3.4** → **3.6**, because for the later macrolactonisation, the *p*-methoxyphenyl ring on the 1,3-dioxane must be in the $\alpha$-position (i.e., equatorial). The desired transformation was achieved with freshly recrystallised DDQ by kinetic control, since the (*E*)-oxonium ion **3.5** was generated in the conformation **5.1** (Scheme 7.5) with syn H atoms as a result of allylic strain[39]*. As soon as equilibration became possible, acetal **3.6** isomerised to the more thermodynamically stable isomer (probably twist boat structure **5.2**)[8] with the

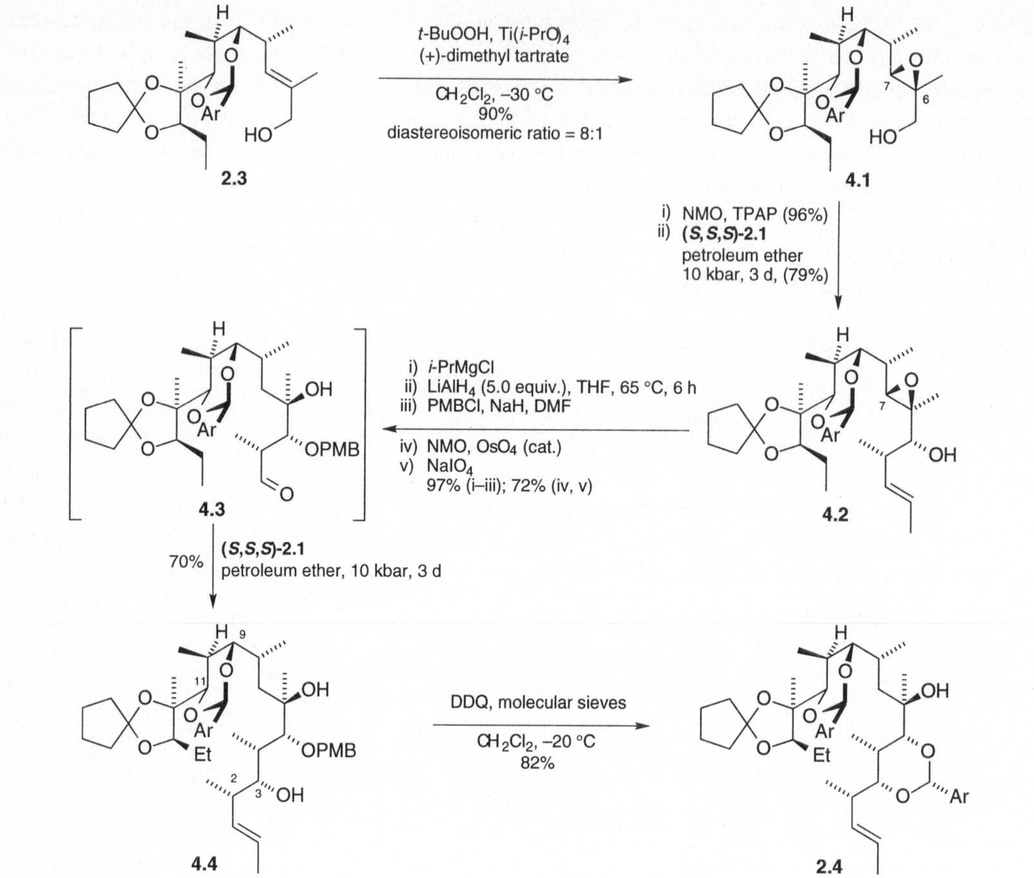

2.3

$t$-BuOOH, Ti($i$-PrO)$_4$
(+)-dimethyl tartrate

$CH_2Cl_2$, –30 °C
90%
diastereoisomeric ratio = 8:1

4.1

i) NMO, TPAP (96%)
ii) **($S$,$S$,$S$)-2.1**
petroleum ether
10 kbar, 3 d, (79%)

4.2

i) $i$-PrMgCl
ii) $LiAlH_4$ (5.0 equiv.), THF, 65 °C, 6 h
iii) PMBCl, NaH, DMF

iv) NMO, $OsO_4$ (cat.)
v) $NaIO_4$
97% (i–iii); 72% (iv, v)

4.3

70% **($S$,$S$,$S$)-2.1**
petroleum ether, 10 kbar, 3 d

4.4

DDQ, molecular sieves

$CH_2Cl_2$, –20 °C
82%

2.4

**Scheme 7.4**

5.1
($E$)-oxonium ion

Kinetic Product

rotation

($Z$)-oxonium ion

5.2
Thermodynamic Product

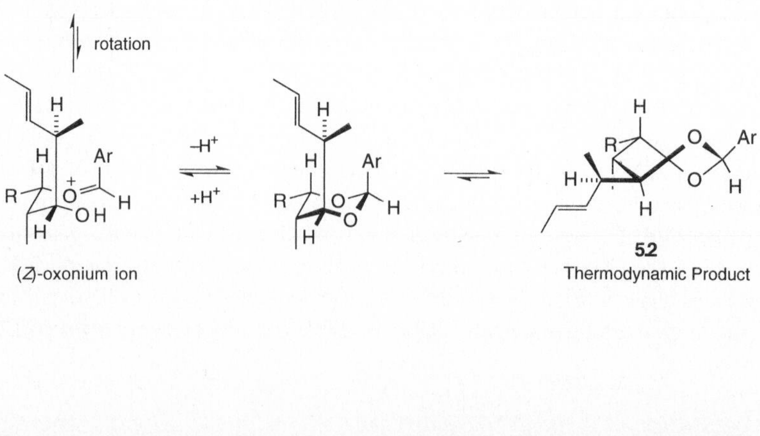

**Scheme 7.5**

aryl substituent in the $\beta$-position and both pendant chains in a *pseudo*-equatorial position, a compound that was worthless for elaboration to the target[1,26]. This happened frequently during the second DDQ oxidation **4.4** → **2.4**, because the acidic dichlorodicyanohydroquinone triggered the epimerisation of the strained C9/C11 dioxane. A solution to the problem was eventually found: by oxidation of **4.4** with DDQ *coated onto molecular sieves,* dioxane **2.4** could be generated without epimerisation. The most difficult step was still to come.

*Phase 3.* Oxidation of the alkene in **2.4** gave the acid **6.1** (Scheme 7.6) which decomposed upon attempted isolation owing to cleavage of the labile C9/C11 and C3/C5 dioxane rings. Worse still, the cyclopentylidene acetal, whose well known acid-lability was a key design feature of the protecting group strategy, could not be hydrolysed selectively in the presence of the *p*-methoxybenzylidene acetals even though this transformation had been successful in model compounds. To start again with another protecting group for the C12/C13 hydroxyls was out of the question; therefore, a way had to be found to slow down the hydrolysis of the *p*-methoxybenzylidene acetals selectively. While the inclusion of the aryl rings in cyclodextrin did not bring about the desired result, a charge-transfer deactivation of the electron-rich *p*-methoxyphenyl rings by trinitrotoluene (10 equiv.) in a two-phase system permitted selective hydrolysis of the cyclopentylidene group in **6.1**. Hydrolysis of the *p*-methoxybenzylidene acetals was suppressed because the positively charged aryl ring in the charge-transfer complex thwarted protonation of the acetal oxygens and the consequent cleavage. Necessity is, indeed, the mother of invention!

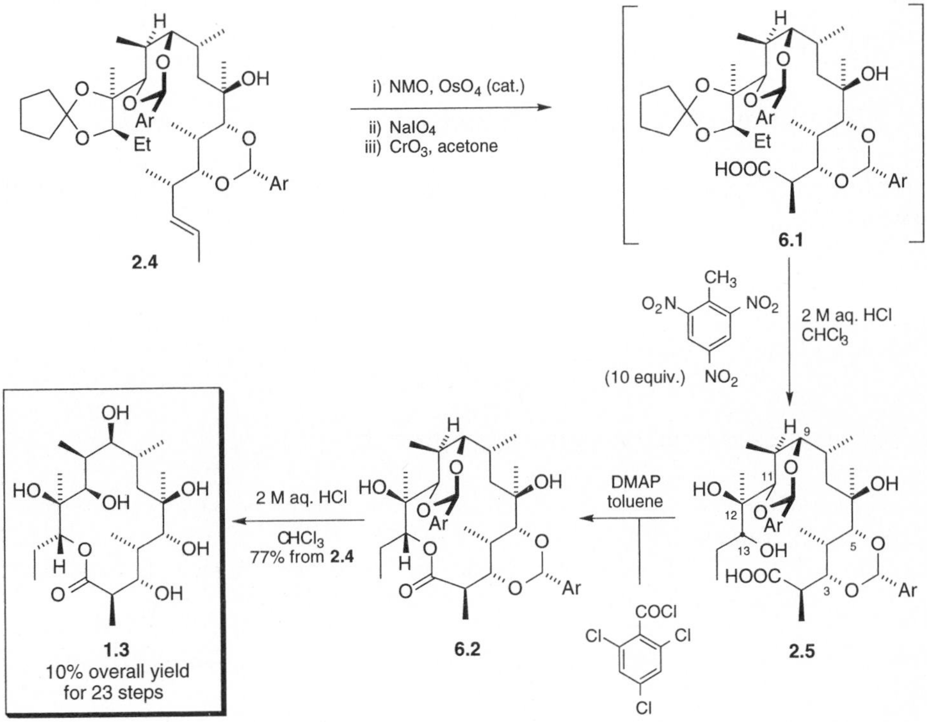

2.4

i) NMO, $OsO_4$ (cat.)
ii) $NaIO_4$
iii) $CrO_3$, acetone

6.1

(10 equiv.) trinitrotoluene; 2 M aq. HCl, $CHCl_3$

2.5

DMAP, toluene; 2,4,6-trichlorobenzoyl chloride

6.2

2 M aq. HCl, $CHCl_3$, 77% from **2.4**

**1.3**
10% overall yield
for 23 steps

**Scheme 7.6**

*Phase 4.* Having caused so many problems *en route*, it was now time to enjoy the intended conformational benefits of the *p*-methoxybenzylidene acetals. Yamaguchi lactonisation (see section 4.2.2) of the seco acid **2.5** (Scheme 7.6) occurred without incident and verified the steric proximity of the hydroxyl and carboxyl groups imparted by the C9/C11 dioxane ring. Finally, removal of the *p*-methoxybenzylidene acetals from the lactone **6.2** gave (9*S*)-Dihydroerythronolide in 77% overall yield from **2.4**.

The Marburg synthesis of (9*S*)-Dihydroerythronolide A was conspicuous for its combination of brevity, economy, and stereocontrol. That the synthesis was brought to a successful conclusion in the face of mounting difficulties, is a reflection of good design, perseverance, luck, and creativity. The synthesis could not have been completed without the help of protecting groups and the fact that they were also the root cause of so many later difficulties is a central theme of this book. It is a dilemma that all synthetic chemists must embrace with grace if not with gusto. We began our survey with a quote from Dr. Johnson. Let us now end it that way too.

> *Difficulties embarrass, uncertainty perplexes, opposition retards, censure exasperates, or neglect depresses. We proceed because we have begun; we complete our design, that the labour already spent may not be in vain: but as expectation gradually dies away, the gay smile of alacrity disappears, we are compelled to implore severer powers, and trust the event to patience and constancy.*
>
> *Samuel Johnson, The Rambler, Tuesday, 10 March, 1752*

# 7.3 Reviews

## 7.3.1 Reviews Concerning the Synthesis and Chemistry of Macrolides


1 *Macrolide Antibiotics: Chemistry, Biology, and Practice.* Omura, S., Ed.; Academic Press: Orlando, 1984.

2 Recent Developments in the Synthesis of Macrolide Antibiotics. Paterson, I.; Mansuri, M. M. *Tetrahedron* **1985**, *41*, 3569.

3 Total Synthesis of Macrolide Antibiotics. Tatsuta, K. In *Recent Progress in the Chemical Synthesis of Antibiotics* Lukacs, G.; Ohno, M., Eds.; Springer: Berlin, 1990; p1.


## 7.3.2 Reviews Concerning Asymmetric Epoxidation


1 Asymmetric Epoxidation of Allylic Alcohols: The Sharpless Asymmetric Epoxidation. Pfenniger, A. Synthesis **1986**, 89.

2 Synthetic Aspects and Applications of Asymmetric Epoxidation. Rossiter, B.E. In *Asymmetric Synthesis* Morrison, J. D. Ed.; Academic Press: Orlando, 1988; p 194.

3 On the Mechanism of Asymmetric Epoxidation with Titanium–Tartrate Complexes. Finn, M.G.; Sharpless, K. B. In *Asymmetric Synthesis* Morrison, J. D. Ed.; Academic Press: Orlando, 1988; p 247.

4 Addition Reactions with Formation of Carbon-Oxygen Bonds: (ii) Asymmetric Methods of Epoxidation. Johnson, R.A.; Sharpless, K. B. In *Comprehensive Organic Synthesis* Trost, B.; Fleming, I., Eds.; Pergamon: Oxford, 1991; **7**, p 389.

### 7.3.3 Reviews Concerning Addition of Crotylmetal Derivatives to Aldehydes

1 Diastereogenic Addition of Crotylmetal Compounds to Aldehydes. Hoffmann, R.W. *Angew. Chem. Int. Ed. Engl.* **1982**, *21*, 555.
2 α-Chiral Boronates: Reagents for Asymmetric Synthesis. Hoffmann, R.W. *Pure Appl. Chem.* **1988**, *60*, 123.
3 Allyl Organometallics. Roush, W. R. In *Comprehensive Organic Synthesis* Trost, B.; Fleming, I., Eds.; Pergamon: Oxford, 1991; **2**, p 1

### 7.3.4 Reviews Concerning 1,3-Allylic Strain

1 Allylic Strain in Six-Membered Rings. Johnson, F. *Chem. Rev.* **1968**, *68*, 375.
2 Allylic 1,3-Strain as a Controlling Factor in Stereoselective Transformations. Hoffmann, R.W. *Chem. Rev.* **1990**, *90*, 1841.

## References


1 Woodward, R. B.; Logusch, E.; Nambiar, K. B.; Sakan, K.; Ward, D.; Au-Yeung, D. W.; Balaram, P.; Browne, L. J.; Card, P. J.; Chen, C. H.; Chenvert, R. B.; Fliri, A.; Frobel, K.; Gais, H. J.; Garrat, D. G.; Hayakawa, K.; Heggie, W.; Hesson, D. P.; Hoppe, D.; Hoppe, I.; Hyatt, J. A.; Ikeda, D.; Jacobi, P. A.; Kim, K. S.; Kobuke, Y.; Kojima, K.; Krowicki, K.; Lee, V. J.; Leutert, T.; Malchenko, S.; Martens, J.; Matthews, R. S.; Ong, B. S.; Press, J. B.; Rajan Babu, T. V.; Rousseay, G.; Sauter, H. M.; Suzuki, M.; Tatsuta, K.; Tolbert, L. M.; Truesdale, E. A.; Uchida, I.; Ueda, Y.; Ueyara, T.; Vasella, A. T.; Vladuchik, W. C.; Wade, P. A.; Williams, R. M.; Wong, H. N. C. *J. Am. Chem. Soc.* **1981**, *103*, 3210.
2 Corey, E. J.; Hopkins, P. B.; Kim, S.; Yoo, S.; Nambiar, K. B.; Falck, J. R. *J. Am. Chem. Soc.* **1979**, *101*, 7131.
3 Kinoshita, M.; Ohsawa, N.; Gomi, S. *Carbohydr. Res.* **1982**, *109*, 5.
4 Kinoshita, M.; Arai, M.; Tomooka, K.; Nakata, M. *Tetrahedron Lett.* **1982**, *27*, 1811.
5 Kinoshita, M.; Arai, M.; Ohsawa, N.; Nakata, M. *Tetrahedron Lett.* **1982**, *27*, 1815.
6 Kochetkov, N. K.; Sviridov, A. F.; Ermolenko, M. S.; Yashunsky, D. V.; Borodkin, V. S. *Tetrahedron* **1989**, *45*, 5109.
7 Sviridov, A. F.; Borodkin, V. S.; Ermolenko, M. S.; Yashunsky, D. V.; Kochetkov, N. K. *Tetrahedron* **1991**, *47*, 2291.
8 Hikota, M.; Tone, H.; Horita, K.; Yonemitsu, O. *Tetrahedron* **1990**, *46*, 4613.
9 Corey, E. J.; Trybulski, E. J.; Melvin, L. S.; Nicolaou, K. C.; Secrist, J. A.; Lett, R.; Sheldrake, P. W.; Falck, J. R.; Brunelle, D. J.; Haslanger, M. F.; Kim, S.; Yoo, S. *J. Am. Chem. Soc.* **1978**, *100*, 4618.
10 Corey, E. J.; Kim, S.; Yoo, S.; Nicolaou, K. C.; Melvin, L. S.; Brunelle, D. J.; Falck, J., R.; Trybulski, E. J.; Lett, R.; Sheldrake, P. W. *J. Am. Chem. Soc.* **1978**, *100*, 4620.
11 Kochetkov, N. K.; Sviridov, A. F.; Ermolenko, M. S.; Yashunsky, D. V.; Borodkin, V. S. *Tetrahedron Lett.* **1987**, *28*, 3835.
12 Mulzer, J.; Kirstein, H. M.; Buschmann, J.; Lehmann, C.; Luger, P. *J. Am. Chem. Soc.* **1991**, *113*, 910.
13 Masamune, S.; Hirama, M.; Mori, S.; Ali, S. A.; Garvey, D. S. *J. Am. Chem. Soc.* **1981**, *103*, 1568.
14 Myles, D. C.; Danishefsky, S. J.; Schulte, G. *J. Org. Chem.* **1990**, *55*, 1636.
15 Totah, N. I.; Schreiber, S. L. *J. Org. Chem.* **1991**, *56*, 6255.
16 Oikawa, Y.; Nishi, T.; Yonemitsu, O. *J. Chem. Soc., Perkin Trans.1* **1985**, 1.
17 Heathcock, C. H.; Young, S. D.; Hagen, J. P.; Pilli, R.; Badertscher, U. *J. Org. Chem.* **1985**, *50*, 2095.
18 Kobayashi, Y.; Uchiyama, H.; Kanbara, H.; Sato, F. *J. Am. Chem. Soc.* **1985**, *107*, 5541.
19 Burke, S. D.; Schoenen, F. W.; Murtiashaw, C. W. *Tetrahedron Lett.* **1986**, *27*, 449.
20 Burke, S. D.; Chandler, A. C.; Nair, M. S.; Campopiano, O. *Tetrahedron Lett.* **1987**, *28*, 4147.
21 Burke, S. D.; Schoenen, F. J.; Nair, M. S. *Tetrahedron Lett.* **1987**, *28*, 4143.
22 Hoagland, S.; Morita, Y.; Bai, D. L.; Märki; H.-P.; Kees, K.; Brown, L.; Heathcock, C. H. *J. Org. Chem.* **1988**, *53*, 4730.
23 Born, M. C.; Tamm, C. *Tetrahedron Lett.* **1989**, *30*, 2083.

24 Kim, Y. G.; Whang, K.; Cooke, R. J.; Cha, J. K. *Tetrahedron Lett.* **1990**, *31*, 3275.
25 Schlessinger, R. H.; Mialli, A. M. M.; Adams, A. D.; Springer, J. P.; Hoogsteen, K. *J. Org. Chem.* **1992**, *57*, 2992.
26 Stork, G.; Rychnovsky, S. D. *J. Am. Chem. Soc.* **1987**, *109*, 1565.
27 Yonemitsu, O.; Tone, H.; Nishi, T.; Oikawa, Y.; Hikota, M. *Tetrahedron Lett.* **1987**, *28*, 4569.
28 Paterson, I.; Rawson, D. J. *Tetrahedron Lett.* **1989**, *30*, 7463.
29 Martin, S. F.; Pacofsky, G. J.; Gist, R. P.; Lee, W.-C. *J. Am. Chem. Soc.* **1989**, *111*, 7634.
30 Bernet, B.; Bishop, P. M.; Caron, M.; Kawamata, T.; Roy, B. L.; Ruest, L.; Sauvé, G.; Soucy, P.; Deslongchamps, P. *Can. J. Chem.* **1985**, *63*, 2810.
31 Hanessian, S.; Rancourt, G. *Can. J. Chem.* **1977**, *55*, 111.
32 Hanessian, S.; Rancourt, G.; Guindon, Y. *Can. J. Chem.* **1978**, *56*, 1843.
33 Nakata, T.; Fukui, M.; Oishi, T. *Tetrahedron Lett.* **1988**, *29*, 2219.
34 Chamberlin, A. R.; Dezube, M.; Reich, S. H.; Sall, D. J. *J. Am. Chem. Soc.* **1989**, *111*, 6247.
35 Woodward, R. B. In *Perspectives in Organic Chemistry*; A. Todd, Ed.; Interscience: New York, 1956; pp 155.
36 Stürmer, R.; Ritter, K.; Hoffmann, R. W. *Angew. Chem. Int. Ed. Engl.* **1993**, *32*, 101.
37 Hoffmann, R. W.; Ditrich, K.; Köster, G.; Stürmer, R. *Chem. Ber.* **1989**, *122*, 1783.
38 Oikawa, Y.; Nishi, T.; Yonemitsu, O. *Tetrahedron Lett.* **1983**, *24*, 4037.
39 Broeker, J. L.; Hoffmann, R. W.; Houk, K. N. *J. Am. Chem. Soc.* **1991**, *113*, 5006.

# Index